RELIABILITY WEAROUT MECHANISMS IN ADVANCED CMOS TECHNOLOGIES

RELIABILITY WEAROUT MECHANISMS IN ADVANCED CMOS TECHNOLOGIES

Alvin W. Strong
Ernest Y. Wu
Rolf-Peter Vollertsen
Jordi Suñé
Giuseppe La Rosa
Stewart E. Rauch, III
Timothy D. Sullivan

IEEE Solid-State Circuits Society, *Sponsor*
IEEE Press Series on Microelectronic Systems
Stuart K. Tewksbury and Joe E. Brewer, *Series Editors*

IEEE PRESS

A JOHN WILEY & SONS, INC., PUBLICATION

CONTENTS

2 DIELECTRIC CHARACTERIZATION AND RELIABILITY METHODOLOGY 71

Ernest Y. Wu, Rolf-Peter Vollertsen, and Jordi Suñé

3 DIELECTRIC BREAKDOWN OF GATE OXIDES: PHYSICS AND EXPERIMENTS

209

Ernest Y. Wu, Rolf-Peter Vollertsen, and Jordi Suñé

5 HOT CARRIERS   441

Stewart E. Rauch, III

PREFACE

As consumers today we care a great deal about the useful life of a car, a computer, or a pacemaker. Folks of an earlier century were concerned about the life of their carriages and we will return to this later. Manufacturers have always needed to understand the length of the useful life of their product. In addition to potential warranty costs, there are customer satisfaction issues and there may even be liability and ethical ramifications. Reliability, then, is justifiably of great concern to all. It must answer the questions: What is the useful lifetime of a given product and how can one verify that lifetime? In a CMOS world, the operating criteria for products can be very different. Some are used rarely and so may require only a few minutes of useful life. Singing greeting cards and yodeling stuffed animals would fall into this category. Other CMOS products, the pacemaker for example, would be an unmitigated disaster if it failed within minutes of implantation. CMOS applications in space can have both very long expected operational lifetimes as well as very severe environmental conditions. The wide range of CMOS applicability makes the reliability questions all the more interesting, and challenging.

The purpose of this book is to present in one place the physics, stress and analysis techniques, and models necessary to correctly determine the time of wearout for the major reliability mechanisms associated with CMOS technology. Given these tools and the product application and specifications, the engineer will be able to accurately predict when wearout will start to occur for any given product application. The engineer will also be able to verify that there are no surprises lurking to cause an early failure. Most previous books covering CMOS reliability have focused on only one or two of the technology mechanisms, have generally been more product application specific, and have not delved as deeply into the physics of the mechanisms as we have done here.

Reliability Wearout Mechanisms in Advanced CMOS Technologies has been written at a beginning graduate, or senior undergraduate, level and assumes some solid state physics background. It is designed to teach the physics of the major CMOS reliability mechanisms, the impact those mechanisms have on the device and circuits, and how to calculate that impact. The book assumes that the engineer has little or no reliability education or experience. The engineer that masters this book should have a good understanding of CMOS reliability physics, be able to design and conduct appropriate reliability experiments, analyze data, and accurately make the resultant reliability projections.

The book is divided into seven relatively independent chapters. The first chapter, the Introduction, discusses the assumptions necessary for any reliability work, describes the various types of CMOS reliability mechanisms, and at a very basic level introduces the statistics necessary for CMOS reliability analysis. The second chapter, Gate Dielectric Characterization and Methodology, describes the techniques available to understand the properties of the dielectric in question and discusses the failure distribution models for that dielectric. The third chapter, the Dielectric Breakdown of Gate oxides, describes the physics behind dielectric breakdown and provides the experimental and analytical steps required to perform dielectric breakdown projections. The fourth chapter, the Negative Bias Temperature Instability, introduces transistor device behavior as it impacts device reliability; discusses the device configurations and process variations that are important to the effect; describes the physics that controls the degradation mechanism; and finally gives experimental procedures to measure, analyze, and project NBTI lifetimes. The fifth chapter, Hot Carriers, describes the sensitive operational configurations for the HC effect and why they are important, discusses the physics and models of the effect, how the effect degrades the transistor performance, and describes how to measure HC degradation and project device lifetimes that are limited by that degradation. The sixth chapter, Stress-Induced Voiding, moves to the interconnect levels and introduces the theory and models that apply when a constrained system undergoes temperature changes; describes the impact of those temperature changes to the metal layers, vias, and their interfaces; and presents the analysis techniques to estimate the lifetime of the metallurgy. The seventh and last chapter, Electromigration, addresses the lifetime of current-carrying metal components; discusses the physics of failure for those components; and gives the experimental procedures, models and analysis techniques for projecting lifetimes limited by electromigration.

Because of the depth and breadth of CMOS technology reliability itself, we do not discuss electrostatic discharge, latchup, radiation-induced soft error rates, package reliability, or the reliability of the package and chip interactions.

The following poem returns us to those folks of that earlier century who were also interested in reliability — the reliability of a carriage, as recorded by Oliver Wendell Holmes.

ALVIN W. STRONG

Essex Junction, Vermont
July 2009

The Deacon's Masterpiece
or
The Wonderful "One-Hoss Shay"

Have you heard of the wonderful one-hoss shay,
    That was built in such a logical way
      It ran a hundred years to a day,
  And then of a sudden it — ah, but stay,
I'll tell you what happened without delay,
      Scaring the parson into fits,
    Frightening people out of their wits, –
    Have you ever heard of that, I say?

    Seventeen hundred and fifty-five.
    Georgius Secundus was then alive,

Snuffy old drone from the German hive.
    That was the year when Lisbon-town
Saw the earth open and gulp her down,
And Braddock's army was done so brown,
    Left without a scalp to its crown.
    It was on that terrible Earthquake-day
That the Deacon finished the one-hoss shay.

Now in building of shaises, I tell you what,
    There is always a weakest spot, –
    In hub, tire, felloe, in spring or thill,
    In panel or crossbar, or floor, or sill,
In screw, bolt, throughbrace, — lurking still,
    Find it somewhere you must and will, –
    Above or below, or within or without, –

And that's the reason, beyond a doubt,
That a chaise breaks down, but doesn't wear out.

But the Deacon swore (as deacons do,
With an "I dew vum," or an "I tell yeou")
He would build one shay to beat the taown
'N' the keounty N all the kentry raoun';
It should be so built that it couldn' break daown:
"Fer," said the Deacon, "t's mighty plain
Thut the weakes' place mus' stan' the strain;
N the way t' fix it, uz I maintain, is only jest
'T' make that place uz strong uz the rest."

So the Deacon inquired of the village folk
Where he could find the strongest oak,
That couldn't be split nor bent nor broke, –
That was for spokes and floor and sills;
He sent for lancewood to make the thills;
The crossbars were ash, from the the straightest
trees
The panels of whitewood, that cuts like cheese,
But lasts like iron for things like these;
The hubs of logs from the "Settler's ellum," –
Last of its timber, — they couldn't sell 'em,
Never no axe had seen their chips,
And the wedges flew from between their lips,
Their blunt ends frizzled like celery-tips;
Step and prop-iron, bolt and screw,
Spring, tire, axle, and linchpin too,
Steel of the finest, bright and blue;
Throughbrace bison-skin, thick and wide;
Boot, top, dasher, from tough old hide
Found in the pit when the tanner died.
That was the way he "put her through,"
"There!" said the Deacon, "naow she'll dew!"

Do! I tell you, I rather guess
She was a wonder, and nothing less!
Colts grew horses, beards turned gray,
Deacon and deaconess dropped away,
Children and grandchildren — where were they?
But there stood the stout old one-hoss shay
As fresh as on Lisbon-earthquake-day!

EIGHTEEN HUNDRED; — it came and found
The Deacon's masterpiece strong and sound.
Eighteen hundred increased by ten; –
"Hahnsum kerridge" they called it then.
Eighteen hundred and twenty came; –
Running as usual; much the same.
Thirty and forty at last arive,
And then come fifty and FIFTY-FIVE.

Little of of all we value here
Wakes on the morn of its hundredth year
Without both feeling and looking queer.
In fact, there's nothing that keeps its youth,
So far as I know, but a tree and truth.
(This is a moral that runs at large;
Take it. — You're welcome. — No extra charge.)

FIRST OF NOVEMBER, — the Earthquake-
day, –
There are traces of age in the one-hoss shay,
A general flavor of mild decay,
But nothing local, as one may say.
There couldn't be, — for the Deacon's art
Had made it so like in every part
That there wasn't a chance for one to start.
For the wheels were just as strong as the thills
And the floor was just as strong as the sills,
And the panels just as strong as the floor,
And the whippletree neither less or more,
And the back-crossbar as strong as the fore,
And the spring and axle and hub encore.
And yet, as a whole, it is past a doubt
In another hour it will be worn out!
First of November, fifty-five!
This morning the parson takes a drive.
Now, small boys get out of the way!
Here comes the wonderful one-hoss shay,
Drawn by a rat-tailed, ewe-necked bay.
"Huddup!" said the parson. — Off went they.

The parson was working his Sunday's text, –
Had got to fifthly, and stopped perplexed
At what the — Moses — was coming next.
All at once the horse stood still,
Close by the meet'n'-house on the hill.
First a shiver, and then a thrill,
Then something decidedly like a spill, –
And the parson was sitting upon a rock,
At half past nine by the meet'n'-house clock, –
Just the hour of the earthquake shock!

What do you think the parson found,
When he got up and stared around?
The poor old chaise in a heap or mound,
As if it had been to the mill and ground!
You see, of course, if you're not a dunce,
How it went to pieces all at once, –
All at once, and nothing first, –
Just as bubbles do when they burst.

End of the wonderful one-hoss shay.
Logic is logic. That's all I say.

1

INTRODUCTION

Alvin W. Strong

1.1 BOOK PHILOSOPHY

This CMOS technology reliability book has been written at a beginning graduate level or senior undergraduate level and assumes some solid state physics background.

The book is divided into seven relatively independent chapters consisting of an introduction, gate dielectric characterization, gate dielectric physics and breakdown, negative bias temperature instability or just NBTI reliability, hot carrier injection or hot electron reliability, stress-induced voiding or stress migration reliability, and electromigration reliability. The chapters describe the reliability mechanisms and the physics associated with them. They then take that understanding as the framework to build the bridge between the accelerated mechanism and the product mechanism.

For a CMOS reliability course or understanding focused only on one of the mechanisms, the authors expect that the material covered would include most of the first chapter and that focus chapter.

Several mechanisms are occasionally considered with reliability mechanisms, but these are not included here. Examples of these include latch-up [1], electrostatic discharge (ESD) [2, 3], and the radiation-induced soft-error rate (SER) [4].

Reliability Wearout Mechanisms in Advanced CMOS Technologies. By Strong, Wu, Vollertsen, Suñé, LaRosa, Rauch, and Sullivan

1.2   LIFETIME AND ACCELERATION CONCEPTS

It is a fact of life that every human-devised system has a finite lifetime before the catastrophic failure of the system occurs. However, most systems have a reasonably well-defined lifetime, and the catastrophic failure, or wearout, occurs well past that expected lifetime. That system has met our expectation and the customer is satisfied. Wearout is best thought of in terms of all of the systems or subsystems failing within one or two orders of magnitude in time. For example, a computer system with an expected lifetime of 10 years should experience no significant wearout before 10 years. However, all of the systems could be expected to wear out sometime between 20-plus years and 200-plus years.

1.2.1   Reliability Purpose

The purpose then of reliability is to ensure that the life of the system will be longer than the target life and that the failure rate during the normal operating life of the system will be below the target failure rate. The reliability of the product must be known when the product is sold so that the operating-life warranty costs can be quantified and customer satisfaction protected. Ensuring these objectives are met means that each failure mechanism must be quantified so that its impact during normal operating life can be predicted and the time at which it starts to cause the system to wear out can be predicted as well.

The length of time one has to do the reliability stressing and make the predictions is dependent on the state of the program. As a new technology is being developed, reliability engineers should be generating reliability data to help guide the program in the appropriate design, cost, and reliability tradeoffs. This work may occur over the course of several months to a few years. However, feedback on any given experiment needs to be given as quickly as possible. Once the technology is ready for implementation, it would typically undergo a "qualification" of no more than three months in duration. If a problem is discovered after qualification, that is, during manufacturing, it is all the more crucial to give feedback quickly.

The concept of an accelerated life is necessary for reliability stressing to have meaning. That is, it must be possible to find some condition or conditions that will allow one to shrink a 10-year product life down to a three month period, or less, so that the reliability of the system can be investigated and guaranteed in that three months. The conditions used to accelerate a given mechanism usually cannot be applied to the whole system (in our case, the semiconductor product chip). In this case, a test structure must typically be built that will replicate the behavior of the element in the product chip, but allow one to apply an accelerating condition. Hence, with the concept of accelerated life, we also need to posit the concept of a representative test structure.

It must be noted that in all of the above discussion, the product is assumed to operate perfectly when it is first turned on, for example, at time zero.

Once the reliability of each element has been investigated, understood, and modeled, an additional step should be feedback for the next design pass so that the product team can design in reliability. A simple example of this design for reliability would be to use minimal groundrules only where necessary.

1.2.2 Accelerated Life

An accelerated life concept must include several features to be useful. In addition to the requirement that it can be used to accelerate a particular reliability mechanism, it must be possible to quantify how much that condition actually accelerates the reliability mechanism. It must be possible to build a bridge between the accelerated stress conditions and the use condition so that it is possible to quantify the degree or amount of acceleration. This quantification is also necessary to ensure that no mechanism is introduced with the accelerating condition that does not exist at the use condition of the product. One must understand if the behavior of the mechanism is uniform and consistent from the use condition to the accelerating conditions.

Once an appropriate accelerating condition has been determined, the acceleration between the stress conditions and the use condition can be determined. First, data from at least two different values of accelerating conditions are measured. The cumulative fails from those conditions are then plotted on the y axis, versus time on the x axis. This plot is done on a set of axes that have been transformed in such a way that the resulting plot is a straight line. The methodology for doing this for the most common distributions used in semiconductor reliability is discussed in detail later in this chapter. The two distributions that are most commonly used in semiconductor technology reliability are the two-parameter lognormal and Weibull distributions because each of these is very flexible and can be used to describe many different types of behaviors. These distributions provide a functional form with which a distribution can be characterized so that it can then be treated analytically. The details of these distributions, and their axes transformations, are the topics of Section 1.4. A simple, although somewhat unrealistic, example would be a distribution whose cumulative failures were linear with the log of time. If the cumulative fails were to be plotted against time, a nonlinear curve would result. However, a transformation of the x axis by taking the log of the time and then plotting the cumulative failures against that log of time would result in a straight line. These transformations are necessary so that the distributions and the slopes remain invariant across all accelerating conditions and down to the use conditions. All transformations have been made for the example in Figure 1.1 so that the plot is linear. Two sets of voltage data are shown plotted on the top left of Figure 1.1. At least three different values of each accelerating condition are preferred but only two are shown on Figure 1.1 for simplicity. The data from these accelerated conditions are used to calculate an acceleration factor, which is in turn used to calculate the acceleration time between the lowest accelerated condition and the use condition.

One uses the lowest accelerated condition to minimize projection error. Obviously if any new mechanism is introduced due to the accelerated condition, or a nonlinearity in the expected mechanism is introduced, this acceleration time

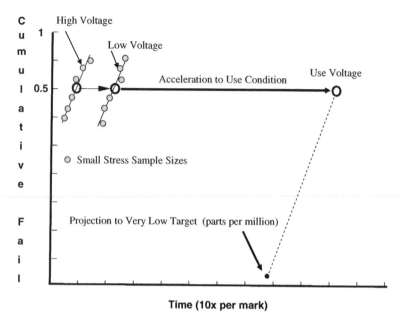

Figure 1.1. Acceleration example with two voltage conditions with the first projection to use condition at a 50% life and the final projection to low (ppm) failure rate target.

would not be valid. The transformation equations for the x and y axes of the plot will depend on the particular reliability mechanism in question and the probability distribution function that is most appropriate for that mechanism. The sample sizes for the stresses are typically very small, whereas the wearout target is typically expressed in terms of parts per million (ppm) or less. This means that once the acceleration between the stress conditions and the use condition is calculated, a second projection must then be made from that value, which typically is at about the 50% fallout point of each of the accelerated curves, to a very small percent fallout for that second projection curve. Note that the 50% value is used only as a convenient example here. The exact value at which the acceleration calculations will be made will depend on the distribution that is to be used and is discussed in detail later in this chapter. The intent here is to give a broad overview and avoid losing the reader in the detail. This extrapolation is done using the same slope found during at the stress conditions, since the axes have been transformed so that they remain invariant across all of the conditions of interest. Thus, an error in the acceleration factor causes the use condition to be incorrectly located in time, and any errors in determining the correct slope causes the projection to the small fraction fail target to have an additional error. Often this last error, the error due to an incorrect slope determination, can cause the largest error in the resultant projection. It should be highlighted that we are not speaking of graphing errors here since the calculations can all be performed using computer software. If done graphically, those errors would be in addition to the errors mentioned previously.

Three plots are shown in Figure 1.1. The two plots on the left show the two different accelerated voltage conditions with all other conditions held constant. The dotted line plot on the right shows the projection from the 50% failure time for the use condition to the failure time associated with the target failure rate for that mechanism. One other potential factor that is not shown in Figure 1.1 is a test-structure scaling factor. This factor will be discussed in each of the chapters for which it is applicable. Although there are many accelerating conditions as shown in Section 1.2.3, by far the two most common accelerating conditions are voltage and temperature. The minimum experimental design that can yield a voltage accelera-tion factor is the two conditions as shown in Figure 1.1. For example, assume that the stress voltages in Figure 1.1 are 4.37 V and 4 V. The lines on the left side of Figure 1.1 represent fits to data taken at these accelerated voltage conditions. The slopes of the two stress conditions are shown as equal. The slope fit would normally be accomplished by a fitting program, which could force the best fit to all of the accelerated curves simultaneously. For this case we will assume an Eyring accel-eration model [5] applies that has the form $Acc = t_2/t_1 = \exp\{(\Delta H/k)\{(1/T_2) -(1/T_1)\}\} \exp\{-\beta_V (V_2-V_1)\}$. For the acceleration due to voltage, $Acc_{VOLTstress} = t_2/t_1 = \exp\{-\beta_V(V_2-V_1)\}$, where V is voltage, t is time, and β_V is the voltage acceleration factor for this Eyring model. The temperature acceleration model by itself has the form $Acc_{TEMPstress} = t_2/t_1 = \exp\{(\Delta H/k)\{(1/T_2)-(1/T_1)\}\}$ where k is Boltzmann's constant and is also known as an Arrhenius model. These models will be discussed in more detail throughout this book but are introduced here to give the reader an early qualitative introduction to the acceleration concepts. Observation of the first two curves will reveal a time difference or acceleration of about $30 \times$. The large circles in Figure 1.1 represent a mean life of the hardware under stress and as mentioned above are used as a convenient example. As will be discussed later, the points at which the acceleration calculations will be made are the most accurate values for the distribution under consideration. If the voltage used for the first curve on the left is $V = 4.37$ and the voltage for the second curve is $V = 4$, then β_V may be calculated, given $Acc_{VOLTstress} = 30$, as $\beta_V = (\ln Acc_{VOLTstress})/(V_1-V_2) = 9.2$. This value for β_V is then used to project from $V = 4$ to the $V = 2$ use condition as $Acc_{VOLTuse} = \exp\{9.2 \times (4-2)\} = 10^8$. Having made this calculation, one then needs to consider whether or not the value calculated is reasonable based on comparable data both from the reliability analyst's prior work, as well as literature values. A similar procedure would be used to calculate any acceleration including, for example, a temperature acceleration. This example should give the reader a better understanding of the actual process of stressing and then projecting to use conditions using acceleration concepts. Obviously the stress conditions must be appropriately chosen and the experiment appropriately designed to achieve useful results. Note that if too small of difference is used between two accelerating conditions then the experimental error and the statistical variation in the two sets of data may cause enough overlap of data such that the acceleration factor between the two sets of data cannot be calculated. On the other hand if the difference between the two sets of data is too large, the failure times of the lower condition may be longer than the time designated for the stress. The discussion of the extrapolation to very small failing percentage targets will commence in Section 1.4.5.

We now return to a more general discussion of acceleration. The mean life from Figure 1.1 is plotted against one of the accelerating conditions in Figure 1.2. Figure 1.2 presents a picture of the progress of the state of the art of reliability stressing across the last 20-plus years. Each circle represents a change of approximately 40 × in time.

More than 20 years ago, all reliability stressing was done with the reliability test structures wire-bonded onto a die carrier or module. This structure, which was contained within the package, was then put into a stress apparatus, which typically applied stress temperature and voltage between weeks and months, depending on the mechanism under investigation. The readouts were made at preset values, typically on the order of two or three times per decade. The test structures had to be removed from the stress apparatus and physically transported to a tester for each readout. This stressing is represented in Figure 1.2 by the second circle from the left. Note that the left most circle represents the useful life of the structure, typically 10 years. For mechanisms like ionic contamination, which will relax unless the voltage is continuously applied, it was necessary to have large batteries connected to keep the hardware at stress voltage while transporting the hardware

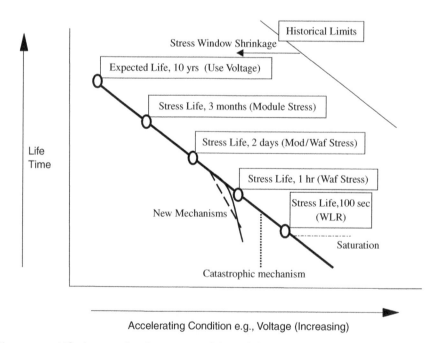

Figure 1.2. Lifetime projection curve with each large circle representing the 50% life for the accelerating or use condition and showing various nonlinearities that can compromise a highly accelerated stress.

to the tester and awaiting test. Even for mechanisms that do not relax when not under bias, this method has the disadvantage that no data can be obtained for the first few weeks after the hardware becomes available because the wafers are being built onto the chip carriers or modules and this build typically takes several weeks. The advantage of this method, even today, is that the stress equipment is relatively inexpensive and the test equipment is general enough to be used for readouts for several mechanisms. This amortizes the test equipment across all of these stresses and further decreases the cost of ownership.

However, for accuracy and simplicity, it is desirable to use the same equipment for the stress application and the readouts. This also minimizes handling damage and human error. Data acquisition improvements during the past 20 years have allowed the detection of exact times-to-fail even for the long three-month stresses. Now whether the stress is a long, three-month stress or a very short stress, the exact times-to-fail can be obtained with the same equipment.

In addition, advances in the state of the art in hot carrier stressing, in dielectric stressing, and in electromigration stressing have moved the leading practice to the far right two points on Figure 1.2, that is, to stresses of hours and minutes. Typically, stressing with seconds of duration is used in conjunction with at least one additional stress of longer duration. For example, in the case of dielectric stressing, optimally three voltages are stressed with the shortest stress duration having a median fallout on the order of 10 to 100 sec and the longest (lowest voltage) having a duration of 1000 to 10000 sec. This has been practiced for the last 5 to 10 years for dielectric stressing but as the state of the art thickness approaches 1 nm, it may actually become necessary to return to the relatively long stresses of several months. In the case of electromigration, the quantitative bridge for stressing on the order of seconds was only demonstrated a few years ago [6–8].

One of the obvious points that should be explicitly made is that for a three-month stress, the extrapolation to a 10-year use life is only a factor of 40. While for a 100 sec stress, the extrapolation is a factor of more than 1E6. Much more care concerning the projection error must be exercised when one is extrapolating six orders of magnitude, than when one is extrapolating only a little more than one order of magnitude.

Also one has to investigate very carefully whether any change in the accelerating condition of the mechanism in question has occurred or can occur under any reasonable set of conditions. This is depicted graphically in Figure 1.2 as new mechanisms, which may occur above a certain stress level. If any such mechanisms exist, they may be either linear or nonlinear as shown, and they would preclude exceeding that stress level since no straightforward model or bridge to use conditions would be possible in that case. Another possibility is that the accelerating or stress condition saturates above a certain level. That is, a further increase in the accelerating condition causes no resultant decrease in the lifetime. Again, a mechanism of this nature would limit the accelerating condition to a value no higher than just below its saturation value, and even there, the physics would need to be well understood.

The final concept that Figure 1.2 attempts to depict is that the window available for stressing is shrinking as the technology features continue to shrink. In the past, a very significant margin existed in many of the mechanisms. Often a calculated acceleration would demonstrate that the stress had gone well beyond a 10-year life but no wearout for that mechanism had been observed. Dielectric stressing is an excellent example of this. For 12 nm oxides, dielectric stressing on the order of three months could not detect any indication of wear out, and the stress focus was just on the extrinsic or defect part of the curve. Today, models are constructed to understand whether fractions of nanometers can be shaved from the oxide thickness and still meet the end of life targets.

1.2.3 Accelerating Condition

Acceleration concepts have been discussed and we now turn our attention to accelerating conditions. What types of external forces can be applied to the semiconductor test structure in such a way as to cause the end of life, say 10 years, to be reached in a time period that is significantly shorter than that 10-year life. In principle, the shorter the stress time the better, as long as one can still bridge to the use conditions. Examples of accelerating conditions for semiconductors are shown below. This extensive list includes all of the common accelerating conditions and a list of pertinent mechanisms with chapter references where applicable.

- Voltage (DC)
  - Dielectric breakdown (3.4)
  - Electromigration {indirectly} (7.3)
  - Hot carrier (5.2)
  - Temperature bias stability (4.3–4.5)
  - Interconnect opens and shorts (7.3)
  - Ionic contamination
  - Energetic particle-induced soft error mechanisms
  - Variable retention time mechanisms
  - Leakage mechanisms
- Voltage Change (AC)
  - Conducting hot carrier mechanisms (5.2)
- Temperature
  - Dielectric breakdown (3.4)
  - Electromigration {indirectly} (7.3)
  - Stress migration (6.2–6.5)
  - Interconnect shorts and opens (6.2–6.5, 7.3)
  - Hot carrier (5.2)
  - Temperature bias stability (4.3–4.5)

- ○ Ionic contamination
- ○ Variable retention time mechanisms
- ○ Leakage mechanisms
- • Temperature Change
- ○ Interconnect opens
- • Temperature Change Rate
- ○ Interconnect opens
- • Current Density
- ○ Dielectric breakdown (3.2–3.4)
- ○ Electromigration (7.3)
- • Humidity
- ○ Corrosion
- • Humidity and Pressure
- ○ Corrosion
- • Harsh environment
- ○ Corrosion
- • Mechanical pull tests
- ○ Mechanical strength of interconnects and adhesives
- • Radiation
- ○ Some dielectric breakdown concerns
- ○ Soft error rate (SER) for certain flash memory

Note: The SER effect does not get worse with time for most CMOS devices.

1.3 MECHANISM TYPES

1.3.1 Parametric or Deterministic Mechanisms

A parametric or deterministic mechanism is defined, albeit somewhat arbitrarily, as any mechanism that impacts all identical structures nearly equally. A stress for this type of mechanism will always cause the parameter under question to shift. And, even if many samples are stressed, the shifts will all be very close to the same value assuming all of the stressed structures are identical. For this reason, very small sample sizes can successfully be used to characterize a parametric mechanism. Most of the variation of the shifts observed for parametric mechanisms is caused by variations of the controlling parameters and not by random statistical variation.

The hot carrier (HC) mechanism is one example of a parametric mechanism. While a field effect transistor (FET) is turning on or off, the gate current has a peak value resulting from channel hot electron injection. These electrons gain enough energy to surmount the Si/SiO_2 interface without suffering energy-losing collisions in the channel. The electrons are trapped and result in FET performance degradation. This mechanism is uniform and parametric in the sense that for a set of FETs that are all structurally identical, the shifts resulting from the above stress will be almost identical across all of the devices stressed; that is, the shifts will be determined by their parameter values, not by the random variation. In practice, if chips from several wafers or lots are stressed, variation will be seen but that variation will be a function of slight differences in the structures of the FETs across the wafers and lots.

Electromigration is an example of a mechanism that has aspects of a parametric mechanism. A current flowing through a line will cause atomic motion in that line. If that line is aluminum, significant atomic motion will occur at higher current densities and will cause the line resistance to increase and ultimately open. This is fundamental to the structure and the metallurgy. For high enough current densities, electromigration will always occur for that aluminum line. It is not caused by a defect although it can be exacerbated by a defect. Although the physics cannot be changed, sometimes it is possible to mitigate the problem. For example, if redundant layers of certain other metals are used in conjunction with aluminum, the sandwich line structure will increase in time-zero resistance if the overall cross-sectional area of the line remains constant, but electromigration typically will only cause a resistance increase and not an open under the same high current-density stress. For some metallurgies, no electromigration will occur even at higher current densities. However, it must be pointed out that in the case of electromigration, there are also aspects of a random mechanism because the grain structure of the line is random. And this randomness is true for metallurgy that is identical in processing. Typically, larger sample sizes are necessary when stressing mechanisms that have a greater degree of randomness.

Obviously it is crucial to understand the fundamental physics for a parametric mechanism. Once the physics is understood, strategies can be put into place to mitigate the effect or to eliminate the problem by structural or operating-point changes. Sometimes mitigation is possible and sometimes it is not. For the electromigration example, tungsten is sometimes used for the lower levels of wiring where the distances are small and the higher time-zero line resistance is tolerable. For the longer wiring levels, the resistivity of tungsten is too large and aluminum or copper must be used and other strategies invoked to decrease the impact of electromigration.

Once the physics is understood, so that all of the controlling parameters are identified and each of their impacts quantified, it is possible to address elimination and mitigation strategies. To be able to quantify the impact of a given parameter, it is usually necessary to characterize the impact of that parameter on a test structure where individual control of all of the terminals is possible. If, for example, the physics of the mechanism is related to a parasitic edge transistor in parallel to the bulk transistor, the decision must be made as to whether to change the process to eliminate the parasitic transistor, or to simply mitigate its impact on the circuit. A problem may occur only at one extreme of the normal processing window or set of biases and tolerances. HC is one example since it is worst at the shortest channel lengths for a given set of stress conditions. In some cases the strategy may be to run the process to a tighter manufacturing limit. Because this type of mechanism equally affects all structures with the identical process, only a few structures need to be investigated to reasonably well characterize a parametric mechanism. However, these devices under tests (DUTs) must all be structurally identical.

From this previous discussion it should be obvious that it is necessary to investigate parametric mechanisms at all salient process window extremes to ensure that no undesired effects occur. Again, each process window investigation point requires only a small sample size.

Below are some examples of parametric mechanisms and one or two strategies for controlling or eliminating the effect. In most of the cases, there are other strategies that could also be invoked. Applicable chapter references are shown.

- *Hot carrier*: design point change, e.g., lower operating voltage the device experiences
- *Bias temperature stress or (negative bias temperature instability)*: design point change, e.g., decrease operating voltage
- *Ionic contamination*: discovery and removal of contamination source
- *Stress induced leakage current*: design point change, e.g., decrease operating voltage or thicken gate oxide
- *Electromigration*: design point change, decrease current density
- *Soft error rate (radiation induced)*: design point change, increase critical charge of pertinent cells or decrease charge collection efficiency. This is not discussed further in this book

1.3.2 Structural Mechanisms

Structural mechanisms are those mechanisms for which the fails physically occur in the same place. The distinction here from the structurally induced parametric fails is that these fails are only a function of a structural artifact. Although these definitions are all somewhat arbitrary, they help in understanding the sample size differences recommended in the later chapters. Usually significant failure analysis is required to determine that a particular failure type has a structural, systematic signature. Often this signature only occurs at one of the process extremes so that it does not occur on every wafer or lot. Sometimes it is even more difficult to identify because not only does it only occur at one process extreme, it may also require a certain set of process biases and/or tolerances to align in just a "right" way for the failure to occur. This may take the form of one part of the wafer having an acute susceptibility, or it may be tool dependent. In some of these cases, it may appear random, while in fact, the fail is part of a manufacturing defect or process window tail. This can usually be avoided if a large enough sample is investigated and if at least part of that sample comes from the salient process extremes. If the failure analysis then identifies a particular feature failing more than once, that feature should undergo very careful scrutiny.

Sampling is very important since the problem may not impact all lots or wafers or die equally. The sampling must gauge all process variations unless one process extreme can be identified as the worst case for the given mechanism. A minimum of three manufacturing lots is recommended with one produced at the identified critical extreme. For this type of mechanism, sampling for random statistical variation is less important than sampling for the pertinent process extremes.

Once this type of mechanism is understood, it can often be mitigated with a strict application of statistical process control (SPC). However, no structural fails should be acceptable within the normal process limits. Otherwise this would represent a technology weakness that, if accepted, would likely result in an inordinate number of customer failures even with tight SPC. It is always better in the long term, to fix a problem rather than to try control it. Fixing the problem can take the form of structural modifications or a redefinition of the process limits. Especially for hardware made later in a program, the normal exercise of SPC, once the process line is full of hardware, may eliminate the possibility of a problem.

Process improvements made during manufacturing can inadvertently introduce new structural mechanisms. An effective method to avoid this is to sample a large number of chips and wafers looking for changes even in time-zero characteristics. Changes in the time-zero characteristics will not always flag a change in a reliability mechanism, but a change in the time-zero characteristic should be carefully investigated especially if a significant database exists for the normal properties of the parameter. Wafer-level reliability (WLR) is an even better gauge as to the impact of process improvements on reliability. And in fact, occasionally the time-zero properties have been changed through process changes only to make the reliability worse in a direct tradeoff between yield and reliability. All of the examples for this type of mechanism are very technology/process dependent.

1.3.3 Statistical Mechanisms

Statistical mechanisms are defined as those mechanisms that are primarily random. Thus the more susceptible area to a particular statistical mechanism, the more likely that mechanism will cause a chip fail. The occurrence of the fail will be totally random within that susceptible area. This is in contrast to a structural fail, which will always occur at a given feature within a structure. It is also in contrast to the parametric or deterministic mechanism for which the process variation or process extreme will have a larger impact on the result than does the random statistical variation within a given process point. One must be careful at this point because the statistical mechanisms are also caused by fundamental physics unless the discussion is limited to defects. The distinction is more focused on the impact that the random statistical variation has on the investigation of the mechanism.

For the statistical mechanisms, it is critical that a significant sample be stressed to understand the failure behavior. Hardware made at the process extremes is only important in as much as it has an impact on the occurrence of the mechanism. For example, a thinner oxide will have a higher failure rate for a given set of conditions than the thicker oxide. In this case a significant sample should be stressed at the process minimum thickness. The fails will be randomly distributed throughout the area in both cases.

Aspects of dielectric breakdown and electromigration are two examples of statistical mechanisms. The aspect of electromigration that is statistical in nature is the grain structure of the line. Although the fabrication conditions very clearly impact the grain structure, that grain structure will still have random variation even when the fabrication conditions are identical. And first dielectric breakdown is generally accepted to be random and has been shown to follow a weakest-link behavior resulting in a Weibull distribution for its cumulative fail distribution. This will be discussed in Section 1.4.7.

For a statistical mechanism, a much larger sample is required than for a parametric mechanism. The reason is that there is a given randomness in samples that have identical structures, even for identical processing in as much as possible. Thus for a given stress on parts that are identical as far as can be known, there will be a distribution of results instead of a single-valued result. The sample size must be large enough to ensure that some mean or characteristic life is truly representative of the population from which the sample is drawn and not just caused by random statistical variations. To further elucidate this point: if five processes were being compared and five small samples were chosen for each case, very different results could be obtained just based on the random statistical variation, and it could be possible, and in fact likely, to not choose the best process due to this random variation. Obviously it must also be assumed in this case, as well as the other cases, that the samples being investigated are representative not only of the population from which they were drawn but also from the entire production population.

A discussion of statistical mechanisms must include a general discussion of defects. There are four classifications of random defects depending on the time of occurrence. The first class of defects is screened out at time zero. These are obviously the yield fails. The next class of defects is the infant defects. These pass all time-zero testing but fail very early during stressing. These first two categories of random defects typically reflect the manufacturing defect level. The third class of defects is the operating life defects. These defects reflect the ultimate manufacturing and process capability and cause failure throughout the useful life of the product. The final class of defects is wearout. This category of random defects reflects the technology capability and was the subject of the preceding paragraphs.

Figure 1.3 shows the instantaneous failure rate, or hazard function, versus time for the last three reliability classifications depicted as the so-called bathtub curve for reliability. The first stage on the left is the early or infant mortality region. The instantaneous failure rate is here characterized as rapidly falling, as the parts are in the very early stage of use and the weaker parts are still failing at a

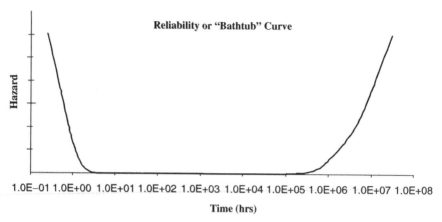

1.0E−01 1.0E+00 1.0E+01 1.0E+02 1.0E+03 1.0E+04 1.0E+05 1.0E+06 1.0E+07 1.0E+08

Time (hrs)

Figure 1.3. Reliability bathtub curve, a hazard plot or instantaneous failure rate plot in units of fails per time.

relatively rapid rate. A short, highly accelerated stress, called a burn-in stress, can often protect the customer because that stress has the possibility of moving the hardware past the infant region before any of it is shipped to the customer. Thus the customer only receives parts that are already in the operating life region where the failure rate is very low. Although shown constant for convenience above, typically the failure rate is slightly decreasing during the operating life for semiconductors. Finally, ultimately the hardware will start to wear out. The depiction here is of a wearout target that is 10 years or about 100 K hours.

1.3.4   Infant Defects

Infant defects are those that pass time-zero tests without fail, but fail shortly thereafter. Often failure analysis of the infant fails reveals structures that were extremely marginal but somehow survived the time-zero or yield tests. These defects should be entirely due to manufacturing defects. As the process matures for a new technology, the level of these defects should decrease to a low number bounded by the capability of that manufacturing facility.

There should be no structural fails contained within the infant defect population. The existence of structural fails could indicate a technology problem that still needs to be addressed or a process window problem that needs to be eliminated.

Burn-in is the primary method of removing infant defects once the product is completed. Burn-in for semiconductors is almost always done with a short temperature and voltage stress. Sometimes both a wafer burn-in and a module burn-in are performed. The wafer burn-in will typically occur at elevated temperatures and very high voltages compared to the use conditions, and will last only a few seconds. The module burn-in conditions will include an elevated temperature and a voltage that is higher than use condition, but lower than the wafer burn-in voltage, and will be applied for several hours. As the process matures and the manufacturing line defects decrease, the burn-in times and/or conditions will also typically be decreased.

There are several types of module burn-in. The most effective burn-in is also the most costly. The simplest but least effective burn-in is a DC stress where the parts are not exercised during burn-in and hence, neither are they tested while under stress. They are only tested before and after the burn-in stress. This method suffers from the fact that not all circuits are being exercised, and therefore the burn-in coverage is less than 100% and possibly very significantly less than 100%. It also suffers from the fact one cannot be sure that all of the parts really received the voltage stress. This could happen for several reasons but the parts that do not receive burn-in are called escapes and will be the most likely to fail in the field. The next level of burn-in is for the parts to be exercised but not tested during burn-in. Again they are only tested before and after the burn-in stress. This method still suffers in that parts may be incorrectly inserted during stress, hence one cannot be sure that all of the parts really received the voltage stress. The most complex, costly, and effective burn-in is for the parts to be both exercised and tested during burn-in as well as before and after the burn-in stress. This last option ensures that all of the parts that are put into the burn-in chamber will be flagged as fails at both the stress and measurement conditions. This last option also minimizes escapes since the chip responds in the oven to the testing while it is at the stress conditions. However, even for this last option a few escapes can still occur.

Infant defects can be minimized by attention to general line cleanliness and particulate control and monitoring. Again, the defect level typically decreases as the technology maturity increases on a given fabrication line.

Good design-for-reliability practices also improve the apparent defect learning. One of the more common practices is to use the minimum groundrules only when absolutely necessary. Another practice is to use special features only for those circuits where there is great leverage.

Infant defects can impact many features and their nature depends on the details of the technology. If the infant defects are not adequately eliminated either through strict process control and line maturity, by burn-in, or both, then the product will have high very early fallout rate when the customer starts using it.

1.3.5 Operating Life Defects

Operating life defects are those defects that occur, as the name suggests, during the operational life of the product. The instantaneous failure rate should be small and must be contained within the target to ensure that the product does not fail at greater than the expected rate. To meet a given specification, accelerated life modeling is used to predict the product defect level and to bridge that to the given specification. The sample size used to determine the level of operating life defects must be relatively large since this is a random mechanism and the operating-life defect level is very small.

1.3.6   Wearout

Ultimately, some part of any system will cause wearout. The time to wearout is dependent on many factors. The first factor is the technology itself. In addition to the technology, examples of other factors include the manufacturing process, the temperature, and the voltage conditions. The frequency of use or duty cycle is also very important. For example, a redundancy check circuit that is only used when the chip is powered up can tolerate much more degradation per cycle than can a circuit that is always operating when the chip in use. Other factors also have an impact on the onset of wearout including the circuit itself.

The objective of modeling in this case is to ensure that the wearout does not start until well after whatever the lifetime specification is for the product. Again, because this is a random statistical mechanism, sample size is important.

One good example of a random mechanism that causes wearout is the intrinsic dielectric mechanism of relatively thick oxides.

1.4   RELIABILITY STATISTICS

1.4.1   Introduction

The following treatment of reliability statistics is very abbreviated since the primary focus of this book is the physics of the CMOS reliability mechanisms, and there are many excellent texts on the subject of reliability statistics [9–13]. Ideally, the reader will already have some background in reliability statistics. However, since that will not be the case for everyone, this abbreviated treatment is included.

The author's experience, from teaching a course based on this material, is that some students raise objections to any treatment of statistics because they do not understand the necessity of even a minimal background. By the end of the class, however, the students have an appreciation for this background. This material is necessary to understand the following chapters, and in fact, further treatment of these concepts is given in the remaining chapters of this book.

1.4.2   Assumptions

Many assumptions, and indeed compromises, must be made in the exercise of semiconductor reliability. The first assumption is that the variation in the stress and test results is due to just the random statistical variation and/or random process variations. This assumption can usually be met, but care must be taken in the design of the experiment to ensure that the test site is appropriately designed, that appropriate equipment is being used for the stress and test, and that the equipment is in calibration.

The second assumption is that the sample stressed is representative of the population. Clearly, if the sample is not representative of the population, one can question what the results really mean and how pertinent they are to what will happen to the product in the field. This assumption is typically only weakly satisfied in the case of technology qualifications. Fortunately, technology qualifications offer special opportunities that provide relief from only weakly meeting this assumption. Quality control (QC) sampling is very different and can typically satisfy this assumption quite well.

A third assumption is that each fail is independent of all other fails and therefore has no impact on those other fails. This assumption significantly simplifies model generation and can generally be considered true for reliability defects. However this is typically not true for yield defects and sometimes there is a direct relationship between yield and reliability fails [14–16]. For some special fail cases one can decrease yield fails at the expense of increasing reliability fails and vice versa, with appropriate process changes. Please note that we are talking about process changes and not burn-in tradeoffs here. It is well known that yield fails are not usually randomly distributed on a wafer. Stapper [17, 18] and others have demonstrated that defects often follow a negative binomial distribution. Simplistically this means that defects are clustered about areas on the wafer instead of being independently distributed about the wafer area. Thus one is more likely to find a second defect in the general area that already has a defect than near an area on the wafer that has no defects. Yield issues and the negative binomial distribution are both beyond the scope of this book, but they are mentioned because of the occasional relationship between yield and reliability fails.

A fourth assumption is that it is possible to represent the discrete results obtained during reliability stressing with continuous functions. Again, this assumption is necessary in order to be able to generate models with predictive capability. Much debate and agonizing can surround this assumption depending on how well the data are behaved and how well they fit the continuous function and which continuous function fits best. Often the time range of data is limited to such a narrow window so as to allow the use of more than one continuous function to characterize the data. Thick dielectric data are a classic example of this case and the appropriate model was debated for many years. Sometimes very early or late fails cloud the fit, but much of the time these fails can be explained due to known phenomena, albeit often only after the stress. Care must be taken to ensure no new regime or mechanism has been introduced during the stressing if some fails behave significantly differently than the majority of the population of fails. If this is the case, the stress conditions may have been too extreme and the fails are no longer representative of the use conditions. Even if all of the fails follow the same distribution, care must be taken not to stress at an extreme that is not representative of the use conditions as described previously.

Two or more modes can be present with two different characteristic life times and two different characteristic variation parameters. Here a bimodal fit can often be achieved. An example would be dielectric breakdown for a relatively thick dielectric of 5–10 nm. In this case both the characteristic life and shape parameter are very different for the intrinsic or wearout population and the extrinsic or defect population. There is a clean break in the curve between the two populations and nearly all of the fails can be clearly attributed to one population or the other.

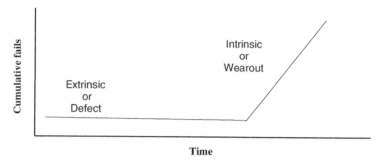

Figure 1.4. Example showing the extrinsic, or defect failure typically having a shallow slope, and the very late intrinsic, or wearout failure having a steep slope.

This example is shown in Figure 1.4 for the case of a relatively thick dielectric. (The case where the characteristic life is the only variable for two dielectric distributions will be discussed further in the dielectric chapter in relationship to the Weibull distribution.) This has direct applications to other distributions as well by inference. One would hope and expect to be able to identify two different failing types through failure analysis when two distributions are seen. This is usually the case, although it is not always true possibly due to the obvious limitations of failure analysis resource and the total number of fails. In the end, some continuous function must be determined to be able to make the reliability predictions.

A final comment about sample size is that there are some applications where the attempt is so futile that other considerations must be used. For example, the number of parts required for space applications is typically small. In these cases, the qualification would need to bridge to similar hardware, use very long-term product stressing, or possibly some other method.

1.4.3 Sampling and Variability

One can have variability in results due to raw material variations, process variations, test equipment variations, stress equipment variations, or random statistical variations. The process norm and its variations as well as the random statistical variations are the subjects of interest. The other variations should be minimized as much as possible.

Test and stress equipment variations can be minimized by timely and scheduled calibration, and detected by using standards during the testing and stressing. The standards are parts that have not undergone stress and therefore should show no variation from readout to readout. For *in situ* test equipment, that is, stress equipment that not only applies the stress but also performs the readouts, standards cannot be as readily used, so more care should be taken to ensure that equipment is calibrated. It is important to establish the variation of each piece of stress equipment. For example, ovens should have a temperature profile done for each oven and that profile should be done when the oven is fully loaded with DUTs. Ideally, a thermocouple would be monitored in each (DUT) position. For most stresses, one would want to know the temperature to within a few degrees or better. A fully loaded oven will change the airflow characteristics and could result in hot or cold spots during the stress. Each stress should be considered in a similar fashion. Note that here we are not talking about variation that would necessarily be random in these cases. If the loaded oven interior is running five degrees high, the projections will all have an additional error caused by that source. If the projections are far enough from the norm so that the ovens are checked after the fact, appropriate corrections can be made then; however, it is much more satisfying to the customer and the responsible engineer to have the equipment calibrated before the stress. A more extreme but yet typical example is the case where several tiers of DUTs are put into an oven. Depending on the airflow, each tier may have a temperature offset and even the DUTs in one tier could be different. Here is a case where temperature profiling is crucial because normally one would not keep track of the DUT position within the chamber. Thus it would not be possible to apply correct temperatures after a stress even if the oven were to then be profiled. The ovens provide a clear example of the importance of using calibrated stress and test equipment to avoid nonrandom variations. They also provide a clear example of a nonrandom variation that could be interpreted as random variation if it were unknown as in the case of temperature variation within each tier and even DUT. Another example of nonrandom variation would be dielectric stresses where one needs to ensure that the interconnect wiring has a low enough resistance, such that when leakage current is present, either due to tunneling currents or stress induced leakage currents, that the current does not cause voltage drops along the wiring and especially that the current does not cause nonuniform voltage drops along the wires. In that worst case scenario, each measurement and stress would have its own set of variations that would have to first be understood and then considered when doing the projection. It is obviously highly desirable to minimize these nonrandom variations everywhere possible. One must consciously think through each experiment to ensure that the experiment is correctly designed using appropriate and calibrated equipment. One last example of the use of appropriate equipment includes the detailed consideration of the how the measurements will be taken. If the equipment has been calibrated but it is to be used in an auto-range mode, care must be taken that there is enough time in that auto-range mode for the measurement to be returned to the computer. Even just a step up to the next higher range might cause an additional delay time. If the time required for the measurement was set at one scale without thought to these cases, errors could occur that would be nonrandom and very confusing. If this type of variation is present and not detected, it will add to the overall error. This error could cause the projections to be skewed due a number of factors including, for example, a constant offset or additional data scatter. In some cases the projections would be pessimistic and in others the projection would be optimistic. However, without recognizing the error in the system, the reliability analyst would not even know to doubt the projection. Sections in each of the following major chapters will address how to avoid many of these pitfalls for each of the mechanisms in terms of test site design, stress, and test.

One also has variation due to the process differences within each chip, wafer, and lot. Ideally, the sample chosen for stress would contain all process iterations possible in the manufacturing line.

Note that the challenge here is in the early qualification of new technologies and not in the ongoing inspection, or quality control, of the manufacturing line. These stresses are done at a much lower stress level and a shorter time than the qualification stresses since the hardware must still be shippable after these stresses. That is, a significant portion of the intrinsic life of the product cannot be used during the QC stressing. It is also true of burn-in (BI) that a significant portion of the intrinsic life of the product cannot be used during BI. For this QC issue, one has a continuous flow of parts through a short stress and a test. The good news is that a very large sample is typically stressed so one has a representative sample of the population for that time period. This stressing, taken in conjunction with the qualification stressing, does give a good overall picture of the population. Thus for QC, as yield learning occurs and the appropriate process changes are made, hardware with those changes are available for the QC-type stressing and testing. Typically in this case, no special process window lots are fabricated but the hardware currently being made on the line can be stressed to the QC times and conditions.

Now let us return to the variation within each chip, wafer, and lot due to the process differences. Our focus here is now the early qualification of new technologies. Care needs to be taken to have samples that are as representative of the standard process as possible. As discussed previously for random statistical mechanisms, the larger the representative sample from the population, the better the statistics of the result. This is true for lots, wafers, and test structures. However, typically three to five lots of hardware with several wafers from each lot is about as much hardware that the reliability engineer can expect to obtain for a process qualification. Most specifications from JEDEC, a standards governing body used by most of the semiconductor industry, require a minimum of three lots. Recall that the assumption is that the sample is randomly selected from the total population. For the ongoing quality inspections this can be true; however, for the early qualification work, this is not possible. The first set of the lots that have the final approved process will most often be used for the qualification work. This hardware is also necessarily from a single snapshot in time. Because it is a technology qualification, the hardware must be sampled very early in the manufacturing cycle, and hence, before significant additional process learning has had time to occur. This hardware should be reasonably representative of that early vintage hardware, but without due diligence during the continuing process learning, the process may change in such a way that the reliability becomes unacceptable.

The sampling should always focus on the final expected process since the processing order of two processes can be critical even when they are ostensibly independent of each other [16]. This is not usually a problem but the engineer should be aware that it could possibly make a difference. A careful design of the experiment is always a requisite to avoid confusing or misleading results.

The point of the above brief discussion is to point out that even with the best one can do to ensure that the sample is representative of the entire population, realistically one can only weakly satisfy this assumption in the semiconductor reliability world during the early qualification phases. Other strategies must be instituted to mitigate this weakness and they are discussed below. However, it is possible to do a much better job at satisfying this assumption in the ongoing quality control during the life of the product.

Returning to a focus on the types of mechanisms, if the lots are chosen wisely and if the experiments are carefully designed including the test structures, then one can investigate and qualify a new technology with a very high degree of confidence even though the sampling is extremely limited. The stress and test equipment to be used in the experiment must be carefully considered and calibrated, so that the variability in the experiment can be limited to just the process and statistically random variability that one is trying to understand.

It is at this point that previously given mechanism definitions, that is, parametric, structural, and statistical mechanisms, should be applied. The design of the experiment and sampling plan must consider each type of mechanism uniquely and separately. For the parametric mechanism, if only three lots are available for stressing, at least one of the three lots must be manufactured in such a way so as to guarantee the pertinent extreme of the process variation. For example in the case of gate dielectric, the thinnest gate permitted by the process definition must be investigated. The thickest gate permitted does not typically need to be investigated from a technology reliability perspective since the lifetime of the thicker gate will usually be greater, and depending on the variation, possibly much greater than the lifetime of the thinnest dielectric. That is not to say that no one cares how thick this gate dielectric gets. At its thickest extreme, device performance may be compromised so that extreme must be investigated for product performance related issues but not for reliability issues. Sometimes all of the pertinent process extremes can be fabricated on a single lot; however, often at least two lots are needed to obtain all of the process extremes necessary to address each of the parametric mechanisms.

Investigations to find structural mechanisms can also be nicely satisfied with lots made with processing to the extreme edge of the allowed manufacturing range. In this case it is not quite so obvious which extremes are required and lots made with nominal processing should also be investigated. However large samples are not required since by the definition given previously, if a chip is made with this particular set of features, it will fail. The challenge then is to ensure that a chip is made with those features. But, of course, unless one has already seen a fail, one does not know what set of features would cause a fail. Here, past experience is the best guide for finding that set of process conditions that might result in structural failures.

As suggested previously, large, representative sample sizes are needed for the purely statistical mechanism. For today's technology few mechanisms are purely statistical in nature. That is the good news in terms of sample size. The other good news is that what one lacks in terms of representative lots and wafers, one can often compensate through the use of good test site designs. Also appropriate test site designs can provide a sufficient effective sample size even though the total wafer count may be limited.

There are several other types of variables in the realm of sampling that should be mentioned for the reader's awareness. Sampling wafers from a tool that has just gone back into production after a preventative maintenance cycle can cause problems, as can wafers from new chemical suppliers even though they have been approved.

1.4.4   Criteria, Censoring, and Plotting Points

Data can be collected either in terms of specific shifts or in terms of pass/fail criteria. Electromigration, for example, will cause resistance increases. Depending on the process details, the lines undergoing electromigration may or may not actually go to a high resistance state or an open. The data taken and recorded may simply be those lines that shifted by more than a given value or even just the number of lines shifting by more than the given value. The data taken and recorded may include the resistance shift of each line. Clearly, if the resistance shift of each line is taken, the data can later be expressed in terms of a pass/fail limit.

For technology qualifications, it is usually desirable to record as much data as possible including the actual shifts. These data are useful at several levels including failure analysis decision making and isolating wafer and lot dependencies. The total shift data are vital for the parametric mechanisms but are also desirable for the structural and statistical mechanisms. Once the product is in manufacturing and minimal stressing and testing can be done, often pass/fail criteria are adequate as part of the shipping qualifications. Even during the wafer stressing and testing for the shipping qualifications, chip failure location on the wafer can be used to great advantage and flag process problems early.

One manufacturer [14] has shown excellent results using product yield as an indicator of defect density for neighboring chips. They divide the chip into an edge chip region and a center chip region. Chips in a "bad" neighborhood, in either region, are 10 times more likely to fail during burn-in than those chips from a "good" neighborhood in either region. This work covered one hundred thousand to one million chips and included technologies from 0.25 μm down to 0.09 μm as well as both aluminum and copper. They give a quantitative expression for the number of chips that constitute a "bad" neighborhood and show excellent correlation between their modeled behavior and actual burn-in data. This paper also provides a nice set of references for earlier work in this area.

Most stress equipment built today has *in situ* capability. That is, it has the ability to both stress the DUT as well as to test that device. Hence, obtaining exact times-to-fail is simple. For example, the exact time for a 20% shift in the line resistance due to electromigration can be obtained from the test equipment. In this case the time for each fail or predetermined-shift is known uniquely and can be plotted uniquely.

Occasionally, times are known only for groups of fails. This may happen if older stress equipment is used or if stresses are done in ovens because of large sample sizes. Here the parts are stressed for a given period of time, removed from stress, taken to a tester, tested, and then put back on stress. For reference, this equipment is typically referred to as *ex situ* equipment. Now only those parts failing after the given time period are known. The cumulative population failing at that time is typically then plotted against that time. However, depending on the design of the experiment and the specific issues being investigated, it is sometimes preferable to break any experiment into intervals and plot the cumulative fallout in the middle of that interval be it on a log or linear scale. Plotting positions are discussed in more detail below.

One of the challenges of *ex situ* testing is that care must be taken that the mechanism does not relax. For example if one were stressing for ionic contamination in *ex situ* equipment, one would need to apply a bias to the parts as they came off the *ex situ* equipment so that the parts did not recover while awaiting test. Historically large automotive batteries were used for this task when *ex situ* equipment was used to stress mechanisms that relaxed.

Data may be censored for many reasons. There are statistical procedures for handling nearly every type of censoring. The most common reason for censoring is likely lack of stress time. Parts can only be stressed for a given time period and it is expected that some part of the sample will survive beyond that time. Normally for parametric mechanisms, one would have a model that predicted how long the product should last given the shift, the stress conditions and the time under stress. The model may allow the entire sample to be used.

If most DUTs failed within a relatively narrow time range but several did not, two questions arise. Were these DUTs significantly better than the rest and potentially not subject to the mechanism that caused the majority to fail, or was the stress or test somehow compromised on these parts? Again, failure could be defined as a certain shift or an actual open, short, or cessation of operation. The nonfailing parts might be candidates for censoring if the stress was not fully applied, or they might be part of a second and better population so that a bimodal distribution should be used. There are at least three common ways of handling these cases depending on the exact circumstances. One characterization method would be to characterize the failing distribution with a bimodal distribution. Another way would be to censor the unfailing or late-failing parts, subtract them from the sample, and recalculate the failing distribution and times based on the new sample size. A third method would be to leave them in the sample. It is clearly highly desirable to know the reason for the preponderance of early or late fails before making the decision as to how to characterize them. However, the statistics in doing so are straightforward. Another pertinent example is those DUTs that pass the initial test but fail during the first application of the stress. These are not yield fails, nor are they real reliability fails since they failed as the initial stress was being applied. These would actually be in the ship product quality level (SPQL) category of fails. This is the class of fails that cause a product to look good as shipped from the supplier, but when the customer first turns on the machine, it fails to function. This is obviously a very serious class of fails but can show up on reliability plots and unless handled correctly can compromise the reliability conclusions.

Step stressing is a technique by which multiple stress conditions are obtained on the same set of hardware. These techniques are advantageous because a smaller overall sample is then required since the same parts are used for every condition. Usually, but not always, the steps have increasing acceleration as outlined below, and if done carefully, step stressing is powerful. One of the attributes of a well designed experiment is equal representation across all cells so that if one stress condition or cell gives results unlike the other stress cells, it cannot be due to one wafer or lot dominating that cell or lacking in that cell.

Historically, the key to step stressing is that each new applied condition must result in an acceleration of about $100 \times$ in time from the previous cell. With a $100 \times$ acceleration, it is possible to calculate all of the acceleration parameters to the same degree as if two independent cells had been stressed. If this cannot be achieved, the separation in the results may not be adequate to interpret the data, compromising the entire experiment. If a step stress is contemplated, a focus on experimental design is necessary. This type of step stress is discussed further in Section 2.3 with a dielectric example.

1.4.4.1 Plotting Position.

The simplest and most straightforward method of plotting the cumulative fails would be to simply plot the fail percent against the time of fail. For example, assume that a stress of four parts resulted in one fail at the one-hour readout, a second fail at the second-hour readout, a third fail at the third-hour readout and finally the fourth fail at the fourth-hour readout. The concept of a plotting position is to plot the fail in the most representative position in the interval of failure. Thus instead of plotting the first fail at 25% in our example, we plot it someplace between 0 and 25%. One option is to plot the fails at the midpoints of the intervals such the fails would be plotted at the 12.5%, 37.5%, 62.5% and 87.5% points on the y axis corresponding to 1, 2, 3, and 4 hours respectively on the x axis. Conceptually, this y-interval plotting position may be the simplest of the plotting positions. This plotting position is shown for 4 data points and 20 data points in Figure 1.5. There can be several choices for plotting positions and good statistical arguments for each. Some of the more common plotting positions are given below, but a thorough treatment of the statistical rationale behind each choice is beyond the scope of this book. The interested reader is referred to the statistics texts. In addition to the statistical arguments suggesting the requirement of a plotting position, there would be the practical problem that plotting simplistically as mentioned above, would require plotting a point at 100% fallout which for log plots does not exist.

The smaller the sample, the more crucial is the use of an appropriate plotting position. Several authors have suggested different plotting positions, any one of which may be best under a given set of circumstances. Two common plotting positions are $F(t_i) = i/(n+1)$ or $F(t_i) = (-1/2)/n$ [19] and $F(t_i) = (-0.3)/(n+0.4)$

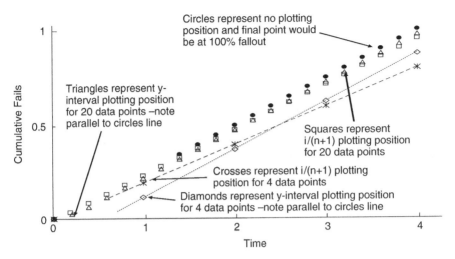

Figure 1.5. Plot comparing *y*-interval and *i*/(*n*+1) plotting positions with no plotting position for 4 and 20 data points.

[20]. Other authors have suggested various alternatives to the above plotting positions. It is a simple matter to compare the possibilities with one's actual data especially in today's world where the data is typically analyzed using a computer.

Note that the above discussion is only applicable and pertinent when graphical procedures are used. Typically the analysis and estimation procedures will be done using computer procedures.

As can be seen by Figure 1.5, there is little difference between the values of the *y*-interval plotting position and the $F(t_i) = i/(n+1)$ plotting positions for samples of 20 or greater, although the slopes are somewhat different. There are significant differences both in value and slopes for both of the plotting positions shown on Figure 1.5 for the case of only four points.

1.4.5 Definitions (Normal)

1.4.5.1 Failure Figures-of-Merit.

The first set of definitions that are given are for the generic failure language that is used within the reliability field. Wearout, intrinsic failures, or end-of-life failures are defined at the end of the expected life of the product. Defect or extrinsic failures are typically expressed both in terms of end-of-life failures and various points of time throughout the life of the product especially after one year. As seen below, these expressions may either be in terms of the cumulative fails or the failure rates of the hazard function. The terminology used for cumulative fails is typically parts per million (ppm), and the terminology used for failure rates may be fails per 1000 hr or fails per billion device hours (FITs).

1.4.5.2 General. The independent variable for nearly all practical cases within reliability statistics is time. Sometimes the distribution parameters will be used as the independent variables when comparing distributions or models. This might take the form of plots comparing the mean and standard deviation for various lognormal distributions. (See Section 1.4.8) For Weibull distributions this might take the form of a comparison of the various shape parameters at the characteristic life for each distribution (See Section 1.4.7).

Recall that our first assumption above was that the variation in the results was due only to random variations. A ramification of this definition is that the actual value is not known at any precise point in the domain of interest (nearly always time for reliability work) due to the random statistical variation. Within the framework of reliability modeling, a distribution function is sought to describe the pattern of values that are most likely to occur.

1.4.5.3 Probability Density Function (PDF). The probability distribution function or probability density function is that function that describes the fails in each period of time. This is nothing more than a histogram with time intervals as the abscissa and the percentage of fails as the ordinate or *y*-axis value. Since this function gives the fails that have occurred in the previous time steps, if an analytical expression can be found that accurately describes those past times steps, that expression can then be used to predict the fails that would be expected in future time steps. For example, if a stress results in all samples failing of 50 samples on stress, and the data is then grouped within subregions, the discrete distribution in Figure 1.6 results. Note that these 50 data points were generated by a simulation with a mean fail time of $\mu = 10$ hrs and a standard deviation of $\sigma = 2$, and hence, the data are representative of some mechanism that causes the population to fail with the above characteristics. The intent of this plot is that it represents a small sample from a much larger population. The next charts will also present data representative only of the empirical sample, not of the true population. These charts will highlight the randomness of the resulting distributions when only small samples are possible.

The assumption is that the true mean and standard deviation of the population is known. The charts then depict that variation for the distributions associated only with choosing a small sample. There are several features to notice about this empirical distribution before continuing. The first is that it does indeed have the general appearance of a normal distribution with a mean of 10 hr. The second observation is that it has significant aberrations from the expected values both for the 3.5–4.5 time interval and the 6.5–7.5 time interval. But this is just a small sample from a true normal distribution so the aberrations are not a consequence of data collection issues; they are just a function of the random variation to be expected because the sample size chosen was only 50. Note that if this would have been a real experiment, the early fail at about four hours might have been attributed to something other than the primary fail mechanism, which acts between 6 and 14 hours and hence, considered as a nonrepresentative fail and eliminated from the population. Obviously, this could be the case in a real experiment; however, here it is an integral part of the normal fail distribution and is equivalent in all respects, except its time to fail, to the later fails. As such it cannot be eliminated from the sample without compromising the experimental conclusions. The same observations could be made for some of the fails in the 6.5–7.5 time interval. Here two to three fails were expected and seven were observed for this simulation. One could wonder about power spikes and any number of other experimental problems during this time interval. However, the large number in this interval in our case is again purely random statistical variation and must be accepted as experimentally valid.

The point of the above discussion is twofold. First, when dealing with statistical mechanisms, it is crucial to have large sample sizes. Otherwise, one may not be able to differentiate between experimental data issues and purely random statistical variation for those points that appear to be far from the expected distributions. The second point is that data points should only be removed from the sample when there are very clear experimental reasons to justify it.

When the time interval approaches zero in the limit, the above discussion applies to continuous functions as well. If the discrete function is well behaved, that discrete function may be represented by a continuous distribution. The continuous PDF, which was used to generate the simulation shown in Figure 1.6, is given in Figure 1.7. Five such distributions were simulated and averaged together and are plotted in Figure 1.8 after normalization. Note that the distribution parameters were identical in all five of the simulations used to generate Figure 1.8, but each individual simulation of 50 points varied significantly from the curve shown in Figure 1.7. The combination of the five simulations of 50 points shown in Figure 1.8 after normalization, much more closely resembles the continuous distribution shown in Figure 1.7. The difference between Figures 1.6 and 1.8 is due purely to random statistical variation.

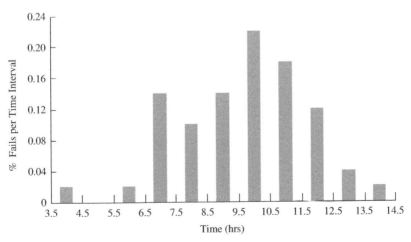

Figure 1.6. A histogram corresponding to a sample from a normal distribution with $\mu = 10$ and $\sigma = 2$; one simulation with 50 points.

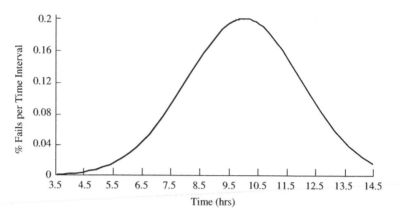

Figure 1.7. Continuous normal distribution with $\mu = 10$ and $\sigma = 2$.

Figure 1.9 shows a simulation using the same starting distribution but with 250 data points normalized to 50 data points instead of just a simulation of 50 data points. Here there is still significant deviation from the normal distribution. This is most obvious in the 8.5–9.5 and 10.5–11.5 intervals. Figures 1.8 and 1.9 show deviations because the sampling is still limited. Figure 1.10 shows that same simulated distribution but now with 1000 data points and again normalized. This curve is now very true to a normal distribution especially at the values close to the mean. The shape of all of the curves in Figures 1.6–1.10 are the classic "bell curves." But if only 50 points or less are plotted, serious deviation from the pure normal distribution should be expected.

Mathematically, one can allow the subregions to become smaller and smaller, and to in fact approach zero. This is the case shown in Figure 1.7 where the PDF becomes the continuous function, $f(t)$ and is defined as the probability that the

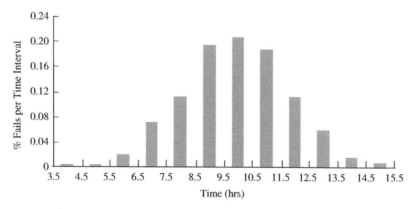

Figure 1.8. A histogram corresponding to a sample from a normal distribution with $\mu = 10$ and $\sigma = 2$ for 5 simulations each with 50 points after normalization.

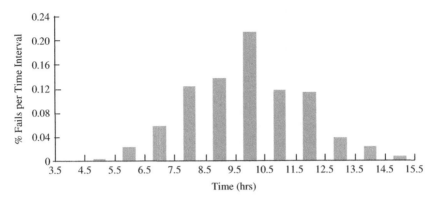

Figure 1.9. A histogram corresponding to a sample from a normal distribution with $\mu = 10$ and $\sigma = 2$ for a single simulation with 250 points normalized to 50 points.

values of t lie between $(t-0.5dt)$ and $(t+0.5dt)$, where $0 < t < \infty$. The equation for the PDF for the normal distribution is given Equation 1.1. Note in the general case the mean may have any value whereas the standard deviation must have a positive value.

$$f(t) = \frac{1}{\sigma\sqrt{2\pi}} \exp\left[-(t-\mu)^2/2\sigma^2\right] \quad \text{where } 0 < t < \infty \qquad (1.1)$$

where μ is the mean time to fail and σ is the standard deviation.

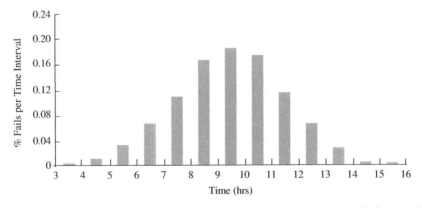

Figure 1.10. A histogram corresponding to a sample from a normal distribution with $\mu = 10$ and $\sigma = 2$ for a single simulation with 1000 points normalized to 50 points.

1.4.5.4 Cumulative Distribution Function (CDF). The cumulative distribution function (CDF) is the cumulative sum from each subregion of the failing population for a discrete function. The CDF, or $F(t)$, is that fraction of the population that has failed by time t. Stated another way, $F(t)$ is the probability that a part will fail by time t. Hence, as one adds the failures from each time interval, the CDF increases from zero to one. The CDF for a continuous function is then simply the integral of the PDF.

The CDF will typically be the first plot that the reliability engineer makes after taking the experimental data. Notice that for the CDF, only the fails are plotted. This has the important ramification that only the failing samples will be used to determine the distribution. If the reliability engineer has designed an experiment with 1000 samples for a given time duration, and after the end of that time, only five samples have failed, only five points can be plotted and only the tail of the distribution will be displayed. The remaining 995 samples are censored before failure occurs. As we shall see later, a large variation would be expected in the results at the tail of a distribution due to large confidence bounds. Hence, very little about the sample distribution could be expected to be accurately determined in that case even though a large sample was stressed. At the other extreme, if only five samples were stressed and all failed, there would again be only five fails to plot but that would represent the entire CDF for this sample. Now because the total sample size was so small, the confidence bounds would be very large, and the distribution parameters of the population would have a such wide range of possible values for a given set of confidence bounds that little would be known about the actual distribution of the population, even though it is known for our sample of five.

The CDFs that are shown in Figures 1.11 and 1.12 are related to the discrete and continuous PDFs given in Figures 1.6 and 1.7, respectively. We need to highlight that these are empirical or sample CDFs and not those for the population. As expected, since these data were generated by a simulation of a

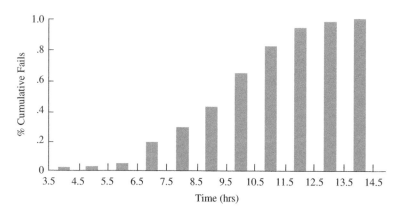

Figure 1.11. CDF for discrete normal distribution with $\mu = 10$ and $\sigma = 2$ for a single simulation with 50 points.

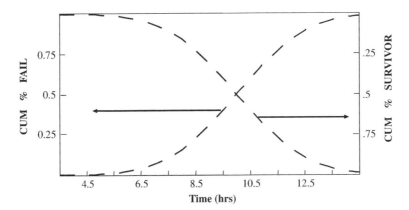

Figure 1.12. CDF and reliability function for continuous normal distribution with $\mu = 10$ and $\sigma = 2$ for a single simulation with 50 points.

normal distribution with a mean of 10, half of the population has failed by 10 hours. Equivalently, the probability of failure by 10 hours is 50%. Expressed mathematically, the probability of fail at a given time is the area expressed by

$$F(t) = \int_0^t f(x)dx \qquad (1.2)$$

From the PDF the most general equation for the CDF is given as Equation 1.3 below. The reliability subset of this equation limits t to the positive time values.

$$F(x) = \int_0^t \frac{1}{\sigma\sqrt{2\pi}} \exp\left[-(x - \mu)^2/2\sigma^2\right] dx = \Phi[(t - \mu)/\sigma] \qquad (1.3)$$

where Φ is the normal distribution function and $0 < t < \infty$. The probability that t lies within the range from 0 to $+\infty$, is equal to one since in our case that range includes all possible values of t. Hence, the value of the integral must be 1 representing 100% failure.

1.4.5.5 Reliability Function R(t).
Whereas the CDF is the cumulative failing population, the reliability function, R, is the cumulative surviving population. It is obtained by subtracting the cumulative fails (CDF) from 1, i.e., $R(t) = 1 - F(t)$. Just as the CDF must equal one after the last fail, the reliability function must equal 0 since there are by definition no survivors. The equivalent continuous reliability distribution is also plotted in Figure 1.12 where the reliability scale is on the right. Obviously for the R, the point at which 50% of the population has survived is 10 hours or the mean of the normal distribution.

1.4.5.6 Instantaneous Failure Rate (IFR) or Hazard Function, h(t).
Another important statistical concept that is used within the reliability community is that of the instantaneous failure rate (IFR) or hazard function $h(t)$. The $h(t)$ is

defined as the probability that those parts that have survived until a time, t, will fail in the next increment of time, Δt. The $h(t)$ is then a probability divided by the incremental time Δt which converts it to a rate. Hence, $h(t)$ is a failure rate expressed either in terms of fails per time or fraction failing per time. The $h(t)$ or hazard function is defined mathematically as

$$h(t) = f(t)/[1 - F(t)] = f(t)/R(t) \qquad (1.4)$$

Derivations for the $h(t)$ are given in references [12] and [5]. Those derivations are beyond the scope of this book although the reader may find that a review of those derivations may be helpful in better understanding the hazard function.

This figure of merit is most useful in describing the case of fallout during the normal lifetime of the product. That fallout is also known as extrinsic fallout or defect fallout. The $h(t)$ can be integrated in a manner similar to the PDF or $f(t)$ to obtain $H(t)$ sometimes referred to as the cumulative hazard function. For more information on the cumulative hazard function than given below, the interested reader is directed to the books previously referenced and especially reference [10].

The discrete $h(t)$ related to Figures 1.6 and 1.11 is shown in Figure 1.13 and the continuous $h(t)$ related to Figures 1.7 and 1.12 is shown in Figure 1.14. For the case of this normal distribution, the $h(t)$ is an increasing function. The discrete $h(t)$ shown in Figure 1.13 has a general similarity to the continuous $h(t)$ in Figure 1.14 but because of random statistical variation is far from identical.

Because the CDF is the summation or integral of the PDF, the statistical variation can appear to be mitigated. This could be the conclusion of a casual comparison of Figures 1.6, 1.11, and 1.13 as compared to Figures 1.7, 1.12, 1.14 respectively. However, a careful inspection will still reveal the variation.

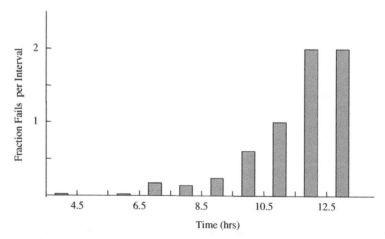

Figure 1.13. Hazard function of discrete normal distribution with $\mu = 10$ and $\sigma = 2$ for a single simulation with 50 points.

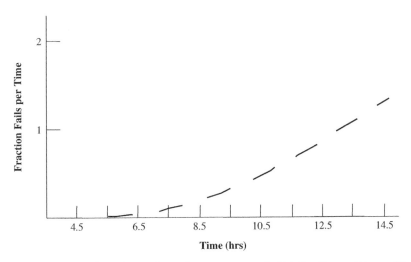

Figure 1.14. Hazard function of continuous normal distribution with $\mu = 10$ and $\sigma = 2$ for a single simulation with 50 points.

1.4.5.7 Cumulative Hazard, H(t). The cumulative hazard function is the integral of the hazard function as shown in Equation 1.5 and is another function that can provide useful insight. The cumulative hazard is directly related to the CDF as shown in Equations 1.4 and 1.5. The $H(t)$ can be more convenient for some types of censored data when plotting manually than is the CDF. However, for semiconductor technology reliability and in today's computer world, it is rarely used and is presented here for completeness.

$$H(t) = \int_0^t h(x)dx \qquad (1.5)$$

Another potentially useful relationship is $H(t) = -\ln R(t)$. This equation may be verified by taking the derivative of both sides to obtain Equation 1.5. This relationship may be expressed in several different forms.

1.4.5.8 Average Failure Rate (AFR). The average failure rate (AFR) is the total number of fails within a given time interval expressed as a rate. This is a useful figure of merit because the time interval can be defined and then a single number used to characterize the reliability. Note that the hazard function is variable so the value depends on the chosen time. One could get the same value of AFR for equivalent product from two vendors using the hazard function, but the shape of the curves in arriving at that value might cause one product to be acceptable and the other one unacceptable. Because the AFR is an average, it is a simpler figure of merit and is an acceptable figure of merit in many cases. It is typically the figure of merit of choice to compare failure rates after one year and sometimes after 5 or 10 years as well. The one-year AFR is not necessarily set exactly at one year or 8.76 khr, sometimes it is approximated as 10 khr. The AFR is just the integral of $h(t)$ between a specific time interval, divided by that time interval. Mathematically it is represented as

$$AFR(t_1, t_2) = \left(\frac{1}{t_2 - t_1}\right) \int_{t_1}^{t_2} h(t)dt. \qquad (1.6)$$

1.4.5.9 Moments. Although moments of a probability density function may be defined that will characterize a PDF, the accepted practice in reliability engineering is to use the distribution attributes themselves, such as the characteristic life and shape factor for a Weibull distribution, rather than the moments of the PDFs. A given moment may be described as the relationship of every value of $f(x)$ with respect to some fixed value.

A detailed discussion of the moments of distributions is beyond the scope of this book and is generally not necessary for reliability engineering. However, the interested reader is directed to Green and Bourne [12], for a rather complete discussion on the various moments for the major distributions. The moments of most interest in semiconductor reliability are 1) the first moment about the origin which is the mean for the normal distribution and the characteristic life for the exponential distribution; and 2) the second moment about the mean which is the variance for the normal distribution and the square of the characteristic life for the exponential distribution.

First moment about some constant x_0:

$$M_1 = \int_{-\infty}^{\infty} (x - x_0)f(x)dx. \qquad (1.7)$$

Mth moment: $$M_m = \int_{-\infty}^{\infty} (x - x_0)^m f(x)dx. \qquad (1.8)$$

Then the first moment about the origin ($x_0 = 0$) is the mean:

$$M_1 = \int_{-\infty}^{\infty} xf(x)dx. \qquad (1.9)$$

The second moment about the mean is the variance:

$$M_2 = \int_{-\infty}^{\infty} (x - M_1)^2 f(x)dx. \qquad (1.10)$$

1.4.5.10 Fractile or Quantile Function. The p fractile, or equivalently the p quantile, is the time at which that fraction, in percent, of the sample fails. Hence, $p = 5\%$ represents the 5% failure point of the sample and $p = 50\%$ represents the 50% failure point, and $p = F(t_f)$. It is sometimes the case that the time position of the fifth, tenth, or twentieth percentile of the lifetime distribution is specified as the fail criterion. This is especially true when the tails of the population are the primary cause of failure. The time to fail of the p fractile is then just $t_f = F^{-1}(p)$ or the inverse of the CDF. Thus for the example in Figure 1.12, the time of fail for the 25 fractile or 25% failing portion of the sample is about 8.75 hrs. Note that the 25, 50, and 75 fractiles or quantiles are also known as the quartiles.

1.4.5.11 Reliability Projections and Closure. The normal distribution was used to introduce the various reliability functions because everyone typically has had either direct or indirect experience with the normal or Gaussian distribution.

One final concept is needed before moving to the other distributions that will be commonly used in reliability statistics. That concept is the one of transforming the distribution of choice in such a manner that the CDF is linear when plotted.

The CDF plot for a continuous normal distribution was given in Figure 1.12. However, one could not make any graphical projections using that chart since the plot is not a straight line. The axis needs to be modified to achieve a straight line. Alternatively the equivalent set of equations could be numerically solved to make the projections. In practice both are typically done. A computer is used to solve equations, fit data, and do all of the calculations, and it is also used to plot the results on the appropriate axis so that one may have a visual demonstration of the solution and of the projection.

Before continuing the endeavor to make the CDF of the normal distribution a linear plot on some axis, the concept of the standard normal distribution needs to be considered. Equation 1.3 is not soluble in closed form. Originally, tabular values were used to solve these equations. A standard normal distribution was introduced so that only one set of values would be necessary and all variations of a normal distribution could be converted to that standard normal distribution. In today's world of computers and numerical integration, this procedure would not be necessary, but it is instructive and should give the reader a better intuitive understanding. Given a distribution where x is the random variable of a normal distribution that has a mean of μ and standard deviation of σ, the transformation to the standard normal distribution is given by

$$z = (x - \mu)/\sigma. \tag{1.11}$$

With this transformation and noting that for a standard normal distribution $\sigma = 1$ by definition, the PDF given in Equation 1.1 and the CDF given in Equation 1.3 become

$$\text{PDF: } \phi(z) = \frac{1}{\sqrt{2\pi}} \exp\left[\frac{-z^2}{2}\right] \quad \text{where } -\infty < z < \infty \tag{1.12}$$

and

$$\text{CDF: } \Phi(z) = \int_{-\infty}^{z} \left[\frac{1}{\sqrt{2\pi}} \exp\left(\frac{-w^2}{2}\right)\right] dw \quad \text{where } -\infty < z < \infty. \tag{1.13}$$

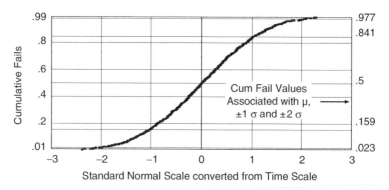

Figure 1.15. Plot of an empirical CDF of 1000 points $\mu=0$ and $\sigma=1$ with "typical" axis.

As mentioned above, historically the values for the standard normal CDF were tabulated and all normal distributions converted to the standard normal distribution for integration based on that tabulation. Most, if not all statistics books, contain tables showing the solution of Equation 1.13, including the referenced statistics books. Plots of an empirical, normal CDF are shown in Figures 1.15 and 1.16 for a standard normal distribution for the 1000 point experiment with the points generated by simulation assuming the standard normal distribution. Note that for the standard normal distribution, the PDF is symmetric about zero so that indeed $\mu=0$ and hence, at zero, half of the population will have failed.

We return to the second part of the normal CDF in Equation 1.3 to obtain the y-axis transformation for the normal plot that will yield a linear CDF plot if the distribution is normal. Mathematically, this may be expressed from Equation 1.3

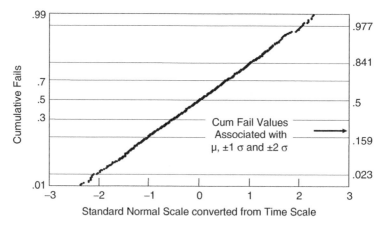

Figure 1.16. Plot of an empirical CDF of 1000 points $\mu=0$ and $\sigma=1$ with axis transformation to yield a linear CDF plot if it fits a normal distribution.

for the normal CDF as $F(x_p) = \Phi(\{x_p - \mu\}/\sigma)$. Solving this equation for x_p by taking the inverse of the function $F(x_p)$, one obtains $x_p = \mu + \sigma \Phi^{-1}(x_p)$. If x_p is then plotted versus $\Phi^{-1}(x_p)$, a straight line will result if the distribution in question can be represented with a normal distribution. Note that each of the quantile pairs is plotted equidistant from the mean on this new axis. Thus $+/-\sigma$ points are equal distant from the mean as are the 0.01 and 0.99 points whereas the distance on the y axis between 0.01 and 0.023 is nearly equal to that between 0.3 and 0.5. Note that the right y-axis scale could have chosen to show the equivalent σ value instead of the actual cum fail percent. Given this choice of percentile values, the right y-axis equivalent values are 2, 1, 0, −1, and −2 σ, respectively, from top to bottom. Thus Figure 1.16 has an axis that transforms the CDF of a normal distribution into a linear function.

Note the power of this transformation. The CDF curve is now linear since the data follows a normal distribution and while the tails of the distribution vary from the straight-line projection, one could project from the data back to the time at which 1 ppm or 0.0001% would be predicted to fallout, or to any other desired fallout. Obviously this projection assumes that no new mechanisms were encountered as per the assumptions previously stated in Sections 1.2.2 and 1.4.2.

It is instructive to observe these two plots when only 50 experimental data points are available. We have already discussed the importance of sample size for at least statistical mechanisms, but since a picture is worth a thousand words, the picture is shown below. Clearly the slope and mean of the curve for Figure 1.16 are known to a much higher confidence than that for Figure 1.17, or equivalently Figure 1.18, which has the axis transformation.

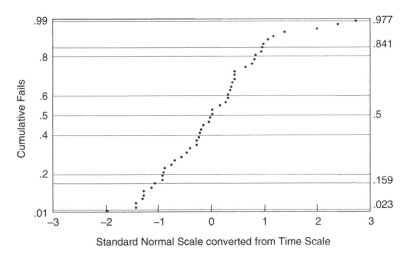

Standard Normal Scale converted from Time Scale

Figure 1.17. Plot of an empirical CDF for normal distribution of 50 points for $\mu = 0$ and $\sigma = 1$ with "typical" axis.

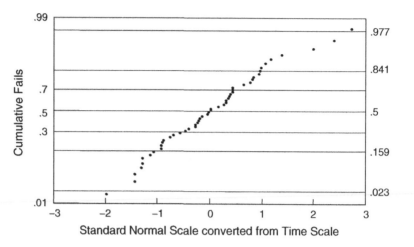

Figure 1.18. Plot of an empirical CDF for normal distribution of 50 points $\mu = 0$ and $\sigma = 1$ with axis transformation to yield a linear CDF plot if it fits a normal distribution.

The reliability concepts have been introduced using the normal distribution since most are familiar with it and its bell-shaped PDF. The normal distribution will be discussed further in terms of the lognormal distribution in Section 1.4.8 below as well as in the electromigration chapter in Sections 7.4.2 and 7.4.3. We will now move to the distributions that will be most commonly used in the remainder of this book. Those distributions are primarily the lognormal distribution and the Weibull distribution which are both two-parameter distributions as typically used in semiconductor reliability, although a third parameter could be introduced for either. It has in fact been shown that a third parameter for the lognormal distribution can more effectively describe electromigration behavior because it can be used to model an incubation period [21]. The cost is a much larger sample size to determine all three parameters than that required to determine just two parameters so that the more complex model has historically rarely been used. As features continue to shrink it may become advantageous to move to a three-parameter distribution to gain the additional accuracy that could provide additional reliability margin.

1.4.6 Exponential Distribution

The exponential distribution is a single-valued distribution and has the very important attribute that $h(t)$, the instantaneous failure rate, is a constant The PDF, CDF, and $h(t)$ are given in Equations 1.14, 1.15, and 1.16, respectively, where time, $t \geq 0$.

$$\text{PDF: } f(t) = \lambda \exp(-\lambda t) \tag{1.14}$$

$$\text{CDF}: F(t) = \int_0^t f(x)dx = 1 - \exp(-\lambda t) \qquad (1.15)$$

$$\text{IFR}: h(t) = f(t)/[1 - F(t)] = \lambda \qquad (1.16)$$

Because $h(t)$ for the exponential distribution is a constant, the probability of failure in the next time increment is independent of the length of time the product has already been in use or on stress. The ramification of this is that the part has no aging that impacts its failure rate. A new part has the same failure rate as does the part that is still functioning having already survived many years of operation. Another way of stating this is that the part has no memory of its past operation.

The first moment about the origin is $1/\lambda$ and the second moment about the mean is $1/\lambda^2$. These moments are given for comparison to the normal distribution and are not frequently used in semiconductor technology reliability.

We show a set of curves for the PDF, CDF, and $h(t)$ for the exponential distribution where λ has the values of 0.5, 1, and 2. Figures 1.19 and 1.22 depict the PDF and $h(t)$, respectively. The exponential CDFs are plotted in Figures 1.20 and 1.21.

Note that the vertical axis has been modified in Figure 1.21 so that the CDF for the exponential distribution will plot linearly. The procedure here is much simpler than for the normal distribution, although in principle it is similar. Equation 1.15 is solved for λt yielding $-\ln(1-F(t))$. The scale transformation is possibly most recognizable for $\lambda = 1$, since at $t = 1$, the exponential argument value is 1 and the cumulative percent failed is 63%.

Thus far we have been discussing the exponential distribution in terms of a single parameter. The exponential distribution can also be utilized with two parameters and the PDF, CDF, and $h(t)$ are shown in Equations 1.17–1.19 for

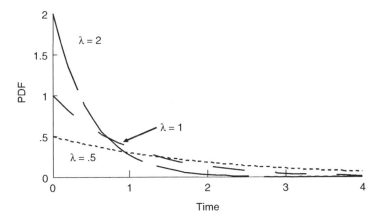

Figure 1.19. PDF of exponential distribution for three values of the distribution parameter, λ, plotted on linear axis.

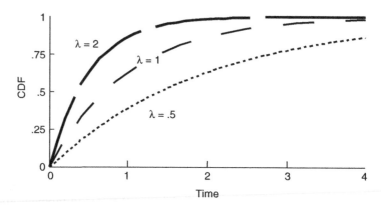

Figure 1.20. CDF of exponential distribution for three values of the distribution parameter, λ, plotted on linear axis.

that case. The second distribution parameter, t_0, would typically be a timescale shift in the case of semiconductor reliability although in some cases its interpretation might more appropriately be a threshold parameter.

$$\text{PDF: } f(t) = \lambda \exp[-\lambda(t - t_0)] \tag{1.17}$$

$$\text{CDF: } F(t) = \int_0^t f(x)dx = 1 - \exp[-\lambda(t - t_0)] \tag{1.18}$$

$$\text{IFR: } h(t) = f(t)/[1 - F(t)] = \lambda \tag{1.19}$$

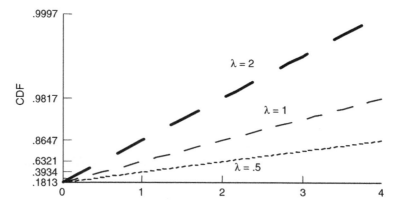

Figure 1.21. CDF of exponential distribution for three values of the distribution parameter, λ, a plotted on a transformed axis to yield linear CDFs.

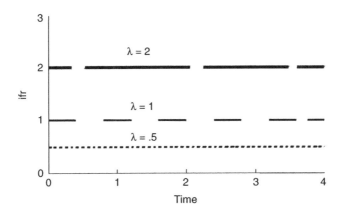

Figure 1.22. $h(t)$ of exponential distribution for three values of the distribution parameter, λ, plotted on linear axis.

1.4.7 Smallest Extreme Value and Weibull Distributions

The smallest extreme value distribution is included primarily to demonstrate its relationship to the Weibull distribution; however, it does have some application in its own right. These applications are not in general use in semiconductor reliability today and will not be discussed in this book. The smallest extreme distribution is applicable for $-\infty < x < \infty$ where κ is the location parameter and may be positive or negative and ζ is the scale parameter which must be positive. As in the Weibull distribution, 63.2% of the population has failed at the value $F(x = \kappa)$.

$$\text{PDF: } f(x) = \frac{1}{\zeta} \exp \left[\frac{x - \kappa}{\zeta} \right] \exp \left[-\exp \left(\frac{x - \kappa}{\zeta} \right) \right] \qquad (1.20)$$

$$\text{CDF: } F(x) = 1 - \exp \left[-\exp \frac{x - \kappa}{\zeta} \right] \qquad (1.21)$$

$$\text{IFR: } h(x) = (1/\zeta) \exp[x - \kappa/\zeta] \qquad (1.22)$$

To obtain the Weibull CDF one defines $x = \ln y$ and $\kappa = \ln \varphi$, and $\zeta = 1/\beta$, then substituting in Equation 1.21 one obtains

$$F(y) = 1 - \exp[-\exp(\ln y - \ln \theta/1/\beta)] = 1 - \exp \left[-(y/\varphi)^\beta \right], \qquad (1.23)$$

where y is now constrained to be $y > 0$.

The result in Equation 1.23 is identical to the Weibull CDF given in Equation 1.25. It is left as a student exercise to show that this special case of the smallest extreme value function reduces to the Weibull distribution for the PDF and $h(t)$. A more complete discussion is given by Nelson [10].

The Weibull distribution, as a two-parameter distribution, has a characteristic life, φ, and a shape parameter, β. The PDF, CDF, and $h(t)$ are given in Equations 1.24, 1.25, and 1.26, respectively.

$$\text{PDF}: f(t) = (\beta/t)(t/\varphi)^{\beta} \exp\left[-(t/\varphi)^{\beta}\right] \qquad (1.24)$$

$$\text{CDF}: F(t) = \int_0^t f(x)dx = 1 - \exp\left[-(t/\varphi)^{\beta}\right] \qquad (1.25)$$

$$\text{IFR}: h(t) = f(t)/[1 - F(t)] = (\beta/t)(t/\varphi)^{\beta} \qquad (1.26)$$

Obviously for semiconductor reliability purposes, these equations are applicable only for $t \geq 0$.

Notice first the case where the shape parameter is equal to 1, that is $\beta = 1$. This is the case in which the Weibull distribution collapses to the exponential distribution. In this case, Equations 1.24–1.26 reduce to Equations 1.14–1.16, respectively, where $\lambda = 1/\varphi$.

The next observation gives the characteristic life its meaning. Substituting φ for t in the CDF, we get $F(t) = 1 - \exp(-[\varphi/\varphi]^{\beta}) = 1 - e^{-1} = 0.632$. Hence, 63.2% of the population fails by the characteristic life, φ, independent of the shape parameter β. This makes the characteristic life a powerful figure of merit for the Weibull distribution. Therefore when discussing distributions which follow a Weibull distribution, it is the 63.2 percentile that is the desired figure of merit, not the fallout at the 50 percentile as is the appropriate figure of merit for the normal distribution.

The shape parameter, β, gives the slope of the Weibull distribution. Usually in reliability projections, the slope has a larger impact on the final projection than the characteristic life since the projections are typically extrapolated across several or even many orders of magnitude. Any error in the shape parameter is then magnified by that extrapolation.

Many reliability systems can be and are modeled using a Weibull distribution. As the figures depicting the Weibull distribution below will show, it is a very flexible distribution. This flexibility is one of the reasons it is very useful for reliability engineers. The Weibull distribution is the distribution of choice for systems that have many, identical competing elements that can each cause a fail and for which the first element to fail causes the entire system to fail. This is discussed briefly in Section 1.4.8 and more fully in the dielectric chapters, Sections 2.1.2 and 2.4. Several authors have discussed the theoretical justification for using the Weibull distribution and how it follows from the extreme value theory when used in conjunction with the weakest link model [22–25].

The first moment about the origin is $\varphi\Gamma\{(1 + \beta)/\beta\}$ for a Weibull distribution. The second moment about the mean for a Weibull distribution is $\varphi^2\Gamma\{(2 + \beta)/\beta\} - \Gamma^2\{(1 + \beta)/\beta\}]$ where Γ is the gamma function. These moments are not

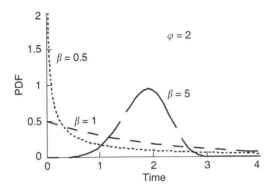

Figure 1.23. PDF for Weibull distribution with three values of the shape parameter β and a characteristic life, $\varphi = 2$.

frequently used in semiconductor reliability work but are given here for completeness and for comparison to the normal distribution.

The next topic for the Weibull discussion is the conversion of the axes for the CDF plot to those that will result in a straight line plot for the Weibull CDF. Using the same general procedure as for the exponential linearization Equation 1.25 is solved for "t" by subtraction and taking the natural logarithm twice obtaining

$$\ln[-\ln[1 - F(t)]] = \beta \ln t - \beta \ln \varphi. \tag{1.27}$$

The term on the left has been defined as a Weibit, W, where $W = \ln(-\ln(1-F(t)))$. If we then plot W versus $\ln t$ on the x axis, a linear plot for the CDF is achieved if the distribution follows a Weibull distribution.

Figures 1.23, 1.24, 1.26 are plotted on a linear scale and show the PDF, CDF and $h(t)$, respectively, for the Weibull distribution. For Figure 1.25 the Weibull CDF is plotted on a scale having Weibits as the y axis and $\ln t$ as the x axis.

Notice that in Figures 1.24 and 1.25, the CDF plots for Weibull distributions having a characteristic life of two, the cumulative failure for each of the Weibull distributions is 63.2% at that characteristic life of two, regardless of the shape parameter. Also note that widely differing distributions occur for the shape factors chosen.

The $h(t)$ for a shape parameter, $\beta = 1$ is a constant as discussed above and is demonstrated in Figure 1.26. Returning to Figure 1.23, note the drastic difference in the PDF for Weibull distributions having shape parameters of 0.5, 1, and 5. For Weibull distributions having values of β less than 1, the PDF starts high at time zero and is a decreasing function. This is typical of a defect mechanism and indeed, historically, for oxides thicker than about 5 nm, $\beta \leq 1$, was typically interpreted as extrinsic or defect fallout while for values of $\beta > 1$, the interpretation was that the Weibull distribution represented the intrinsic fallout or wearout. This is a very

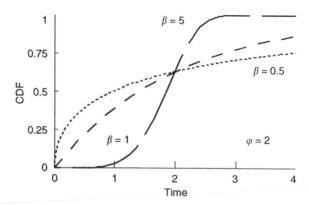

Figure 1.24. CDF for Weibull distribution for three values of the shape parameter β and for a characteristic life $\varphi = 2$.

graphic example of why the Weibull distribution is so useful in reliability modeling. Many experiments can be successfully modeled using a Weibull distribution; however, it is most powerful when a theoretical reason exists for its use to describe a particular mechanism.

The three-parameter Weibull distribution is given in the Equations 1.28–1.31. Again, the third parameter would typically be a time-shift parameter for semiconductor reliability modeling although it could be cast as a threshold parameter especially in the more general cases. Note that as semiconductors are

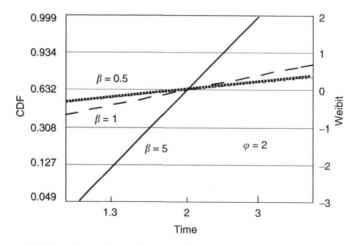

Figure 1.25. CDF for Weibull distribution for three values of the shape parameter β and for a characteristic life $\varphi = 2$ plotted on a transformed axis to yield a linear CDF plot.

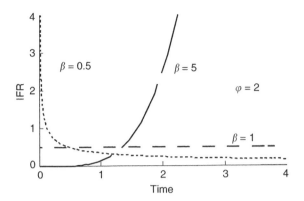

Figure 1.26. $h(t)$ for Weibull distribution with three values of the shape parameter β and a characteristic life, $\varphi = 2$.

driven to the limit, it may become necessary to include more terms in the modeling distributions to more accurately characterize the end of life predictions.

$$\text{PDF}: f(t) = (\beta/t - t_0)(t - t_0/\varphi)^\beta \exp\left[-(t - t_0/\varphi)^\beta\right] \qquad (1.28)$$

This equation can be written somewhat more simply if it is recast as

$$\text{PDF}: f(t) = (\beta/\varphi^\beta)(t - t_0)^{\beta-1} \exp\left[-(t - t_0/\varphi)^\beta\right] \qquad (1.29)$$

$$\text{CDF}: F(t) = \int_0^t f(x)dx = 1 - \exp\left[-(t - t_0/\varphi)^\beta\right] \qquad (1.30)$$

$$\text{IFR}: h(t) = f(t)/[1 - F(t)] = (\beta/\varphi^\beta)(t - t_0)^{\beta-1} \qquad (1.31)$$

1.4.8 Lognormal Distribution

The lognormal distribution, as a two-parameter distribution, has a median parameter, $\mu_{\ln}$, and a shape parameter, $\sigma_{\ln}$. $\mu_{\ln}$ is also called the log mean in some statistics texts. It is important to understand the relationship between the lognormal distribution and the normal distribution. The median parameter, $\mu_{\ln}$, is the mean of the natural log of the lifetime and the shape parameter, $\sigma_{\ln}$, is the standard deviation of the natural log of lifetime. Thus if Y is a random variable with a normal distribution having a mean of μ, and a standard deviation of σ, the lognormal distribution with a random variable of Z which would be derived from that normal distribution is given by $Z = \exp Y$. For this lognormal distribution the shape parameter $\sigma_{\ln}$, is equal to the normal standard deviation σ, and the median parameter is related to the normal mean by $\mu_{\ln} = \exp \mu$.

The PDF, and CDF equations for the lognormal distribution are shown in Equations 1.32 and 1.33 where $0 < t < \infty$.

$$\text{PDF}: f(t) = \left(\sigma_{\ln} t \sqrt{2\pi}\right)^{-1} \exp\left[-((\ln t - \ln \mu_{\ln})^2 / 2\sigma_{\ln}^2)\right] \qquad (1.32)$$

$$\text{CDF}: F(t) = \int_0^t \left[\left(\sigma_{\ln} x \sqrt{2\pi}\right)^{-1} \exp\left[(-(\ln x - \ln \mu_{\ln})^2 / 2\sigma_{\ln}^2)\right]\right] dx \qquad (1.33)$$

The median parameter above is designated as '$\mu_{\ln}$' to emphasize the relationship to the normal distributions but it is often designated as T_{50}. The $h(t)$ would be numerically calculated and is given by the usual $h(t) = f(t)/[1-F(t)]$ formula. Note that the hazard rate of the lognormal is not monotonic as per Figure 1.30. The mean or first moment about the origin for a lognormal distribution is $\exp[\ln(\mu_{\ln}) + 0.5\sigma_{\ln}^2]$. The variance or second moment about the mean for a lognormal distribution is $\exp[2 \ln(\mu_{\ln}) + \sigma_{\ln}^2]\{\exp(\sigma_{\ln}^2)-1\}$. Again these moments are given only for comparative purposes as they are rarely used in semiconductor reliability.

Figures 1.27, 1.28 and 1.30 depict the lognormal PDF, CDF, and $h(t)$ respectively. Figure 1.29 depicts the CDF after the transformation to the linear scale. Observe that in Figures 1.27–1.30 the lognormal distribution also has a great deal of flexibility as to the shape of the distributions that it can model. Hence, the lognormal distribution can also be used to model a large number of phenomena. Also note that for the CDF, the median life time-to-fail, $\mu_{\ln}$, is independent of the shape parameter, $\sigma_{\ln}$.

The lognormal scale for the CDF plot may be transformed in a manner similar to the normal scale except now instead of starting with $F(t_q) = \Phi(\{t_q - \mu\}/\sigma)$ as for the normal distribution, we start with $F(t_q) = \Phi(\{\log (t_q) - \ln(\mu_{\ln})\}/\sigma_{\ln})$. Solving for t_q by taking the inverse of the function, $\ln(t_q) = \ln \mu_{\ln} + \sigma_{\ln} \Phi^{-1}(t_q)$ is

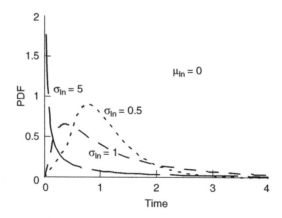

Figure 1.27. PDF for lognormal distribution with three values of the shape parameter $\sigma_{\ln}$ and a median parameter, $\mu_{\ln} = 0$.

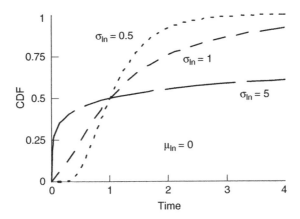

Figure 1.28. CDF for lognormal distribution with three values of the shape parameter σ_{ln} and a median parameter $\mu_{ln} = 0$.

obtained. Hence, for the lognormal, a plot of $\ln(t_q)$ versus $\Phi^{-1}(t_q)$ will yield a straight line if the distribution can be modeled by a lognormal distribution. This is depicted in Figure 1.29.

Because of the flexibilities of both the lognormal distribution and the Weibull distribution it is often possible to model data using either distribution. However, typically data can only be collected across two to three orders of magnitude and projections must be made another two or three orders of magnitude in time beyond the data collection time. For example, a typical stress will last at most 100–1000 hrs and the product must last for 100 Khr. Also, the number of parts that can be stressed is usually very limited because of early hardware delivery limitations as well as stress facility limitations. The lack of stress time beyond a three month maximum

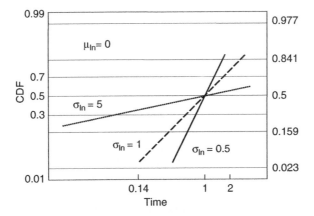

Figure 1.29. CDF for lognormal distribution with three values of the shape parameter σ_{ln} and a median parameter, $\mu_{ln} = 0$ and with axis transformed to yield a linear CDF.

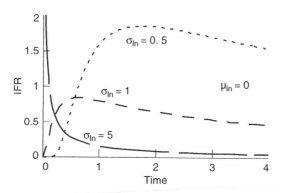

Figure 1.30. $h(t)$ for lognormal distribution with three values of the shape parameter, σ_{ln}, and a median parameter, $\mu_{ln} = 0$.

precludes several serial stresses. Thus if only 50 parts are stressed, the first fail already represents the 2% fallout point. Based on the amount of data and the range of data, one may not be able to experimentally determine which distribution fits best and indeed, they may fit the data equally well. The results of projecting across the two to three orders of magnitude, however, could well mean the difference between a product that passes the requirement by an order of magnitude or more and one that fails the requirement by an order of magnitude or more. Recall that in addition to projecting across those orders of magnitude of time one is also usually projecting to ppm failure rates with the last data point at typically 1% or 2%.

For that reason, one prefers to appeal to a theoretical reason for choosing one distribution over the other whenever possible. It will be shown in Chapter 2, Section 2.4 that the Weibull distribution should be used when modeling phenomena that exhibit an extreme value behavior. Extreme value distributions are used to characterize systems where there are many competing mechanisms in parallel, and any one of which failing, can cause the whole system or product to fail. Dielectric breakdown is a prime example this behavior and the theory in Section 2.4 is applied to dielectric breakdown. The reader is advised to review that chapter as well as the aforementioned references to fully understand the theoretical reasons for choosing a Weibull distribution, to observe graphically the difference between choosing the lognormal and Weibull distributions, and finally to gain an appreciation for the sample size and time required to experimentally determine which distribution to use if no theoretical reason can be determined.

One basis for the use of a lognormal distribution is a proportional growth or multiplicative model [26, 27]. The model of failure in this instance is one where a shift, or a change, or a degradation starts ever so slowly, and then with time, that change grows or multiplies. Electromigration is typical, and will be modeled in this book, with a lognormal model. In the electromigration case, a void is formed, but that void growth starts with a single metallic atom exit. As the void grows from the molecular size, the volume surrounding the void increases and more atomic or molecular motion can contribute to the increasing void growth. Ultimately the void becomes such a large percent of the line that the resistance increases to the point of failure. Crack growth and fatigue are also examples of mechanisms typically modeled by lognormal distributions as are diffusion and chemical reactive processes. Thus failures which result from small degradation processes that continue to grow until failure occurs are often best modeled by the lognormal distribution. As mentioned above, electromigration is caused by material transport and both the normal and lognormal distributions will be further developed in the support of the electromigration work in Section 7.4. The relationship of the material transport to the lognormal distribution will also be treated further there. Note that an understanding of the physical mechanism should provide significant direction as to the choice between the lognormal distribution model and the Weibull distribution model. In fact, based on the inability to take enough data to distinguish between the distribution models, the physical mechanism is typically a much better starting point to determine the correct model. The difficulty of choosing a model based on small amounts of empirical data will be clearly demonstrated in Section 2.4.3.

This introduces a further reason for understanding the physics of the mechanism in question that was not discussed in Section 1.3. Without a good understanding of the physics behind the mechanism, one might incorrectly choose a distribution with which to model the behavior and, as a consequence, be either radically optimistic or radically pessimistic without even knowing one's bias.

Sometimes it is useful to introduce a third parameter into the lognormal distribution to reflect either an incubation period, or more generally, an additional feature of the distribution. As mentioned above, this was proposed [21] for electromigration and was very successfully used to model an incubation period for electromigration and is discussed further in Section 7.4.5.

The PDF and CDF equations for the three-parameter lognormal distribution are shown in Equations 1.34 and 1.35 where $0 < t < \infty$.

$$\text{PDF: } f(t) = \left(\sigma_{\ln} t \sqrt{2\pi}\right)^{-1} \exp\left[-\left(\frac{[(\ln t - \mu_{\ln})/t_0]^2}{2\sigma_{\ln}^2}\right)\right] \qquad (1.34)$$

$$\text{CDF: } F(t) = \int_0^t \left[\left(\sigma_{\ln} t \sqrt{2\pi}\right)^{-1} \exp\left[\frac{-[(\ln x - \mu_{\ln})/t_0]^2}{2\sigma_{\ln}^2}\right]\right] dx \qquad (1.35)$$

The CDF for a three-parameter lognormal distribution is shown in Figure 1.31, where the third parameter is the location parameter which allows the direct modeling of an incubation period or of some threshold.

1.4.9   Poisson Distribution

The next distribution covered will be the Poisson distribution. It will be the distribution of choice for modeling some of the device negative bias temperature instability (NBTI) effects. More generally the Poisson distribution is often used to model recurrence data.

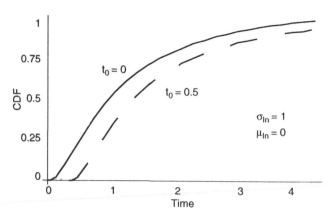

Figure 1.31. CDF for lognormal distribution having three distribution parameters with the shape parameter $\sigma_{ln} = 1$, the median parameter $\mu_{ln} = 0$, and location parameter $t_0 = 0$ and 0.5.

The Poisson distribution is very different from any of the distributions considered earlier because it belongs to the family of discrete distributions. Until now, we have modeled data with continuous distributions even though the data have been discrete. The discrete distribution gives a tool by which one can model the number of failures, or shifts, above a certain minimum shift or fail point.

The probability density function of a continuous distribution describes the fails in each period of time which can be plotted as a histogram with the time interval as the abscissa and the number of fails as the ordinate or y-axis value. A discrete distribution has an equivalent function that is called the probability function (pf). It consists of the probabilities of all of the occurrences that can occur in a given system. A simple discrete function is the geometric distribution and its probability function is given by $f(x) = p\,(1-p)^{(x-1)}$, where p is a probability and hence, $0 < p < 1$, x is any one of the total set of observations, and $f(x)$ is then the probability of that observation or outcome occurring.

The cumulative distribution function of a discrete function is the summation of the probabilities of all possible observations or outcomes and must equal one as x approaches $+\infty$, that is, as all of the probabilities of all possible outcomes are included in the summation. As x approaches $-\infty$, the cumulative distribution function must approach zero since the probability of none of the outcomes would be included in the summation. For the geometric and Poisson distributions, the observation or outcome factor must be an integer; however, this is not true of discrete distributions in general. A much more complete discussion of discrete distributions is given by Nelson [10].

The pf and CDF for the Poisson distribution are shown in Equations 1.36 and 1.37, where λ_P is the observation or outcome rate factor with $\lambda_P > 0$, $n \geq 0$ and is the number of observations or events, and t is the observation variable which in the case of semiconductor reliability would typically be time but could, in principle

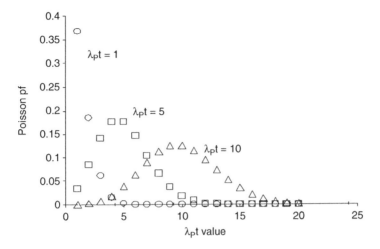

Figure 1.32. Poisson pf for λt values = 1, 5, and 10.

be any variable of interest. The pf and CDF are plotted for three values of λ_P in Figures 1.32 and 1.33, respectively.

$$\text{pf: } f(nt) = ((\lambda_P)^n/n!)\exp[-\lambda_P] \tag{1.36}$$

$$\text{CDF: } F(n) = P(N \leq n) = \sum_{i=0}^{n}\left((\lambda_P)^i/i!\right)\exp[-\lambda_P] \tag{1.37}$$

The mean and variance for the Poisson distribution are equal to each other, and each is equal to λ_P. Again the reader is referred to Nelson [10], and Meeker [9], for a much more complete discussion on the Poisson distribution as well as Poisson analysis (Figure 1.33).

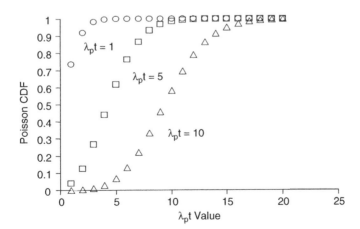

Figure 1.33. Poisson CDF for λt values = 1, 5, and 10.

1.5 CHI-SQUARE AND STUDENT t DISTRIBUTIONS

This section introduces those distributions that will be used not to characterize the physics of a mechanism, but rather to provide the statistical framework to test the applicability of the results.

1.5.1 Gamma and Chi-Square Distributions

The chi-square test is one of the primary metrics used to determine if the distribution chosen to characterize the mechanism is consistent with the sample data. Hence, the chi-square test is used to determine the goodness of fit. Because it is a special case of the gamma distribution, the gamma distribution is shown first. The PDF and CDF for the gamma distribution are given in Equations 1.38–1.40.

1.5.1.1 Gamma Distribution

$$\text{PDF}: f(t; \alpha_G, \beta_G) = \left(t^{\alpha_G - 1}/\beta_G^{\alpha_G}\Gamma(\alpha_G)\right)\exp[-t/\beta_G] \quad \text{for } t > 0. \tag{1.38}$$

$$\text{CDF}: F(t; \alpha_G, \beta_G) = (1/\Gamma(\alpha_G))\int_0^t \left(x^{\alpha_G-1}/\beta_G^{\alpha_G}\right)\exp[-x/\beta_G]dx. \tag{1.39}$$

$\Gamma(\alpha_G)$ is the gamma function given by

$$\Gamma(\alpha_G) = \int_0^\infty x^{\alpha_G-1}e^{-x}dx. \tag{1.40}$$

If $\beta_G = 2$ and $\alpha_G = v/2$ where v is an integer representing the number of the degrees of freedom of the distribution, then the gamma distribution reduces to the chi-square distribution. The chi-square distribution PDF and CDF are given in Equations 1.41 and 1.42. The PDF and CDF for chi-square distributions of one and five degrees of freedom are shown in Figure 1.34.

1.5.1.2 Chi-Square Distribution

$$\text{PDF}: f(t) = \left(t^{(v/2)-1}/2^{(v/2)}\Gamma(v/2)\right)\exp[-t/2] \quad \text{for } t > 0 \tag{1.41}$$

$$\text{CDF}: F(t) = (1/\Gamma(v/2))\int_0^t \left(x^{(v/2)-1}/2^{(v/2)}\right)\exp[-x/2]dx \tag{1.42}$$

It should also be noted that for the special case of two degrees of freedom ($v = 2$), the chi-square distribution itself reduces to an exponential distribution with a mean equal to 2. This can be seen from the PDF of Equation 1.41. $\Gamma(v)\} = \{1/\Gamma(v/2)\}$ $\{x^{(v/2-1)}/2^{v/2}\}\exp(-x/2) = \Gamma(1)\}\{1/2\}\exp(-x/2) = \{1/2\}\exp(-x/2)$. The

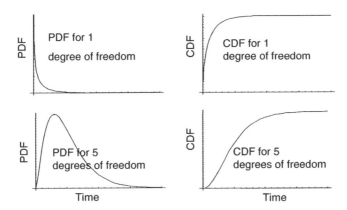

Figure 1.34. PDF and CDF for chi square distribution for 1 and 5 degrees of freedom.

gamma function could also be expressed in terms of a simple factorial expression since the gamma function argument υ has an integer value, e.g., $\Gamma(v) = (v-1)! = 1$ in the case of $v = 2$. Further discussion of the chi-square distribution is given in Green and Bourne [12].

1.5.2   Student *t* Distribution

The Student *t* test can be used to draw inferences about the sample, the population, and the statistical significance of the results. It has many uses within the field of semiconductors but its development is beyond our scope here [28]. The PDF and CDF for the Student *t* distribution is given in Equations 1.43 and 1.44, respectively.

1.5.2.1   Student *t* Distribution

$$\text{PDF: } f(x) = \frac{(\Gamma[(v+1)/2]/\Gamma[v/2])}{\left(\{1+(x^2/v)\}^{(v+1)/2}\sqrt{\pi v}\right)} \quad \text{where } -\infty < x < \infty \quad (1.43)$$

$$\text{CDF: } F(x) = \frac{(\Gamma[(v+1)/2]/\Gamma[v/2])}{(\sqrt{\pi v})} \int_{-\infty}^{x} \left(\{1+(y^2/v)\}\right)^{-(v+1)/2} dy \quad (1.44)$$

where Γ is the gamma function and v is the degrees of freedom.

The Student *t* distribution is symmetrical about the mean and it converges to a normal distribution as v approaches infinity. Student *t* distributions are shown in Figure 1.35 for 2 and 20 degrees of freedom.

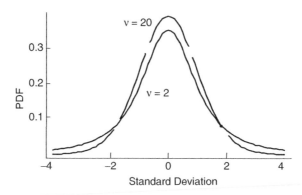

Figure 1.35. Student *t* PDF for 2 and 20 degrees of freedom.

1.6 APPLICATION

1.6.1 Readouts Versus "Exact" Time-To-Fail

Data censoring is often necessary even given the experimenter's best effort to avoid it. Various types of data censoring are discussed below. Our focus, as throughout this book, will be on the practical application to semiconductor technology reliability. Historically, one had to use separate test and stress equipment sometimes called *ex situ* stress equipment. That is still necessary sometimes today, but more often one piece of equipment can be used for both the stressing and testing of the DUTs. This is likewise sometimes referred to as *in situ* stressing. This equipment will give the "exact" time-to-fail of each DUT. *In situ* test equipment typically will be able to determine the time-to-fail within milliseconds and although this is not mathematically exact, it can be considered exact for any realistic reliability stress that is at least a few seconds or greater in duration. All of the semiconductor technology reliability stressing that is practiced today including very fast wafer-level reliability stressing is at least a few seconds in stress duration.

When *ex situ* stressing is necessary, the sample is first tested and then put into the stress chamber. Readouts should usually be done on a log-of-time basis. As an example, consider a stress sample that is to be 500 hr in duration. Assume that each test of the total sample takes one hour on the tester with a five-hour overhead of removing the sample from stress, transporting it to and from the tester, and returning the sample on stress. Further assume that either no relaxation occurs or that the samples are held at a voltage while awaiting their turn on the test equipment. The next issue to be addressed is the tradeoff between data collection, the time at which most of the failures or shifts are expected, and the total time of the stress activity. If the prime concern is the percent of the sample that survives until 500 hr, very few, intermediate readouts may be chosen especially if time is of great concern, which is often the case. As the reliability engineer, one wants to be very cautious in this case. If the hardware all passes at the end of the 500 hr, there is no issue. However, if there are significant fails, the customer will then more than likely desire additional information. For the example requiring 6 hours off stress, the readout schedule might be 10, 30, 100, and 500 hours. Obviously if only two or three readouts were made, this becomes problematic. The best experimental design can only be achieved when the behavior is reasonably well known beforehand, so that the readouts can be taken in approximately equal steps of transformed time. Note that the first plotted fail on any CDF plot will not occur at time zero, but at the first nonzero readout. Depending on how quickly fails or shifts are expected, the reliability engineer might even choose to do a two-hour readout in addition to the above readouts or even both one- and two-hour readouts.

For the case of *in situ* stressing, a data-retention strategy is typically chosen by the equipment manufacturer. Here the issue becomes too much data when only a little information is desired. If the first significant shift or fail occurs after 18.554 hr for one of the samples on stress and the equipment has been making a readout once every 0.001 hr, someplace in the equipment storage 18,553 entries exist that present no information. Typically equipment today will give the engineer the ability to save measurements that have a shift greater than a per cent change from the last measurement or beyond some minimum threshold change. This is obviously both desirable and necessary.

The actions of the reliability engineer are the same in both cases after the data is taken. The CDF will be plotted on the appropriate axes. The difference is that for the *ex situ* case, the only points that will be plotted will be at the precise readout times or a plotting position appropriate to those times. However, what is really known about each fail in this distribution is only its time-to-fail within a certain time range. For example, for a fail that occurred at the 50-hour readout, it is only known that that fail occurred sometime between the previous readout, e.g., 20 and 50 hr. Obviously, all of the fails that occurred between 20 and 50 hr are grouped together and plotted at a single time. For this reason this type of data is called grouped data or readout data. Because these fails are only known within a range of time values, this is also called interval censoring. For the in situ case, values will be plotted at the "exact" times-to-fail, again considering the best plotting position strategy.

1.6.2 Additional Types of Censoring

Two of the most common reasons for censoring are lack of time or lack of stress equipment or both. If a stress is terminated before all of the parts have failed, it is called time censoring or Type I censoring. The times-to-fail are not known for the samples failing after the stress termination. Time censoring is very common and often necessary. The challenge for the reliability engineer is to then design the experiment in such a fashion that most of the population fails in the allotted time. Type II censoring occurs when the sample is removed from stress after a given percent of the samples have failed. This may also be called failure censoring. Failure censoring has the advantage of guaranteeing that a given, cumulative fallout can be plotted on the CDF plot. Often the tail of the distribution may be significantly more robust than the main population so it may survive another decade or more in time. But in terms of the modeling and projection, this tail would be appropriately discounted, so time can be saved without any significant information lost in this special case of failure censoring. It is assumed that for either time censoring or failure censoring, the stress plan is decided beforehand and that the censoring is hence, unbiased. If time or fail censoring decisions are made during the stress that are based on the stress results, a bias, either favorable or unfavorable, could be introduced into the results.

Other reasons for censoring might include equipment failure or voltage surges due to power outage and return. In these cases the compromised samples would need to be removed from stress. If one were stressing on two separate stressors and only one suffered from the problem, those samples would be removed at that time, and as the stress continued on only the uncompromised equipment, the percent fails from that time on would be calculated based on the new, smaller sample.

Another reason for censoring would be the need to obtain failure analysis on an early fail or shift to determine if the signature was the expected fail signature. For example, if for an electromigration stress 20% resistance shift was considered a failure, a DUT might be pulled from the stress and sent to failure analysis if it shifted 10% much more quickly than expected. Here the data point would be sacrificed for an immediate determination as to whether a new mechanism was at work. In principle the same could hold true for a stress that uncovered several mechanisms, but for which only one was of concern. Although this would not be a common case in semiconductor reliability, it can happen.

1.6.3   Least-Squares Fit and Application

It is expected that the engineer will be using numerical techniques for plotting and analyzing the data. Hence, we do not give computer programs, nomographs, tables of calculated values for special functions, or charts showing continuous values of special functions. Once the data has been plotted, preferably by numerical means, a fitting routine can be run to give a best fit line through those values.

One of the simpler choices would be a least-squares fit to the data. In this case, the square of y-axis distance from each point, to the fitted line is summed and minimized to achieve the least-squares fit. This is shown graphically in Figure 1.36.

As discussed in Section 1.4, we choose axes which cause the function that is to be plotted to have a linear form, thus the least-squares-fit function is also a linear function.

$$y_i(a, b) = a + bt_i \tag{1.45}$$

$$Q^2(a, b) = \sum_{i=1}^{n} \left[(y_i - (a + bt_i))^2 \right] \tag{1.46}$$

Equation 1.46 is minimized when the derivative is set equal to zero. Since Equation 1.46 is a function of both 'a' and 'b,' each partial derivative must be

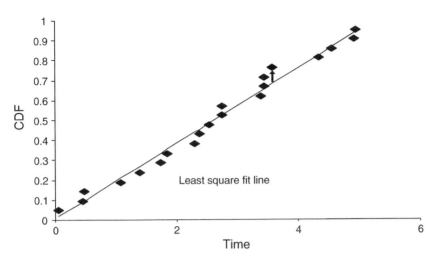

Figure 1.36. Least-squares-fit line to 20 point data set plotted with $i/(n+1)$ plotting position.

individually set to zero. Note that Q could be a function of many parameters for the nonlinear case, and in that more general case, the equivalent of Equation 1.46 would be set to zero for all partial derivatives. The derivation of the equations below is beyond the scope of this book and the reader is referred to any theoretical statistics book for a complete derivation and more complete description of these equations. The reader is directed to the end of this discussion for an example that walks through each of the following steps. Under almost every circumstance in today's world, any of several excellent software packages would be used to do these calculations; however, it is important for the reliability engineers to understand the basis of those calculations even if they have not derived each equation themselves. We start with the least-squares fit.

$$\partial Q^2(a,b)/\partial a = -2\sum_{i=1}^{n}[(y_i - (a + bt_i))] = 0. \tag{1.47}$$

$$\partial Q^2(a,b)/\partial b = -2\sum_{i=1}^{n}[(y_i - (a + bt_i))t_i] = 0. \tag{1.48}$$

The estimates of the regression coefficients, 'a' and 'b,' are given by the following equations. The estimate of the 'b' coefficient, $\hat{b}$, is the estimate of the slope of the least-squares-fit line and is given by

$$\hat{b} = \frac{\sum_{i=1}^{n} y_i t_i - [(\sum_{i=1}^{n} y_i)(\sum_{i=1}^{n} t_i)/n]}{[\sum_{i=1}^{n} t_i^2 - (\sum_{i=1}^{n} t_i)^2/n]} \tag{1.49}$$

The estimate of the 'a' coefficient, or estimate of the intercept of the least-squares-fit line, is given by:

$$\hat{a} = \left(\sum_{i=1}^{n} y_i \right) \bigg/ n - \hat{b} \left(\sum_{i=1}^{n} t_i \right) \bigg/ n \qquad (1.50)$$

Two figures of merit for the least-squares fit include the correlation coefficient, ρ_C shown in Equation 1.51, and a data-based estimator for the variance of the error s^2, shown in Equation 1.52. This estimator s^2 will be used throughout this chapter since it is usually true in semiconductor reliability that the variance of the population is not known. The correlation coefficient is a measure of the strength of the linear relationship between the two variables under study. ρ_C is dimensionless and ranges from -1 to 0 to $+1$, meaning an excellent negative correlation, no correlation, or a positive correlation (Figure 1.37). This is sometimes called the Pearson product moment of correlation. Although this would not normally be an issue in semiconductor reliability, note that excellent correlation does not imply a causal relationship, but only that a linear relationship exists between the variables in question. The estimator of the variance, s^2, gives a measure of the dispersion of the actual y values about the y values as fitted to the least-squares line. If the estimator of the variance was very large, it would be possible that little, if any, correlation existed even if the correlation coefficient was close to a positive or negative one. The correlation coefficient ρ_C and the variance estimator s^2 are given by:

$$\rho_C = \frac{\left[\sum_{i=1}^{n} y_i t_i - \left(\sum_{i=1}^{n} y_i \right) \left(\sum_{i=1}^{n} t_i \right) / n \right]}{\sqrt{\left[\sum_{i=1}^{n} y_i^2 - \left(\sum_{i=1}^{n} y_i \right)^2 / n \right]} \sqrt{\left[\sum_{i=1}^{n} t_i^2 - \left(\sum_{i=1}^{n} t_i \right)^2 / n \right]}}. \qquad (1.51)$$

$$s^2 = \left\{ \frac{\left[\sum_{i=1}^{n} y_i^2 - \left(\sum_{i=1}^{n} y_i \right)^2 / n \right] - b \left[\sum_{i=1}^{n} y_i t_i - \left(\sum_{i=1}^{n} y_i \right) \left(\sum_{i=1}^{n} t_i \right)^2 / n \right]}{n - 2} \right\}. $$

$$(1.52)$$

One very significant attribute should be noted about the least-squares fit. The error between the actual point and the fitted point contained within the line is squared. The sum of these squares is then minimized. Another and possibly preferable option would be to simply minimize the sum of the errors instead of the sum of the square of the errors. However, absolute values are not analytically soluble. The result of minimizing the sum of the squares is that points further from the fitted line have a larger contribution and are thus more heavily weighted than the closer points. Usually, but not always, the end points of the distribution, the first readouts and the last readouts, will have the largest error. These early and late points will typically be the most distant from their equivalent points on the fitted line. Hence, these early and late readouts have a greater influence on the resulting fitted line than do most of the other readouts. One would typically want to minimize the influence of these points rather than maximize it. The early points

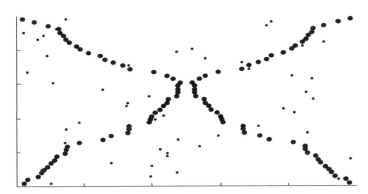

Figure 1.37. Positive correlation (increasing slope with larger dots) and negative correlation (decreasing slope with larger dots) where correlation coefficient, ρ_c, would be relatively close to $+$ or -1, respectively, and no correlation (smaller dots) where the correlation coefficient, $\rho_c \sim 0$.

may be only slightly longer than the minimum response time of the stress/test system. The late points may not be representative of the main population due to any number of stress/test issues, or they may truly be representative of a population that has significantly longer life.

The confidence interval, at the $100(1-\alpha)$ percentile for the slope $\hat{b}$ of our straight line fit is given by Equation 1.53. Note that this will yield a range of slopes that are acceptable within our specified confidence interval. As alpha gets smaller, the range of acceptable slopes will increase. Hence, there is a tradeoff, the tighter the confidence interval desired, the larger will be the spread of values for $\hat{b}$ that must be accepted to meet that increased confidence level. This is quantified in the example at the end of this chapter. The confidence interval for the slope, C_b, is given by

$$C_b = \hat{b} \pm t_{\alpha/2} s \left\{ \sum_{i=1}^{n} t_i^2 - \left(\sum_{i=1}^{n} t_i \right)^2 \bigg/ n \right\}, \qquad (1.53)$$

where $\hat{b}$ is the estimator for the slope and is given in Equation 1.49, $t_{\alpha/2}$ is taken from Student distribution tables and is also a function of the degrees of freedom, s is the square root of the variance estimator given in Equation 1.52, and n is the sample size.

The next figure of merit is for the $100(1-\alpha)$ percent confidence interval at a given value of time, t, for the mean value of y which in this case is the CDF_i. The reader is referred to any of the referenced statistics books for more detail. The confidence interval, C_I, is then given by

$$C_I = \hat{y} \pm t_{\alpha/2} s \sqrt{ n^{-1} + (t - \bar{t})^2 \bigg/ \left(\left(\sum_{i=1}^{n} t_i^2 - \left(\sum_{i=1}^{n} t_i \right)^2 \right) \bigg/ n \right) }, \qquad (1.54)$$

where $\hat{y}$ corresponds to time t and is given by $\hat{y}(i) = a + \hat{b} t_i$.

Cautionary note: The confidence intervals, *t*-tests, and hypothesis tests shown above are only exact for the case where the E_i are also normal random variables. The transformations discussed in Section 1.4, which are required to cast the various failure distributions in a linear form, unfortunately cause the above regression model to be not strictly valid after the transformations. In particular, the noise terms E_i, are not independently distributed, having a tendency to be larger towards the edges of the data. Hence the maximum likelihood estimators are much preferred for the evaluation of confidence intervals, *t* tests and hypothesis testing and although they are beyond the scope of this book the reader is directed to any of the referenced statistics texts for thorough discussions of these topics. Understanding the weakness of this analyzed regression model we will continue using it for simple graphical illustrative purposes. Also note that its validity could be tested by using simulations if it were to be used instead of the maximum likelihood estimator. As already mentioned, the expectation is that one of the many computer programs for confidence interval and hypothesis testing would be used to generate the results, and that in fact a maximum likelihood estimator program would be used.

Example 1: A simple linear example of the application of Equations 1.45 through 1.54 will be given next. The simplest case to illustrate the applications of the equations and avoid getting lost in the transformation equations is used. Note that in Section 1.4 the transformations that were shown allow the normal, exponential, Weibull, and lognormal distributions to be plotted linearly after the axis transformations. The $i/(n+1)$ plotting position is chosen in this example for simplicity and there are 20 data points. The actual least-squares fit is shown in Figure 1.36. It was generated with software, but we now go back to reconstruct the method of doing it manually. Table 1.1 gives the time and CDF values as well as the additional terms required for the calculations.

First the slope and intercept estimates are calculated for the least-squares fit line drawn in Figure 1.36.

From Equation 1.45: $\hat{b} = \dfrac{34.01 - [(10)(52.34)/20]}{[178.5 - (52.34)^2/20]} = 0.189$

From Equation 1.46: $\hat{a} = 10/20 - (0.189)(52.34)/20 = 0.00571$

Hence, the formula for the line drawn in Figure 1.36 is $\hat{y} = 0.00571 + 0.189\ t$.

The correlation coefficient ρ_C is calculated from Equation 1.51 and yields $\rho_C = 0.991$. The value is very close to $+1$, indicating a very strong positive correlation. This indicates that there is a very strong linear relationship between these two variables.

$$\rho_C = \frac{[34.026 - (52.367)(10)/20]}{\left[\sqrt{6.508 - (10)(10)/20}\right]\left[\sqrt{(178.64) - (52.367)(52.367)/20}\right]} = 0.991$$

TABLE 1.1. Table Showing Time, CDF, Squares, Cross Products, and Sums for Least-Squares-Fit Procedure

	Time	Time^2	CDF	CDF^2	Time $\times$ CDF
	0.064	0.004	0.048	0.002	0.003
	0.446	0.199	0.095	0.009	0.042
	0.467	0.218	0.143	0.020	0.067
	1.096	1.200	0.190	0.036	0.209
	1.393	1.941	0.238	0.057	0.332
	1.734	3.005	0.286	0.082	0.495
	1.858	3.451	0.333	0.111	0.619
	2.306	5.317	0.381	0.145	0.878
	2.366	5.599	0.429	0.184	1.014
	2.537	6.572	0.476	0.227	1.221
	2.743	7.523	0.524	0.274	1.437
	2.753	7.582	0.571	0.327	1.573
	3.386	11.467	0.619	0.383	2.096
	3.426	11.735	0.667	0.444	2.284
	3.431	11.774	0.714	0.510	2.451
	3.572	12.759	0.762	0.581	2.722
	4.323	18.684	0.810	0.655	3.499
	4.556	20.760	0.857	0.735	3.905
	4.929	24.297	0.905	0.819	4.460
	4.955	24.553	0.095	0.907	4.719
SUM	**52.340**	**178.640**	**9.143**	**6.508**	**34.027**

The variance estimator then comes from Equation 1.52 and the square root of the variance estimator for this example is, $s = 0.0384$.

$$s = \left\{ \sqrt{\frac{[6.508 - (10)(10)/20] - [0.189(34.026 - (10)(52.367)/20]}{18}} \right\} = 0.0384$$

The final two metrics investigated in this chapter will be the confidence intervals on the value of the slope and on the mean value of $\hat{y}$ or the CDF estimator at a fixed time. The 95% confidence interval for the slope, C_b, is given by Equation 1.53 and the 95% confidence interval for the mean value of $\hat{y}$ at a fixed time is given by Equation 1.54. In terms of the reliability projection, the value of the slope is by far more critical than the mean value of $\hat{y}$ at a fixed time. That is because any slope error is magnified by the reliability projection across at least several orders of magnitude of time in the course of data being taken at highly accelerated conditions while the projection must typically reach 10 years as demonstrated in our first figure, Figure 1.1. Any error in mean value of $\hat{y}$ at a fixed time remains constant on a percentage basis as the projection extends across the decades. These figures of merit are shown on Figure 1.38 and Figure 1.39 respectively.

$$C_b = 0.189 \pm (2.101)(.0384)/((178.5 - (52.34)^2/20))^{0.5} = 0.189 \pm .037$$

$$C_I = \hat{y} \pm (2.101)(0.0384)\sqrt{\frac{1}{20} + \frac{(t - 2.617)^2}{178.5 - (52.34)^2/20}}$$

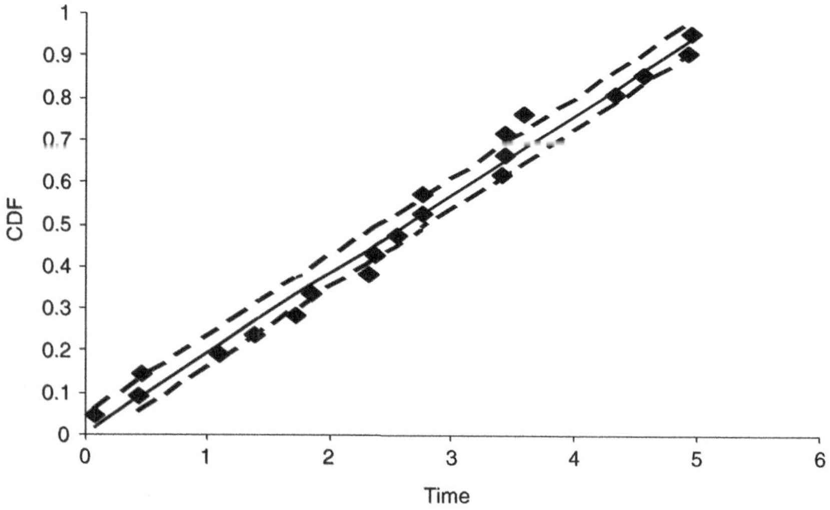

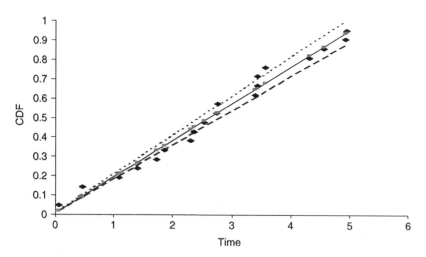

Figure 1.38. Least-squares-fit line to a 20 point data set, plotted showing point-wise confidence intervals computed for each value of *t*, using Equation 1.54.

Figure 1.39. Least-squares-fit line with 20 data points plotted with confidence limits calculated based on Equation 1.53.

For the case of $t = 0$, from Equation 1.44: $\hat{y}(\hat{a}, \hat{b}) = 0.00571 + 0.189t = 0.00571$ and above

$$C_I = (0.00571 + 0.189\,t) \pm (2.101)(0.0384)\sqrt{\frac{1}{20} + \frac{(t - 2.617)^2}{178.5 - (52.34)^2/20}}$$

$$= 0.00571 \pm 0.03736$$

Hence 95% confidence interval for the mean value of $\hat{y}$ at a fixed time of zero lies between the values of -0.0316 and $+0.043$. Obviously for CDF, the value must always be positive. The 95% confidence interval for the slope lies between 0.176 and 0.201.

A casual observation of Figures 1.38 and 1.39 may not be convincing that the conservative value for the slope has a greater impact on the final reliability projection than the conservative value of the least-squares fit due to the mean value of $\hat{y}$ at a fixed time for the 95% confidence intervals. Note that in these figures no projection has been made. Again we must emphasize that all of the results for Example 1 are only approximate because the noise terms are not independently distributed. Hence the results on the figures are also only approximate. The intent here is to give the reader a simple approximation. The more accurate computer generated results for a maximum likelihood estimator are much preferred; however, it is also good to have a simple approximate method as a sanity check.

Figure 1.40 is a qualitative depiction of three errors terms that conspire against the reliability engineer. Two terms were discussed above with respect to the linear regression model even though severe weaknesses were highlighted for its use in determining confidence bounds. Those two error terms are shown qualitatively as the slope error and the mean error in Figure 1.40. The third error is the error in the acceleration factor itself. Limited sample sizes, random statistical variation, and random process variation all limit the accuracy of the acceleration calculation. Often the models used in semiconductor reliability have an exponent that controls the acceleration so that a very small error in that exponent is magnified many times. Here we have assumed that there are no experimental or equipment errors. The slope and mean error indicators are only shown on the worst case side of the two acceleration error indicators. The data points are practically lost on a chart such as this and one can see that the total error bar is approximately an order of magnitude as it is depicted in this figure. Although this is nothing more than a sketch, it is very realistic to expect these three errors to result in an uncertainty of that magnitude. If the time is to be a given, then the uncertainty is in the value of the failure rate and again that can be on the order of a factor of ten.

Finally compare this figure to our starting point of Figure 1.1. Hopefully at this point the reader understands not only the procedure for generating plots like Figure 1.1 or 1.40, but also understands the precautions that must be considered as one goes about designing an experiment, taking the data, determining the

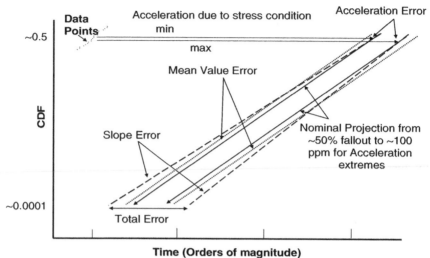

Figure 1.40. Sketch of 20 data points showing errors due to acceleration uncertainties, slope uncertainties, and mean value uncertainties, projecting from the 50% fallout to ~100 ppm.

appropriate distribution function, calculating the acceleration factors, and finally making the plots either physically or with a computer.

1.6.4   Chi-Square Goodness of Fit Application

The chi-square goodness of fit test may be used to test the hypothesis that data, sampled from a population, fit an assumed distribution. The concept behind this test is to first divide the data into intervals or cells, and to then compare the expected value in each interval based on the assumed distribution, to the observed value in that interval. The test is sensitive to both the size of the intervals and the data frequency within each interval. Note that if the expected data frequency is less than five, some intervals may need to be combined.

$$\chi^2 = \sum_{i=1}^{c} \{(Observed_i - Expected_i)^2 / Expected_i\} \qquad (1.55)$$

In Equation 1.55, c is the number of the intervals or cells or groups of data and for Equation 1.55 to approach a true chi-square distribution, the sample size, n, must approach ∞. However, as long as n is large enough so that each grouping of data has a minimum of five observations, Equation 1.55 can be used for a chi-square goodness of fit test. Note that the chi-square statistic has $c-1$ degrees of freedom if no parameters need to be estimated.

Typically the parameter values will be estimated from the data in the case of semiconductor reliability. For that case, Equation 1.55 may be rewritten explicitly in terms of the sample size and the probability of the samples falling within a given grouping. The student is referred to Mann [11] for a detailed discussion of this subject.

$$\chi^2 = \sum_{i=1}^{c} \{(Observed_i - nP_i)^2/nP_i\} \tag{1.56}$$

where c is the number of the intervals or cells or groups of data and again the sample size, n, is sufficiently large for each cell to have at least five observations.

The first step in applying the chi square goodness of fit test then, is to divide the data into groups so that a histogram of the actual data may be plotted. Typically the null hypothesis to be tested in this case will be that the data can be described by the assumed distribution and the alternative hypothesis is that the data cannot be described by the assumed distribution. The chi square figure of merit is defined by Equation 1.56. This test assumes the chi square distribution has $c-p$ degrees of freedom where c is the number of nonempty cells, and p is the number of unspecified parameters of the distribution plus one. For the two-parameter Weibull distribution in Equation 1.25 or the two parameter lognormal distribution in Equation 1.33, $p = 3$. The hypothesis that the data can be described by the assumed distribution is rejected if Equation 1.57 is true. The term $\chi^2_{(1-\alpha,m-p)}$ is calculated from the chi square CDF given in Equation 1.42.

$$\chi^2 > \chi^2_{(1-\alpha,m-p)} \tag{1.57}$$

where α is the level of significance and m–p are the degrees of freedom.

An application of the chi square test using these equations is given below. The fail times are given in Table 1.1 and the chi square test will now be used to determine whether or not these fail times belong to a uniform distribution with $0 \le t \le 5$. Three cells are chosen between 0 and 5 such that the cell boundaries are at $c_1 = 1.4$, $c_2 = 2.8$ and $c_3 = 5$ with the cells having 5, 7, and 8 observed fails respectively. Also note that since we have discrete fail times, the cell separations, c_i, must be chosen so as to avoid those fail times. The expected values for the three intervals or cells may be calculated using the equation: $P_i(t) = (t_i - t_{i-1})/(t_2 - t_1)$ where the distribution is defined between the values of t_1 and t_2, and which in this case are 0 and 5, respectively. Those results are: $P_1 = 0.28$, $P_2 = 0.28$, and $P_3 = 0.44$. After multiplying by the sample size, the estimated number of observations in each cell becomes 6, 6, and 9, respectively. Note that the requirement of a minimum of five expected occurrences per cell is met in this example. Substitution of these results into Equation 1.56 then yields:

$$\chi^2 = (5-6)^2/6 + (7-6)^2/6 + (8-9)^2/9 = 0.444$$

The hypothesis that the data is sampled from a uniform distribution is not rejected for $\alpha = 0.05$, since $\chi^2_{0.95}(2) = 5.99$.

1.6.5   Maximum Likelihood Estimation (MLE)

Conceptually, the objective of MLE is to determine that set of distribution parameters that will maximize the likelihood of representing the sample data. MLE is both a powerful and versatile method for fitting statistical distribution models to sample data. MLE is very useful for semiconductor reliability both because it has the power to fit all of the distributions commonly used in reliability to sample data and because it provides a best fit for all of the distribution parameters across all of the cells of a stress. MLE can also be used for hypothesis testing. Although the computation required for MLE is typically complex, this is not an issue given today's software packages.

The likelihood function, L, is the joint probability shown in Equation 1.58.

$$L = \prod_{i=1}^{n} f_{xi}(x_i; \theta_1, \theta_2, ..., \theta_k) \tag{1.58}$$

where $f_{xi}(x_i; \theta)$ is the probability of all of the x_is occurring together given the distribution parameters $\theta_1, \theta_2, ..., \theta_k$, f is the probability density function, x_i is a random sample of size n, and θ is a vector consisting of the set of unknown distribution parameters.

Discussion of the MLE theory and technique is beyond the scope of this book and the reader is referred to, for example, Mann [11], for an excellent discussion of the MLE technique as applied to a three-parameter Weibull distribution. Li et al. recently published two excellent articles on the MLE theory and techniques for the lognormal distribution [29, 30]. Nelson [10] also addresses MLE in a comprehensive manner.

1.6.6   Closure

This brief introduction to statistics should be adequate for the reader to navigate the remainder of this book. It has been abbreviated because the focus of this book is on the reliability mechanisms themselves, not on reliability statistics. The development of the Weibull and lognormal distributions will be discussed further in Chapters 2 and 7, respectively. Also in Chapter 2, additional examples of confidence bounds are given. However, none of these statistics topics have been treated rigorously since this book is focused on CMOS reliability. Furthermore, some very important topics, for example MLE, have only been mentioned. The reliability engineer that clearly understands statistics is in the best position to both understand what information is in the data as well as advance the understanding of the physics of the mechanism. In no way should the brevity of this treatment be construed as diminishing the importance of reliability statistics. Both introductory reliability statistics books as well as advanced reliability statistics books have been referenced and the readers are encouraged to avail themselves of these references as well as any of the other of the many statistics books on the market.

Many other areas of statistics have not even been mentioned, including some the statistical considerations that go into the design of the experiments. Although several reasons for censoring have been highlighted, we stopped short of providing the statistics to do so. The reader should also be aware that the field of statistics includes the appropriate formalisms to treat replacement parts and repairable parts. One can think of fitting multiple cells and the statistical treatment required for factorial experiments. All of these have more or less applicability to semiconductor reliability depending on the particular issues and experiments, and again the reader is referred to the full books on these subjects. Several good texts not already referenced include [31, 32].

We have discussed in some detail the mechanics of a projection. Once the mechanics are mastered, the real engineering begins. A Weibull distribution was shown to have theoretical justification for modeling a physical system that behaves as a 'weakest link' system, as discussed and referenced in Section 1.4.7. In this case the theoretical support for the Weibull distribution, a subset of an extreme value distribution, is very strong. One theoretical basis for lognormal distribution is the proportional growth or multiplicative model and is typically the case for material transport. Knowing the mechanism is an important part of choosing the correct modeling distribution.

One final word of caution for today's world where most, if not all, of reliability analyses are done by machine—this is truly progress; however, always do a sanity check on the results. Try to estimate the result so that you will not be blindsided by input errors that may have missed an order of magnitude or used inconsistent units. At the risk of nostalgia, the one and only advantage of the ancient instrument called a slide-rule was that it gave the result only to three significant digits and it did not show the decimal point. The three significant digits could be refined but the engineer had to independently calculate the decimal point. The engineer was forced to look at the problem and result in enough detail to notice an order of magnitude error.

REFERENCES

1. Ron R. Troutman. *Latchup In CMOS Technology: The Problem and Its Cure*. Kluwer Academic Publishers, Norwell, MA: 1986.

2. Steven Howard, Voldman. *ESD Physics and Devices*. John Wiley and Sons, Hoboken, NJ: 2004.

3. Ajith E. Amerasekera, Charvaka, Duvvury. *ESD in Silicon Integrated Circuits*. John Wiley and Sons, Chichester: 2002.

4. Special SER Edition. IBM J. of Res. and Dev., 40(1): Jan 1996, pp 1–136.

5. Tobias, Paul A., Trindade, David C. *Applied Reliability*. Van Nostrand Reinholdt, New York: 1986.

6. Lee, C. Tom, Deborah, Tibel, Sullivan, Timothy D., Forhan, Sheri. Comparison of isothermal, constant current and SWEAT wafer level EM testing methods. IRW: 2001, pp 194–199.

7. Tom C. Lee, Michael, Ruprecht, Deborah, Tibel, Timothy D. Sullivan, Wen. Electromigration study of Al and Cu metallization using WLR isothermal method. IRPS: 2002, pp 327–335.

8. A. E. Zitzelsberger et al. On the use of highly accelerated electromigration tests (SWEAT) on copper. IRPS: 2003, pp 161–165.

9. W. Q. Meeker, L. A. Escobar. *Statistical Methods for Reliability Data*. John Wiley and Sons, New York: 1998.

10. Wayne, Nelson. *Applied Life Data Analysis*. John Wiley and Sons, New York: 1982.

11. Nancy R. Mann, Ray E. Schafer, Singpurwalla, D. Nozer. *Methods for Statistical Analysis of Reliability and Life Data*. John Wiley and Sons, New York: 1974.

12. A. E. Green, A. J. Bourne. *Reliability Technology*. John Wiley and Sons, London: 1972.

13. Terry, Sincich. *Statistics by Example*. Dellen Publishing Co., San Francisco: 1982.

14. W. C. Riordan, R. Miller, E. R. St. Pierre. Reliability improvement and burn in optimization through the use of die level predictive modeling. IRPS: 2005, pp 435–445.

15. W. C. Riordan, R. Miller, Sherman, J. M. Hicks. Microprocessor reliability performance as a function of die location for a 0.25 µm five layer metal CMOS logic process. IRPS: 1999, pp 1–11.

16. A. W. Strong et al. Gate dielectric integrity and reliability in 0.5 µm CMOS technology," IRPS: 1993, pp 18–21.

17. C. H. Stapper. LSI yield modeling and process modeling. IBM J. of Res. and Dev., 20: 1976, pp 228–234.

18. C. H. Stapper, A. N. McLaren, M. Dreckmann. Yield model for productivity optimization of VLSI memory chips with redundancy and partially good product. IBM J. Res. and Dev., 24, 1980, pp 398–109.

19. G. H. Hahn, S. S. Shapiro. *Statistical Models in Engineering*. John Wiley and Sons, New York: 1967.

20. Johnson, Leonard G. The median ranks of sample values in their population with an application to certain fatigue studies. Industrial Mathematics, 2: 1951.

21. R. G. Filippi et al. Paradoxial predictions and minimum failure time in electromigration. App. Phys. Letters, 66(16): 1995, pp 1897–1899.

22. D. R. Wolters, J. F. Verwey. Breakdown and wear-out phenomena in SiO2 films. In: *Instabilities in Silicon Devices*, edited by Barbottin and Vapaille. North-Holland: 1986, pp 315–362.

23. R. P. Vollertsen, W. G. Kleppmann. Dependence of dielectric time to breakdown distributions on test structures. Proc. of IEEE 1991 Int. Conf. on Microelectronic Test Structures, 4: 1991, pp 75–80.

24. E. Y. Wu, J. H. Stathis, L.-K. Han. Ultra-thin oxide reliability for ULSI application. Semiconductor Sci. and Tech., 15: 2000, pp 425–435.

25. Yashchin, Emmanuel. Modeling and analyzing breakdown phenomena in insulators–A stochasitic approach. Research Thesis, Israel Institute of Technology, Haifa: May 1981.

26. D. S. Peck. Semiconductor reliability predictions from life distribution data. Proc. of the AGET Conf. on Reliability of Semiconductor Devices: 1961.

27. B. T. Howard, G. A. Dodson. High stress aging to failure of semiconductor devices. Proc. of the Seventh National Symposium on Reliablity and Quality Control: 1961.

28. T. T. Soong. *Fundamental of Probability and Statistics for Engineers*. Wiley and Sons, Hoboken, NJ: 2004.

29. B. Li, Yashchin, E. Christiansen C., J. Gill, R. Filippi, T. Sullivan. Application of three-parameter lognormal distribution in EM data analysis. Microelectronics Rel, 46: 2006, pp 2055–2049.

30. B. Li, C. Christiansen, J. Gill, T. Sullivan, E. Yashchin, R. Filippi Threshold electromigration failure time and its statistics for copper interconnects. J. of Appl. Phys., 100: 2006, pp 114516-1–10.

31. C. R. Hicks. *Fundamental Concepts in the Design of Experiments.* Holt, Rinehart and Winston, New York: 1973.

32. William J. Diamond. *Practical Experimental Designs for Engineers and Scientists.* Van Nostrand Reinholt, New York: 1989.

2

DIELECTRIC CHARACTERIZATION AND RELIABILITY METHODOLOGY

Ernest Y. Wu, Rolf-Peter Vollertsen, and Jordi Suñé

2.1 INTRODUCTION

This chapter summarizes the fundamental aspects of characterization methodology and statistical models for dielectric reliability with the focus on silicon dioxide (SiO_2) and/or SiO_2-based dielectrics, the key ingredient of CMOS. These dielectrics can be found in many flavors in today's integrated circuits. There are the thick inter- and intrametal dielectrics with thicknesses of several hundred nanometers. On the other hand, the film thickness approaches one nanometer for high performance CMOS technology in which the predominant use of SiO_2 as gate dielectric is to passivate the silicon surface. SiO_2-based dielectrics have owned this primary place in the microelectronics revolution due to the unique properties of SiO_2, which are summarized in Table 2.1.

Because of SiO_2's predominant use in microelectronics technologies, this material has been studied most intensively over the last 40 years. In comparison with other insulators, the excellent quality of SiO_2 grown on a Si substrate (SiO_2/Si) interface with optimized fabrication processes has yielded very low defect density. The large barrier height ($\sim 3\,eV$) of the SiO_2/Si interface associated with the large SiO_2 bandgap ($\sim 9\,eV$) leads to much reduced leakage current for sufficiently thick films and results in highly resistive insulators. The high melting temperature of SiO_2 makes it very compatible with CMOS processing steps performed after the formation of SiO_2-based gate dielectrics.

Reliability Wearout Mechanisms in Advanced CMOS Technologies. By Strong, Wu, Vollertsen, Suñé, LaRosa, Rauch, and Sullivan
Copyright © 2009 the Institute of Electrical and Electronics Engineers, Inc.

TABLE 2.1. Overview of Physical and Chemical Properties of SiO$_2$ at Room Temperature

Structure	amorphous
Molecular weight	60.1 g/mole
Density	2.19 g/cm^3
Molecules/cm^3	2.2×10^{22}/cm^3
Melting point	$\sim 1600^\circ$C
Thermal expansion coefficient	$5 \times 10^{-7}{}^\circ$C^{-1}
Young's modulus	6.6×10^{10} N/m^2
Poission's ratio	0.17
Thermal conductivity	0.014 W/cmK
Relative dielectric constant	~ 3.9
Energy gap	~ 9 eV
DC resistivity	10^{14}–10^{16} Ohm-cm
Infrared absorption band	9.3 μm
Index of refraction	1.459
Electron mobility	20–40 cm^2/Vs
Hole mobility	$\sim 2 \times 10^{-5}$ cm/Vs

Sources: A. S. Grove, *Physics and Technology of Semiconductor Devices*, Wiley, New York (1967); S. Sze, *Physics of Semiconductor Devices*, 2nd ed., Wiley, New York, (1981); W. R. Runyan and K. E. Bean, *Semiconductor Integrated Circuit Processing Technology*, Addison-Wesley Publishing Company, New York, (1990).

These unique and excellent properties have enabled the continuous scaling of microelectronics technology for ever increased circuit performance and circuit density. From a reliability viewpoint, it is very important to understand whether these excellent properties, and the related electrical device performance character- istics, can be maintained through the useful lifetime of CMOS products, particularly as the film thickness of SiO$_2$ is continuously reduced. Besides SiO$_2$-based gate dielectrics, other dielectrics have been used in CMOS technologies. For example, silicon nitride (Si$_3$N$_4$) or multilayers of SiO$_2$/Si$_3$N$_4$ and SiO$_2$/Si$_3$N$_4$/SiO$_2$ structures are used as the storage capacitors in some memories. SiO$_2$ and Si$_3$N$_4$ dielectrics are also used as the insulating materials between metal layers in the so-called MIM capacitors used in analog and radio frequency (RF) applications. In recent years, dielectric materials with high permittivity (the so-called high-K dielectrics) have been intensively investigated to replace SiO$_2$-based dielectrics for leakage–current reduction to achieve low power consumption. The family of hafnium-oxide-based materials (HfO$_2$, HfSi$_x$O$_y$, HfO$_x$N$_y$, and HfS$_x$O$_y$N$_z$) emerges as the leading candidate. However, none of the materials have been found to possess the same quality as SiO$_2$ films. The contrary happens in the metallization levels where the conventional SiO$_2$ insulators are replaced by dielectric materials with low permittivity (the so-called low-K dielectrics) as interconnect insulators in order to reduce the time delay in the back-end-of-line (BEOL) of CMOS techno- logies. Although the charge conduction and breakdown (BD) mechanisms might be quite different or unique in different types of dielectrics, many fundamentals of the characterization methodology and the main principles of the theoretical formulations discussed in this chapter remain more or less the same. Furthermore, even the physical models and concepts obtained from the state-of-the-art SiO_2 reliability physics have already been extended to characterize these different types of dielectrics. Therefore, many parts of this chapter can be beneficial even for readers dealing with different kinds of dielectrics.

2.1.1 Application and Fabrication of Silicon Dioxide-Based Dielectrics

As already mentioned above, silicon dioxide had been first used as the gate dielectric in metal–oxide–silicon field-effect transistors (MOSFET) some decades ago, when the gate material was the metal aluminum. In the meanwhile, the gate employs a highly doped polysilicon. The simple MOS capacitor structure shown in Figure 2.1 (which is also used in products) is the main test structure used for dielectric reliability assessment. The capacitor is in most cases bounded by a field oxide-like shallow trench isolation (STI) on all four sides. The transistor gate is bounded by diffusions along two sides and STI on the other two sides. Before STI was used for isolation, field oxide produced by other methods was grown. The best known is probably the local oxidation of silicon (LOCOS).

The MOSFET depicted in Figure 2.1 serves as a fundamental building block in a wide variety of semiconductor products. The best examples are probably the microprocessor and the memory. Memories come in many varieties: There are dynamic random access memories (DRAMs), static random access memories (SRAMs), and nonvolatile memories (NVRAMs or NVMs). Some of those memories can be embedded in application specific integrated circuits (ASICs).

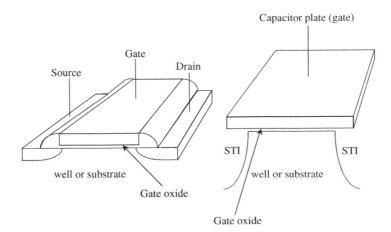

Figure 2.1. MOS transistor (left) and capacitor (right). The transistor is bordered by two diffusions (source and drain) and two STI (shallow trench isolation) areas (not shown due to view angle). The capacitor is STI bounded on all four sides.

But the variety of CMOS products is much wider and includes power CMOS, radio frequency (RF) devices, and much more. Each of these products uses circuits that are comprised of MOSFETs or MOS capacitors for which the gate dielectric integrity needs to be maintained throughout the lifetime for continued functionality at the specification level. The gate can lose the control of the MOSFET current depending on the leakage current increase through the dielectric due to localized dielectric failure, as we will discuss later in this chapter. In the case of memory applications such as storage capacitors, a dielectric failure can result in a loss of information due to its high sensitivity to even a small increase of leakage current from the specified functional level.

Silicon gate oxides can be fabricated using a variety of techniques based on thermal oxidation, chemical deposition, or physical deposition. Historically, the gate oxide of MOS devices is thermally grown by exposing Si to O_2 at elevated temperatures. Because of its processing simplicity and the nearly perfect quality of the Si/SiO_2 interface, this method has been used to obtain very thick SiO_2 films in a range of several tens or hundreds of nanometers. In the early days of the microelectronic industry, ionic contaminations were major problems so that reliability aspects only came into focus after the interface and bulk properties were significantly improved. The optimized silicon dioxide gate has been a fundamental component for the success of CMOS technology.

Thermal oxidation can be accomplished in a wide range of temperatures in dry oxygen or in an oxygen–water ambient. By combining the oxidation processes in dry and wet ambient, oxidation time can be sufficiently reduced for a given film thickness; at the same time, a better reliability is often achieved. As oxide thickness has been reduced below 3 nm for technology generations of 180 nm and beyond, new serious concerns are raised: 1) high leakage current due to direct tunneling between the gate and substrate; 2) out-diffusion from the heavily boron-doped polysilicon gate in pMOSFETs and penetration of this boron into the channel of the device; and 3) associated reliability issues such as breakdown (BD) for these thin dielectric films [1].

The incorporation of amounts of nitrogen as small as $<5\%$ in SiO_2 films was found to be very effective to reduce the boron penetration. These SiO_2 films with nitrogen composition are usually referred to as silicon oxynitrides (Si–O–N), or nitrided oxides, or SiO_2-based dielectrics. Nitrogen can be incorporated by using a thermal oxynitridation process and substituting O_2 with NH_3, N_2O, or NO gas. Silicon dioxides and silicon oxynitrides can be deposited by chemical deposition techniques such as rapid thermal chemical vapor deposition (RTCVD) or by physical deposition techniques such as oxygen or nitrogen ion implantation. In the latter case, nitrogen is incorporated by implanting it in the Si substrate, followed by an oxidation process.

Usually, the interfacial properties of gate dielectrics deposited by these techniques are inferior to thermal oxidation. A post-deposition annealing at temperatures higher than deposition temperature is required, although a low thermal budget was initially intended when using these techniques. The incorporation of a small amount of nitrogen only represents a slight modification of pure SiO_2 films. As film thickness continues to scale down below 1.5 nm, a drastic increase in direct tunneling current becomes a serious concern. In addition, the nitrogen profile within SiO_2 films was found to be very critical and affects both device performance and reliability. In particular, high nitrogen concentration at a Si/SiO_2 interface can lead to detrimental effects in reliability. Therefore, an ideal nitrogen profile requires a high concentration at the poly-Si interface to suppress boron penetration and a small enough concentration at the Si/SiO_2 interface to control channel hot electron degradation. With these limits, thermal techniques are no longer able to produce high nitrogen content and good control of the nitrogen profile. Plasma-assisted deposition techniques, in which the nitrogen-containing species are plasma-excited, can not only produce sufficient nitrogen concentration but also spatially selective nitrogen profiles that meet many demands required for high device performance.

To minimize the damage to the Si substrates by energetic particles in the deposition process, a remote plasma nitridation (RPN) method has been developed. Furthermore, the addition of sufficient nitrogen (5 to 15%) into SiO_2 films effectively increases the permittivity of the silicon oxynitride films since the dielectric constant of silicon nitride is approximately twice that of SiO_2. This means that the physical thickness of heavily nitrided oxide can be greater than that of a single SiO_2 layer of the same equivalent thickness. Consequently, direct tunneling currents are reduced with respect to a pure SiO_2 layer in this range of thickness. This RPN deposition process or the decoupled plasma nitridation (DPN), a modified version of RPN process, is widely used nowadays in the most advanced high performance MOSFETs for film thickness around one nanometer. Figure 2.2 shows a cross-section image obtained from high resolution

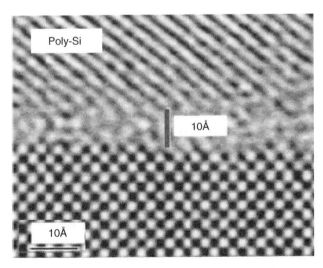

Figure 2.2. The lattice images of transmission electron microscopy for a MOS transistor with polysilicon gate and silicon substrate with a SiO_2 film of 1.0 nm thickness [2]. Reprint Courtesy of International Business Machines Corporation, copyright 2002 © International Business Machines Corporation.

transmission electron microscopy (X-TEM) for 1.0 nm silicon oxynitride [2]. In this thickness range, an oxide film consists of a few molecular layers only (distance of O–Si–O bridge ~ 0.4 nm). As discussed later in the subsection of oxide scaling, this certainly affects the insulating properties and the power-consumption constraints due to the increased direct tunneling current.

2.1.2 Failure Modes of Gate Oxide and Reliability Requirements

2.1.2.1 *Gate Oxide Breakdown.* Dielectric breakdown is the irreversible local change of the dielectric isolation property. In other words, within the dielectric area a tiny spot develops with increased conductivity compared to the rest of the dielectric area that remains nearly unchanged. Although the conducting spot is very small compared to the capacitor or device area, it now dominates the current flow and therefore changes the electric characteristics (current–voltage curve) from a behavior before breakdown to a clearly different characteristic after breakdown. A breakdown happens after a certain amount of time during which the oxide is subjected to an electrical stress at product operation or elevated conditions. This local change of properties causes, in most cases, the product to stop functioning as intended. The details of dielectric breakdown will be discussed in Section 2.3 and Section 3.2.

In most sections the so-called intrinsic breakdown is considered. This kind of breakdown is an inherent material property of good and flawless dielectrics, which nevertheless wear out over time and finally fail at the moment of breakdown. Being related to a wearout process, intrinsic breakdown exhibits an increasing failure rate. In practice, this means that within a relative short period of time after the intrinsic breakdown of the first of the samples, the entire population that was put on stress (or in operation) at the same time and condition will fail. This leads to steep statistical failure distributions as discussed in Sections 2.4 and 3.2.

However, some failures can also occur much earlier than expected for intrinsic breakdown. Since these fails occur early, they are called early or defect-driven fails. They usually occur distributed over a wide time range before the intrinsic breakdown, causing a wide or shallow distribution before the intrinsic distribution. Those flat distributions are called extrinsic distributions and typically exhibit a decreasing failure rate that may be improved by burn-in. The dielectric structures within this category that fail do so not from wearout but from flaws in the dielectric. As an example, the dielectric could be thinner at one spot or contain a particle that almost penetrates the entire dielectric. In such cases, the weak spot degrades faster than the remaining, flawless dielectric and breaks down before the intrinsic breakdown at any other spot is reached. The flaws are usually defects introduced by process irregularities. It should be kept in mind that intrinsic fails reflect wearout and material properties, whose models and physics are discussed in the following sections. The results derived for intrinsic fails may not, in all cases, be transferred to extrinsic, defect-driven fails. Nevertheless this is often attempted due to the lack of models for this failure mode and the much larger sample size required for proper investigation of extrinsic mechanisms.

2.1.2.2 A Historical Perspective. Electrical breakdown of dielectrics by lightning was one of the first known electrical phenomena. The first laboratory experiments on breakdown of glass were carried out in 1799 [3]. Since then, the investigation of dielectric breakdown has had two main motivations: the scientific interest for the fundamental understanding of this phenomenon and the practical application of dielectrics as insulating materials. Much of the work on breakdown done in the early days was nicely summarized in the book by O'Dwyer [4].

The experimental work on dielectric breakdown, especially for silicon dioxide, can be traced back to nearly 40 years ago. In the early 1970s, the experimental studies were focused on the dielectric breakdown caused by the presence of mobile ions (particularly Na^+) in contaminated SiO_2 films [5–8] while other investigations were aimed at studying the reduction in breakdown strength due to the radiation damage created by ion implantation [9]. These studies clearly demonstrated an enhanced electric conduction with a reduced barrier height due to Na^+ ions drift in SiO_2. The results can be well correlated to the reduction of breakdown strength and degradation of TDDB measurements. The presence of these defects in SiO_2 films presented a fundamental limit for the application of SiO_2 films in MOS technology. The investigation of the intrinsic properties of oxide breakdown became possible only when the oxide processing technology improved to the extent that mobile ion contamination was drastically reduced or eliminated.

Early experimental investigations of oxide breakdown focused on voltage and temperature acceleration as well as on area scaling characteristics for thick films with oxide thickness varying from 10 to 70 nm [10–14]. On the other hand, similar work was conducted using fast voltage/current ramping techniques [14–17]. Among these studies, an exponential field dependence of time-to-breakdown was generally reported while an Arrhenius thermal activation process was shown to be compatible with the experimental temperature dependence of time-to-breakdown. At the same time, efforts to understand the relation between the generation of defects and the breakdown mechanism were launched [18, 19] that tried to correlate defect generation and its impact on electron conduction mechanisms with oxide breakdown. The thermochemical model was first proposed in 1985 to explain the exponential dependence of time-to-breakdown (t_{BD} or T_{BD}) on oxide field based on the generalized Eyring model [13]. Other researchers considered that an electron avalanche process was at the origin of the breakdown of SiO_2 dielectrics [19–22]. In these early investigations (even in those based on impact ionization and electron avalanche processes), the oxide electric field was almost exclusively considered as the driving force for oxide breakdown. It was not until the end of the 1980s that electron heating resulting from the interaction between the oxide field and electron lattice was shown to be a key element in the defect generation process leading to oxide breakdown [23].

On the other hand, as far as the breakdown statistics are concerned, there was quite a general presumption among these early studies to use the log normal function as the best choice to model the cumulative t_{BD} failure distribution, although some authors questioned the validity of this assumption [14]. Some theoretical justification for the log-normal distribution was given by assuming that the instantaneous breakdown probability is proportional to the FN current density in combination with an enhanced cathode field due to hole trapping [24, 25]. However, Wolters et al. experimentally demonstrated that the Weibull distribution provides the best description for the oxide breakdown [26] for 40 nm thick oxides using the grouping experimentation method. This was subsequently verified for intermediate thick and ultra-thin oxide films [27, 28].

At present, it is widely accepted that the breakdown is triggered by the generation of microscopic defects in the oxide which finally lead to the formation of defect-assisted conduction paths that trigger the oxide breakdown. These microscopic oxide defects should not be confused with the earlier mentioned extrinsic defects that are rather macroscopic. The relation between stress-induced defect generation and breakdown statistics is already well established in the so-called percolation models of oxide breakdown [29–32]. On the contrary, there is no general agreement about the physical mechanisms involved in the generation of these defects, and a certain controversy still exists among different models. In Sections 2.4 and 3.2, we discuss the statistical properties of oxide breakdown in detail, while in Section 3.3 some of the most important degradation and breakdown models are presented and compared.

2.1.2.3 Gate Oxide Reliability Requirements. Intrinsic breakdown should not happen in a product during the specified operating lifetime. Extrinsic fails may not be completely avoidable in a product, but their occurrence must be reduced to a tolerably low rate. This leads to the definition of the following reliability requirements: All circuits of a chip must operate in their initial manner for a certain period of time (usually 10–15 years, depending on the application), i.e., no catastrophic breakdown of the gate dielectric is tolerable. It must be assured that during the intended period of operation, the fallout remains below a certain tolerable percentage (e.g., 10–100 ppm for intrinsic fails, depending on the product).

A chip could contain circuits operated at different voltages and with different duty cycles. Thus for each circuit or each group of similar circuits, the failure rate at the end of the intended chip lifetime needs to be assessed. The combined failure rate of all circuits, or groups of circuits, in the chip must not exceed the tolerable failure rate. The tolerable failure rate is a fraction of the total tolerable chip failure rate because it needs to consider contributions from other mechanisms such as interconnects, hot carrier degradation, and soft error rate (SER). In consequence, for the gate oxide reliability assessment we need to know the criteria listed below:

- The gate oxide-related fraction of the total tolerable failure rate at the end of the intended lifetime (estimated from product life tests)
- The operation conditions of the various circuits (voltage, temperature, duty factor, and state during operation)
- The burn-in conditions (if applicable)
- The total number of devices under the same stress condition for each circuit to determine the gate oxide area for that condition and circuit

- A definition of what kind of dielectric failure causes a circuit to no longer function (relevant failure criterion, e.g., a maximum leakage current for the failing transistor)

All of this information is necessary to define the target and some details for a product-relevant lifetime projection. This is an input from product design and/or product reliability sources. But for the lifetime estimate, much more is needed. One needs to know the following information.

- The time-to-breakdown distribution type and the oxide thickness and stress condition specific parameters
- The voltage acceleration model and dielectric-technology-specific acceleration parameter
- The temperature activation model and the technology-specific activation energy
- The area scaling model

The above can be determined in part with the help of literature, but most of the results must come from the reliability engineer's own measurements. These measurements are necessary because most of the parameters of the models are technology-specific within some margin. Assumed literature parameters or models may lead to technology/product conclusions that are too optimistic or conservative. The details of these models are the subject of Section 2.4 and Sections 3.2–3.4. For a relevant reliability assessment, appropriate characterization tools are also essential. These tools are the test structures that deliver representative results, the stress equipment, and the algorithms for both stress and evaluation that represent the state-of-the-art. Another set of tools are measurement methods for basic characterization, e.g., for thickness measurement, electrical failure analysis, interface characterization, and so on. Many of these tools are described and discussed in Sections 2.2–2.3.

2.1.2.4 The Reliability Projection Procedure. As discussed above, an appropriate projection procedure and methodology is required to verify the gate dielectric reliability requirements are fulfilled. Figure 2.3 gives a schematic illustration of the series of steps involved in a realistic reliability projection. The stress data are usually collected at accelerated voltages and elevated temperatures on test structures (capacitors or devices). In a typical lifetime projection for a given oxide thickness, the time-to-breakdown is extrapolated from the stress voltages to a lower voltage which is representative of the use condition. Similarly, a projection to the temperature at use condition can be done. Since the stresses are often performed on test structures with small area, a projection to the total gate area of a chip is required. Finally, the lifetime at higher failure percentiles must be projected to the lower cumulative failure percentiles to satisfy the specified requirements. For this extrapolation to low percentiles, a good knowledge of the shape of the time-to-breakdown distribution is of paramount importance.

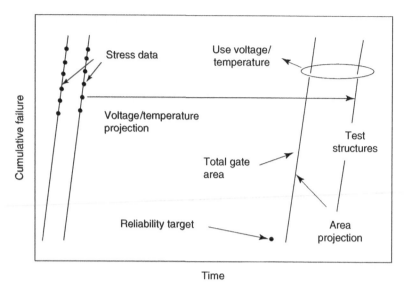

Figure 2.3. A schematic illustration of projection procedure used in gate dielectric reliability assessment. The symbols represent stress data at high voltages. The cumulative failure is given in percentiles along the y ordinate. The reliability target is specified at a given time (10 years) and a specified cumulative percentage.

In summary, there are four main steps involved in the entire projection: the measurement of the time-to-breakdown distribution at different accelerated voltage and temperature conditions; the voltage and temperature projection; extrapolation to the chip area; and the projection to the low failure percentile target to ensure highly reliable products. For both area extrapolation and percentile projections, a proper choice of breakdown distributions must be made (discussed in Sections 2.4 and 3.2). Furthermore, a reasonable acceleration model is required for voltage and temperature extrapolation. Three voltage acceleration models, widely discussed in the literature, are the exponential law $T_{BD} \sim \exp(-\gamma V)$, the power law $T_{BD} \sim V^{-n}$, and the exponential law with reciprocal voltage dependence $T_{BD} \sim \exp(C/V)$, a simplified version of the $T_{BD} \sim \exp(G/F_{OX})$ model as discussed in Section 3.4. Depending on the model choice, the projected lifetimes can vary by several orders of magnitude, although within a certain voltage range all three models can equally fit the experimental data obtained from a limited time window. Further differences can result from applying an inadequate temperature acceleration model and from picking a wrong distribution type that would lead to wrong area and percentile extrapolations. Sections 3.1, 3.2, and 3.3 are intended to provide the readers with elements of knowledge and understanding with the final goal of helping in the critical process of choosing the reasonable models for reliability qualification.

2.1.3 Impact of Oxide Scaling

The primary reason of the downscaling of the device dimensions is to gain high device performance and ever increased device density of integrated circuits in microelectronics technologies. The shrinkage of oxide thickness increases the gate capacitance as required to maintain high drive current for scaled MOSFETs and helps keep short-channel effects under control. This has pushed SiO_2-based gate dielectrics to reach the atomic scale as already revealed in Figure 2.2. Figure 2.4 (a) shows the reduction of gate oxide thickness with device scaling of various technology generations according to the ITRS roadmaps. Notice that a change in the ITRS roadmap from 1999 to 2000 results in a shift of oxide thickness by nearly 25%. On the other hand, the operational voltage does not decrease as rapidly as the reduction in oxide thickness, particularly in 2006 ITRS roadmap as shown in Figure 2.4 (b). However, having reached the 1 nm thickness range, the downscaling has slowed considerably in 2006 ITRS roadmap. The reason is that the gate leakage is comparable or larger than the device-off-current and can even get close to the device-on-current. This means an enormous increase of the standby power which is not tolerable for mobile applications and requires tremendous cooling efforts in other applications. While thicker oxides are typically made of pure silicon dioxide, ultra-thin oxides are nitrided oxides or oxinitrides because the nitridation of SiO_2 reduces the leakage somewhat and also reduces boron penetration. The reason behind the observed increase of the leakage current is that the oxide thickness is so thin that direct quantum-mechanical tunneling primarily controlled by film thickness can take place as considered in detail in Section 2.2.

The downscaling of oxide thickness (T_{OX}) has also a significant impact on oxide reliability and characterization [33]. In this regard, we must point out that the reduction of oxide thickness has a relevant impact on almost every aspect related to the oxide degradation and breakdown. The properties of the breakdown mechanism responsible for the oxide degradation change with T_{OX}. The electrical consequences of the generated defects and hence the way we can indirectly measure their density also change drastically from thick to thin oxides. The reduction of T_{OX} changes the main properties of the statistical distribution of oxide breakdown as well as the voltage acceleration. As a consequence of all these changes, the oxide lifetime is strongly affected by oxide thickness scaling. Let us now briefly consider all these issues one by one. First of all, it must be recognized that the physical mechanisms responsible for the degradation and breakdown change drastically with T_{OX}. In this regard, for example, stress voltages could be so high in thicker oxides that impact ionization could take place [19–23, 34, 35] and eventually be responsible for the generation of oxide defects. However, this mechanism is completely impossible in ultra-thin oxides and other mechanisms must be taken into account as described in Sections 3.1 and 3.2.

In general, although all oxides break down when subjected to prolonged accelerated stress, the mechanisms by which this phenomenon is triggered depend on the oxide properties (mainly thickness) and the stress conditions. Another

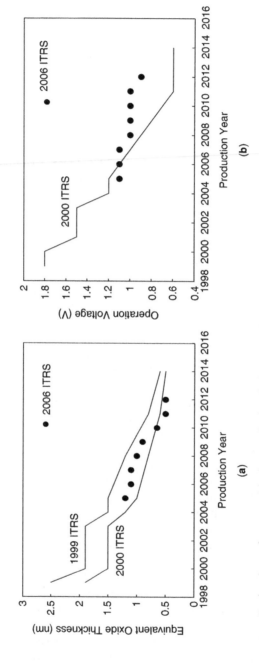

Figure 2.4. Reduction of gate oxide (or equivalent) thickness (a) and decrease of operation voltage (b) with device scaling according to technology generations (adapted from ITRS) [1].

important issue related to the characterization of the degradation that leads to breakdown is the electrical effects of the stress-induced oxide defects. In this regard, oxide thickness scaling produces important changes. While charge trapping in the oxide bulk occurs during stress in thicker oxides and has been used to monitor oxide degradation, this disappears at around 4 nm because the oxides are too thin to trap charges in significant amounts (because the defects are easily discharged to the electrodes by direct tunneling). In thin oxides, the main electrical signature of the generated defects is an increase of the current with stress time (particularly at low voltages) which is related to a mechanism of trap-assisted tunneling. This stress-induced leakage current (SILC) becomes noticeable at low voltages before breakdown occurs for oxide thickness below 10 nm [36] as discussed in Section 3.2. The measurement of the evolution of SILC during stress has also become a quite common way to monitor oxide degradation in thin oxides.

From the viewpoint of reliability, the most important fact is that when T_{OX} is downscaled, the time-to-breakdown is drastically reduced if the gate voltage is maintained. For example, it has been reported that the scaling of T_{OX} from 3.5 to 1.0 nm at fixed voltage causes the mean time-to-breakdown to decrease about 25 orders of magnitude [37]. On the other hand, the scatter of the times to breakdown (the width of the statistical breakdown distribution) significantly increases as the oxide thickness is scaled down [30–32]. While for thick oxides the intrinsic time-to-breakdown distributions are steep and tight, they widen and become flatter for thinner oxides, as discussed in Sections 2.4 and 3.2. This forces a larger sample size to be used for reliability assessment of ultra-thin oxides to maintain the same tightness of the confidence bounds as for thicker oxides. On the other hand, it causes a relevant reduction (about five to six additional orders of magnitude in scaling from 3.5 to 1.0 nm) of breakdown lifetime related to the extrapolation to low failure percentiles. Fortunately, a positive aspect of the thickness reduction is that the voltage acceleration becomes stronger for thinner oxides [38] so that it helps to partially compensate the reduction of reliability when the gate voltage is also reduced. Voltage acceleration plays one of the most important roles in affecting the projected lifetime due to oxide breakdown.

For product development and circuit design it is important to know the possible operation voltage. Using the projection procedure provided above, Figure 2.5 compares the projected maximum operational voltages as a function of oxide thickness using different voltage acceleration models at a specified failure rate of 100 parts per million (ppm). According to the projection using power-law voltage acceleration, the projection shows that the required V_{MAX} by 1999 ITRS can be satisfied while the use of exponential law for projection suggests that the V_{MAX} requirement can not be sustained.

This comparison reveals how important it is to determine the correct voltage acceleration model to be used in reliability projection. We will discuss different voltage acceleration models and their implications in detail in Sections 3.3 and 3.4. Nevertheless, the trend in Figure 2.5 clearly indicates that when using either projection model, as oxide thickness decreases the maximum allowed operational voltage also decreases. This has motivated the interest for the study of

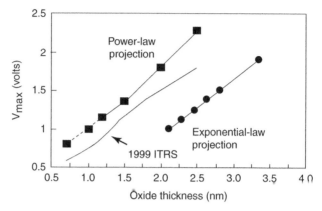

Figure 2.5. The projected supply voltage as a function of oxide thickness for 100 ppm and 0.1 cm² at 110°C for dielectric failure [37]. The dashed lines are extended from the projected maximum voltages for oxide thickness ∼1 nm. The dielectric failure is defined as the first event of dielectric breakdown regardless of the magnitude of leakage current at the moment of breakdown. Reprinted with permission from Ref. 37, Copyright 2003, Elsevier.

post-breakdown properties and of the real impact of breakdown on device and circuit performance [39–54], as briefly considered in Section 3.5. In this regard, it is worth mentioning that thick oxides break down hard, i.e., they exhibit mostly a low resistive ohmic characteristics after breakdown. This is due to the fact that for thick oxides, much higher stress voltages are required to cause the breakdown because of experimental time constraints. On the other hand, thinner oxides (<5 nm) break down soft, i.e., they exhibit a rather low-current nonlinear post-breakdown IV characteristic and, as discussed before, may not even cause the circuit to fail [41].

Soft breakdowns [39–42] and progressive breakdown (a steadily increasing leakage current after the initial breakdown path formation) [43, 44] have been reported before a hard breakdown eventually occurs. Although a final hard breakdown is usually observed after a sufficiently long stress time, the question is whether this hard breakdown or a preceding soft breakdown ends the functionality of a circuit, and whether the extra time elapsed from the first breakdown event to the final device failure gives a significant additional reliability margin. Since the distinction between breakdown and device/circuit failure might be relevant for reliability assessment, several groups have investigated the impact of ultra-thin oxide breakdown on devices [41, 45–49], and others have reported the effect of thin oxide breakdown on digital circuits [50, 51] with focus on inverters and RF circuits [52–55] or SRAM cells [56–58]. Section 3.5 considers some of these issues with particular emphasis on the statistical description of post-breakdown reliability while further details about the prediction of reliability from the system point of view can be found in a recent review paper in reference [59].

2.2 FUNDAMENTALS OF INSULATOR PHYSICS AND CHARACTERIZATION

The knowledge of the electric properties of dielectric films in metal–oxide–semiconductor (MOS) structures is one of the most important fundamentals to develop a complete understanding of dielectric reliability. In particular, the reliability evaluation at accelerated conditions requires a comprehensive understanding of both device operation and physical effects, which may or may not be present at low-voltage conditions. These effects become more pronounced as the thickness of dielectric film is reduced. In this section, we will discuss some of these topics and, in particular, the capacitance–voltage characteristics, the carrier injection phenomena, and the determination of oxide film thickness, since these are the most relevant for the characterization of the dielectric breakdown process.

2.2.1 Capacitance–Voltage Characteristics

Capacitance versus voltage measurements are basic for the characterization of a MOS structure as already discussed in many excellent textbooks [60–63]. Unlike conventional capacitors, a MOS structure consists of an insulator layer sandwiched between a metal or a polysilicon gate-electrode and a semiconductor substrate, as schematically shown in Figure 2.6. This configuration gives rise to a large spectrum of electrical phenomena in the capacitance–voltage (CV) characteristics. The capacitance measurements can provide various means to determine many electrical parameters, which can be related to the physical parameters in the fabrication process and to the bulk material or the interface properties of the MOS structure. In addition, capacitance measurements can be applied to

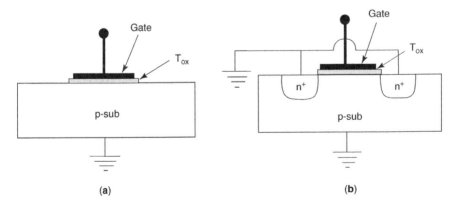

(a)   (b)

Figure 2.6. Schematic diagrams of the bias configurations used for the current-voltage measurements and capacitance-voltage measurements. (a) a MOS capacitor as often used for the accumulation mode with a negative bias on the gate electrode and (b) a MOS transistor as often used for the inversion mode with a positive bias on the gate electrode.

reveal the dynamic behavior of defect generation that results from the application of electrical stress or the damage induced by other external means such as irradiation. Therefore, the capacitance–voltage characteristics are a key element in understanding the degradation process and the dielectric breakdown in the context of microelectronics reliability.

2.2.1.1 MOS Capacitor Structures.
A simple MOS capacitor with a metal gate and a p-type substrate is schematically shown in Figure 2.6 (a). When a negative gate voltage is applied to the gate, electrons in the gate electrode and holes (majority carriers in this case) in the substrate are accumulated at the metal/ SiO_2 and SiO_2/Si interfaces, respectively. This bias condition is commonly referred to as NFET accumulation mode and the corresponding energy-band diagram is given in Figure 2.7 (a). As the gate voltage is swept from zero to the positive direction, the silicon surface close to the SiO_2 layer is first depleted and then eventually inverted, depending on the availability of electrons. Minority carriers (electrons in this case) in the surface channel can be thermally generated within a p-type substrate. However, this method of generating minority carriers is not effective due to the long characteristic times associated to the kinetics of the generation–recombination processes.

One way for more efficient generation of minority carriers is to expose the MOS structure to intense illumination so that electron hole pairs can be generated by absorption of photon energy. Another commonly used technique is to supply electrons from an additional electrode as shown in Figure 2.6 (b). In this case, the auxiliary electrode is formed by the n^+-type diffusion, which acts like an electron reservoir. Thus, electrons are constantly supplied by this electrode connecting to the external circuitry. Once sufficient minority carriers are generated or supplied, the surface is said to be inverted and the bias stress condition is often called NFET

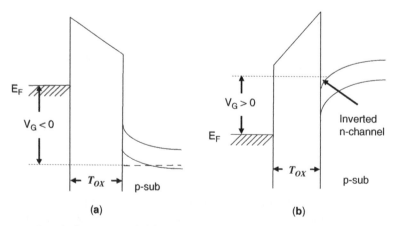

Figure 2.7. Band diagrams of (a) accumulation mode (gate injection) and (b) inversion (substrate injection) for an MOS structure with a metal gate and p-type Si substrate.

inversion mode. The energy-band diagram corresponding to this stress mode is shown in Figure 2.7 (b). Similar bias configurations and their energy-band diagrams can be obtained for MOS structures with a p^+-type polysilicon gate and an n-type substrate for a PFET device. Conventionally, the capacitance measurements are carried out at low voltages where the leakage currents are negligibly low, especially for thick oxides and in either the accumulation or inversion modes. However, for thin oxides, large direct tunneling currents can significantly alter the capacitance measurements, as we will see in Section 2.2.2.

In reliability evaluation, much higher voltages are applied to the MOS structures to cause dielectric breakdown within a reasonable time window. In typical accelerated conditions, appreciable leakage currents ($\sim 10\,\mu A$ or lower) can be measured in the gate electrode. These leakage currents arise from the injection of electrons or holes across the SiO_2 barrier by quantum mechanical (QM) tunneling. For instance, in NFET accumulation mode, the electrons are injected from the gate to the substrate, so that this injection mode is called gate injection. When electrons are injected from either an inverted Si surface in NFET inversion mode or from an accumulated n-type substrate, this injection mode is called substrate injection. Note that in the bias configuration of Figure 2.6 (a), sufficient amount of electrons can not be supplied at low temperatures without using illumination or without the connection to an external reservoir. If carriers are not efficiently supplied to the Si surface to maintain the inversion layer, the substrate is in a so-called deep depletion condition, with the most of applied voltage being dropped in the substrate. Therefore, the measured gate currents become very low and tend to saturate.

2.2.1.2 Conventional CV Characteristics.

Let us first consider a MOS structure with a p-type substrate and a metal gate as shown in Figure 2.6 (a). In a biased MOS structure, the gate voltage is divided into a voltage drop across the oxide layer and a voltage drop over the surface layer of the silicon substrate, as it is expressed in the following equation:

$$V_G = V_{FB} + V_{OX} + \Psi_S = V_{FB} + \frac{-Q_S}{C_{OX}} + \Psi_S, \qquad (2.1)$$

where V_{OX} stands for the oxide voltage, Ψ_S is the surface potential at the silicon surface, and V_{FB} is the flatband voltage that is equal to the workfunction difference ($\Phi_{GS} = \Phi_G - \Phi_S$) between the gate electrode (Φ_G) and the Si substrate (Φ_S) in the simplest case where the charges in oxides can be ignored. Q_S is the surface-charge density induced in the silicon substrate and C_{OX} is the oxide capacitance per unit area, which is given by

$$C_{OX} = \frac{\varepsilon_{OX}}{T_{OX}}, \qquad (2.2)$$

with ε_{OX} being the dielectric constant of the SiO_2 film and T_{OX} is the oxide thickness. The gate capacitance per unit area of a MOS capacitor with a metal gate can be conveniently expressed as a series combination of the oxide capacitance and the Si surface capacitance:

$$\frac{1}{C_G} = \frac{1}{C_{OX}} + \frac{1}{C_S}, \tag{2.3}$$

where the silicon surface capacitance, C_S, is a measure of surface charge density responding to the change of surface potential

$$C_S = -\frac{dQ_S}{d\Psi_S}. \tag{2.4}$$

The schematics of the equivalent circuit of this MOS system are given in Figure 2.8. The oxide capacitance (C_{OX}) is fixed, but the gate voltage changes the surface potential and this in turn changes the gate capacitance. If the surface charge density is known as a function of surface potential, $Q_S(\Psi s)$, the gate capacitance $C_G(V_G)$ can be readily determined as a function of the gate voltage and this can be compared with the experimental capacitance measurements. For the classical mechanics treatment of the MOS system assumed here, the surface charge density function, $Q_S(\Psi s)$, can be analytically obtained by solving the Poisson equation [62].

The calculated capacitance–voltage characteristics for a p-type MOS structure are given in Figure 2.9 [63]. As the gate voltage (V_G) takes large negative values in strong accumulation mode, the total capacitance approaches the capacitance of the oxide film since the contribution of C_S to the total capacitance is negligibly small. This is indeed the basis for the use of the measurement of the capacitance in accumulation to extract the oxide film thickness in an ideal situation. However, as one will see later, this straight forward extraction method becomes difficult due to the substantial contribution of C_S and its strong voltage dependence, even for

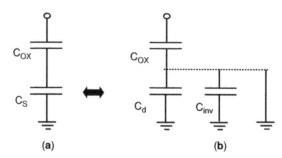

(a)   **(b)**

Figure 2.8. Equivalent circuits of an MOS capacitor with a metal gate electrode. (a) C_S represents the contributions from all capacitances of the Si surface. (b) C_S represents either a depletion-layer capacitance prior to the formation of an inversion-layer or an inversion-layer capacitance measured in low frequency or high frequency with externally supplied carriers. The solid line over the equivalent silicon capacitor C_S represents the ideal situation in which C_S is shunted in full inversion.

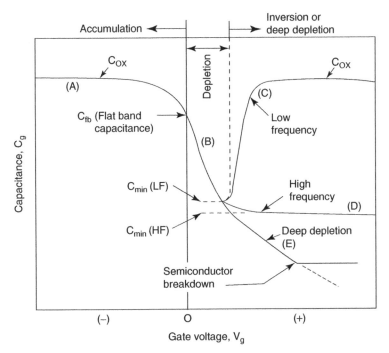

Figure 2.9. Calculated gate capacitance versus gate voltage for an MOS capacitor with a metal gate and p-type Si: (A) accumulation, (B) depletion, (C) inversion measured with low frequency. (D) Inversion-layer cannot form with high frequency. (E) Deep depletion. Reprinted with permission from Ref. 63, Copyright 1963, John Wiley & Sons, Inc.

thick oxides. When V_G increases towards positive values, the MOS structure enters into the depletion region, and the gate capacitance continuously decreases until the threshold voltage is reached. In the depletion region, the contribution of C_S can not be neglected because the depletion width increases with the increasing surface potential (Ψ_S).

The depletion capacitance decreases as its depletion region widens and leads to a decrease in the total gate capacitance. This is illustrated in Figure 2.8 (b) as C_S is replaced with the correspondent Si depletion capacitance (C_d). As the gate voltage increases beyond the threshold voltage, the formation of an inversion layer, (i.e., n-type channel) at the silicon surface becomes feasible. However, whether minority carriers (electrons) can populate the n-type channel depends on the availability of electrons as discussed above. If the electrons are mainly supplied by the thermal generation mechanism along the silicon surface, these minority carriers can not be generated fast enough to follow high frequency signals. In this case, the gate capacitance decreases to a minimum value as the silicon depletion layer approaches its maximum width, and the corresponding minimum surface capacitance $C_S = C_D(min)$, as shown in Figure 2.9. If the DC bias is changed too fast, the substrate enters into the so-called deep depletion condition in which the depletion width increases until it can provide enough minority carriers for the formation of an inversion layer (if this is at all possible). At this point, the gate capacitance saturates to a value lower than $C_D(min)$.

If the variation of AC signal is sufficiently slow (low frequency condition), the thermally generated electrons can populate and depopulate the inversion layer, following the AC voltage. Thus, the total MOS capacitance would start to increase as indicated in Figure 2.9. In this case, the main contribution to the C_S is the capacitance of the inverted silicon layer C_{inv}, as illustrated in Figure 2.8 (b). When the surface is fully inverted at high positive V_G, the gate capacitance eventually reaches the value of C_{OX} because the inversion layer shunts the depletion layer, as schematically shown by the dotted line in Figure 2.8 (b). Due to the low frequency used to ensure the supply of minority carriers by thermal generation processes, this capacitance measurement is often referred as the quasi-static capacitance measurement. As shown in Figure 2.6 (b), the use of n^+-type diffusions directly connected to the surface channel can also supply the minority carriers. The quasi-static CV characteristic can also be measured under high frequency conditions using this type of auxiliary electrodes or using MOSFET devices with source and drain grounded since, in this case, source and drain connect the channel to the external circuit and efficiently supply the required minority carriers.

2.2.1.3 *Polysilicon Depletion and Surface Quantization Effects.* In the present MOS technologies, highly doped polycrystalline silicon gates are used due to technological advantage such as self-alignment [64]. However, if the polysilicon doping is not heavy enough to ensure a degenerate electron gas, electrons or holes tend to be deleted in the polysilicon gate when the device is biased in inversion [65]. This results in an additional voltage drop at the interface between the polysilicon gate and the SiO_2 film. More importantly, an additional capacitance component associated to the depleted polysilicon gate comes into play in series with the oxide capacitance and the substrate capacitance, leading to a reduction of the total gate capacitance of the MOS structure. Figure 2.10 shows the equivalent circuit of a MOS structure with a polysilicon gate [66] while the energy-band diagram for an n^+-type polysilicon gate with a p-type substrate is given in Figure 2.11 (a). Thus, in the case of a polygate, the total gate capacitance is given by

$$\frac{1}{C_G} = \frac{1}{C_{OX}} + \frac{1}{C_S} + \frac{1}{C_{Poly}}, \qquad (2.5)$$

where

$$C_{Poly} = -\frac{dQ_{Poly}}{d\Psi_{Poly}} \qquad (2.6)$$

and Q_{Poly} is the charge density in the polysilicon depletion layer and Ψ_{Poly} is the surface potential at the poly-Si/SiO_2 interface. Equation 2.1 also needs a

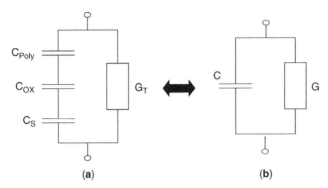

Figure 2.10. Equivalent circuits for an MOS device with n$^+$-type polysilicon gate. (a) The polysilicon capacitance, C_{Poly}, oxide capacitance, C_{OX}, and silicon capacitance, C_S, all in series with the tunnelling conduction, G_T, parallel and (b) the measured capacitance and conductance as equivalent circuit. Reprinted with permission from Ref. 66, Copyright 1995 Elsevier.

modification to include the voltage drop in the polysilicon layer and becomes

$$V_G = V_{FB} + V_{OX} + \Psi_S - \Psi_{Poly}. \tag{2.7}$$

As CMOS devices continue to shrink for high performance, the electric field at the Si/SiO$_2$ interface can reach very high values exceeding $5 \times 10^6\,\text{V/cm} = 5\,\text{MV/cm}$. Under these conditions, the classical theory is no longer adequate for the description of the physics of the MOS structures because quantum-mechanical effects

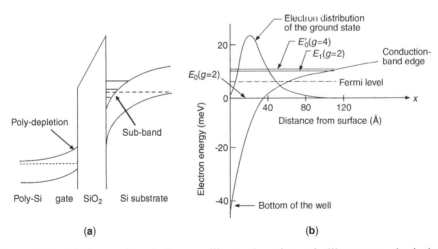

Figure 2.11. (a) Energy band diagram illustrating the polysilicon gate depletion effect and surface quantization effect at the Si surface when a positive V_G is applied to an NFET device. (b) Expanded view of energy sub bands in the potential well with electron distribution. Reprinted with permission from Ref. 67, Copyright 1967 the American Physical Society.

become pronounced. Consider the case of an NFET device in the inversion mode (shown in Fig. 2.11a); when high electric fields are present at the interface, a narrow potential well is formed as the bottom of the conduction band is bent downwards with a roughly triangular shape. As a consequence, the motion of electrons in this well is confined in the normal direction to the Si surface.

According to the QM theory, the electron energy associated to perpendicular transport can only take a series of discrete energy levels rather than the continuum of levels considered by the semi-classical theory [62, 66]. However, the electron's motion is not confined in the transverse direction in which their energy spectrum remains continuous. Therefore, the behavior of electrons should be regarded as a series of two-dimensional (2D) electron gases (quasi-2D sub bands) as depicted in Figure 2.11 (a). An example of the electron distribution is given in Figure 2.11 (b) for an enlarged view of the potential well under a high surface field. The electron distribution approaches zero at the interface with the assumption of an infinite oxide barrier for $x < 0$. The bent potential well is approximated by a triangular potential, $V(x) = qF_Sx$ for $x > 0$. In contrast with classical theory, the electron distribution reaches a maximum away from the interface (at about 20 Å in this particular case) and then gradually decays into the Si substrate. This shift from the interface causes another capacitance component to appear in series with the oxide capacitance and represents another fundamental limitation to the desired increase of the gate capacitance. Note that this effect will remain even if

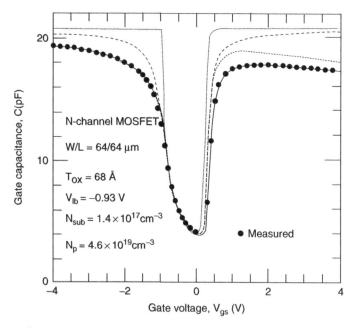

Figure 2.12. The measured gate capacitance in comparison with the results of classical models and the quantum mechanical model. See the text for the detailed discussion. Reprinted with permission from Ref. 68, © 1995 IEEE.

polydepletion effect can be eliminated by moving to metal-gate technologies in the near future.

A complete quantum-mechanical description of either the inversion or the accumulation layer at the Si interface involves a numerical self-consistent solution of the coupled Poisson's and Schrödinger's equations [67]. The quantum mechanical behavior of confined carriers, either electrons or holes, at the Si/SiO$_2$ interface is found to affect device operation because of an increase in threshold voltage and a decrease in the gate capacitance [62]. In the context of dielectric reliability evaluation, the QM effects can also affect the accurate determination of oxide voltage because of the electron motion quantization causing a larger voltage drop in the Si substrate, as we will discuss in Section 2.2.3. It is important to note that the surface quantization effect appears both when the MOS structure is biased in inversion and in accumulation regardless of the type of minority carriers, either electrons or holes.

Figure 2.12 displays the measured capacitance as a function of gate voltage for a MOS device structure with an n$^+$-type polysilicon gate and a p-type Si substrate with an oxide thickness of 6.8 nm [68]. The inversion capacitance was obtained by the quasi-static CV measurement. Several theoretical calculations of gate capacitance based on the different assumptions are included for comparison. The dotted line represents the calculated gate capacitance using a constant surface potential model of $\Psi_S = 2\Psi_B$, where Ψ_B is the so-called bulk potential, defined as separation of the Fermi potential with respect to the midgap in the substrate bulk [60]. The dashed line shows the calculated results using the solution corresponding

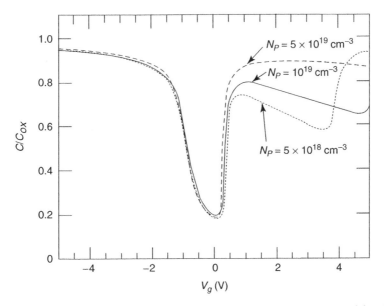

Figure 2.13. Low frequency CV curves of a p-type MOS capacitor with n$^+$poly-silicon gate doped with different doping concentrations. Reprinted with permission from Ref. 69, © 1994 IEEE.

to a classical treatment of the surface charge density as a function of surface potential. The dotted-dashed line represents the classical solution but includes the classic treatment of the polydepletion effect. In these theoretical calculations, the total gate capacitance approaches the value of oxide capacitance at high V_G in accumulation mode, but the difference is still appreciable even in strong accumulation. All these calculations significantly overestimate the gate capacitance in both accumulation and inversion region. Only when the surface quantization effect is properly considered along with the polydepletion effect, the theoretical results agree well with the experimental capacitance measurement in the entire voltage range (solid line in Fig. 2.12) [68].

When the active doping concentration is too low, the polysilicon interface can become inverted if the applied gate voltage is sufficiently high [65, 69, 70]. Figure 2.13 shows CV measurements with different doping concentrations [69]. In this particular case, for low doping concentrations, the capacitance at 2 V is nearly 30% lower than that of C_{OX}. It continues to decrease with increasing gate voltage but rises beyond 4 V for $N_{Poly} = 5 \times 10^{18} \, cm^{-3}$. This recovery of the capacitance is the first indication of inversion in polysilicon gate. The capacitance recovery is identified to be caused by the thermal generation [71]. As oxide thickness continues to shrink below $\sim 2 \, nm$, the accurate measurement of capacitance becomes impossible even at low voltages. This is because the direct tunneling in ultra-thin

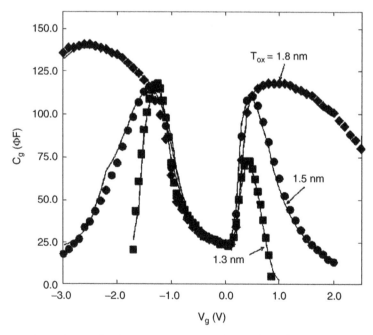

Figure 2.14. Experimental CV curves measured on ultra-thin oxides. The sharp decline of capacitances in accumulation and inversion regions is due to high direct tunneling currents. Reprinted with permission from Ref. 72, © 2000 IEEE.

oxide significantly contributes to the conductance of the dielectric system. Any series resistance associated with the channel and source/drain can further magnify this problem. Figure 2.14 displays the total gate capacitance measured for three different oxide thicknesses of 1.3 m, 1.5 m, and 1.8 nm [72]. It is seen that the capacitance at both inversion and accumulation attenuates rapidly with increasing gate voltages in either direction. The reduction in capacitances is far below the theoretical prediction including both surface quantization and polydepletion effects [72]. Sophisticated techniques have been developed to reconstruct the CV curves and the readers are referred to the literature for further reading [72].

2.2.2 Carrier Tunneling and Injection Mechanisms in MOS Structures

Dielectric media are insulating materials by definition. However, this property does not necessarily mean that it is impossible to have current flow through a dielectric film. The mobility of electrons in the oxide conduction band is of the same order of magnitude as that in semiconductors so that the current would be measurably large if carriers were efficiently injected to the oxide conduction band. In any case, the current is usually very low because the conduction band is difficult to populate with electrons. The carrier injection through an insulator layer such as SiO_2 is well known to play an important role in oxide degradation and fabrication processes and has been studied for many years. Either electrons or holes can be injected in the MOS structure by many different means. For simplicity, we focus on electron injection in the following discussion. In many cases, the methods for electron injection can be equally applied to the case of hole injection with proper consideration of energy band diagrams and the signs of applied voltages.

The most common electron injection mechanism is quantum mechanical tunneling through the classically forbidden energy barrier formed by the discontinuity between the silicon and the silicon dioxide conduction bands. The tunneling process is usually considered as an elastic process. Unlike carrier injection mechanisms such as hot carrier injection or avalanche injection (considered in Chapter 5), tunneling does not involve carrier injection over the energy barrier. In general, tunneling probability depends exponentially on both applied voltage and oxide thickness. For this reason, tunneling currents across oxide barrier were historically considered negligible and not considered a limiting factor in microelectronics technology when much thicker oxide films (>7 nm) were employed. However, as oxide thickness is reduced to 7.0 nm and down to ∼1.0 nm, quantum mechanical tunneling becomes appreciable and is responsible for many of the limitations for technology scaling such as power-consumption and reliability constraints. In the following we will discuss tunneling phenomena in the MOS system in detail.

2.2.2.1 Fowler–Nordheim Tunneling. The quantum mechanical tunneling mechanism was well understood in the early development of quantum physics at the beginning of the twentieth century. The tunneling mechanism was first considered by Fowler and Nordheim (FN) in 1928 to study the field emission of

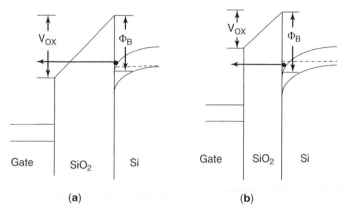

Figure 2.15. The energy band diagrams for electron tunneling process: (a) Fowler-Nordheim tunneling across the triangular barrier ($V_{OX} > \Phi_B/q$) and (b) direct tunneling through a trapezoidal barrier ($V_{OX} < \Phi_B/q$).

electrons through the triangular barrier that forms when a field is applied to a tungsten electrode in vacuum [73]. The tunneling mechanism was also found to play an important role in the conduction and breakdown properties of semiconductor junctions [74, 75]. Ever since silicon technologies were introduced in the 1950s, tunneling phenomena through metal–oxide–semiconductor (MOS) transistor structures has received much attention. The energy band diagram for FN tunneling in a MOS structure with a metal-gate electrode and a p-type substrate is given in Figure 2.15 (a). The simplest expression for the FN current is derived using the Wentzel–Krammers–Brillouin (WKB) approximation [73]:

$$J_{FN} = AF_{OX}^2 \exp\left(-\frac{B}{F_{OX}}\right), \qquad (2.8)$$

where F_{OX} is the electric field across the insulator layer, often called the oxide field. The prefactors, A, and the slope, C, are given by

$$A = \frac{q^3}{16\pi^2 \hbar m_{ox}^* \Phi_B}$$

$$= 1.54 \times 10^{-6} \frac{1}{\Phi_B} (\text{Amp/volt}^2) \qquad (2.9)$$

$$B = \frac{4}{3} \frac{(2m_{ox}^*)^{1/2}}{q\hbar} \Phi_B^{3/2}$$

$$= 6.83 \times 10^7 \left(\frac{m_{ox}^*}{m_0}\right)^{1/2} \Phi_B^{3/2} (\text{volt/cm}) \qquad (2.10)$$

where q is the electron charge, Φ_B is the barrier height, and $\hbar$ is Planck's constant, respectively. In turn, m_{ox}^* is the effective electron mass in the forbidden gap of the oxide, which is expressed in terms of electron mass as $m_{ox}^* = m_{ox}m_0$, so that m_{ox} is a dimensionless parameter. m_0 is the electron mass in free space. In the FN theory, the tunneling current density shows a strong exponential dependence on the oxide field. The two most important parameters are the barrier height (Φ_B) and the electron effective mass (m_{ox}). Note that the tunneling currents can be strongly affected by these two parameters because of the exponential dependence of FN currents on their values. It is often convenient to express Equation 2.8 in the following alternative form:

$$Ln\left(\frac{J_{FN}}{F_{OX}^2}\right) = -\frac{B}{F_{OX}} + Ln(A) \qquad (2.11)$$

because plotting $Ln(J_{FN}/F_{OX}^2)$ versus $1/F_{OX}$ yields a straight line if the current is due to FN tunneling. This graphical representation is the so-called Fowler–Nordheim plot. B is also called the FN-plot slope and it has the same units as the oxide field.

In 1969, Lenzlinger and Snow first reported that the measured gate current densities follow the FN tunneling theory as illustrated in Figure 2.16 for three different gate electrodes (polycrystalline Si, aluminum, and magnesium for relative comparison) [76] although the absolute magnitude of currents was later considered to be too low due to measurement artifacts [77]. This figure reveals that plotting the experimental tunneling currents in the FN plot yields straight lines as predicted by the FN tunneling theory. The change of gate electrode from polycrystalline Si to aluminum, and to magnesium leads to an increase in the FN tunneling currents measured at the same field. This is attributed to a decrease

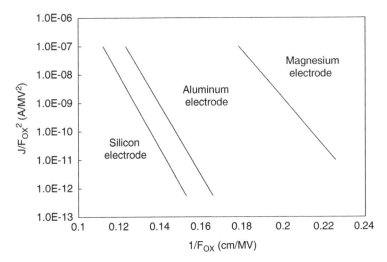

Figure 2.16. Fowler-Nordheim plot of tunneling current versus voltage for thermal oxides. Reprinted with permission from Ref. 76, Copyright 1969, American Institute of Physics.

in the barrier height for aluminum and magnesium electrodes in comparison with polysilicon gate as shown in Figure 2.16.

The FN tunneling currents are now commonly observed in the MOS structures of modern microelectronics technologies. Instead of using the WKB approximation, one might refer to work done by Gundlach in 1966 that obtained an exact solution for the electron tunneling current by calculating the transmission probability though a triangular barrier [78]. His work predicted that electron tunneling current would exhibit an oscillatory behavior due the quantum-mechanical interference effects related to electron-wave reflections at the conduction-band discontinuity of the Si/SiO_2 anode interface. This oscillation behavior was experimentally confirmed by Maserjian [79]. In this case, the expression for electron tunneling current is modified as

$$J = B_0 J_{FN} \tag{2.12}$$

where J_{FN} is given by Equation 2.8 and B_0 is the oscillatory component of tunneling currents in Gundlach's analytic solution [78].

The logarithm of the normalized experimental current density, i.e., $Ln(J/J_{FN})$, plotted as a function of the oxide voltage across a 4.8 nm oxide [80] reveals this remarkable oscillatory behavior as shown in Figure 2.17. In principle, the quantum interference should occur for all oxide thicknesses in the FN regime. However, for thick oxides above 6.0 nm, the oscillation amplitude damps due to electron scattering with phonons in the oxide conduction band so that the oscillatory behavior cannot be observed in conventional experiments. The quantum oscillation can be readily observed for oxide thickness thinner than 6.0 nm because electron transport becomes ballistic due to the reduced scattering. The quantum interference effect has also found many practical applications for the characterization of oxide films such as the extraction of the film thickness and the study of the interface roughness and transition layer as well as index of refraction [81–83] as will be further discussed in Section 2.2.4. In addition, the existence of quantum oscillation provides a clear proof of ballistic electron transport in MOS structures [84].

2.2.2.2 Direct Tunneling. If the SiO_2 layer is very thin, electrons can directly tunnel from the cathode to the anode through the forbidden gap of a SiO_2 film. This phenomenon is usually referred to as direct tunneling (DT). It occurs when the oxide voltage is less than the barrier height ($V_{OX} < \Phi_B/q$), as shown in Figure 2.15 (b), i.e., when the shape of the tunneling barrier is trapezoidal. Direct tunneling current is known to strongly depend on oxide thickness as shown in Figure 2.18. For thicker oxides, direct tunneling is very much reduced so that it cannot easily be observed using conventional experimental setups. However, in recent years, the advance of microelectronics technology has pushed oxide thickness down to the 1.0 nm range so that direct tunneling phenomenon becomes a routine measurement in laboratories.

Using WKB approximation, Simmons and Stratton published the earlier studies of direct tunneling phenomena through thin dielectric films [85, 86].

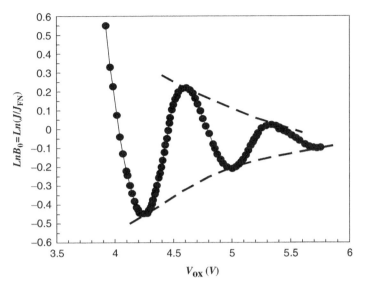

Figure 2.17. Quantum interference oscillations (LnB_0) of the tunneling currents as a function of oxide voltage, V_{OX} for $T_{OX} = 4.83$ nm. Reprinted with permission from Ref. 80, Copyright 1975, American Institute of Physics.

In these studies, a physical picture of independent particles and elastic electron tunneling with one-band parabolic dispersion were assumed. However, at that time, the experimental data was scarce and limited to one thickness, so that the work was mainly theoretical. Moreover, those studies focused on the metal–oxide–metal system rather than on the MOS structures of current interest. In recent years, similar studies based on the same assumption were carried out to study the direct tunneling currents measured on SiO_2 films fabricated with the state-of-the-art microelectronics technology [66, 67, 87–89]. Under the WKB approximation, direct tunneling current can be expressed as [66, 67, 79].

$$J_{DT} \propto \frac{1}{T_{OX}^2} \exp(-2\kappa T_{OX}) \qquad (2.13)$$

where T_{OX} is the oxide thickness and κ the magnitude of the electron wave vector in the forbidden band-gap of SiO_2 that can be expressed as

$$\kappa = \left(\frac{2m_{ox}^* q[\Phi_B - (V_{OX}/2)]}{\hbar^2} \right). \qquad (2.14)$$

Despite the simplicity of the WKB approximation, this equation shows that direct tunneling currents exponentially depend on oxide voltage and thickness. Thickness dependence of direct tunneling currents is given in Figure 2.18 at a fixed gate voltage of ± 1.5 V for NFET and PFET inversion, respectively. This figure shows

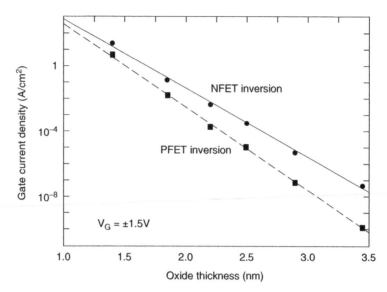

Figure 2.18. Thickness dependence of direct tunneling current densities at $V_G = \pm 1.5V$ at 30°C for n^+-type polysilicon/NFET and p^+-type poly-silicon/PNFET under inversion.

that, in the case of NFET inversion mode, the direct tunneling current exponentially increases at a rate of ~ 4.5 decade for every nanometer of oxide thickness reduction (~ 4.5 dec/nm). Although the DT current starts with a much reduced value for thick-oxide PFETs stressed in inversion, it rises rapidly with a steeper slope and becomes comparable to the NFET current for thinner oxides. This result is also consistent with the prediction of Equation 2.13 if we take into account that the barrier height is different for both types of devices. In the case of PFETs biased in inversion, the gate current is controlled by the direct tunneling of valence-band holes with a barrier height $\Phi_B \sim 4.9$ eV while for the NFET inversion mode, the gate current is dominated by direct tunneling of conduction-band electrons with $\Phi_B \sim 3.0$ eV.

Up to this point, we have only dealt with the treatment of tunneling based on the Wentzel–Kramers–Brillouin (WKB) approximation. However, more advanced treatments are also relevant from the technology point of view. These treatments require the numerical solution of the coupled Schrödinger and Poisson equations to properly treat the surface quantization effects [90–93], as already discussed in Section 2.2.1. Quantum mechanical treatment was first applied to calculate the tunneling current from an accumulation layer in the FN regime, showing excellent agreement with experiments [90, 91]. Later, QM simulation of tunneling currents was applied to the direct tunneling regime [92, 93]. Figure 2.19 compares the experimentally measured DT currents with the QM calculation, producing excellent agreement [93]. At low voltages, the curvature of DT

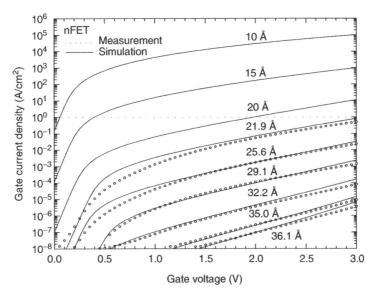

Figure 2.19. Measured IV characteristics under n⁺poly/NFET inversion with various values of oxide thickness. The lines represent the simulated IV curves of solution to the Schrödinger and Poisson equations. Reprinted with permission from Ref. 93, © 1997 IEEE.

tunneling currents can be reasonably captured by QM solution while the WKB approximation fails to do so.

While the self-consistent QM simulation of the tunneling current is a more rigorous approach, it has the drawback of being is computationally expensive. Thus, many researchers have also attempted to perform the tunneling current calculation using the analytical approaches while retaining the accuracy achieved by QM simulation. To resolve the discrepancy between the WKB model and experiments at low voltages, Register et al. developed a scheme to correct the conventional WKB approach and obtain an analytical solution for the direct tunneling current in good agreement with the experimental IV measurements [95]. A semiclassical compact model has been proposed including a long list of parameters for the simulation of conduction-band electron tunneling, valence band electron tunneling, and valence band hole tunneling in the DT regime [96]. The modified WKB approximation was extended to thinner oxides down to ~1.5 nm [97] and dual-layer MOS structure such as those based on high-K dielectrics [97, 98].

Until now, we have separately discussed tunneling in the FN regime (i.e., through a triangular barrier) and the DT regime (i.e., through a trapezoidal barrier). However, it is instructive to examine the tunneling phenomena in a more global scale and examine the experimental data for both regimes in a single plot as that of Figure 2.20, which depicts the IV characteristics for oxides of different

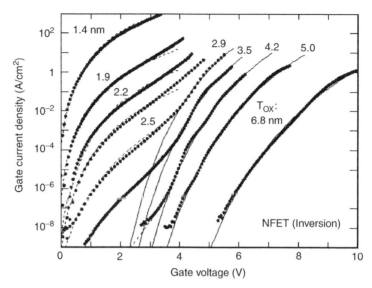

Figure 2.20. Measured tunneling currents for a wide range of oxide thickness from 1.4 nm to 6.8 nm at 30°C in n⁺-type polysilicon/NFET inversion. The solid and dashed lines represent the calculation using the FN expression (Eq. 2.8) and using the semi-classical approach [96] for DT currents, respectively.

thickness. Above $V_G \sim 4\,V$ ($V_{OX} \sim 3\,V$), FN tunneling dominates with a much steeper slope for the current–voltage dependence. Below $V_G \sim 4\,V$ ($V_{OX} \sim 3\,V$), direct tunneling dominates with a weaker voltage dependence but a much faster increase with decreasing oxide thickness. These results suggest that in the DT regime the oxide thickness plays a relatively more important role, while the barrier height along with the tunneling distance, which indirectly depends on oxide voltage and thickness, dominates the current in the FN regime.

While the strong thickness dependence of the direct tunneling current was already predicted by Simmons [85] and Stratton [86] more than 40 years ago, its impact on technology scaling was only appreciated in recent years. For CMOS technology applications, various scaling limits have been projected in numerous manners over the last decade. It has been revealed that the collapse of electronic structure due to the penetration of electron waves from the opposite electrodes would eventually cause SiO_2 to lose its insulating property as oxide thickness is reduced to about 0.7 nm, with a practical limit of about 1.2 nm considering interface roughness [99]. In recent years, the power consumption for CMOS technology applications has imposed a serious limitation caused by the drastic increase of the direct tunneling current. An initial evaluation study indicated that the thickness limit due to tunneling current is about 1.5 nm [100]. However, oxide scaling has further proceeded down to about 1 nm for high performance applications.

2.2.2.3 Temperature Dependence and Tunneling Parameters. It is well
known that tunneling itself is a temperature-independent process; however,
electrons incident on the barrier do not necessarily reach it with the same energy.
In principle, the FN expression given in Equation 2.8 strictly corresponds to the
zero temperature limit; for other cases, thus, it represents a low-temperature
approximation. Until now, only limited studies on the temperature dependence of
FN tunneling currents have been reported [76, 101–105]. Figure 2.21 (a) shows the
FN plots for 7.8 nm oxides at various temperatures [104]. The linearity of the FN
plots is evident for all the temperatures under investigation. A certain impact of
temperature on the FN current is evident, especially for temperatures above
250°C. The experimental data shows a gradual increase at low temperatures but
much stronger dependence at the highest temperatures. At lower fields, the
temperature dependence is much stronger than at high fields. As shown in Figure
2.21 (a), the calculations considering a Fermi–Dirac distribution with a constant
barrier height show good agreement with the tunneling data only at low
temperatures but deviate significantly at higher temperatures. A satisfactory
agreement can be achieved as depicted in Figure 2.21 (b) when a linear
temperature-dependence of the barrier height is assumed. This conclusion appears

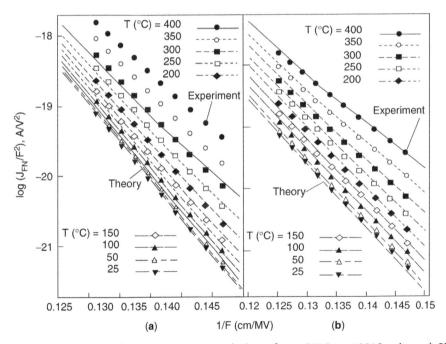

Figure 2.21. FN plots for temperature variations from 25°C to 400°C using n+-Si
substrate in the accumulation mode with $T_{OX} = 7.8$ nm using a temperature
independent barrier height (a) and an assumed linear temperature-dependent
barrier height (b). Reprinted with permission from Ref. 104. Copyright 1995,
American Institute of Physics.

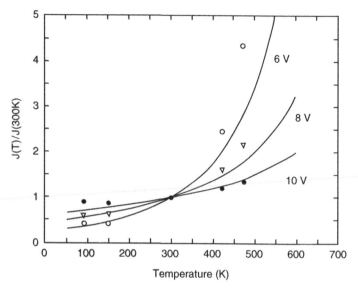

Figure 2.22. Temperature dependence of FN tunnelling currents at fixed voltages. Symbols for experimental data and the lines from QM simulation. Reprinted with permission from Ref. 107, © 1991 IEEE.

to be inconsistent with the results of internal photoemission measurements from which it was concluded that the barrier height is relatively insensitive to temperature variation [101].

In the above studies, quantum effects such as the formation of sub-bands at the SiO_2/Si interface were neglected in the analysis of the temperature dependence of the FN current. A comparison between the experimentally measured current and the results of the self-consistent solution of the Schrödinger and the Poisson equations is given in Figure 2.22 [103]. This type of QM calculation is capable of capturing the main features in the experimental temperature dependence. In this QM picture, the most important contribution to the temperature dependence comes from the temperature-dependent occupation of excited energy sub-bands above the ground-state since the contribution of the lowest sub-band is mainly temperature-independent. At low temperatures, the effects of high energy sub-bands above the Fermi level are negligible but substantially increase with increasing temperature due to the exponential form of the Fermi–Dirac distribution. This explains the small temperature dependence at low temperatures and the stronger dependence at high temperatures. The stronger temperature dependence of tunneling current at low fields can also be explained by the fact that separation between sub-bands is smaller at low fields so that the occupation of high energy sub-bands is larger. In other words, at low fields the transmission probability is small; the current is mainly affected by the occupancy probability. This explains why at high fields the transmission probability is larger and the occupancy has much less effect on the current [103]. In summary, it appears that the QM simulation gives a natural explanation for the main experimental trends of the temperature dependence of the FN current without arbitrarily changing the injection barrier height.

A detailed understanding of the tunneling mechanisms requires a firm knowledge of the involved parameters. The key parameters used to accurately characterize or calculate the FN or DT currents are the barrier height and the effective electron mass. Due to the exponential dependence of tunneling currents on these parameters, their values are crucial for accurate calculations. Unfortunately, the accurate determination of these two parameters is a complicated topic. Due to the limited space, we only provide a qualitative discussion here. In general, there are two approaches to determine these two parameters: the electrical extraction method and the direct physical method. If the electric field can be determined independently as discussed in Section 2.2.3, the prefactor and the slope of FN plot ascribed in Equation 2.11 can be used to determine the barrier height and effective mass.

The physical techniques such as internal photoemission technique and X-ray photoelectron spectroscopy (XPS) provide a direct extraction of the tunneling barrier height without using any theoretical model. However, it is questionable to presume that the physically measured tunneling barrier height will be the same as that obtained from the electrical extraction method. Moreover, the energy sub-band formation due to surface quantization and the image force effect can play a

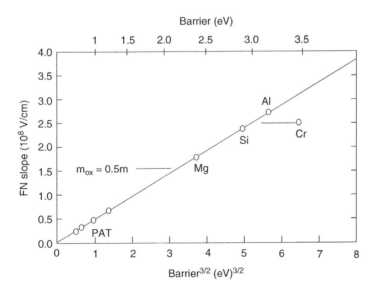

Figure 2.23. Experimental FN slopes as a function of the barrier height expressed in $(\Phi_B)^{3/2}$. The solid line is from the FN theory using Equation 2.3 with $m_{ox}^* = 0.5m_0$. PAT refers to photo assisted tunneling from Al with the barrier, $\Phi_0 = \Phi_{Al} - \hbar\omega$ where $\hbar\omega$ is the photo energy. Reprinted with permission from Ref. 106, Copyright 1982, American Institute of Physics.

significant role in determining the effective barrier height for carrier tunneling injection. The discussion of these topics is beyond the scope of this book, and the readers are referred to the specialized literature for further reading [106–109]. Despite many difficulties in determining effective electron mass and barrier height, when the experimentally determined FN slopes are plotted against the barrier heights measured for various gate materials using the data reported (by different groups as seen in Fig. 2.23), a single value of $m_{ox}^* = 0.5m_0$ provides a surprisingly good description for all the extracted barrier heights [106]. Based on these results, most of the researchers have systematically assumed this value for the tunneling effective electron mass.

2.2.2.4 Thermionic (or Schottky) Emission. Besides the QM tunneling mechanism that describes the transmission of electrons through the insulator barrier (passing through a classically forbidden region), electrons can also be injected over the barrier as illustrated in Figure 2.24 (a). This process is often referred as thermionic emission since it strongly depends on temperature. The thermionic (or Schottky) emission current [110] can be calculated as

$$J = \frac{qm_I^* k_B^2 T^2}{2\pi^2 \hbar^3} \exp\left(\frac{-\Phi_B + (qF_I/4\pi\varepsilon_0\varepsilon_I)^{1/2}}{k_B T}\right), \tag{2.15}$$

where F_I is the electric field across the insulator, q the electronic charge, Φ_B the barrier height in eV as shown in Figure 2.24 (a), ε_0 the permittivity of free space, ε_I the dielectric constant of the insulator film, m_I^* the effective electron mass and $\hbar$ the Planck's constant, k_B the Boltzman constant, and T the temperature in Kelvin (K). Notice that the square-root field-dependence of thermionic currents arises from the consideration of the image force effect, which is extensively discussed in literature

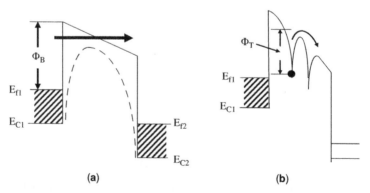

(a)   **(b)**

Figure 2.24. Schematic diagrams for (a) thermionic (Schottky) emission and (b) Frenkel–Poole emission. The dashed line in (a) represents the barrier lowering due to the image-force. The barrier height (Φ_T) in (b) stands for the depth of a trap potential well.

[108, 109, 111]. In addition, a linear plot of $LN(J/T2)$ versus $1/T$ can provide a means to verifying whether this mechanism is consistent with the experimental data. The data published in the 1960s for the insulating layers of SiO_2 deposited under vacuum conditions can be well described by this thermionic emission process [112, 113]. The derived barrier height and dielectric constant were found to be 0.39 eV and 3.2 eV, respectively. While the dielectric constant seems to be plausible in comparison with the value of 3.9 eV widely used today, the extracted barrier height of 0.39 eV is much smaller than the value of 2.9–3.15 eV reported nowadays for the state-of-the-art thermal SiO_2 or SiO_2-based dielectrics with different levels of nitridation. For this reason, these results obtained in the 1960s suggest those oxides show much worse quality. This carrier injection process is not considered to play a significant role in modern microelectronics technologies that use SiO_2 films for silicon surface passivation in normal stress and operation conditions.

2.2.2.5 Frenkel–Poole Injection. If many kinds of traps are present in an insulator, the electron injection mechanism can be different from the tunneling processes described above. In this case, when electrons are injected into such an insulator they can easily get trapped. However the current conduction can still take place when an electron can pass from one trapping site to another as schematically described in Figure 2.24(b). This is often referred to as the Frenkel–Poole mechanism [114, 115]. The thermionic emission process discussed above is a cathode-limited process. However, the Frenkel–Poole injection is a bulk-limited mechanism since the motion of electrons through bulk trapping sites controls the conduction process. This bulk-limited current is given by

$$J = J_0 F_I \exp\left(\frac{-\Phi_T + (qF_I/\pi\varepsilon_0\varepsilon_I)^{1/2}}{k_B T}\right) \qquad (2.16)$$

The proportionality constant, J_0, is a function of density of the trapping sites. Notice that the exponential functions of Equations 2.15 and 2.16 are identical. However, the physical origin is slightly different. In the Frenkel–Poole (FP) emission, the potential barrier is modulated by the trapped charges in the sites while the image force effect is considered for the thermionic (or Schottky) emission. Furthermore, the barrier height (Φ_T) in Equation 2.16 refers to the depth of a trapping site from the bottom of the insulator conduction band, as shown in Figure 2.24 (b). This equation has been commonly used to describe the experimental current–voltage characteristic for the conduction process through silicon nitride films fabricated by chemical vapor deposition (CVD) and also in other dielectric films. An example of this conduction mechanism was reported for an Au-Si_3N_4-Mo structure with a film thickness of 290 nm [115].

The temperature dependence of the FP current arises from the fact that this process involves a field-enhanced thermal excitation of trapped electrons into the conduction band. Based on the experimental measurements of field- and temperature-dependence of the FP current densities, the barrier height and dynamic dielectric constant are derived to be (1.3 ± 0.2) eV for Φ_T and 5.5 ± 1

for ε_I, respectively [115]. The value for the dielectric constant (ε_I) is close to the dynamic dielectric constant since the measured dielectric constant ranges from 4 expected for the visible range to 7 for the static dielectric constant of Si_3N_4. This was claimed to be a crucial verification of this conduction mechanism for CVD Si_3N_4 films [115]. On the other hand, for silicon nitride fabricated using jet vapor deposition (JVD) process, the level of as-fabricated defects or traps of these Si_3N_4 films is much reduced; hence, the electron conduction is dominated by FN tunneling rather than by the FP mechanism [116].

2.2.2.6 Substrate Hot Electron Injection. The conduction currents based on tunneling or other injection mechanisms described above are exclusively controlled by the applied field across the dielectric layer at a fixed temperature. For the investigation of the dielectric breakdown mechanism, it is sometimes desirable to separate the effects of applied field and current density. Special test structures like the one schematically represented in Figure 2.25 (a) are used to show this. These structures consist of an NFET device with a separate N^+P injector. Source and drain are connected together to the ground. The gate voltage is varied to change the oxide field while keeping the transistor in inversion. The electrons are supplied into the p-type substrate from the forward-biased injector. Gaining sufficient energy from the electric field in the depletion layer of the reverse-biased substrate, the electrons are injected over the barrier into the conduction-band of the dielectric film as described in the energy band diagram Figure 2.25 (b). This injection mechanism is called substrate hot electron (SHE) injection. The injector voltage does not modify the oxide field but changes the number of injected carriers and consequently the gate current. Furthermore, the substrate terminal can be biased to change the energy distribution of the injected electrons, as shown schematically in Figure 2.25 (b).

Figure 2.26 displays the gate currents as a function of the gate voltage measured using these special structures with a substrate bias of $-8.25\,\mathrm{V}$ and several injector biases [117, 118]. At low gate voltages, the measured gate current densities due to SHE injection are several orders of magnitude larger than those obtained by the conventional tunneling mechanisms at these voltages. Only at very high gate voltages, the gate currents measured under SHE and tunneling bias conditions become comparable due to the dominant role of tunneling effect. Note the gate currents using SHE injection method show almost no dependence on gate voltage and are mainly modulated by the injector voltage. The SHE technique is also useful to force a given current density (by changing the injector bias) for different oxide thicknesses while keeping the same oxide field at the same time.

The injection mechanisms discussed above are often referred to as uniform injection mechanisms since the carriers are uniformly injected through the insulator barrier across the entire area. Carrier injection can be also carried out nonuniformly in a MOS structure (as discussed in Chapter 5). For example, channel hot electron injection is an example of nonuniform carrier injection. In this case, electrons are mainly injected into the gate oxide in the overlap region of the channel and drain regions using a high drain bias to accelerate the carriers.

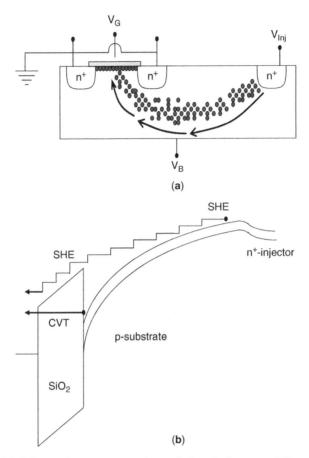

Figure 2.25. (a) Schematic representation of the devices used for substrate hot electron (SHE) injection experiments. Drain and source are short-circuited to ground. The gate voltage determines the oxide field provided that the inversion channel is formed. The substrate bias (V_B) changes the distributions of the energy of the carriers while the injector voltage (V_{Inj}) determines the current through the oxide. (b) Schematic spatial band diagram showing the effects of substrate polarization on the energy of injected carriers. CVT represents the constant voltage tunneling current component and SHE, the substrate hot electron current component.

In addition, other injection mechanisms such as avalanche injection, thermal- or optical-excited substrate hot carrier injections have been reported for the characterization of gate oxides. The readers can find an excellent review of these topics given by Ning [119]. In summary of this subsection, we have discussed the most important and relevant mechanisms of carrier injection through an insulator layer used in microelectronics. The understanding of carrier injection processes is

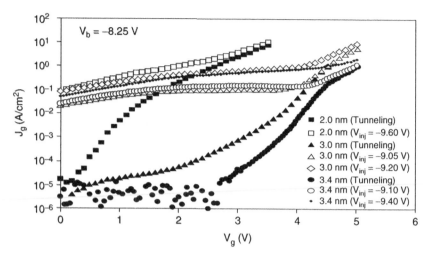

Figure 2.26. Gate current density versus gate voltages with a substrate bias of $-8.25\,V$ and several injector voltages for 2.0, 3.0, and 3.4 nm oxides. Reprinted with permission from Ref. 117, © 2000 IEEE.

very important in reliability evaluation, especially in subsequent sections when we deal with the degradation and breakdown phenomena under accelerated stress conditions.

2.2.3 Oxide Voltage (Field) and Electron Energy at Anode

2.2.3.1 Determination of Oxide Voltage (Field). The proper determination of oxide field or voltage is important, for instance, when the time-to-breakdown data is compared with a theoretical model. When dealing with extremely thick oxides with a metal gate, the gate voltage has been simply assumed to be the voltage across the oxide layer, i.e., $F_{OX} \cong V_G/T_{OX}$ or $F_{OX} \cong (V_G - V_{FB})/T_{OX}$. However, even in these extreme cases, this assumption is only approximately correct, particularly when the voltage is low. In an MOS structure with a poly-Si gate and a thin oxide, the situation can be substantially more complicated as already seen in Figure 2.12 for 6.8 nm oxides. Substantial errors can occur if surface quantization effect and poly-depletion effects are not properly accounted for.

The determination of oxide voltage can be accomplished either experimentally or theoretically. In the experimental approach, only electrical methods are available which are based on the determination of the CV and IV characteristics. One of the electric methods consists in using the FN tunneling current to derive the oxide field with assumed values of Φ_B and m_{ox}. Historically, the FN plots such as those in Figure 2.16 which are traditionally used to derive the values of Φ_B and m_{ox}, can also be used to determine the oxide field. For example, in a MOS system with a metal gate and a very thick oxide, $\geq 10\,nm$, the measured tunneling currents can be well described by the simple FN expression (Eq. 2.8). Although the precise values of the barrier height and the electron effective mass still remain somewhat unresolved, a reasonable knowledge for the values of these two parameters has been established after many years of research as discussed above. Therefore, with the assumed values for Φ_B and m_{ox}, the FN expression can be directly used to extract the oxide field as a function of V_G from the measured tunneling currents. However, for thinner oxides, especially between 3.0 nm and 6.0 nm, the direct application of Equation 2.8 is not possible because the gate current is modulated by the QM oscillation effects. Recently, an alternative method has been developed to determine the oxide field by finding the bias voltages corresponding to the nodes where the oscillation diminishes, that is, the bias points where $B_0 = 1$ in Equation 2.12, or equivalently $LnB_0 = 0$ [81]. In this procedure, the measured gate current is first used to calculate the field values by inverting Equation 2.8. Because the gate currents deviate from the FN current described by Equation 2.8, these calculated field values (E_{cal}) do not necessarily correspond to the true oxide fields except for those at which $B_0 = 1$. The fields at the nodes are true oxide fields and can be determined by numerically differentiating the curve of E_{cal} versus V_G curve. Thus, the oxide field as a function of gate voltage can be obtained [81, 120].

Another method is to use the CV characteristics as originally proposed by Berglund [121]. This is generally a useful method, provided there is no capacitance attenuation effect due to the tunneling current increase (Fig. 2.14). Thus, this method is limited to oxide thickness above ~ 2 nm. The analysis given by Berglund was extended by Depas et al. to include the contribution of polysilicon capacitance, C_{Poly} [66]. By combining Equations 2.4 to 2.7 and applying the requirement of charge neutrality $dQ_{Poly} = -dQ_S$, we obtain the following equation for oxide field:

$$F_{OX} = \frac{1}{\varepsilon_0 \varepsilon_{OX} A_{OX}} \int_{V_{FB}}^{V_G} C(V_G)dV, \qquad (2.17)$$

where the oxide voltage can be determined from $V_{OX} = F_{OX} \times T_{OX}$ if T_{OX} is known. Also note that this method assumes a prior knowledge of the gate oxide area.

It is worthwhile to point out that the derivation of Equation 2.17 is only based on the analysis of the equivalent circuit. Therefore, the results are general and can be applied to all the cases even if we deal with MOS systems using ultra-thin oxides in which QM effects are pronounced. In practice, Equation 2.17 can directly be used to obtain oxide field by integrating the measured capacitance as a function of gate voltage. A prior knowledge of the flatband voltage and oxide area is required. The flatband voltage can be determined independently [61, 122] or obtained from a QM simulator. In order to obtain the oxide voltage, it is necessary to know the oxide thickness accurately. We will defer the discussion of oxide thickness measurement to Section 2.2.4. Figure 2.27 compares the V_{OX} versus V_G relation determined by the IV and CV methods [120]. The theoretical calculation

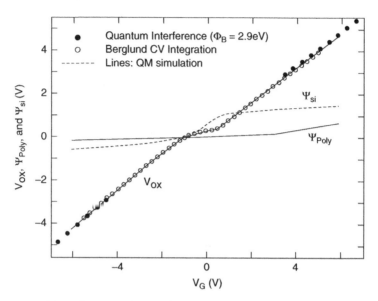

Figure 2.27. Oxide voltage, voltage drop (Ψ_{Poly}) in n$^+$-type polysilicon gate and surface potential (Ψ_{Si}) in Si surface as a function of gate voltage in comparison with quantum interference technique and the calculation of quantum simulation. The substrate is p$^+$-type and the oxide thickness is 3.7 nm. Reprinted with permission from Ref. 120, © 1997 IEEE.

using the QM simulator is also included for comparison. The excellent agreement among these three methods demonstrates the success of the model that includes the QM effects for characterizing the MOS system. Moreover, the results in Figure 2.27 suggest that a simple linear relation is sufficient to describe the V_{OX}–V_G (or F_{OX}–V_G) relation.

2.2.3.2 Electron Energy at the Anode Interface. To gain deep understanding of the physics of electron transport, oxide degradation, and other applications, it is most useful to have an accurate knowledge of the value of the electron energy at the anode interface. Figure 2.28 (a) and (b) display the energy band diagrams of electron tunneling in FN and DT regimes. As shown in Figure 2.28 (a) for the FN regime, when an electron travels through the conduction band of SiO$_2$, it gains kinetic energy from the electric field and, on the other hand, it loses part of this energy due to the inelastic interaction with phonons. Since stochastic processes are involved in the electron transport, not all the electrons reach the anode interface with the same energy. The determination of the shape of the energy distribution requires the solution of the Boltzman equation using the Monte Carlo simulation, as we will briefly discuss in Section 3.3. However, for the purpose of general understanding, a phenomenological approach was developed as an adequate approximation [123]. Within this approach, the average energy of

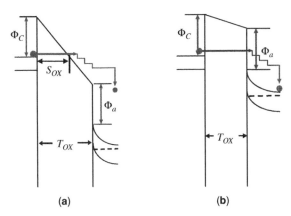

(a)   (b)

Figure 2.28. Band diagrams for calculation of electron energy at the anode (a) FN tunneling regime and (b) direct tunnelling regime. Reprinted with permission from Ref. 123, Copyright 1985, American Institute of Physics.

electrons in the SiO_2 conduction band is the solution of the following differential equation:

$$\frac{dE}{dx} = qF_{OX} - \frac{E}{\lambda}, \qquad (2.18)$$

where E is the electron energy, F_{OX} is the oxide field, λ is the mean free path of electron energy loss, and q is the electron charge. The term on the left hand side of Equation 2.18 represents the net energy gained by electrons per unit distance. The first term of the right-hand side is the kinetic energy gained by electrons from the oxide field while the second term represents the energy loss due to the scattering. Assuming that the mean free path is independent of energy as a first-order approximation, the solution of Equation 2.18 can be obtained by integration:

$$E(x) = qF_{OX}\lambda\left[1 - \exp\left(-\frac{(x - S_{OX})}{\lambda}\right)\right], \qquad (2.19)$$

S_{OX} being the position at which the electron enters the oxide conduction band ($x = 0$ and $x = T_{OX}$ are the positions of the cathode and anode interfaces of the oxide, respectively). The total energy of the electron entering into the Si-anode is

$$E_{GAIN} = q\Phi_a + qF_{OX}\lambda\left[1 - \exp\left(-\frac{1}{\lambda}\left\{T_{OX} - \frac{\Phi_c}{F_{OX}}\right\}\right)\right], \qquad (2.20)$$

where Φ_c and Φ_a are the electron barrier heights at the cathode and anode, respectively. If both cathode and anode materials are made of n-type silicon, a single value of Φ_B can be used to simplify the notation, i.e., $\Phi_B \equiv \Phi_a \equiv \Phi_c$. The second term of the RHS is the net kinetic energy gained by the electron from the oxide field as given by Equation 2.19 for $E(x)$ at $x = T_{OX}$. The first term ($q\Phi_a$)

represents the potential energy gained by the electron when it crosses the SiO_2-Si anode interface. Therefore, E_{GAIN} represents the total energy of the electron with respect to the bottom of the conduction band at the silicon anode. For direct tunneling electrons, the situation is much simpler as depicted in Figure 2.28 (b), so that the total energy of the electron is just given

$$E_{GAIN} = qV_{OX} \quad \text{for} \quad V_{OX} < \Phi_c. \tag{2.21}$$

Figure 2.29 depicts the calculated average electron energy using Equations 2.20 and 2.21 as a function of oxide voltage for several values of oxide thickness. The reported values of λ range from 1.3 to 4.0 nm [80, 123]. The bold solid line represents the kinetic energy of the electron gained by the field without suffering any scattering event, i.e., the so-called ballistic limit. As seen in Figure 2.29, when the oxide field (or voltage) increases, the electron energy initially increases very fast and then gradually saturates due to the effects of scattering, which is particularly significant at high fields (voltages). For very thick oxides ($\gg \lambda$), the dampening effect on average electron energy is much more pronounced due to the long distance that the electrons travel in the oxide conduction band and to the reduction of the oxide field. On the other hand, for thinner oxides, the energy loss due to scattering is less effective so that the kinetic energy continues to increase. This is particularly true for ultra-thin oxides with thickness approaching the mean free path ($T_{OX} \to \lambda$), the average electron energy converges toward the limit of ballistic tunneling without any scattering.

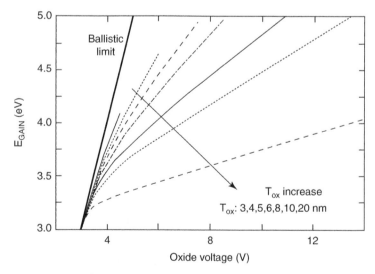

Figure 2.29. Average electron energy versus oxide voltage for several oxide thicknesses using Equation 20. A value of 1.5 nm was used for the mean free path (λ) as discussed in the text.

2.2.4   Determination of Oxide Thickness

Oxide thickness is one of the most important parameters in the characterization of a MOSFET device. We will dedicate this subsection to providing an overview of various techniques used for determining the oxide thickness. The techniques can be classified into three categories: physical, electrical, and theoretical methods. It is important to recognize that these techniques are based on the interpretation of the physical principles involved in each method. For example, the electrical techniques involve the physical models which are assumed for the description of the MOS system. There are different levels of complexity in those models which allow different levels of accuracy in the description. The limitations of these physical models are certainly inherited by the thickness extraction procedures. The accuracy of the measurement strongly depends on the validity of the model assumptions as a description of the realistic measurement environment. Furthermore, because of different physical principles, thickness values extracted using various different techniques should not be regarded as the same physical identities, even if an excellent agreement is achieved. In this regard, for example, the thickness extracted from a CV measurement might be slightly different from the value obtained from an IV measurement because the impact of atomic-scale variations in thickness (due to roughness of interfaces) on the average capacitance and on the average current might also be different.

2.2.4.1   Electrical Extraction Based on Tunneling Current Measurement. Electrical methods for oxide thickness extraction are based on either current or capacitance measurements. For thickness extraction using a current measurement, the use of simple FN tunneling theory (Eq. 2.28) with inclusion of temperature effect has been proposed [124, 125]. This method employs only a pair of current–voltage measurements and an iterative solution of the FN equation with the assumed barrier height and effective electron mass. Moreover, the applied voltage is also assumed to be equal to oxide voltage. Nevertheless, a good correlation between this IV method and the CV extraction was obtained for oxide thickness around ~ 22 nm [125]. Unfortunately, this simple and fast method is limited to very thick oxides using metal gates without gate-depletion effects. Because of the exponential dependence of the FN current on T_{OX}, any localized oxide thinning can significantly contribute to the measured gate current and prevent the use of this method for small area devices. On the other hand, as already discussed in Section 2.2.2, for thin oxides between 3.0 nm and 6.0 nm the oxide field can be extracted from the QM oscillations of tunneling current using the technique proposed by Zafar et al. [81]. Thus, after identifying the maxima and minima of the current oscillation pattern (LnB$_0$) like the one shown in Figure 2.17, the oxide thickness can be readily determined using the following relation:

$$T_{OX} = K_n \left[\frac{\hbar^2}{2qm_{ob}F_{OX,n}} \right]^{1/3} + \frac{\Phi_B}{qF_{OX,n}}, \tag{2.22}$$

where $F_{OX,n}$ is the oxide field corresponding to the nth maximum (minimum) and K_n is the nth zero of the Airy function (derivative of the Airy function) [81]. m_{ob} is the average effective mass of the electron in conduction band of SiO_2. Note this is different from m_{ox}^*, which is the effective mass in the forbidden gap of SiO_2 as given in Equation 2.10.

2.2.4.2 Electrical Extraction Method Based on Capacitance Measurement.

As already stated above, the oxide thickness can also be extracted from capacitance measurements. Strictly speaking, this is a measure of an equivalent thickness of a MOS capacitor, rather than the actual physical thickness of the SiO_2 film, because a different stoichiometry certainly exists at the Si/SiO_2 interface which has some impact on the local material permitivity. Let us consider the simple case of a MOS structure with a metal gate and a semiconductor substrate as shown in Figure 2.6. The total gate capacitance is given in Equation 2.3 as the sum of the reciprocal of the gate insulator capacitance and silicon depletion capacitance. Theoretically, the gate capacitance only equals to the oxide capacitance in strong accumulation or inversion regions as seen in Figure 2.9. In reality, the gate capacitance does not saturate in strong accumulation mode due to the significant contribution of the silicon capacitance (C_S) as shown, for example, in Figure 2.12. Therefore, many classical techniques based on CV measurements have attempted to obtain the silicon capacitance by using physical models for the Si surface charge density, $Q_S(\psi_S)$, so that silicon capacitance can be expressed as a form of the measurable quantity such as the derivative of the gate capacitance dC_G/dV_G. Many classical techniques are principally based on this approach but vary in the details depending on the physical assumptions of each technique [126–129]. The first CV extraction method was proposed by Maserjian et al. [126] for the case of degenerate Si in the strong accumulation regime. An approximation for including quantum effects is used to derive the following analytic equation:

$$\frac{1}{C_G} = \frac{1}{C_{OX}} + \beta \left(\frac{dC_G^{-2}}{dV_G}\right)^{1/4}, \tag{2.23}$$

where β is a constant that depends on the area and the effective electron mass in the accumulation layer. A plot of $1/C_G$ versus dC_G^{-2}/dV_G would give a straight line in which the intercept is $1/C_{OX}$. It can be seen that the Si capacitance (C_S) in Equation 2.23 is expressed in terms of the derivative of the gate capacitance $C_G(V_G)$. However, this equation is only valid in the strong accumulation limit. Subsequently, several researchers have developed similar procedures based on the fitting to the measured CV curves or on some extrapolations of the experimental results [130–133]. In some cases, it is found that a combination of the Marserjian's approach and the new procedure [131] can produce the best agreement with the thickness extracted from ellipsometric techniques if the functional form assumed for the charge statistics can be tuned [132]. A similar approach also using some graphical extrapolation procedure was developed in the same timeframe by McNutt and Sah [127]. Several extensions of this method were also reported some time later [134, 135]. Another relevant method consists in exploring the CV properties near the flatband condition [128, 129]. In Lehovec's graphical approach [128], an initial estimate of oxide capacitance is required whereas the technique proposed by Ricco et al. involves a second derivative of CV measurements [129]. In practice, these extraction algorithms require particular care since these methods are very sensitive to the derivative (and second derivative) of the experimental CV characteristics and, hence, they are prone to errors. In principle, these CV methods do not treat the quantum effects adequately, thus presenting a fundamental limitation for the accurate characterization of dielectric film thickness as we will discuss later in the comparison of the results obtained with these different techniques.

2.2.4.3 *Electrical Extraction Method Based on Quantum Mechanical Simulation.*

The thickness extraction methods based on capacitance measurements largely rely on the validity of the physical models assumed to describe the carrier properties at the Si/SiO_2 interface. When the classical theory is applicable, an analytical solution can usually be obtained and this facilitates the comparison with the experimental measurements either for capacitances or tunneling currents. However, as the aggressive scaling of microelectronics has pushed oxide thickness down to about 1.0 nm; the effects associated with poly-depletion and surface quantization can not be ignored in the evaluation of MOS parameters. Therefore, the extraction of oxide thickness would inevitably require a good understanding of quantum effects in the MOS system. Quantum mechanical treatment of the MOS system is well established and the readers are referred to the vast literature for further reading. Here, we can only summarize the key results in terms of oxide-thickness extraction.

In the QM simulation, a guessed value for oxide thickness is a first input to the QM simulator, then the coupled Schrödinger and Poisson equations are solved numerically to obtain the self-consistent solution. The calculated capacitance results are compared to the experimental measured CV curves. Upon the comparison, an updated value of oxide thickness is generated as the new input to the QM simulator. This iteration process continues until a satisfactory match is achieved between the simulation and the experimental CV measurements. Fortunately, it is found that the gate capacitance in strong accumulation is rather insensitive to the doping concentrations in either polysilicon or substrate both for n^+-type poly/NFET or p^+-type poly/FET devices [93, 94]. This feature renders the self-consistent QM simulator a well defined tool to accurately extract the oxide thickness as illustrated in Figure 2.30. In this figure, the physical thickness in the y ordinate is obtained from the calculated QM calculation after its satisfactory convergence with the gate capacitance data while the equivalent thickness in the x ordinate is obtained by the gate capacitance measurement at $V_G = -2.5$ V for NFET devices. It can be seen that the direct gate capacitance yields an equivalent thickness of about 0.6 nm thicker than the physical thickness (QM CV-extraction). Two different values for polysilicon doping concentration are included and no significant differences. Two sets of curves corresponding to the classical theory using Maxwell–Boltzman and Fermi–Dirac statistics are included for comparison

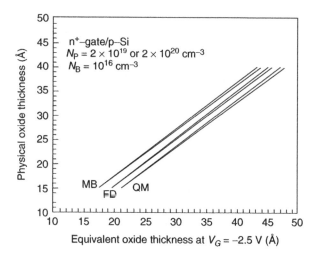

Figure 2.30. The extracted physical oxide thickness using the QM simulation versus the thickness derived at $V_G = -2.5$ V using different theories: classic theory with Maxwell-Boltzman (MB); classic theory with Fermi–Dirac statistics (FD); and quantum theory (QM) [94]. Reprinted with courtesy of International Business Machines Corporation, copyright 1999 © International Business Machines Corporation.

in Fig. 2.30. It is evident that the classical theory overestimates oxide thickness by 2–4 Å in comparison with QM simulation. Similar to the classical CV methods for thickness extraction, any uncertainties in dielectric constants can give rise to an error in the thickness determined from the use of the QM CV-extraction method.

2.2.4.4 *Physical Extraction Method Based on Ellipsometry.* The most common optical technique for oxide thickness extraction is based on ellipsometry measurements of film thickness and refractive index. In general, this is a technique that measures the changes in the polarization state of a monochromatic light beam reflected from the surface of a dielectric film as described in detail in [136, 137]. This is a fast, contactless, and nondestructive technique which is widely used for in-line monitoring of film thickness in the microelectronics industry. Based on the model of a single homogeneous dielectric layer on a substrate with known refractive index, the thickness and refractive index of the dielectric layer can be calculated from measured optical parameters.

Unfortunately, the model of a single homogeneous layer is too idealized because the stoichiometry at the Si/SiO_2 interface is different from that of the bulk material. In addition, the interface roughness is well known to modulate the interface properties of bulk materials. For film thickness below 10 nm, refractive index values are found to be higher for thinner films and approach the values of the bulk for SiO_2 layers prepared both by thermal and CVD process [138], suggesting that the interfacial region is optically different from the bulk region. Moreover, the random measurement errors in addition to truncation or round-off errors can propagate and result in significant errors in extracted SiO_2 refractive

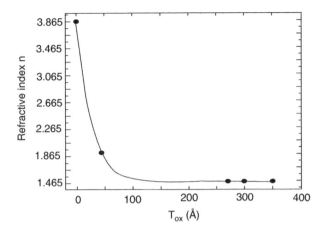

Figure 2.31. The refractory index versus oxide thickness. The line represents a fit using the exponential dependence. The value of 1.864 for 4.4 nm oxides is extracted from ellipsometry using the knowledge of oxide thickness based on quantum interference technique. Reprinted with permission from Ref. 139, Copyright 1996, The Electrochemical Society.

index [139]. This makes the determination of film thickness nearly impossible for film thickness less than 10 nm. For this reason, many researchers used an a priori assumed value for the bulk refractive index (1.462 ± 0.005 for 6328 Å) to extract the film thickness for oxides thinner than 10 nm. On the other hand, Herbert et al. have used the oxide thickness determined from the quantum oscillation method as an *a priori* assumption to extract the refractive indexes from the ellipsometry measurements. Using this method, they found an average value of 1.894 ± 0.110 for an oxide of 4.4 nm [139]. Figure 2.31 displays the refractive index as a function of oxide thickness [139]. The best fit gives an exponential dependence of refractive index on oxide thickness. In modern microelectronics technologies, the silicon dioxide layer is heavily nitrided to reduce the gate tunneling current and the boron penetration for ultra-thin films approaching 1.0 nm. It is well known that the refractive index increases with the nitrogen concentration. These complexities add to the already discussed uncertainties for the ellipsometric-based extraction of ultra-thin oxide thickness. Thus, the accuracy of this technique is seriously compromised in the particular case of ultra-thin oxides.

2.2.4.5 Physical Extraction Method Based on Transmission Electron Microscopy.

Another useful method for film thickness measurement is based on transmission electron microscopy (TEM), sometimes also called high-resolution TEM (HRTEM). This technique is not only used to measure the thickness but also to investigate properties such as interface roughness and other structural and morphological features of the dielectric or semiconductor film. Typical examples of cross-sectional high-resolution TEM images are shown in Figure 2.32 for 2.4 nm oxide films in MOS structures, respectively. The atomic lattice spacing of

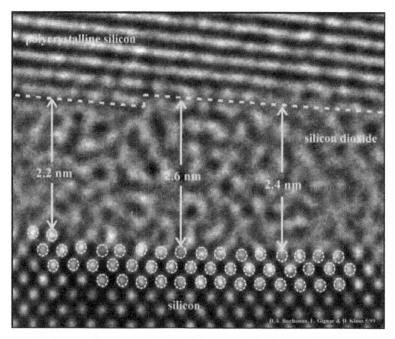

Figure 2.32. The lattice images of transmission electron microscopy for a MOS capacitor with polysilicon gate and silicon substrate with a SiO_2 film of 2.4 nm. Notice the atomic step at polysilicon/SiO_2 interface as discussed in the text [140]. Reprinted courtesy of International Business Machines Corporation, copyright 1999 © International Business Machines Corporation.

the Si substrate is clearly visible and can be used for measurement calibration, which is the advantage of the TEM technique. For the 2.4 nm film of Figure 2.32, both Si/SiO_2 and poly-Si/SiO_2 interfaces can be identified. A thickness variation of 0.2–0.3 nm at either interface can be related to the atomic silicon steps which have a similar value ($a_0 = 3.138$ Å). The amorphous nature of SiO_2 films is revealed in both films. The HRTEM technique is very attractive as a direct measurement of film thickness. In contrast to all other techniques, it does not require an *a priori* knowledge of physical parameters of the film such as the refractive index in the case of ellipsometry, the dielectric constant in capacitance-based extraction methods, or the mean free path (or electron attenuation length) in the XPS technique. On the other hand, however, as it can be seen in the case of the 1 nm oxide layer shown in Figure 2.2, the interfaces are sometimes rather blurry, making it difficult to obtain a precise definition of the film thickness. In addition, the projection of the lattice planes for the whole sample thickness and small tilt angles or stresses can also contribute to errors in the TEM measurement. Moreover, this method is time consuming and the measurement is made in a region of atomic-scale dimension which is much smaller than the size of the devices of interest. Thus, the obtained result might not be a good measurement of the average film thickness since there are atomic-scale thickness variations along the device dimensions.

2.2.4.6 Physical Extraction Method Based on X-ray Photoemission Spectroscopy.

X-ray photoemission spectroscopy (XPS) has been demonstrated to provide another physical method for oxide thickness measurement [141–145]. This technique is commonly used to examine the local chemical environment and the structure of the Si/SiO_2 interface, giving much information such as the size and nonstoichiometry of the interface transition layer or the valence-band alignment at the interface. In principle, XPS delivers a high resolution measurement of oxide thickness with fairly good accuracy. In XPS analysis, the photoelectron spectra provide much information about the binding energy of the core electronic levels, Si (2p), for Si atoms in the dielectric film. The film thickness measurement is based on X-ray photoelectron line intensity. The intensity ratio of the Si(2p) peaks of the Si substrate to that of those in the SiO_2 film is given by

$$R = R_0 \frac{\exp(-T_{OX}/(\lambda_0 \mathrm{Sin}\,\theta))}{1 - \exp(-T_{OX}/(\lambda_0 \mathrm{Sin}\,\theta))}, \tag{2.24}$$

where λ_0 is the effective attenuation length of photoelectrons (also called the mean escape depth) which is defined as the distance for which $1/e$ of the photoelectrons are not inelastically scattered; θ is the photoelectron take-off angle; R_0 is the ratio of Si(2p) intensity for an infinitely thick dielectric with respect to that of an oxide-free Si substrate (hydrogen passivated). The XPS technique requires instrumental alignment and hence involves two separate measurements for the determination of R_0. An accurate knowledge of the electron attenuation length (λ_0) is also required. Monte Carlo simulation has shown that this parameter is equivalent to the inelastic scattering mean-free path [146]. In practice, this parameter is determined from a set of samples with known thickness, i.e. measured by means of other techniques such as HRTEM, for example. Therefore, this technique is not readily available in microelectronics manufacturing environment.

2.2.4.7 Comparison of Thickness Measurement Techniques.

We will now compare the oxide thickness extracted with the various techniques presented in the previous sections. Table 2.2 summarizes the results obtained with different techniques for relatively thick oxides [130]. For the ellipsometry measurement, a fixed value of 1.465 was assumed for the refractive index. For the 6.5 nm and 10 nm oxides, all the thickness measurements show reasonable agreement within the estimated errors. In this study, the authors found that thicker films of 13 nm and 20 nm exhibited considerable roughness due to the oxide fabrication processes such as substrate material growth (Czochralski or floating zone), prediffusion cleaning, and oxidizing conditions, as indicated from their TEM images [130]. In the case of 13 nm and 20 nm oxides, the result of CV is not in agreement with that of ellipsometry due to the simple assumption of a single homogeneous layer that neglects the roughness effect. For the 20 nm oxides, the value (16.6 nm) obtained

TABLE 2.2. A Comparison of Different Extraction Methods for Thick Oxide with Thickness in Angstroms

CV	FN	Ellipsometry	TEM	Roughness
64.5 ± 0.15	69 ± 4	66.8 ± 1.5	67.6 ± 3	≤ 3
100 ± 1.5	99 ± 5	100.5 ± 1.7	97 ± 3	≤ 3
127 ± 1.8	121 ± 6	129 ± 1.8	108^*	10
196 ± 1.8	166 ± 10	201 ± 2	160^*	40

Note: The data with an asterisk are the mean values between the maximal and minimal extrema.
Source: Reprinted with permission from Ref. 130, Copyright 1992, Elsevier.

by the FN method is considerably lower than those obtained from CV measurements and ellipsometry. This is most likely caused by the extremely large roughness of 4.0 nm (a 20% of the overall thickness), since FN current is more sensitive to the existence of thin spots in the samples because the tunneling current exponentially decreases with T_{OX} if voltage is kept constant. A good agreement between the CV and FN methods was also reported by other authors for oxides with thickness in the 20–25 nm range, but no information about interface roughness was provided [124, 125, 130].

Table 2.3 compares the results obtained from different CV methods for oxide thicknesses ranging from 2.7 to 5.8 nm [147], with HRTEM results included as a reference. Among all the methods, the thickness values obtained from the QM–CV method yield the best agreement with the TEM measurements. The comparison with the CV methods of Ricco et al. and McNutt et al. yields unsatisfactory results. In this regard, it was suggested that the quantum effects can also play a significant role near the flatband conditions, an issue that was previously considered insignificant [148]. The CV method by McNutt et al. also neglects the quantum effects in the Si substrate; it was derived for metal gate MOS structures in accumulation mode. Thus, its application to a MOS structure with polysilicon gate further magnifies the thickness difference, resulting in a much smaller thickness in comparison with the TEM measurements. The thickness values obtained by using the Maserjian's method are also shown to be lower than those of the TEM and QM–CV methods.

Figure 2.33 compares the extracted thicknesses for thin oxides with T_{OX} ranging from 1.7 to 5.5 nm using five different methods: HRTEM, ellipsometry, QM interference technique, Maserjian's classical CV method, and QM–CV

TABLE 2.3. A Comparison of Different Extraction Methods for Thin Oxide with Thickness in Nanometers

Iterative QM	Riccò et al.	Maserjian et al.	McNutt/Sah	TEM
2.75	3.43	2.45	~ 1.55	2.8 ± 0.2
3.42	4.06	2.68	~ 2.4	3.5 ± 0.2
5.82	6.32	5.07	~ 4.6	5.9 ± 0.2

Source: Reprinted with permission from Ref. 147, © 2001 IEEE.

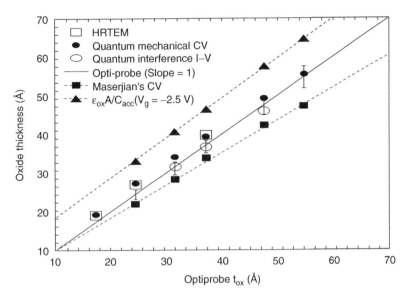

Figure 2.33. Comparison of extracted oxide thickness using several techniques as discussed in the text. Reprinted with permission from Ref. 149, © 2000 IEEE.

extraction method [149]. The thickness derived using the gate capacitance at $V_G = -2.5$ V is also included for comparison, revealing that the error involved in this rough estimation can be as large as 10 Å. In general, except the Maserjian's method, the other thickness extraction techniques yield results which are in reasonable agreement among each other, consistent with the previous findings of other groups [93, 120]. The thickness extracted from the QM–CV method based on the algorithm developed by Hauser et al. [150] is in good agreement with those obtained from HRTEM within an experimental error around 1 Å.

The quantum interference method and the ellipsometry method show nearly the same results. Similar to the results in Table 2.2, the oxide thickness values extracted from Maserjian's method are consistently lower than those of all other methods. As it has been remarked before, the formulation of Equation 2.23 approximately includes the surface quantization effect in the Si substrate but with a metal gate. Thus, for a MOS system with a polysilicon gate, a direct application of the Maserjian's method usually results in a thinner extracted oxide thickness. The values extracted by the ellipsometric method using a fixed refraction index (1.459) are also slightly lower than those of other methods [149].

The authors attribute these differences partly to the nonstoichiometry of the transition layer and partly to the interface roughness since the uncertainties in dielectric constant can affect any thickness extractions using the capacitance. On the other hand, other researchers have found that the thickness values used from the TEM and XPS techniques are smaller than those of ellipsometry using the bulk SiO_2 refraction index. It was shown that a higher refraction index, as shown in Figure 2.31, yields smaller oxide thickness values [146] in better agreement with the other physical

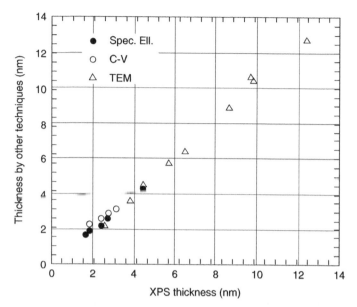

Figure 2.34. Comparison of thickness measurements using XPS technique, TEM imaging method, and ellipsometry technique. CV extraction method is also included for comparison. Reprinted with permission from Ref. 145, Copyright 1997, American Institute of Physics.

methods. Rather than single wavelength ellipsometry, the spectroscopic ellipsometry can allow determination of the refraction index as well as the film thickness.

A comparison of the oxide thickness extracted from three different physical techniques (spectroscopic ellipsometry, TEM, and XPS) is provided in Figure 2.34 [145]. The thickness values derived from the capacitance measurement are also included [145]. These values are slightly larger than those obtained by physical measurement techniques. This is attributed to the fact that quantum effects were not included in this particular CV extraction [145]. The three physical techniques yield reasonable agreement over a wide range of oxide thickness from ~ 12 to 1.8 nm. Through these comparative studies, it becomes clear that QM effects are a reality in the MOS systems with oxide thickness of few nanometers we have to deal with in the characterization of modern microelectronic devices. Only when the QM effects are properly considered, the electrically extracted thicknesses yield satisfactory agreement with those obtained by the physical methods.

2.3 MEASUREMENT OF DIELECTRIC RELIABILITY

2.3.1 Measurement Methods

In Section 2.1.2 it was already pointed out that measuring reliability of a system or part of a system (e.g., the dielectric, which is tested by stressing a product relevant test structure) involves stressing at accelerated conditions in order to cause failure within a practical time. Assessing reliability almost always involves making the parts fail and thus applying an irreversible test, which can not be repeated on the same device. Therefore a statistically sufficient number of fresh samples need to be available for reliability testing (Section 2.1.4). The stressing of such a sample size can be done by stressing one part after another or by connecting all the parts in parallel and by then stressing them simultaneously. For fast tests with very short durations a fully automatic wafer prober is best suited; for longer durations a semiautomatic wafer prober works well, but the alignment of every wafer requires more manpower. Stresses with very long durations are most efficiently done by dicing the wafers and packaging the test structures. The packaged structures are put on stress boards and subjected to the same stress simultaneously.

The necessary accelerators for dielectrics are voltage or current and temperature. The major acceleration is driven by voltage or current, while temperature is less effective. Nevertheless, increasing the temperature for the stress above the specified operation condition of the final product helps to reduce the stress time further. Increasing the temperature too much may change the activation energy and thus may require a more sophisticated method to account for this, as will be described in Section 3.4. A useful stress temperature is the burn-in temperature for the product, because this usually assures that no new degradation mechanisms are activated. Many IC products are burned in at temperatures in the range between 130–150°C, while the specified maximum operation temperature is commonly in the range from 70–125°C. The reliability stress therefore requires a hot chuck that keeps the wafer at the desired temperature during the entire stress. Such a hot chuck is not always available; hence, there are a lot of publications using room temperature for dielectric reliability stressing, compensating for the lacking temperature acceleration by applying a higher voltage or current. The general dielectric stress setup is shown schematically in Figure 2.35 regardless of temperature.

The simplest setup for stressing a MOS capacitor requires a voltage or current source, which depends on the intended stress method. For monitoring the feedback of the system over time a current or voltage meter are necessary, respectively. In most cases the source and the meter are in the same box called source–meter–unit (SMU). SMUs can be operated manually or be controlled by a

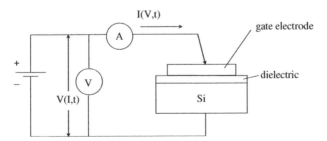

Figure 2.35. Schematic of general MOS capacitor dielectric stress setup. A voltage or current source are needed and, correspondingly, a current or voltage meter to monitor the stress over time. Often this is achieved by involving a source-meter-unit (SMU).

personal computer (PC), which is most useful for the automation of tests for dealing with large sample sizes. In general it is a good practice to perform a leakage test, e.g., at operation condition, before the stress to verify integrity of the sample and also after test to verify breakdown.

In the following sections, the different methods that are applicable for dielectric stressing are described. The method of choice depends on various considerations, like need and availability of acceleration model, consistency of stress with operation conditions or purpose of data acquisition, e.g., qualitative sanity check, reliability modeling, or physical understanding of dielectric breakdown mechanism. It is implied that the stresses described below are computer controlled, i.e., software is required to perform the stress. Possible features of the software are mentioned. One feature is the storage of the measured data for subsequent evaluation. This requires additional analysis and breakdown detection software. The advantage is that the analysis can be very thorough and if necessary can be repeated considering new understanding. This is especially useful for the long term stresses (described below) during process development and qualification. For more mature, well known processes the breakdown detection can be more or even completely automated.

Although with most methods, charge-to-breakdown (q_{BD} or Q_{BD}) can be determined (it should be kept in mind) that especially for thinner dielectrics, q_{BD} depends on stress conditions and the method used to acquire it, i.e., different q_{BD} values are found for the same oxide when applying different methods as well as different stress conditions. On the other hand, it has been shown that for oxides in the thickness range 10–100 nm at consistent conditions the resulting q_{BD} distributions from different stress methods can be the same [151].

Breakdown detection is described with the methods, but discussed in detail in Section 2.3.2. Although the methods are straightforward and rather simple there are a lot of potential influences on the results. For very thick oxides self-healing may cause the first breakdown to be missed if slow equipment is used with highly accelerated stress conditions because the capacitor recovers very fast [152]. Another observation is multiple electrical breakdowns before catastrophic dielectric breakdown [153]. Most of these could have potentially triggered catastrophic dielectric breakdown if sufficient energy would have been supplied. Therefore the impedance of the source affects the breakdown time [153]. Similarly, a low current compliance in a constant voltage stress can affect the soft breakdown of ultra-thin oxides and its post breakdown resistance [154]. Also, inserting an inductance in the measurement circuit increases the probability of soft breakdown with increasing inductance due to its impact on the dynamics during breakdown [155]. More details regarding soft breakdown will be discussed in Section 2.4.4 with respect to test structure and area, in Section 2.3.4 specifically focusing on breakdown detection. In Section 3.2 and 3.5, the physics of breakdown and the post-breakdown behavior are treated, respectively. Reviews of measurement techniques have been published ([151]) or in conjunction with the 1/E acceleration model [156], and with many technical details, references and a historical development [157].

2.3.1.1 Constant Voltage Stress.

There are various reasons for this stress method. In most integrated circuits the dielectrics are operated at a constant voltage, or to be more exact, switched between two fixed voltages, with one of them being 0 V. In addition, various voltage acceleration models are available for lifetime extrapolation. For ultra-thin oxides it has been shown [158, 159] that the constant voltage stress is the relevant stress.

Prior to starting the constant voltage stress, the integrity of each sample is checked by applying a small voltage, e.g., the operation voltage of the circuit in which the oxide is to be used. This simulates the functional test of the circuit and helps to screen defective samples. If the sample passes, the measured current should be in the noise for thick oxides or on the known IV curve for ultra-thin oxides (see Section 2.2.1). To get an even better idea of the oxide state, instead of only one voltage point, several points at different voltages could be taken and stored. This allows one to reconstruct part of the IV curve of every sample after stress. The benefit of this is to be able to determine the onset of Fowler–Nordheim tunneling for thick oxides or to more accurately distinguish between no breakdown and soft breakdown for ultra-thin oxides. An even more sophisticated pretest involves a voltage ramp, followed by a voltage step-stress (such as described in the subsequent section) [160]. Such a pretest is useful if the time resolution needs to be good, which is important for monitoring early, extrinsic fails. However, care must be taken not to cause significant predamage, which would affect the breakdown time of the constant voltage stress, i.e., the maximum voltage during the extended integrity test should be smaller than the voltage applied in the main stress, or the duration of the stress voltage application during the integrity test should be much shorter than the minimum expected breakdown time (as a rule of thumb, more than a factor of 100 difference). The opposite, an improvement of the time-to-breakdown in the constant voltage stress, has been observed in some oxides [161]. In this case a voltage ramp up to a certain electric field was applied to the oxide before the stress. Oxide stressed without the preceding ramp failed earlier than the oxide subjected to the voltage ramp [161, 162], which was explained by initial positive charge trapping. The integrity test can be also critical when dealing with extrinsic distributions and their voltage acceleration. The slope can be altered depending on the screening condition [163]. For all these reasons the integrity check needs to be tailored in a way that it does not affect the stress results.

While stress and characterization of capacitors is simple and straightforward, more options exist for structures like single MOSFETs or MOS transistor arrays. In this case, the bulk or well terminal and the source and drain terminals could simply be connected together resulting in a capacitor structure again. Naturally, they provide several other options for stress and characterization. For a limited electrical failure analysis the location of the breakdown may be determined if the current from gate to bulk or well, source and drain are measured separately. This requires additional current meters or a switching matrix to determine the major leakage paths. Details of locating the breakdown are given in [164, 165].

After the integrity test is passed successfully, the constant voltage stress can start by turning on the pre-selected stress voltage. This voltage is based on experience or estimated from an IV curve. It needs to be high enough to end up with a practical stress time and low enough to be able to measure the time-to-breakdown. For wafer level stress, the time window starts depending on the equipment at some milliseconds or seconds and ends at several thousand seconds, while package level stresses may run several months. From the moment the stress voltage is turned on, the time and current are measured as fast as possible. The measurement frequency determines the accuracy of the measurement especially for short stress durations. The accuracy improves as the sample survives ten or hundred times the time between two subsequent measurements. The time and current values of the readout are stored, and then the next data point is taken and stored and so on. This may lead to large data files. If there is an issue with storage capacity the number of stored data points can be reduced by lowering the measurement frequency, e.g., keeping the measurement interval at 1% of the time that the sample already survived. Another method is to keep the high measurement frequency but store data only if the current changed more than a predefined percentage compared to the last stored current value. Of course the change should be smaller than the change that is defined to mark breakdown (described below), otherwise little or no information about the current transient is stored. In order to record unstable currents but still keep the amount of stored data at a minimum, the percentage necessary for a current change to be stored could be increased after a certain number of data points are stored, e.g., the percentage could be 5% to begin with, after 20 stored data points it could be increased to 20%, after another 30 data points are stored the percentage is increased to 50% and so on. This method works especially well for thicker oxides with a hard breakdown marked by an abrupt current increase over several decades, reaching the milliampere range or compliance at once. The stored current time values of each sample are utilized to detect the breakdown after stress. However, the stress needs to be terminated at some point. For this purpose a time limit can be set, which prevents a stress of a sample (e.g., no contact) to run forever. The time is set based on experience with that particular dielectric and stress condition, including some generous margin that considers possible variations of the breakdown time. For early fails it is useful to stop the stress if a certain condition is met, e.g., the compliance or a certain current level above the preceding stress current, or if the current increased between two subsequent measurements more than a certain amount, e.g., 20 times. Further methods of breakdown detection will be discussed in Section 2.3.2.

A particular problem is the first readout because for a sample failing between the process of switching the voltage on and getting the first current measurement, no accurate failure time is known. Thus it is useful to mark the fails at the first readout by a small fixed time value (e.g., the best achievable time resolution of the equipment) to indicate failure at first readout instead of the recorded time of the first readout. Such a fail is a reliability fail since it occurred after turning on the high stress voltage, given the sample was still functioning at the integrity test. Using a marker-time instead of the real, equipment-response-time-dependent (and therefore random) time allows one to consider these fails in the statistical evaluation while excluding them in the graphical representation, i.e., those fails are censored and contribute only to the failure percentage determined for the first fail with a valid failure time.

If the equipment is fast enough, the test structure has negligible series resistance and the degradation mechanism does not change, then the constant voltage stress can then be performed at a very high acceleration with breakdown times in the one second range. This allows using a constant voltage stress for quantitative inline monitoring with standard equipment [166] or special equipment additions [167].

For data evaluation after stress, it is useful to plot all the data of one wafer in a current versus time plot (Fig. 2.36). This helps to distinguish at first glance typical behavior from non typical behavior. Also a possible thickness difference across the wafer may be detected by the variation of the current at a fixed time. In Figure 2.36 the electron trapping typical for thicker oxides (> 5.5 nm) can be seen. For thinner oxides the current does not change significantly over the stress time. After some time, the current increases drastically and all samples reach the compliance, which in the case of this figure was set to 0.1 mA. For such curves breakdown time determination is easy and a possible current limit is indicated by the dashed line. The so determined times-to-breakdown (t_{BD}) are plotted in a probability net which is explained in Section 2.3.3. Another characteristic parameter that can be extracted from the I–t curves is the charge-to-breakdown (q_{BD}), which represents the charge that was injected in the oxide during stress.

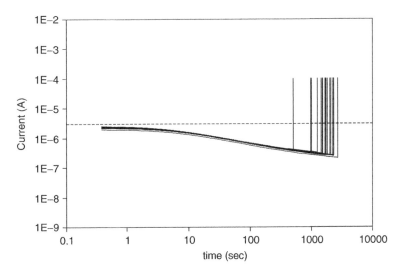

Figure 2.36. Example of a constant voltage stress result: *I–t* curves of a thick oxide that exhibits significant electron trapping. After some time of stress, the current increases dramatically and reaches the preset compliance. This increase marks the breakdown. The dashed line depicts a possible current criterion for breakdown detection.

q_{BD} is calculated by integrating the current between stress start and the last data point before breakdown. The calculated q_{BD} values are also plotted in a probability net, just like the t_{BD} values.

The constant voltage stress is a long term stress because the determination of the voltage acceleration requires stresses at different voltage conditions and also involves lower voltages that run for relatively long times. Such stresses are more efficient if several samples are stressed in parallel. Parallel stresses are also well suited for stressing different structures (see Section 2.4.4) on the same chip simultaneously. Some example setups for parallel constant voltage stressing are depicted in Figure 2.37 (a–d). In Figure 2.37 (a) the most accurate and fastest setup is represented. It requires as many SMUs as channels are involved for parallel stressing. Speed and therefore time resolution is limited only by the SMU and the connection to the PC. This solution is obviously very expensive. An advantage is, however, that each channel can stress its sample at a different voltage or polarity than the remaining channels. A similar but cheaper solution could be implemented using an analog input/output PCI card in combination with a current–voltage converter connected to a PC [167] instead of SMUs.

The setup in Figure 2.37 (b) uses resistors and a scanner matrix to monitor the current of all the channels individually, while the same voltage is applied to all connected samples. Here the speed of the scanner and the serial measurements limit the time resolution, which makes the method more suitable for long term stresses, where the failure times are much larger than the scan and measurement duration. An advantage of this setup is that channels can be easily added without much cost. This makes it simple to add the electrical fail localization after stress for transistor-like structures, as described above.

The setup in Figure 2.37 (c), in which the current is measured directly, without measurement resistances is very similar. It requires, however, more scanner channels in order not to interrupt the stress briefly when the meter is connected in the circuit of a particular sample for measurement. Due to the current measurement, this setup may be more accurate but even slower than the version in Figure 2.37 (b). Instead of the separate voltage source and meter, an SMU could be used. Both setups (Fig. 2.37b,c) are useful for wafer level measurement on a prober as well as insitu measurements of packaged samples in an oven. The last method (Fig. 2.37d) is the cheapest and most useful for measurement of packaged samples in an oven where the wiring does not give access to the individual samples. Instead, only the sum of all currents can be monitored. The resistors in series with the device under test limit the current after breakdown for each sample, ensuring that the broken samples do not draw too much current and thus reduce the stress on the samples that are still to be stressed either through the I–R drop or through pulling down the power supply voltage. Additionally, the current increase of the sum-current is defined by the resistor, i.e., each breakdown causes the same current increase. Therefore this method works well where hard breakdown detection is the only requirement and the tunneling current is sufficiently low. Every increase by the defined amount or a multiple thereof is counted as one or more breakdowns, without having any information about when

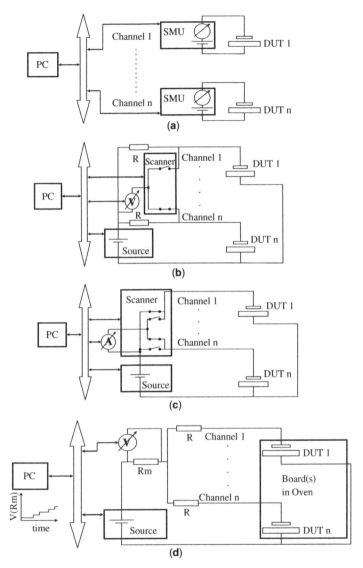

Figure 2.37. Examples of various parallel stress setups: (a) for fast and accurate stress; also suited for very thin oxides (b) and (c) less expensive setup with good time resolution for long term stresses on wafer prober or in oven involving a scanner matrix. Instead of the separate source and meter, an SMU could be used (d) as the least expensive version for long term stresses of packaged samples on stress boards in oven. This is for thicker oxides that exhibit hard breakdown only. The current after breakdown is limited by a resistor, R, hence each breakdown causes a step of defined height allowing to count fails while it is unknown which sample failed.

each sample failed. This information may be collected to some extent by interrupting the stress periodically and doing a readout of all the samples on stress. In fact, running a package-level stress without monitoring the sample behavior on stress (but instead doing readouts at increasing intervals) is also widely practiced. It is, however, labor intensive and involves a lot of handling the devices that are stressed, which may in some cases cause ESD damage or mechanical damage, leading to premature failure or enhanced lifetime due to bad contacts. The last method—as well as most of the described constant voltage stress methods—can be enhanced by addition of a capacitor (e.g., 100 nF) connected in parallel to the device under stress to ensure low resistance breakdown [168] if desired.

2.3.1.2 Voltage Step Stress. The voltage step stress consists of a few constant voltage stresses at different, increasing voltages subsequently applied to the same sample. It is very useful to determine the extrinsic part of a bimodal distribution. For extrinsic fails the stress time can become very long due to the wide distribution and the large sample size required. The idea is to save time, improve the time resolution, and better utilize the available samples and stress equipment by increasing the stress after a sample has survived a predefined time and stress condition. Hence the voltage may be increased three or four times until it is high enough to cause intrinsic breakdown within a short time (Fig. 2.38). However, each stress causes damage in each sample. This damage is not noticeable on good samples, but it causes weak samples to fail considerably earlier in the succeeding step than it would have failed at the same condition without the

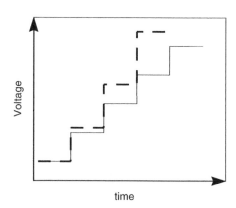

time

Figure 2.38. Two possible designs of voltage step stress are illustrated: solid line using equal voltage steps is likely to cause distorted distributions which need to be corrected by a voltage acceleration model (Fig. 2.39). Dashed line uses increasing voltage steps to shift the predamage to small enough times outside the measurement time window, resulting in undisturbed extrinsic distributions (Fig. 2.40).

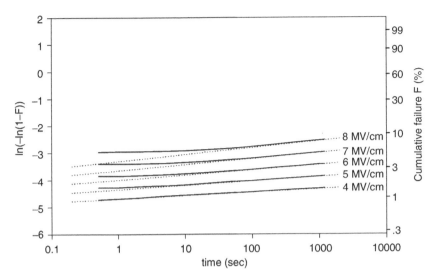

Figure 2.39. Weibull plot of step stress simulation assuming the 1/*E*-voltage acceleration model and no intrinsic fail within the five steps with equal step height. The bold lines represent the step stress result while the dotted lines mark the distributions resulting from constant voltage stresses that use fresh samples for each voltage. An increasing impact of the predamage on the step stress distributions can be observed. This can be corrected if the voltage acceleration model and parameters are known.

preceding stress. It is obvious from Figure 2.39 that at short times, the resulting distributions are not straight lines but start almost horizontally and approach the real extrinsic distribution after some time (much larger than the equivalent stress time from the preceding step). The predamage caused by preceding steps may even increase with each step as shown in Figure 2.39.

If the acceleration model and the corresponding acceleration parameters for those extrinsic fails are known, the predamage can be corrected and conversion of all steps to one voltage is straightforward [169]. If the acceleration model is not known but is to be determined as part of the stress, the predamage needs to be eliminated. The damage from preceding steps cannot be avoided, but it can be shifted to short enough times so that it is negligible within the measurement time window. This is achieved by increasing the voltage so much that the equivalent time calculated from the stress conditions of the preceding step, is much smaller than the time resolution of the subsequent step. The voltage step height needs to increase for each step to accomplish this for all steps (Fig. 2.38, dashed line), which will cause a lot of fails at the first readout of the step, but the subsequently measured times are not affected by the predamage and represent the real extrinsic distribution.

This benefit is illustrated in Figure 2.40 where a simulation of a step stress (bold line) is compared with the regular constant voltage stress (dotted line).

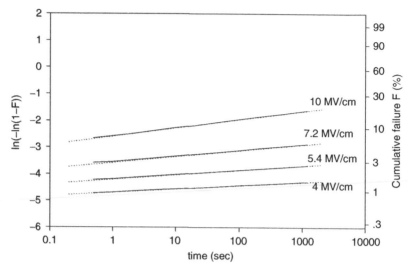

Figure 2.40. Weibull plot of step stress simulation assuming the 1/*E*-voltage acceleration model and no intrinsic fail within the four steps. The bold lines represent the step stress result while the dotted lines mark the distributions resulting from four constant voltage stresses that use fresh samples for each voltage. For two fields a slight deviation from the constant voltage stress can still be seen at times below one second.

For the simulation, intrinsic fails are ignored and the $1/E$-model is assumed as the voltage acceleration model for extrinsic fails [163]. The lowest voltage always yields the same result as a constant voltage stress (although even here deviations from a straight line are possible if the sample is screened aggressively). For the second and third field, a small deviation from the constant voltage stress can be detected at times below one second. Nevertheless, the 20-minute stress per step provides at least three orders of magnitude to determine the distribution parameters. The highest voltage is high enough not to cause detectable deviation to show up in the accessible time window. All four voltage step stress distributions can be used directly in this case to determine the voltage acceleration model and parameters, provided that the sample size was large enough to tighten the confidence bounds sufficiently. Changing the duration of the steps also affects the predamage of subsequent steps, but it is not as efficient as increasing the voltage step [163]. As indicated above, it is desirable to have an acceleration factor of 1000 between the step conditions (voltage and time), and an acceleration factor of at least 100 is absolutely required to be able to clearly interpret a step stress.

2.3.1.3 *Voltage Ramp Stress.* Ramping the voltage and recording the current is a well known method from the electrical characterization of dielectrics. For a reliability assessment it is necessary that the voltage is ramped to the same high, predefined value for all samples. The voltage end point is chosen high enough for all samples to exhibit breakdown during ramping with high probability. The result is a regular IV curve with an abrupt current increase marking the breakdown. Two kinds of voltage ramps have been reported: the linear voltage ramp and the staircase voltage ramp. The descriptor linear or staircase refers to the electric field versus time behavior. Since the current measurement takes some time, most of the ramps on modern test equipment are really staircase like. The steps need to be long enough to allow at least one complete current measurement. The current measurement usually takes more time for low currents and after measurement range changes, which is something that needs to be considered if the timing is critical, e.g., if the exact ramp rate is required. The ramp rate R is generally defined as step height (V_{step}) divided by step duration (t_{step}), i.e., $R = V_{step}/t_{step}$. The step height affects the voltage resolution at breakdown, while the step duration is equipment dependent as discussed above. Both parameters are desired to be as small as possible for the best resolution and approximation of a linear ramp. The ramp rate affects the results of the voltage ramp. A slower ramp causes breakdown at lower voltages because more time is spent stressing the dielectric, while a faster ramp rate allows the dielectric to reach a higher voltage before breakdown occurs. Also, smaller areas cause a larger breakdown voltage. Therefore it is a good practice to state the stress conditions, thickness, and area when presenting breakdown data. For a comparison of results of structures with different areas, the current density instead of the current should be plotted.

The breakdown detection depends on dielectric thickness. For very thick dielectric (>20 nm) the current is generally low and a fixed current limit can be used to detect breakdown. For thinner dielectrics current conduction can reach a high level without representing a breakdown, e.g., due to Fowler–Nordheim tunneling. In this case, the slope between two successive measurements may be used to detect breakdown. For thinner dielectrics (e.g., <5 nm) this method may not work either. In this case breakdown may be detected by lowering the voltage after each step for an integrity measurement [170–172], which increases the total time of the ramp.

A special behavior is observed in thick oxides [34] and thin nitride-oxide dual-layer dielectrics [173], where the current increases by one or more orders of magnitude at a certain critical voltage and then continues to rise normally. Since the dielectric does not break down at the critical voltage, this current jump poses a challenge for automatic breakdown detection.

The output of a voltage ramp is the breakdown voltage if breakdown occurred. In addition, the charge-to-breakdown (q_{BD}) can be calculated by cumulating the injected charge of all steps. Other parameters could include the yield results at predefined voltages. The voltage at a predefined low current above the noise level gives an idea of the uniformity over the wafer or all measured samples. The voltage ramps also provide sufficient resolution to determine the defect density based on the early fails of weak samples. It has been demonstrated for thicker oxides (>10 nm) that the defect density from ramped voltage stress and constant voltage stress is consistent [152]. This still holds for thinner oxides and no contradicting results have been reported.

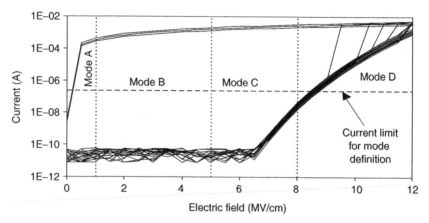

Figure 2.41. Example of an electric field ramp and possible mode assignments. There are immediate shorts (mode A) and early fails in mode D. Intrinsic failure is not reached in this case within mode D.

For comparability of samples with different thicknesses it may be useful to perform an electric field ramp to a maximum field, e.g., 12 MV/cm, with equal field steps (e.g., 0.5 MV/cm and 50 ms duration) instead of a voltage ramp. For this more standardized test the dielectric thickness must be known to determine the voltage increase of the steps and the maximum voltage. It is obvious that the method is sensitive to dielectric thickness variations. The electric field ramp is standardized by JEDEC Group 14.2 [174].

For in-line monitoring of the dielectric quality, an automatic prober is most suitable. For simplified treatment the fails are categorized according to the electric field where the current passes a fail criterion, e.g., 1E-04 A/cm². This criterion requires the knowledge of the dielectric area and may need adjustment for some dielectrics or dielectric thicknesses. As illustrated in Figure 2.41, the ramp may start at zero field to detect early fails in the first interval between 0–1 MV/cm, or for time savings at or close to the operation conditions. Early fails are counted as Mode A and represent fails that would typically be screened in product functionality tests. The next category, Mode B, represents fails between 1 and 5 MV/cm. Those fails are reliability fails and equivalent to burn-in or early life fails of a product. Mode C collects fails between 5 and 8 MV/cm, which are also extrinsic reliability fails potentially during product lifetime. Mode D, the category above 8 MV/cm up to 12 MV/cm, is a mix of fails and good parts for which the current, due to tunneling, reached the failure limit. To distinguish between surviving and failing samples an integrity test after the ramp is necessary. It may happen that all samples failed in Mode D due to intrinsic breakdown. This is the case if the verified fails occurred all at a certain electric field. For more detailed information Mode D could be partitioned in two or more sections. At the end of each section the ramp may be interrupted by a leakage test at a lower field to determine whether the sample is still OK or broken down. Fails in the first section would belong to Mode D and D1 while fails in the second section belong to Mode D and D2 and so on. Thus intrinsic and extrinsic fails could be distinguished more rigorously. From the yields of the various modes the quality of the dielectric is obvious, thus they are useful for SPC cards. In addition, V_{bd} and q_{BD} could be plotted in a probability net (usually a Weibull plot; see Section 2.3.3) and the Weibull parameters could be determined for additional characterization.

The advantages of the ramped electric field test are that it can be a standard, fully automated test procedure that is fairly quick to perform and able to provide much detailed information. Of course methods to project dielectric lifetime by a ramped voltage method have also been proposed based on the E-model [14] and the $1/E$-model [156, 175, 176].

2.3.1.4 Constant Current Stress. The constant current stress (CCS) forces a fixed current through an oxide to assess its robustness. Constant current stress is in most cases rather short due to the high acceleration applied, paired with the fact that electron trapping does not reduce the stress, but rather causes an adjustment of the external voltage in order to keep the current constant. Thus for oxides exhibiting electron trapping, a constant current stress leads to faster breakdown than a constant voltage stress with comparable initial stress conditions. During CCS the voltage is monitored over time and stored for off-line analysis. To limit the amount of stored data, similar algorithms, such as those discussed for the constant voltage stress, can be applied. Generally, the voltage measurement is very fast and provides a very good time resolution. The time-to-breakdown is defined as the time where the voltage decreased more than a predefined percentage (e.g., 20%) of the last measured value. Sometimes it can be observed that instead of the voltage decreasing, it jumps all the way up to compliance. This can happen if, for example, the test structure has too narrow connecting lines, which open as a result of the stress due to electromigration. Such undesired effects need to be screened.

Prior to the constant current stress an integrity test at a low voltage or current is suitable. The same test could be applied after the stress to verify breakdown. The integrity test can be defined analogous to or be even the same as the integrity test of the constant voltage stress.

The output from CCS is the breakdown voltage V_{bd} or the breakdown field E_{bd}, which differs from the initial voltage for dielectrics that exhibit trapping, and the charge-to-breakdown. The product of the forced current density and the time elapsed until the first voltage drop is reached, represents the charge-to-breakdown q_{BD}. q_{BD} depends on stress current density and to some extend on temperature [177, 178], as shown in Figure 2.42. The dependence of Q_{BD} on current density is also a function of oxide thickness. For thin oxides in addition a dependence of Q_{BD} on area was reported [30, 157, 177], which in accordance with the weakest link character of oxide breakdown should also be present for thicker oxides. All those dependencies need to be considered for projections and comparisons of Q_{BD} values. CCS is not suitable for ultra-thin oxides ($<4\,nm$) because Q_{BD} for a fixed current density seems to increase towards thinner oxide thickness, contrary to what is observed from the constant voltage stress for a fixed voltage. The reason

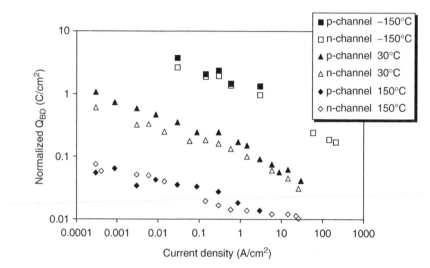

Figure 2.42. Charge-to-breakdown for a 10 nm oxide acquired over a wide range of current densities and three different temperatures. At higher current densities and temperatures lower Q_{BD} values are measured for both channel types [178].

for this is that a reduced voltage is sufficient to drive the same current through thinner oxides. This in turn reduces the defect generation rate dramatically and increases Q_{BD} as discussed later [159]. If CCS is applied to ultra-thin dielectrics appropriate methods for soft breakdown detection need to be introduced, e.g., as proposed for ramped current stresses [179–181].

Constant current stress is sometimes preferred because it is expected to not be sensitive to series resistance. While for a pure and small series resistance this may be true, nonuniformities in the test structure like a distributed series resistance or local potential fluctuations cause current crowding and affect the stress result. Also very high series resistances could cause so much voltage drop that the breakdown detection may be affected. Another source of potential error exists if the dependence on current density is not considered because a higher current density may result in a lower Q_{BD} for the same oxide quality. Q_{BD} is particularly constant for thicker oxides (> 10 nm) over a wide range of current densities, but picking a current density beyond this range—just to cut down the stress duration—may yield an artificially poor estimate of oxide quality.

Previous comments already indicate that CCS is frequently used for the comparison of oxide quality by applying quick stresses. This highlights the advantage of CCS for thick oxides, while for low current densities the duration increases rapidly. Parallel stresses, as described for constant voltage stress, are possible but require a setup with individual current sources while the simple inexpensive setups that are possible for constant voltage stress are not feasible for CCS. Therefore CCS is usually a serial stress method, taking advantage of the high acceleration and the constant stress level. Another drawback of the method is the lack of widely accepted lifetime prediction models for oxides operated at a fixed voltage, although there have been proposals [182]. For the tunneling oxides used in non volatile memories (e.g., Flash, EEPROMs) the current stress is, to the first order, constant and the same for both the product and the test structure. Correlations of Q_{BD} and program/erase cycle times have been reported [183, 184].

For evaluation of the constant current stress the resulting Q_{BD}, V_{bd}, or E_{bd} values of all stressed chips are plotted in a probability plot (e.g., Weibull plot) or in a wafer map to detect regional dependencies.

2.3.1.5 Current Ramp Stress.

Another quick stress method is the current ramp stress. It is the result of an evolution from constant current stress at two or more different stress currents, to get better resolution of early fails. The current ramp forces an exponentially increasing current through the dielectric and is therefore referred to as an exponentially ramped current stress (ERCS). The current is increased either all the way to breakdown or to a maximum current density, which in some cases is kept till breakdown, i.e., an ERCS with subsequent constant current-stress. In some cases, the current density for the subsequent constant current-stress could also be higher to ensure breakdown within a short time. In any case, the current ramp sweeps over a wide range of current densities and (depending on the oxide thickness) over a wide range of Q_{BD}s. Therefore, the q_{BD} distribution from ERCS is steeper than the distribution acquired from CCS or CVS if the dielectric exhibits a dependence of Q_{BD} on the current density. This is important only for quantitative analysis, while for qualitative assessments this is secondary. The ERCS is a widely used fast test because it is relatively simple and fast and has the advantage that breakdown is easier to detect because it is marked by a voltage drop. For the test environment the voltage drop at breakdown is beneficial because the power decreases while it increases in case of a breakdown during the voltage ramp or a constant voltage stress. This could help conserve the lifetime of the probe tips.

For a JEDEC ERCS specification [174], a defined current density marks the starting point of the current ramp. From this current density a current is calculated based on the dielectric area of the structure to be stressed. It is important to comply with a standard because Q_{BD} has been reported to change with the initial current density [185]. The initial current density should only be adjusted if it cannot be reliably forced through the dielectric. Such adjustments should be stated when data are published or used for comparisons to avoid wrong conclusions. The next parameter defining the ERCS is the time spent at each step. This time is usually selected to be short for a quick ramp, but at the same time it needs to be long enough to allow accurate voltage measurement during the steps; thus, it depends on the available equipment. It is useful to do several voltage measurements within one step, e.g., to support noise detection for thinner oxides. The duration of the step is used for calculation of q_{BD} and therefore needs to be maintained as accurate as possible throughout the ramp to minimize errors. Another parameter is the number of steps per current decade. This number affects the resolution of the stress results (e.g., q_{BD}) and the step height as well as the slope of the ramp. About ten steps per decade works well in most cases. Assuming 50 ms per step, the ERCS takes 500 ms per decade and may result in 100 measurement points if the equipment performs one measurement every 5 ms. The total time for the ramp depends on when the breakdown is reached and is in the order of a few seconds.

Breakdown detection during the exponentially ramped current stress is similar as in the constant current stress, i.e., a drop of voltage of 5–10% compared to the preceding value. For thin oxides (< 5 nm), this criterion is usually not sensitive enough to detect soft breakdown. Therefore noise detection is involved as a fast method. Six sequential voltage readings from the same step are sufficient to calculate a slope value [181]. The increase of this value at a soft breakdown is much more significant than the corresponding voltage drop, which means it is a better, more reliable indicator for the soft breakdown detection. If the total time of the current ramp is not an issue, soft breakdown detection could also be performed similar to the voltage ramp by reducing the current after each step to the initial starting current and checking the integrity.

A more advanced method uses a preceding voltage or field ramp to improve the limited resolution for breakdown at low voltages. The preceding field ramp can be started at a level where current cannot be forced reliably and stopped at the current that represents the initial current of the current ramp (Fig. 2.43).

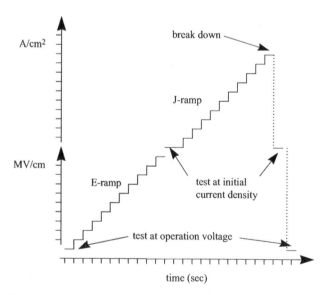

Figure 2.43. Illustration of an ERCS with preceding field ramp. The field ramp stops when it reaches the initial current density of the j-ramp. Two integrity tests are performed: one at operation voltage and one at the initial current density. After breakdown two breakdown verification tests are done at the initial current density and operation voltage, respectively.

In addition integrity tests before and after the ramp can be added similar to the ones already described in previous sections. For the ERCS with preceding voltage ramp, it has been reported that two integrity tests are performed: one at the operation voltage before the voltage ramp and one at the initial current density after the voltage ramp, right before the current ramp. Both tests are repeated for breakdown verification after completion of the current ramp.

The output of an ERCS are the parameters Q_{BD}, the current density J_{bd}, and the voltage or field at breakdown, V_{bd} or E_{bd}, respectively. All of them depend on the ramp rate [186]. Additional parameters can be defined based on the integrity checks, e.g., the current at operation voltage or the voltage at the start current density both determined before and after the ramp. Q_{BD} is in most cases different than the one from CCS due to the increasing current density during the ramp. Therefore conversion of q_{BD} from ERCS to q_{BD} corresponding to CCS has been proposed [186, 187]. Even a method to calculate t_{BD} corresponding to CCS has been published [188].

2.3.1.6 fWLR Stress Methods.

Fast wafer level reliability (fWLR) methods [189] are used for the inline monitoring of intrinsic and extrinsic reliability parameters. The idea is to track the behavior of reliability related parameters such as Q_{BD} over time and detect degraded reliability by comparing the monitored values to reference values from a successful qualification. Such stress methods need to be very fast to allow sufficient sampling of the wafer fabrication or production output to maximize the dielectric information and to minimize the test time. Therefore the ramped voltage and current methods are widely used. Also combined stresses are frequently used to improve the performance and resolution. The ERCS with preceding voltage ramp [181] was described in this section already and it was mentioned that the ERCS may be stopped at a certain current density and succeeded by a constant current stress [190]. The combination ramped and constant stress aims at giving more detailed and accurate information on both the extrinsic and the intrinsic properties of the dielectric.

For some oxides an unexpected behavior with such combined measurements has been observed: In one case the preceding ramp stress improved the reliability of the hardware under the subsequent constant stress [161] depending on the ramp and stress conditions. The cause for this behavior was identified to be initial hole trapping. Thus the combined stress needs to be designed in a way that the preceding ramp neither improves nor significantly degrades the dielectric and consequently does not affect the subsequent constant stress. For thick dielectrics (>25 nm), which exhibit strong charge trapping, it is observed that they fail instantaneously if the stress condition is too high. A pre-stress (e.g., a fast ramp to stress condition) that fills some of the traps prevents the instantaneous fails and enables stressing at the highly accelerated condition. A disadvantage of such combinations may be a long duration. However, with improved measurement speed it is possible to maintain the short duration, e.g., a ramped current stress while completing a short voltage ramp and constant voltage stress [166]. This allows one to determine reliability disposition limits and do a highly accelerated,

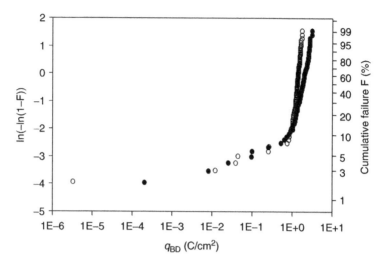

Figure 2.44. Comparison of bimodal q_{BD} distributions acquired on the same wafers by a fast J-ramp (open circles) and a fast constant voltage stress with preceding V-ramp to improve extrinsic resolution (full circles) [166]. The defect density at the intersection point of intrinsic and extrinsic distribution (at slightly below 1C/cm²) remains the same. The slope of the intrinsic distribution from J-ramp is steeper due to the increasing acceleration of this test. Reprinted with permission from Ref. 166, © 2004 IEEE.

quantitative, in-line assessment of the dielectric lifetime. Similar stresses are possible for dielectrics of nonvolatile memory cells using a ramped current and subsequent constant current stress.

The quantitative parameters of those methods represent intrinsic reliability. For extrinsic reliability, the defect density is a quantitative parameter, which can be acquired by all methods that resolve the transition from intrinsic to extrinsic, i.e., defect-driven reliability, as illustrated in Figure 2.44. It has been shown before that the transition between extrinsic and intrinsic distribution is a reproducible point for defect density assessment with all of the methods; hence, ramped stress methods work well, as has been demonstrated for various oxide thicknesses [181, 183, 191]. However, a sufficient sample size or a moving average based on several lots is needed to be able to detect the small defect densities required by today's standards. Furthermore, product related early failure projections have been based on voltage ramp data [192].

2.3.1.7 Constant Current Stress Versus Constant Voltage Stress.
Hokari first proposed the constant current stress (CCS) technique for oxide reliability evaluation [18], especially because this method helps to reveal the role of charge trapping in the breakdown process at a fixed cathode field. Subsequently, this stress methodology had been widely used for evaluation of thick oxides rather than constant voltage stress (CVS). Nevertheless, this methodology already

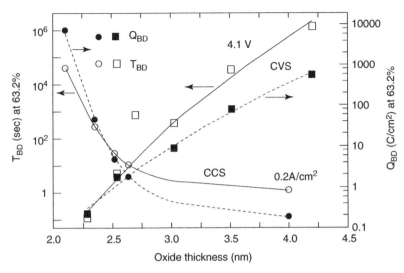

Figure 2.45. T_{BD} and Q_{BD} versus oxide thickness for CVS and CCS. The capacitor structures were stressed in substrate injection with an area of 6.2×10^{-4} cm^2 for CVS mode and an area of 1.0×10^{-4} cm^2 for CCS, respectively. Reprinted with permission from Ref. 159, © 1999 IEEE.

encountered some difficulties in explaining the thickness dependence of oxide breakdown in the range of oxide thickness above 5 nm. Using the CCS method, contradictory results were reported for the thickness dependence of Q_{BD}. At low current densities, thinner oxides (~ 4.0–6.0 nm) tend to show higher Q_{BD} than thick oxides (7.0–33.5 nm) but at high current densities, a reverse trend was observed [193]. In the middle of the 1990s, the concept of energy-driven breakdown was advanced. It was shown that the energy of an electron exiting from the oxides at the anode controls the defect generation and breakdown process [23, 194, 195]. In this scenario, gate voltage plays a more significant role as it is related to energy dissipated by electrons at the anode, especially for oxides stressed in the ballistic FN regime and DT regime.

Figure 2.45 shows the time-to-breakdown (T_{BD}) and the charge-to-breakdown (Q_{BD}) versus oxide thickness for CVS and CCS methods. The CCS results show an increase in T_{BD} and Q_{BD} with decreasing oxide thickness whereas for CVS, these two quantities drastically decrease with oxide thickness. Based on these results using the CCS method, one may conclude that thinner oxides provide much better reliability. On the contrary, the opposite trend is observed in the results using the CVS method. Now, we will use the new breakdown model (Eq. 3.8) developed in Section 3.2 to explain these puzzling results. First, for the CCS method, a much lower voltage is required for thinner oxides due to the strong thickness dependence of direct tunneling currents as shown in Figure 2.20. For a fixed constant current of 0.2 A/cm^2, the stress voltage for 4.2 nm oxide is around 5.8 V whereas for a thinner oxide of 2 nm, a much reduced stress voltage of about

2 V is needed. Hence, the defect generation efficiency is reduced by 8–12 orders of magnitude because of its strong voltage dependence as discussed in Section 3.2. On the other hand, the critical defect density required to trigger a breakdown event only decreases by about four orders of magnitude (Fig. 3.15 and Fig. 3.16). This larger reduction in the defect generation rate compensates for the decrease in the critical defect density. Therefore, for CCS, much higher Q_{BD} and T_{BD} are observed for thinner oxides as compared to thick oxides for a fixed current. However, for CVS, the decrease in Q_{BD} with decreasing oxide thickness directly reflects the reduction in the critical defect density (Figures 3.15 and Figure 3.16).

The CVS method is shown to be capable of resolving the issues due to process effects on ultra-thin oxides while the CCS method reveals contradicting results similar to T_{BD}/Q_{BD} thickness dependence shown in Figure 2.45. [158]. Furthermore, for practical applications, most devices do not operate in CCS mode at use conditions. Based on these considerations, constant voltage stress should be used for oxide breakdown evaluation.

2.3.2 Design of Stress and Test Structures

The design of test structures in general is a very important and complex topic, as demonstrated by the fact that an entire annual conference, the IEEE International Conference on Microelectronic Test Structures (ICMTS), is dedicated to this topic. This section summarizes the basic issues of test structure design for dielectric reliability stressing.

Every test structure needs to be designed to best suit the intended purpose; for investigations of breakdown physics and purely intrinsic breakdown, it is best to work with small structures that do not exhibit any significant edge effects. Small structures are likely to be defect free so that breakdown is really reflecting the dielectric properties. At the same time, edge effects need to be eliminated; these are mostly due to dielectric thinning, e.g., at LOCOS or STI edges, or field enhancement due to sharp gate poly-Si or STI edges [27, 196–198]. Thus the appropriate edge type with the least impact needs to be determined for the technology before designing the test structure. Another purpose of a test structure could be to reflect product and process relevant influences on the dielectric. In this case, the areas and edges of critical circuits in the product are scaled to the test structure, resulting in an array of transistors with the gate terminals connected in parallel and source and drain respectively [27, 199]. Large structures are required if defects and the corresponding defect density is to be monitored [199], considering the low defect densities targeted by today's technologies. Such structures usually take up so much space that they can only be implemented on test chips or special test sites. The advantage of really big test structures is the ability to perform defect monitoring with a relatively small sample size and thus reduced stress time. On the other hand, they may be too sensitive for processes still in development and hence show no intrinsic breakdown at all. The size is limited for test structures that are restricted to the scribe line. This forces a compromise that supports demonstrating intrinsic breakdown properties, as well as extrinsic properties if high defect densities are present or if large sample sizes are used [200]. Another important point to consider for test structure design is the dielectric thickness. Thinner dielectrics exhibit high leakage currents. The overall current may be too high and cloud the soft breakdown that is typical for ultra-thin dielectrics. Hence, for these thin dielectrics, the test structure areas are reduced even to the size of single transistors to ensure that every soft breakdown is detectable [201]. Since it is known that the post breakdown conduction for thin dielectrics is affected by the circuit in which the device is embedded [154], product relevant test structures become more sophisticated; instead of simple transistors or transistor arrays, ring oscillators and inverter chains are used with access to single devices [50].

The design of test structures should be consistent with the design rules of the product as much as possible to avoid effects that are not product relevant. On the other hand, it is a good idea not to use minimum gate length for transistor structures because the poly-Si gate length variations affect the dielectric area directly. The latter can be avoided by designing transistors large enough so that poly-Si variations are negligible, while still maintaining the product relevant ratio between poly-Si and STI edges as well as area. This may change the ratio between channel area and the overlap area of gate and diffusions. Also, the impact that soft breakdown has on a device depends on device length and width [40, 44, 45]. Thus for thin dielectrics (< 5 nm) a compromise is necessary, while for thicker dielectrics that do not show soft breakdown the larger device design works well. Since the area needs to be extrapolated from the test structure to the product, it becomes more important to know the exact test structure area, the larger the area ratio between both of them is. From the above it is obvious that this is more critical for thin dielectrics due to the required smaller test structure area.

A single test structure is usually not sufficient to monitor the dielectric reliability of integrated circuits. Rather, a set of structures is necessary to account for the different oxide thicknesses on the product, for both NMOS and PMOS types as well as for area, poly-Si, or isolation-edge-dominated breakdown. Last but not least, structures for area scaling should be included. While it is obvious why structures for each dielectric thickness and doping type on the product are needed, the other structures require some explanation. For product relevant dielectric reliability testing, arrays of devices are best suited [27, 199]. If they demonstrate the desired reliability, the other structures may not be needed, except for some arrays with different areas to prove that the area scaling of the intrinsic distribution is according to the Poisson model (see Section 2.4). This ensures that there are no parasitic effects [202], and also helps to determine the distribution slope most correctly with the usually limited sample size (Section 2.4.3). The areas of those structures should cover a wide range, e.g., three or more orders of magnitude [202]. If the currents and/or distributions of different areas do not scale, a lifetime projection for the product area does not apply and the root cause needs to be found (e.g., severe thickness variations over the wafer [202]). If bimodal breakdown distributions are observed caused by a high defect density, structures with different edge types and lengths may be stressed. These structures serve as diagnostic structures to help identify the root cause of the deviation from intrinsic distributions. In order to get direct feedback, the diagnostic structures may have the same dielectric area as the array structure. Thus they identify immediately the root cause; if there is a systematic weakness at the isolation edge, the structure with the extensively long isolation edge would show a more severe extrinsic failure distribution or even reduced intrinsic breakdown times [198]. If the weakness is extreme, even the shorter edges might be affected and dominate the breakdown behavior. In this case similar intrinsic distributions may result from all diagnostic structures; however, the area scaling will most likely indicate a problem. This case is probably rare and can occur only if, due to design requirements, every structure has at least a short edge with the weakness (e.g., the connection from the pad usually has to cross an isolation edge, even if the structure is otherwise diffusion bounded—unless the design rules allow contacts to gate poly-Si over active area). Despite the possibility to identify potential weaknesses directly with the test structures, it is a good practice to verify the failure cause by physical failure analysis (PFA) using an optical method that shows whether the fail is at a certain edge or in the dielectric area [198, 203]. For this purpose it is useful to leave the area above the test structure free of metal so that optical methods can detect light emitted at the failure spot.

For diagnosis efficiency and short turn around time it is useful to place the set of test structures at the same pad set, so that with one touch down all or most of the structures can be stressed simultaneously. This avoids the need for several passes and saves time, even if the information from all structures is not always needed. For this purpose the structures should be stressed at conditions that cause approximately the same breakdown time. For different areas the same stress condition should be used unless the acceleration models and their parameters are well known and do not cause artificial deviations from area scaling. An artifact may also be generated if the pads are over thin dielectric or the structure is probed directly on the gate over active area. In such a case, excessive mechanical probe pressure causes shorter breakdown times and skews the breakdown distribution. Therefore pads should always be located over thick isolation oxide.

The most significant issue when designing test structures is avoiding series resistances and nonuniform voltage drops across the dielectric area. It is often not considered that the structure is operated at highly accelerated conditions like elevated temperature and voltage. Any series resistance caused by long and narrow connection lines between the structure and the pad reduces the voltage across the dielectric and hence the resulting time-to-breakdown is longer than it is in a resistance free environment [157, 204, 205]. Also, the slope of the probability distribution is affected by series resistance [202]. Since the series resistance has less effect at lower stress conditions the acceleration determined from such a structure is also usually lower. This may cause the projection to end up artificially low [205]. Thus it is most important to avoid series resistance effects by design. Since the wells and diffusions usually cannot be changed, all the adjustments for reduced series resistance need to be done in the metal levels. The use of wide and short connection lines between pad and structure is essential. In extreme cases narrow lines have been fused during dielectric stress and the open line prevented the collection of information on the post breakdown characteristics [206]. The lines need to be wide for all connecting lines (i.e., the gate connection as well as source, drain and well) to allow stress in both polarities. Sufficient contacts to wells and diffusions as well as to the gate conductor are also necessary. However, if the process exhibits charging during contact etch it may be necessary to optimize the number of contacts to not predamage the dielectric by plasma-induced charging. Plasma-induced charging damage is especially critical for small active dielectric areas (such as single devices) because long connection lines and the standard pads are huge antennas that collect charge and can cause a voltage drop across the small dielectric area, analogous to special test structures designed to monitor plasma-induced charging [207, 208]. Large structures are less sensitive to charging if they are uniform, while nonuniformities or local weaknesses may be damaged and cause shorter breakdown times in the succeeding dielectric stress [209]. Some of the charging can be avoided by design, e.g., connecting the pad at the last metal only, if no stress is intended at a lower level, which due to a higher conductor sheet resistance is questionable for some cases anyway. Another way of protecting the structure would be to connect a protect diode to the gate. The disadvantage is that the structure can then be stressed in one polarity only. Especially at high temperature the connected junction may cause some extra leakage and not allow low current measurements. The protect diode is efficient for one charge type only while the opposite charge may still cause some damage.

If a series resistance is present in a test structure it may be corrected to some extent by methods that have been proposed in the literature [204, 205]. While this may work for some series resistances, it is more sophisticated to correct for distributed resistances that cause nonuniform voltage drops across the structure. The effects are enhanced by the reduction of dielectric resistance at high stress conditions and the growing sheet resistance at high temperature. The effect of the nonuniform voltage drop is that a smaller area of the dielectric is stressed at the intended voltage, while the remaining area experiences a smaller voltage drop [204]. Thus the lifetime is determined by the reduced area and is therefore longer. Both the series resistance and the distributed resistance lead to a deviation from the calculated, theoretical IV curves at high currents. This provides a simple means to check whether the test structures are affected by resistance effects at stress conditions [204].

The most relevant vehicle to test gate dielectrics is fully integrated hardware that contains product and test structure on the same wafer side by side. This type of hardware should be used for qualifications. Even more product relevant are test structures that use product circuits as their basis, e.g., DRAM arrays have been modified by a special wiring of the metal levels to form test structures with large total areas [27, 210]. Such structures are particularly useful for extrinsic dielectric reliability assessment [199]. For technology development, cost and turn around time play an important role. Here so-called short loops could be used. A short loop is hardware used for testing that is produced with only a fraction of the process steps compared to fully integrated hardware. For gate oxide, only a few mask levels are required, but it may be necessary to account for the thermal budget with additional anneals. Short loops can be produced using part of the same mask set as for the product, but using only the gate stack relevant process steps (provided there are test structures next to the product). A short loop can also be a separate test site (without a product on the same wafer) using two or a few mask levels only. This depends on the desired complexity of the structure that addresses gate dielectric reliability issues.

2.3.3 Physical Observations of Dielectric Breakdown

Dielectric breakdown is the irreversible change of physical properties of a local spot in the dielectric from low conduction to higher conduction. It is possible in thicker oxides to observe a light flash with a microscope when this change occurs (if the test structure is not covered by anything absorbing the light). The light flash is due to the overheating of a conducting path between the electrodes that formed as the result of a stress. In the moment this path is formed, energy is released in a very short time (a few ns for 35 nm oxide to < 100 ns for 5.6 nm oxide [152, 211]) in a small, localized spot. This can melt the material in the breakdown channel and locally the electrodes connected to it. Wolters [152] shows scanning electron microscopy (SEM) pictures of capacitors with 40 nm oxide where the poly-Si electrodes exhibit crater shaped damages around the breakdown spot. He explains the self-healing, which is often observed for thick oxide capacitors, by the molten

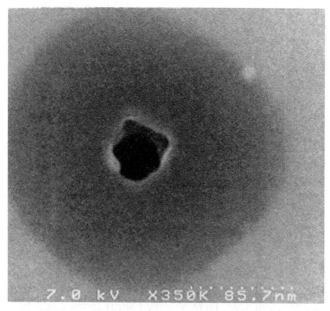

Figure 2.46. Hard breakdown spot in 6 nm gate oxide after poly-Si etch. The etch removes the poly-Si in the breakdown path between the electrodes and some of the silicon below the breakdown spot, which appears as dark gray area around the breakdown spot. Reprinted with permission from Ref. 215, Copyright 2003, Elsevier.

gate material withdrawing from the breakdown channel due to surface tension of the melt. He reports that the size of the breakdown channels is independent of the current density used for stressing. Even the silicon surface can be locally molten as reported for thick oxides [152] as well as for thin oxides [155, 212]. Although for thinner oxides, the hard breakdown does not show any indication in the gate electrode, the breakdown is still causing extreme local heating. A detailed analysis of a breakdown spot cross-section by energy dispersive X-ray spectroscopy (EDX) revealed tungsten at the interface between the undamaged crystalline silicon and the recrystallized silicon under the oxide around the breakdown spot [212]. Due to a visible recrystallization path connecting the center of the breakdown area in the silicon and the tungsten silicide over the poly-Si it is likely that the source of the tungsten at the interface is the tungsten silicide layer on top of the gate poly-Si. A top view of a thin oxide breakdown spot is shown in Figure 2.46 after gate poly-Si removal. The Si-etch also etched out the recrystallized poly-Si in the breakdown channel through the oxide and even etched the silicon under the oxide, causing the darker, concentric area around the hole in the oxide.

Those holes are 50–150 nm in diameter for ∼6 nm oxides, which is consistent with data reported in [155, 211], but clearly less than the 4–5 μm reported for 40 nm oxide [152] stressed at much higher voltages. Worm-shaped holes are another signature, observed from PFA on thin oxide fails that could not be electrically

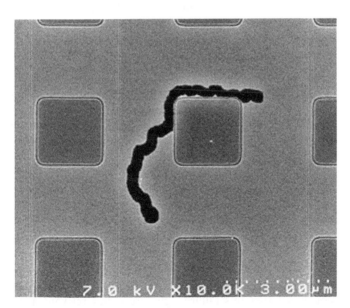

Figure 2.47. Another form of the breakdown is propagated breakdowns consisting of a chain of single breakdowns as the one in Figure 2.46 reaching several μm length. The width of the worm-shaped structure is similar to the diameter of the single breakdown. Part of the breakdown propagated along the STI edge (dark gray squares are STI islands). Reprinted with permission from Ref. 215, Copyright 2003, Elsevier.

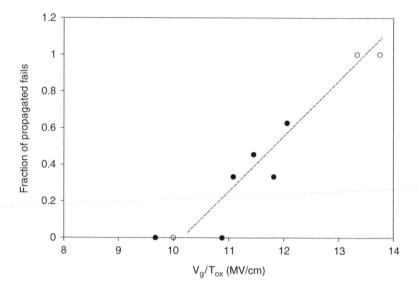

Figure 2.48. Fraction of propagated fails versus nominal electrical field. Below 11 MV/cm no propagated fails were found, thus this signature is not relevant for operation conditions. The fraction of propagated fails from Lombardo's work [211] is assumed based on his statement of good reproducibility of the results: open circles 5.6 nm, 9.3 nm, and 35 nm oxide thickness [211]; full circles 6–7.4 nm oxide thickness [215]; the dashed line is a guide for the eye only. Reprinted with permission from Ref. 215, Copyright 2003, Elsevier.

distinguished from the signature discussed above. An example of such a signature is shown in Figure 2.47. The square-shaped areas are STI islands that provide a long STI edge in this test structure. The worm-shaped oxide breakdown is partly along the STI edge and in the oxide area as well. The width of the worm-like failure signature is about the same as the single hole in Figure 2.46 while the length is several μm. Similar shapes have been reported for the oxide thickness range 5.6–9.3 nm [211], in thick field oxide [213] and in intermetal-level dielectrics [214].

The worm-like shape of the failure signature was explained as propagation of breakdown along a chain of weak spots, which are formed in the oxide during the stress [211]. The propagation speed is very high and the entire structure is created within 100 ns [211]. Furthermore, the defect concentration at breakdown, as well as the propagation speed, increases with oxide thickness [211]. However, for lower stress conditions a lower fraction of breakdowns with the worm-shaped were found and it was concluded that this failure signature is not relevant at the operation conditions [215]. This can be seen in Figure 2.48 where the fraction of worm-shaped fails is plotted versus the nominal electric field. The data of different sources is included and show consistency. No worm-shaped signature was found on samples stressed at electric fields below 10–11 MV/cm, which is consistent with the observation for 35 nm oxides stressed at 10 MV/cm in [211]. For thin dielectrics that exhibit soft breakdown, no such propagated fails have been reported so far.

Another way of observing dielectric breakdown is by employing a lumines-cence method during electrical stress [198]. For thinner oxides photo emission microscopy has been applied to study soft breakdown and its relation to hard breakdown [216, 217]. It was shown that multiple soft breakdowns occur before hard breakdown occurs, all in different locations on the same capacitor [216] with much reduced thermal damage [217]. A hard breakdown can also happen at the same spot where a soft breakdown was observed before, which causes a reduction of the light emission due to the structural change [217].

2.3.4   Considerations for Oxide Breakdown Detection

Oxide breakdown is marked by the discontinuity in stress current or voltage. This is generally referred to as the occurrence of oxide breakdown and also the first breakdown event. The definition of this discontinuity is considerably simpler for very thick oxides than for thin oxides. In the case of very thick oxides stressed in FN tunneling regime at the moment of breakdown, the stress current increases almost instantaneously from stress level to hundreds of micro-ampere as shown in Figure 2.36. In these situations, the detection of breakdown is straightforward. Nevertheless, as mentioned already the proper detection of breakdown can be altered by series-resistance effects and the use of a very large area. Both detection problems aggravate as oxide thickness is reduced. In addition, due to so-called soft breakdown mode or progressive BD mode widely observed in thin oxides, the stress current only shows an incremental change or increase (Fig. 2.49) rather than a sharp increase in stress currents as commonly observed in thick oxides as shown in Figure 2.36. Thus, the larger background direct-tunneling currents can mask the detection of oxide breakdown due to the small jump in gate current as reported in some extreme cases by the use of large area [218]. To circumvent this detection problem, it is more advantageous to repeatedly interrupt the stress to monitor the current at low voltages, often referred to as the stress-induced leakage current (SILC) (for more details see Section 3.2) [159, 219].

Figure 2.49 displays both stress current and SILC as a function of stress time. Because at lower voltages the background tunneling-current is lowered, the change in SILC is much more sensitive to detect the oxide breakdown. In the application of SILC measurements for BD detection, two different approaches can be implemented. 1) A direct use of SILC measurements to replace the stress-current for BD detection. In this method, the SILC measurements should be made frequently enough so that adequate time resolution of t_{BD} can be maintained. The use of low voltage or operational voltage for breakdown detection is proposed as an operational definition of breakdown. It was first proposed for thick oxides to represent a truly irreversible breakdown in the presence of large FN conduction [170]. One of the disadvantages is that this method cannot be used at very short time. 2) The sudden jump in SILC measurements is used to define the time interval in which the BD is subsequently determined by using the stress level. In this case, the time interval for SILC measurements can be made in the logarithmic timescale, not necessarily as frequently as in the previous method.

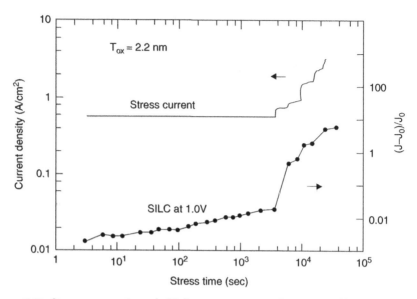

Figure 2.49. Stress current and SILC versus stress time. A diffusion-bounded NMOS capacitor structure was used. Both the diffusion and substrate were grounded with a gate bias of 3.4 V applied. The oxide thickness is determined to be 2.2 nm. Reprinted with permission from Ref. 159, © 1999 IEEE.

In thin oxides below 3.0 nm stressed in the DT regime, the presence of noise or fluctuations in stress currents due to multiple leakage paths prior to breakdown has been known for a long time [220, 221]. Besides the current fluctuations, the current versus time (I–t) characteristics measured on small structures shows relatively large current bursts so-called pre-BD events prior to true breakdown events [180, 222, 223]. This is evidenced by the fact that triggering on these pre-BD events can lead to the t_{BD} distribution not following Poisson area scaling. To overcome these difficulties, many researchers have attempted to develop the criterion and the related algorithm to define the true breakdown [180, 224, 225]. It was first found that the noise increase at the onset of breakdown is much larger than the voltage drop in the case of constant current stress [224]. It is unclear whether the same conclusion can be applied to the CVS method. In order to be able to use the largest possible test area, a normalized change of stress current is proposed as

$$\frac{\Delta I}{I} = \frac{I(t+dt) - I(t)}{I(t)} \tag{2.25}$$

and

$$\frac{RMS}{I} = \frac{RMS(t)}{I(t)} \tag{2.26}$$

This is called local current change because the effects of initial tunneling currents and stress-induced leakage currents are eliminated according to this definition. In actual implementation of this algorithm, a smoothing procedure is included to avoid false triggering of noisy current transients. It was concluded that the use of $\Delta I/I$ trigger is a better BD detection than the noise RMS/I trigger [225]. On the other hand, the breakdown trigger algorithm based on a running-median $\Delta I/I$ is aimed at avoiding the detection of pre-BD as the true BD [180]. Considering a current step as a parameter, it is demonstrated that T_{BD} breakdown distributions measured on different area structures tend to follow the Poisson area scaling in comparison with the cases of pre-BD events being included [180].

As oxide thickness is further reduced down to around 1.5 nm, even individual defects can result in considerable leakage currents prior to breakdown [222]. Thus, even the SILC current due to a single defect can impact the total gate current in extremely small area devices. Furthermore, as stress voltage decreases, the breakdown spot after the formation of breakdown path, i.e., the first BD event, continues to wear out and the breakdown becomes progressive rather than instantaneous. In addition to the large background DT current, the contributions from these effects give rise to many artefacts in the current–time behavior and interfere with the detection of first BD [227]. The issue of breakdown detection becomes more pronounced due to the presence of NBTI defects. Figure 2.50 displays the typical time evolution of the gate current in a p^+ poly/PFET stressed at $V_G = -2.4\,V$.

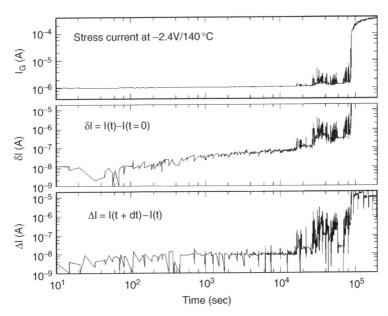

Figure 2.50. Stress current versus time (top). The change of stress current as a function of time in two different definitions: the global current step (δI) and the local current change (ΔI). PFETs of 1.4 nm oxides are stressed. Reprinted with permission from Ref. 226, © 2005 IEEE.

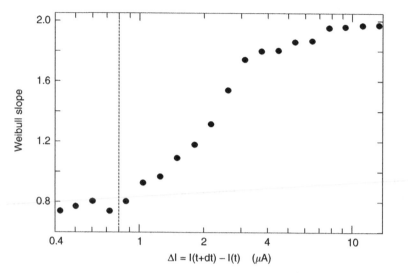

Figure 2.51. Weibull slopes of time-to-creation of leakage path (T_C) versus local current step ΔI for ultra-thin oxides. The dashed line at the onset of rising Weibull slopes is arbitrarily considered for the choice of the first BD event coincident with the formation of a percolation path. Reprinted with permission from Ref. 227, © 2004 IEEE.

At long times, the noisy nature of the stress current can be seen and makes the detection of BD difficult. A new method using the local current step, $\Delta I = I(t+dt)-I(t)$, is recently proposed for BD detection [227]. This method involves the extraction of a series of Weibull distributions for time to the creation of leakage paths (T_C) for all samples as a function of current step (ΔI). By running through a range of ΔI values, several time distributions, (T_C) can be obtained. The Weibull slopes of these time distributions as a function of ΔI values are given in Figure 2.51. It can be seen that at small ΔI values the Weibull slopes remain relatively flat with a plateau region, then they steadily increase with increasing ΔI values as moving into the post-BD phase as discussed in the subsection of progressive breakdown mode in Section 3.5.2. A plateau in β versus ΔI relation in the low ΔI limit is a figure of merit for first BD detection [227]. Specifically, the dashed line in Figure 2.51 between the plateau and the rising part of Weibull slopes is considered the choice for first BD definition. Unfortunately, in many cases this plateau is not always easily identified for ultra-thin oxides.

2.4 FUNDAMENTALS OF DIELECTRIC BREAKDOWN STATISTICS

2.4.1 Weibull Function and Poisson Statistics

The Weibull distribution was introduced in Section 1.4.7. Therefore only its main features are summarized in this section and its relation to Poisson statistics is highlighted. As shown in Section 2.3, oxide breakdown (BD) is marked by a discontinuity in the monitored stress parameter (current or voltage) as a result of the stress. This breakdown is due to increased conduction in a small localized spot and is usually believed to cause the failure of the circuit employing the broken FET or capacitor. The occurrence of oxide BD has a random nature and therefore requires a statistical description by means of a distribution. The choice of a lifetime distribution should not only be based on data fitting, but also on plausibility considerations such as properties and features. As will be shown in Section 2.4.1, distinguishing distributions based on the matching of data requires a large sample size. On the other hand, extrapolating to small, reliability-relevant cumulative failure percentiles causes significant errors if the wrong distribution is chosen. Hence, this section focuses on the properties of the Weibull distribution and the relation to Poisson statistics.

The Weibull distribution is generally accepted in literature as the distribution to describe dielectric breakdown. One reason is that the breakdown spot is very small, much smaller than the device area. In fact, a device can be thought of being composed of a large number of these small and independent subareas. Under stress all of these subareas compete for the weakest one to fail first, causing the entire device to fail. This represents the weakest link principle for which the Weibull distribution was constructed as a general case of the exponential distribution. Waloddi Weibull originally proposed a distribution with three parameters [228]. Since for lifetime assessment only positive times are of interest, the distribution can be reduced to two parameters. This form of the Weibull distribution is described by the following equation:

$$F(t_{BD}) = 1 - \exp\left(-\left(\frac{t_{BD}}{\tau}\right)^{\beta}\right),\qquad(2.27)$$

where $F(t_{BD})$ is the cumulative failure probability and t_{BD} is the time-to-break-down. The characteristic time-to-breakdown τ, is the time-to-breakdown at approximately the 63rd percentile (also called the time scale factor), and β is the Weibull shape factor (also called Weibull slope). It is also common in literature that the upper-case letters such as T_{BD} (or T_{63}) and Q_{BD} (or Q_{63}) are used for the characteristic time-to-breakdown and charge-to-breakdown at 63.2%, respectively. On the other hand, the lower-case letters, t or t_{BD}, (q or q_{BD}), are designated to represent the statistical variables of time-to-breakdown (charge-to-breakdown), except for the case of $t_{63.2}$ or t_{63} which refers to the characteristic time. In the Weibull model, a graphical representation of the data in a $\ln(-\ln(1-F))$ versus $\ln(t_{BD})$ yields a straight line. For this reason, Equation 2.27 may be rewritten to represent the Weibull scale as

$$\ln(-\ln(1 - F(t_{BD}))) = \beta \ln\left(\frac{t_{BD}}{\tau}\right).\qquad(2.28)$$

An important property of the Weibull distribution is that it describes the various failure rates (hazard rates) that occur during product lifetime, i.e., it can be used to express all cases described by the bathtub curve (see Section 1.3). This is a result of the shape factor, which causes the distribution to describe extrinsic, defect-driven, early fails for $\beta < 1$ with decreasing failure rate; random fails for $\beta = 1$ with constant failure rate (exponential distribution); and intrinsic fails or wearout for $\beta > 1$ with increasing failure rate. Thus bimodal distributions, which are often found for oxide breakdown, can be well modeled, as demonstrated in Section 2.4.4. Additionally the Weibull distribution can be derived directly from the failure rate with the postulate of a monotonic behavior and the exponential distribution as starting point. This is an independent derivation of the Weibull distribution and further justification because the failure rates of different distribution types exhibit a different time dependence, which can help to distinguish Weibull (that is, from a log-normal distribution) [229]. Another way to derive the Weibull distribution is based on the extreme value theory. It can be shown that the Weibull distribution is a special case of extreme value statistics, the type III of Gumbel's classification [230, 231]. This explains why the Weibull distribution works successfully for typical weakest link type mechanisms. If the Weibull distribution is an extreme value distribution then the stability postulate must be satisfied [229, 230] and by checking this property other distribution types can be excluded [229]. Besides the fact that the stability postulate was shown to be satisfied [230], it can be checked experimentally by grouping a very large sample size in smaller groups of equal size. The smallest breakdown time of each group represents the new breakdown time of that group. These new breakdown times should be distributed in the same way as the original distribution, but shifted to a lower characteristic time. In fact it has been shown that this requirement is satisfied for various oxide thicknesses [27, 229, 231, 232]. The details can be found in Section 2.4.1. Meeting the stability postulate strongly supports the choice of the Weibull distribution. It also means that the distribution is invariant and preserves its shape if subgroups are formed; in other words, a straight line in the Weibull plot remains a straight line but is shifted after forming groups. Such a grouping of samples is equivalent to testing and comparing samples with different areas. Thus, area scaling provides another important piece of evidence that the Weibull distribution is the appropriate distribution. For this purpose Poisson statistics are applied. The Poisson distribution is used to calculate the probability of an event with the mean number of occurrences m to occur exactly i times as described in Section 1.4.9:

$$P_i = \frac{m^i \times \exp(-m)}{i!} \tag{2.29}$$

For dielectric breakdown it is of interest to know the probability that a device of a certain area will not break down during a given stress (i.e., setting $i = 0$ in Eq. 2.29) which yields the probability $P_0 = \exp(-m)$. Thus the survival probability is defined by the mean number of occurrences m, which can be determined by subjecting several samples of the same type to the same stress. The number of failed samples n divided by the number of stressed samples N yields m if the first breakdown is taken into account only. To consider the area dependence the density of breakdown events D is defined as $D = n/(NA)$, where A represent the dielectric area of the tested device. Thus m becomes $m = n/N = n/(NA) \times A = D \times A$ and $P_0 = \exp(-m) = \exp(-D \times A)$. This represents a particular case of the Poisson distribution and is well known as the Poisson model, which is used for yield calculations in the semiconductor industry (see Section 2.3.4). The yield typically represents the fraction of survivors after a defined short stress. Different stress duration results in a different yield. This time dependence is considered by a more general description, the survival or reliability function $R(t)$ (see Section 1.4.5). The Poisson model relates the reliability function to the exponent of the negative device area A times the density of breakdown events $D(t)$. The survival probability at time t is related to the cumulative failure $F(t)$:

$$F(t) = 1 - R(t) = 1 - \exp(-A \times D(t)) \tag{2.30}$$

The above equation contains the stress duration, which is mostly neglected for yield modeling because the stress is the same for all samples. However, it is important to consider if the stress is not terminated at a fixed point, it instead lasts until breakdown occurs. Thus along a breakdown distribution the breakdown event density is different for each percentile. We can transform a Weibull distribution into a breakdown event density distribution by rearranging Equation 2.28 for the breakdown event density: $D(t) = -\ln(1-F(t))/A$. Taking the logarithm of both sides results in $\ln(D(t)) = \ln(-\ln(1-F(t))) + \ln(1/A)$. This means the logarithm of the event density is the same as the Weibull axis plus a shift that takes care of the area normalization (for an area of $1\,\text{cm}^2$ there is no difference at all,

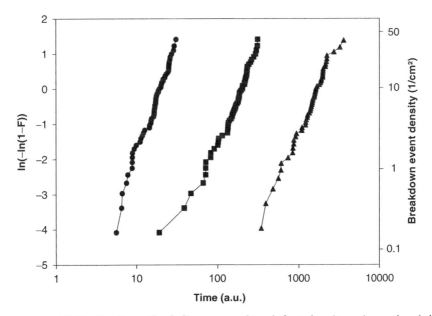

Figure 2.52. Weibull axis on the left compared to defect density axis on the right calculated from the relation given above for one area.

since the unit of the breakdown event density is $1/cm^2$). Replacing the Weibull axis by the event density axis results in no change of the shape of the distributions (Fig. 2.52) and demonstrates the compatibility of Weibull and Poisson statistics, although the underlying models of both are independent of each other. Since the distributions for different stress conditions are parallel it can be concluded that all events are accelerated by the stress in the same way; in other words, the breakdown event density remains the same at each percentile regardless of the stress condition. At this point it should be emphasized that the events are clearly due to defects if extrinsic distributions are discussed—instead of event density the term defect density is usually used. However, for intrinsic distributions, which represent mostly uniform oxides that degrade in a particular way as detailed in Section 3.2, the event density represent purely the statistical nature of oxide breakdown. To avoid confusion between the two different mechanisms the more general term breakdown event density is used.

From Equation 2.30 it is also obvious that for a given event density the cumulative failure depends on area. Increasing the area increases the probability that a faster event is contained in the area and for a given event density the yield for the same stress condition drops, i.e., the failure probability increases. Consequently, each point of the event-density distribution shifts parallel to the Weibull axis. In Figure 2.53 the distributions of three different areas are shown. However, they appear shifted along the time axis rather than shifted along the Weibull axis as expected. This is due to the fact that each of these distributions represents a different event density range, since we usually do not select the samples for the different areas to represent the same event density range. Hence samples with lower event density are added, which represent faster events shifting the distribution of the larger area to shorter breakdown times. If the area difference is not too large, some higher percentiles of the larger area will have the same breakdown times as some lower percentiles of the smaller area. In this overlapping region the event densities are equal as demonstrated in the breakdown event density plot (Fig. 2.53). The distributions of the different areas are shifted onto the same line and overlay where the event density is the same (Fig. 2.53). Deviations from the straight line are, in this case, purely due to statistical variations and can be minimized by further increasing the sample size. The point is that for different areas, a constant event density lies on a vertical line in the Weibull plot (i.e., the same event density causes always the same failure time at a given stress condition) while for a constant area a constant event density is found on a horizontal line (i.e., it causes always the same failure percentile). Based on this the area scaling can be derived from Equation 2.30 by relating the cumulative failure for a given time and different area. The amount of shift caused by the areas A_1 and A_2 where $A_2 = n \times A_1$ is

$$\ln(-\ln(1 - F_{A_1})) - \ln(-\ln(1 - F_{A_2})) = \ln\left(\frac{A_1}{A_2}\right) = \ln\left(\frac{1}{n}\right) \qquad (2.31)$$

This equation provides an important means to verify whether any two distributions obtained for two different areas obey Poisson area scaling law

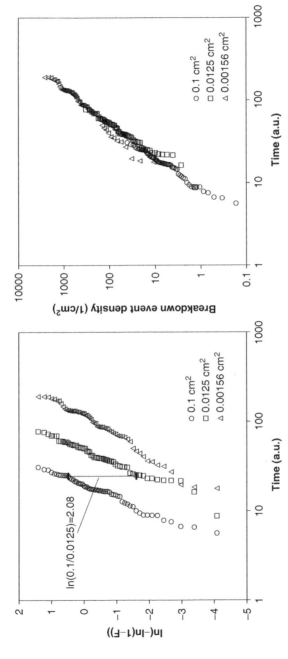

Figure 2.53. Weibull distributions for three different areas and Poisson area scaling on the left compared to defect density distributions of the same data on the right. The shape of the distributions is again the same in both plots, but the area normalization shifts the distributions onto the same line. The smallest area reaches the highest defect density.

without knowing the distribution parameters (characteristic time and slope) and without knowing the event density. The Poisson model is independent of the underlying cumulative failure distribution of the data and can be applied to Weibull and log-normal distributed data in the same manner. However, the Poisson model yields the same result as the formation of subgroups to verify the validity of the stability postulate, hence supporting the Weibull distribution as the appropriate distribution for dielectric breakdown.

For area scaling the Poisson model is applied to each failure percentile regardless of the underlying cumulative failure distribution. However, only in the Weibull net is the shape of the distribution preserved. This demonstrates that the Poisson model can be applied for area scaling of Weibull or extreme value distributions; at the same time, it supports the use of Weibull distributions for oxide reliability assessment. A detailed discussion of Weibull versus log-normal distribution is given in Section 2.4.1, while the area scaling is treated in detail in the next section.

2.4.2   Area Transformation

The area transformation is an important topic for product relevant lifetime prediction. The data on which the projection is based are usually taken on test structures that are (much) smaller than the dielectric area of a VLSI product that is typically comprised of a huge number of individual transistors. Thus an area transformation from the test area to the product area is unavoidable. The area transformation affects the probability to get a fail within a certain time with the failure probability being higher in a larger dielectric area. In other words if samples with a small dielectric area and samples with a large dielectric area are stressed at the same condition, at any given time the smaller area has a lower cumulative failure F than the larger area as shown in the previous section. The Poisson model derived in the previous section is generally accepted for area transformation of intrinsic distributions, but it may also be applied to extrinsic breakdown. This section adds yield models and area transformation of extrinsic distributions but also discusses more details of intrinsic area transformation.

Extrinsic fails are caused by weaknesses in the dielectric. Those weak spots are called defects in yield analysis because they represent unwanted process irregularities that cause a chip to fail at operation condition during initial tests or in early life. For reliability analysis, weaker spots are also called defects as they represent clearly localized spots that are less reliable than other spots in the same dielectric, which is why the fail occurs there (see Section 2.3.1). Since the concept of yield projection [233–236] is a bit easier to understand and represents the historical development, it shall be used here to illustrate the basics of area transformation. The simplification is that a defect is defined by a chip failing during a defined test; no time dependence is considered. Thus the yield is the ratio of functioning chips divided by the total number of tested chips. From the yield, a defect density can be calculated if the relation between yield and defect density is known. This relation depends on the distribution of the defects. Assuming randomly (= uniformly) distributed defects the yield Y is described by the Poisson model [233–238] as derived in previous section ($Y = \exp(-D \times A)$), with D representing the defect density or for intrinsic distributions the breakdown event density and A the area. Initially the area was the total chip area, which was later refined and now A represents the critical area, in this case the dielectric area in the critical circuit(s) of the chip. The area could be even more restricted, that is, the area of one device type (e.g., the NFETs) in a critical circuit, if this device experiences the highest stress during product operation. Sometimes a high stress is compensated by a small area, in this case a device with more relaxed stress conditions but large area could be more critical.

Early on it was observed that the yield did not follow the exponential law, but was much better than predicted by the Poisson model. This was explained by defects that were not uniformly distributed but vary over the wafer and tend to cluster, i.e., it is very likely to find other defects close to a randomly chosen defect and on the other hand larger areas without defects. A general model considering the distribution of defects was proposed by Murphy [239]:

$$Y = \int_0^{\infty} \exp(-D \times A) \times f(D)dD \tag{2.32}$$

with $f(D)$ representing the distribution function of the defect density. Several functions have been used, e.g., a delta function yields the Poisson model as the most conservative model. An exponential function was used by Seeds leading to an optimistic model that represents strong clustering of defects [240]

$$Y = \frac{1}{1 + A \times D}. \tag{2.33}$$

Even more optimistic results are achieved by employing a rectangular or a triangular function for the defect density distribution [233–236] in Equation 2.32. The use of a gamma function [241] leads to negative binomial yield model also called the Stapper model [233–236]. It is a flexible model, which by adjustment of the cluster parameter α is able to describe all of the above models closely.

$$Y = \frac{1}{\left(1 + \frac{A \times D}{\alpha}\right)^{\alpha}} \tag{2.34}$$

The only requirement for the value α is that it must be larger than zero. For α approaching infinity, i.e., randomly distributed defects (no clustering), the Poisson model determines the yield as a lower limit for the Stapper model. With decreasing α, i.e., increasing clustering of defects, the yields becomes more optimistic than predicted by the Poisson model. For $\alpha = 1$ the Seeds model results and for certain values of $\alpha < 1$ the defect density distributions based on the rectangular or the triangular function can be approximated. Therefore the Stapper model is widely used for yield modeling. For reliability projection, the cumulative failure probability F (not the yield) is utilized, which at a given condition is linked to the yield Y at this very condition by $F = 1 - Y$. According to the Poisson model the cumulative failure probability F_B for the area A_B is related to the cumulative failure probability F_T of the test structure with area A_T for any fixed defect density by [27]:

$$F_B = 1 - (1 - F_T)^{\frac{A_B}{A_T}} \tag{2.35}$$

The straight line of a pure Weibull distribution is shifted parallel to the Weibull axis of the Weibull net. The amount of the shift is the logarithm of the area ratio as demonstrated by [152]:

$$\ln(-\ln(1 - F_B)) - \ln(-\ln(1 - F_T)) = \ln\left(\frac{A_B}{A_T}\right) \tag{2.36}$$

From Equations 2.35 and 2.36 it is obvious that the defect density is assumed to remain the same for the different areas and thus cancels out. This assumption is valid only if the different areas were stressed at the same stress condition, i.e., same voltage, same temperature, and same stress duration. Consequently, different cumulative failure probabilities of the same distribution have different defect densities associated with them according to $D = -\ln(Y)/A = -\ln(1-F)/A$, as derived from Poisson model. For higher cumulative failure percentages the defect density is larger due to less severe defects that are present in higher density, causing failure only after extended stress duration. The physical nature of these defects was clear for yield modeling and could be verified easily by physical failure analysis, since yield modeling dealt with extrinsic, defect-driven fails. Based on this experience bimodal failure distributions observed in dielectric reliability were interpreted as a distribution of local oxide thinnings originating from small, local disturbances of the oxide like surface asperities, local barrier lowering, or a locally increased trap generation rate [242]. Later it was reported that intrinsic oxide breakdown is described by a percolation model [29–32, 243] (see Section 3.2.3). According to this model during stress local, conducting connections between the electrodes are generated randomly throughout the oxide area. Breakdown may occur as soon as such a connection has formed anywhere in the stressed area. The statistics of intrinsic breakdown is determined by the probability of such a connection being formed [29–32, 243]. This probability increases for the same oxide thickness with increasing area. Using the early concept of the percolation model [29] it was shown that for intrinsic breakdown, area scaling is well described by the Poisson model [244]. It is generally accepted that the Poisson model is the appropriate and only model for area transformation of intrinsic failure distributions, but it is not restricted to intrinsic distributions.

The area transformation according to Equations 2.33 or 2.34 is independent of any probability distribution (e.g., log-normal or Weibull) and is applied to each cumulative failure percentage F of the distribution. Thus Equation 2.36 can also be used for a raw check whether collected data from two different areas (stressed at the same conditions) follow Poisson area scaling, without knowing the

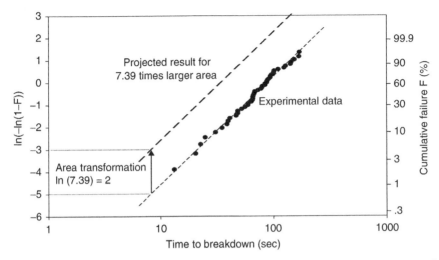

Figure 2.54. Area transformation of an intrinsic breakdown distribution by application of the Poisson model. The shift of the distribution is upward, parallel to the Weibull axis by two units. The shape of the distribution (here a straight line) is maintained only for area transformation using the Poisson model in the Weibull net.

distribution specific parameters. Further examples for application of area transformation are given in Section 2.4.3.

As a practical example for area scaling the Poisson model is applied to data shown in Figure 2.54. The data represent a Weibull distribution as indicated by the straight line. The area transformation to a larger area shifts this line upward, parallel to the Weibull axis. The amount of the shift is calculated by Equation 2.34. A special feature of area transformation in the Weibull net by means of the Poisson model is that the shape of the distribution is maintained, regardless whether it is a straight line (as in this case) or a bimodal distribution due to a mix of intrinsic and extrinsic fails. Due to the preservation of a straight line the area transformation of intrinsic Weibull distributions can also be done along the time axis. By inserting the equation for the Weibull distribution $F = 1 - \exp\left(-\left(\frac{t_{BD}}{\tau}\right)^{\beta}\right)$ in Equation 2.36, it can be shown for every cumulative failure percentile that

$$t_B = t_T \times \left(\frac{A_T}{A_B}\right)^{\frac{1}{\beta}} \tag{2.37}$$

where t_T and t_B are the times to breakdown at the same cumulative failure percentage for the test structure and the new area, respectively. t_B decreases with increasing area as shown in Figure 2.53 already. The exponent is inversely proportional to the Weibull slope. Therefore the actual impact of the area dependence of time-to-breakdown depends on the value of the Weibull slope.

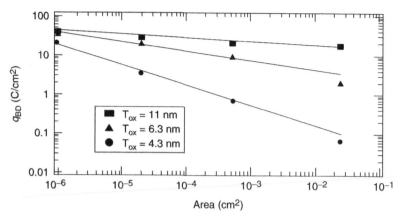

Figure 2.55. Area dependence of Q_{BD} measured for three oxide thicknesses of 4.3, 6.3, and 11 nm. The power law area dependence can be clearly observed. The slope of these lines represents the inverse Weibull slope. Reprinted with permission from Ref. 243, © 1998 IEEE.

Equation 2.37 is also called the power law of area scaling, and it is used in Section 2.4.3 to determine the Weibull slope β of intrinsic Weibull distributions, a possibility that was reported early on [229]. So far the time-to-breakdown (t_{BD}) was used in connection with the Weibull distribution and area scaling, but the same is equally applicable when charge-to-breakdown (q_{BD}) is chosen as the breakdown variable (instead of t_{BD}). Figure 2.55 displays the characteristic Q_{BD} as a function of area, allowing a clear observation of the power-law area dependence of Q_{BD} [30].

The other area transformation models are needed for special cases, when dealing with extrinsic failure distributions. Analogous to the Poisson model, the Stapper model can be derived from the corresponding yield projection model in Equation 2.32 [27]:

$$F_B = 1 - \left(1 + \frac{A_B}{A_T} \times ((1 - F_T)^{-\frac{1}{\alpha}} - 1)\right)^{-\alpha}, \tag{2.38}$$

where for $\alpha = 1$ (severe clustering of defects) yields the Seeds model

$$F_B = 1 - \left(1 + \frac{A_B}{A_T} \times ((1 - F_T)^{-1} - 1)\right)^{-1} \tag{2.39}$$

The effect of Equations 2.38 and 2.39 are discussed using Figure 2.56, which compares the Poisson model, the Stapper, and the Seeds model. The Seeds model clearly changes the shape of the originally straight line due to the reduced effect of area transformation at high cumulative failure percentage as a result of strong clustering. The Stapper model covers a wide range of different clustering severity.

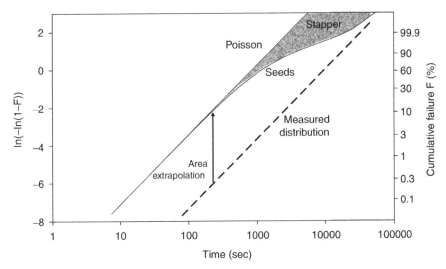

Figure 2.56. Comparison of different area transformation models in the Weibull net. The Poisson model maintains the shape of the distribution, while other models do not. At high cumulative failure percentage the effect of area transformation can be much less than expected from the Poisson model. The Stapper model covers the entire range between Poisson and Seeds model. It can even extend below the Seeds model for extreme clustering.

For less clustering the transformed distribution will be closer to the Poisson model, while for extreme clustering the transformed distribution will coincide with the Seeds model. Since all models are, according to Section 2.3.4, based on a random defect density distribution, they all merge for small cumulative failure percentages with the Poisson model. Thus below 10% the differences between the models become negligible. This means that only for an area transformation for which the resulting cumulative failure percentage is more than $\sim 10\%$ it is necessary to distinguish between the models and to select the correct one. Examples based on experimental data have been reported [27, 242]. Since the targets for products are usually well below 0.1% the Poisson model works in most cases. If area transformations of extrinsic distributions end up in the cumulative failure range in which the Poisson model may be too conservative, then it is necessary to determine the cluster factor α from the area scaling of different areas [27, 238] and chose the model accordingly. For this purpose the distributions of the smaller areas are best transformed to the largest area that has been stressed and exhibits an extrinsic distribution well above 10%. This prevents a reduction of the possible deviation that may occur if all areas are transformed to an even larger, virtual area. If the largest stressed area has an extrinsic distribution below 10%, a combination of areas on the same chip may help to increase the area [27, 238]. In this case, the new area could be the sum of all stressed structures on one chip and the breakdown time is the shortest breakdown time measured on any one of these structures. This results in one new and large area for each chip on a wafer. This method was originally demonstrated to prove the validity of the extreme value statistics for dielectric breakdown [229]. Another circumstance that might contribute to a wrong conclusion regarding clustering is if the Weibull axis spans a very large range, perhaps due to the intrinsic part of distributions from small areas shifting to very high values on the Weibull axis. Fixing the range of the Weibull axis to a regular range (e.g., from -4 to 2) supports distinguishing differences.

Clustering of defects or weak spots should not be confused with systematic, technology-driven weaknesses. A systematic weakness could be a process-induced thinning at the isolation edge, which causes the test structure and the product to fail preferentially at this location instead of randomly anywhere in the test structure. The degree of clustering is not identifiable by the failure location within the test structure, but only by thorough area scaling as described above or physical defect visualization across the wafer. The reason for clustering may be particles being present more densely in some regions of the wafer or even a scratch. Any kind of extrinsic dielectric defect distributed across the wafer nonuniformly causes clustering.

2.4.3 Weibull Versus Lognormal Failure Distributions

Dielectric breakdown is well known to have the characteristics of the weakest link property and Poisson area scaling consistency. In particular, the area scaling property of the intrinsic mode of oxide breakdown follows Poisson statistics as already discussed in Section 2.3.3. Experimentally, the demonstration of the weakest link characteristics of dielectric breakdown is achieved by the subgrouping technique [27, 231, 232] as discussed below. Figure 2.57 shows three t_{BD} distributions corresponding to three areas of 10^{-2}, 10^{-3}, and 10^{-4} cm^2, respectively. The corresponding large sample sizes are ~ 300, ~ 2000, and ~ 4000, respectively. The other two t_{BD} distributions denoted by open symbols were obtained by selecting the minimum t_{BD} drawn from subgroups of equal size. The combination of each subgroup was done by randomly dividing the total population into a certain number of subgroups for a given area. For example, for the case of 4000 capacitors of 10^{-4} cm^2, 40 subgroups of 100 such capacitors are chosen resulting in a total area of 10^{-2} cm^2. As seen in Figure 2.57, the t_{BD} distributions of the combined subgroups show remarkable agreement with those of the larger capacitors with the same areas. These results demonstrate the validity of mathematical combination of sub-area structures as well as the feasibility of using large area structures for reliability evaluation if the series resistance effect and problems detecting the breakdown event can be eliminated. Moreover, these results provided experimental evidence that intrinsic breakdown defects are not only randomly distributed according to the Poisson statistics but also have the character of the extreme-value distribution, the so-called weakest link property. Notice that these t_{BD} distributions for three area capacitors show parallel Weibull slopes. This important result justifies the assumption of area-independent Weibull slopes.

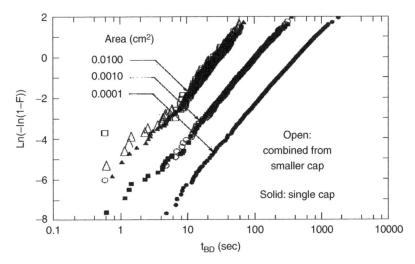

Figure 2.57. Comparison of t_{BD} distributions with different area capacitors. The solid symbols represent t_{BD} of individual capacitors. The open circles show the minimum t_{BD} distributions combined from the smaller capacitors with 10^{-4} cm^2 to get 10^{-3} cm^2. The open squares and triangles stand for t_{BD} of the combined capacitors with 10^{-4} and 10^{-3} cm^2 to obtain an area of 10^{-2} cm^2, respectively. The stress voltage is 4.0 V under substrate injection for p$^+$-poly/PFET in accumulation. Reprinted with permission from Ref. 202, © 2002 IEEE.

Having demonstrated the weakest link characteristic and area dependence of oxide breakdown, we will discuss which failure distribution is suitable for oxide breakdown. Historically, both Weibull and lognormal distributions have been widely used in the literature. It is often argued on practical grounds that both distributions can fit the data. The use of lognormal distribution can still be found in practice in recent years [245], since this distribution can provide much enhanced reliability margin as compared to the Weibull distribution. It was argued that when additional random variation is included, the distribution tends to follow the log-normal function [245].

It was shown that the asymptotic form of the distribution function with the weakest link character is independent of the exact form of the cumulative distribution [230, 246]. Therefore, there is no unique solution for the choice of the cumulative distribution functions at the first sight. A direct demonstration of Weibull distribution for thin oxide breakdown is given in Figure 2.58 for an oxide thickness of 2.67 nm with a standard deviation of 0.0035 nm. It is evident from Figure 2.58 that the log-normal function can hardly fit the data, while the Weibull function excellently matches the data across the whole range with a sample size of about 900 identical capacitors. The large sample size is required because both Weibull and lognormal functions can always fit the data with very limited samples. More importantly, this demonstrates that when the variation of oxide thickness is eliminated, time-to-breakdown is not only an independent random variable but also

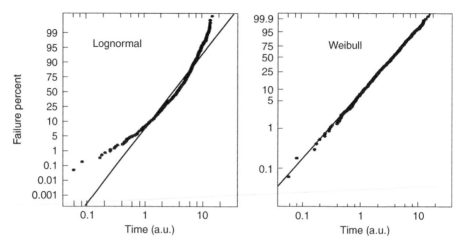

Figure 2.58. Comparison of lognormal and Weibull fit to experimental data. This large sample size of stressed capacitors is selected from a large group of more than 4000 capacitors to obtain high oxide thickness uniformity with a standard deviation of about 0.0035 nm. The capacitors with p$^+$poly gate and n-type substrate were stressed in substrate injection with a gate voltage of 4 V at 140°C. Reprinted with permission from Ref. 28, Copyright 2000, Institute of Physics.

is statistically distributed consistent with the Weibull plot. Thus, the argument that a log-normal distribution can fit a t_{BD} distribution from which thickness or process variation are not eliminated is merely made on practical grounds and fundamentally incorrect despite large sample size. In the presence of such additional random variations, the appropriate treatment of resultant distributions with several statistical variables should be considered as discussed in Section 2.4.4.

As demonstrated above, the Weibull distribution is compatible with Poisson area scaling. In addition Hunter [247] illustrated, that the lognormal distribution is not invariant when Poisson area transformation is applied as shown in Figure 2.59. In other words, if the distribution is a lognormal for one particular area (the straight line in lognormal plot), it would not be a lognormal distribution for other areas (the curved lines in Fig. 2.59). Thus, to be compatible with Poisson area effect, the lognormal distribution must be rejected as the choice to describe breakdown statistical distribution. This again shows that a simple fit to data is not sufficient to select the appropriate distribution. Other indicators and evidence need to be considered. Neglecting a thorough and comprehensive analysis may result in risky predictions.

2.4.4 Estimation of Weibull Parameters

The title of this section already indicates that the Weibull parameters are usually not exactly determined, but due to the statistical nature of dielectric breakdown,

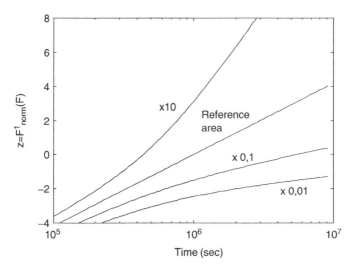

Figure 2.59. The effect of Poisson area transformation on a lognormal distribution. The distribution shows a straight line in the lognormal plot for the reference area, A_{ref}. The other distributions correspond to structures of area $10A_{ref}$, $0.1A_{ref}$, and $0.01A_{ref}$ and are calculated according to $F_2 = 1 - (1 - F_{ref})^{A_2/A_{ref}}$.

estimated based on a limited sample size. This means that there is a certain probability of getting parameters that are different from the true but unknown population parameters. There are several ways to estimate the parameters of the Weibull distribution. The preferred method depends on many factors. These factors include the typical sample size used, whether censoring is involved and if so to what degree, the availability of appropriate statistics software, the degree of accuracy desired, and the depth of statistical considerations applied. Although the trend is clearly towards application of high quality statistics software packages, which deliver the desired parameters as well as the confidence bounds with and without censoring or grouping of data, this section attempts to give a brief overview of what has been or could be done to estimate the Weibull parameters and to give an idea of the complexity involved in the evaluation. Even if the statistical software packages will be used, it is important for the reliability engineer to have some understanding of the calculations done in the package.

First, all of the collected data is plotted in the Weibull net for visualization. This provides a lot of information at one glance, e.g., whether the results are well behaved and the samples have no obvious problems, and whether there are one or multiple failure modes. For such graphical representation, the breakdown times need to be sorted in ascending order and a rank (1, 2, 3...n, where n is the total count of the samples) assigned to each time from which the cumulative failure probability F is determined. Naturally for the first value this plotting position would be $F_1 = 1/n$, the second $F_2 = 2/n$ until the last sample with the highest failure time is reached with $F_n = n/n = 1$. The failure probability of 100% cannot be plotted in the Weibull net and is therefore omitted. This is unsatisfactory, because it means for a small sample size that one data point is missing and since this is the last data point, the longest time was invested to acquire it. Alternately, Gumbel [230] proposed to omit the first data point and plot the last value at $(n-1)/n$. This is also not satisfactory, because the first data point may contain important information, e.g., for dielectrics it may be important for a defect density calculation. Therefore, Gumbel proposed to calculate the probability as $F(t) = i/(n+1)$ with the rank $i = 1...n$. For large n this yields approximately the same as $F(t) = i/n$. Other authors prefer $F(t) = (i-3/8)/(n+1/4)$, or $F(t) = (i-0.5)/n$ or $F(t) = (i-0.5)/(n+0.25)$ (compare also Section 1.4.4). A widely used estimator nowadays is the median rank, which can be closely approximated by $F(t) = (i-0.3)/(n+0.4)$ and is also known as the Bernard estimator [248, 249]. The reader is also referred to Section 1.4.4 and Figure 1.5 for further discussion on plotting positions. Three different methods to determine $F(t)$ for the same set of breakdown times are compared in Figure 2.60. Obviously the tails of the distribution are most affected, especially the lowest percentiles, while the highest percentiles are less affected but the missing data point for the first method is clearly noticed. Results from linear regression indicate that for the different methods, the regression coefficient improves slightly in this case and the slope β becomes noticeably steeper due to the downshift of the lower percentiles. This means that $F(t) = i/n$ may be the

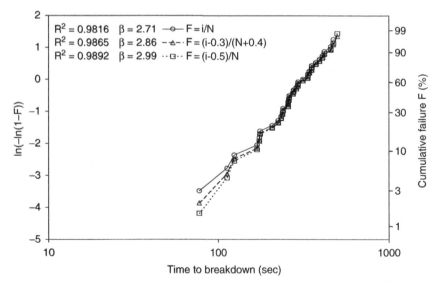

Figure 2.60. Comparison of three different methods for calculating the cumulative failure probability for the same breakdown times. The legend provides results from linear regression that reflect the effect of the downshift of the lowest percentiles.

most conservative method to determine the Weibull slope by linear regression and may compensate for the too optimistic slope reported in studies comparing estimation methods (e.g., [249]). As in those studies, Monte Carlo simulations can be employed to verify which method yields the most accurate results in combination with the parameter estimation method chosen.

Once $F(t)$ is determined, it can be plotted versus the breakdown times using a standard software package. The time zero fails—if not screened out—and the fails at first readout, which typically do not have an exact breakdown time, are usually not plotted. They are contained in the rank number of the first valid breakdown time and hence in $F(t)$, which is how they contribute to the statistics. Samples that did not fail during the stress are treated analogously, i.e., they are not plotted but considered in the total sample size, resulting in a distribution that does not reach the same high cumulative failure percentage as another distribution with the same sample size and 100% fails. From such a plot, the Weibull parameters can be estimated graphically or the data can be evaluated by mathematical procedures.

For some evaluation methods it is useful to exclude data in the distribution tails if they deviate significantly from a straight line and cause an obviously shallower or a steeper fit to the data and if there is an experimental or process reason for excluding those fails. The reason for those excluded fails could be the presence of some extrinsic fails or thickness variations across the samples. One method to extract the distribution parameters is the linear regression as already mentioned (see also Section 1.6.3). It provides a fit to the data that is most consistent with a fit done by eye. This method is well established and yields reasonable results, especially for large and complete sample sizes. It has been shown that it makes a difference, whether the $\ln(-\ln(1-F))$ values are regressed on the times or vice versa [249]. In that work, it was shown that based on Monte Carlo simulations, it is better for the slope parameter to regress $\ln(-\ln(1-F))$ on the times as usually done, while $t_{63.2}$ is more accurate regressing the times on $\ln(-\ln(1-F))$. For large Weibull slopes (>5) there is an almost insignificant advantage for the $t_{63.2}$ estimation by regressing $\ln(-\ln(1-F))$ on the times (or t_{BD}).

A further improved estimate can be achieved by multiple linear regression. This method evaluates the data of distributions from different stress conditions or different areas simultaneously and thus considers more data for the estimate and averages variations. For multiple linear regression it is necessary to know or assume a model that connects the distributions that are to be considered. An often preferred alternative is the maximum likelihood estimate (MLE) (see also Section 1.6.5). This is an iterative method to determine from the data, the parameters that are more likely than other parameters to describe the underlying distribution. One advantage of the maximum likelihood method is that it deals with the breakdown times only and does not rely on the cumulative failure F. The maximum likelihood function is the product of the Weibull probability density function over all samples involved. For more convenient calculation, the natural logarithm of the maximum likelihood function is taken. Then, in order to find the maximum, the first derivative of this equation is obtained by differentiation with respect to $t_{63.2}$ and the slope, β. The resulting equations are equated to zero. After the elimination of $t_{63.2}$ between the two equations and simplification, an equation results that can be solved for β numerically by applying the Newton–Raphson procedure. With this β the $t_{63.2}$ is then estimated by the other equation derived before and rearranged for $t_{63.2}$. The maximum likelihood method requires good starting values, which may be obtained by linear regression. The starting values can influence the results or cause the method to not converge. Also for censored data and bimodal distributions, the math becomes more complex. In addition the MLE of β is biased and requires an unbiasing method [249]. Therefore it is useful to take advantage of high quality statistics software packages. For this reason the reader is referred to text books and literature (e.g., [249–253]) if more details on the MLE method are desired. The MLE method can also be extended to simultaneously consider data from different stress conditions or areas, as in the case of multiple regression.

Comparisons of various estimation methods using Monte Carlo simulations are reported [249, 252]. For small sample sizes and severe censoring, both the least squares and the maximum likelihood method performed poorly [249]. The Bain–Engelhardt estimator and the White estimator, a weighted version of the least square method, outperformed the other techniques [249]. Thus a sample size of more than 30 samples per distribution is mandatory for the usual estimation methods for intrinsic distributions. The sample size needs to be considerably larger for thinner dielectrics with shallower Weibull slopes (see Section 2.4.3) and particularly for extrinsic distributions for which only the failing devices count in the effective sample size. However, using an advanced, high quality software package may provide a competitive advantage with respect to accuracy and sample size. Another possibility to improve the accuracy of the estimated value for the Weibull slope in particular is described in Section 2.4.3.

The estimates from high quality statistics software usually come with confidence intervals. These intervals for $t_{63.2}$ and β indicate the range around the estimated parameter which includes with a certain probability, the true parameter of the sampled population. The confidence level defines the probability at which the true parameter is covered by the confidence interval. The confidence interval is represented by the lower and upper limit, e.g., $t_{63.2 \ low} < t_{63.2} < t_{63.2 \ up}$ and calculated based on the scatter of the data used for the estimate. Other methods determine confidence bounds based on statistical considerations accounting for sample size only as per Section 1.6.3, and cover in many cases a wider interval than the confidence interval that considers the scatter of the analyzed data.

In order to get an idea of what the estimate means, the confidence bounds should be calculated and plotted with the data and the distribution that result from the estimate. The confidence bounds cover with a predefined probability (the confidence level, e.g., 95%) the true but unknown distribution. This means that there is a limited probability of estimating the true distribution parameters based on the data acquired. If sets of data are repeatedly taken at the same stress condition, the extracted parameters vary within a certain range, i.e., they scatter around the true value. Usually only one data set of a given stress condition is taken and for that reason it cannot be expected to always represent the true distribution, instead it will be somewhere close to the real distribution. Therefore, defining a range around the data, which covers the true distribution with a certain degree of confidence, the confidence interval, is very instructive. This can be illustrated by means of the Monte Carlo simulation. A Monte Carlo simulation involves a random number generator that generates a number between 0 and 1, with the same probability for each number. This number is interpreted as the cumulative probability in a breakdown experiment. From this number a break-down time is calculated using the Weibull equation with predefined parameters for $t63$ and slope. This calculated breakdown time represents the result of one stressed sample while the random number is discarded. For each sample a new random number is generated and converted to a time. Once the times for the predefined number of samples are generated they are sorted and evaluated as described above, i.e., a cumulative failure is assigned to each time according to its rank. This yields one distribution. The experiment can be repeated and another distribution can be generated and so on. As an example some distributions from Monte Carlo simulation are shown in Figure 2.61.

Although all distributions were based on the same parameters used for the straight dashed line the variation is significant. Some distributions are shifted towards lower or higher times, while others are steeper or shallower than the true distribution. Some distributions even seem to have extrinsic fails, which in fact belong to a very small percentage of the intrinsic distribution for a huge sample size. For the lowest percentages of the simulated distributions, the spread is the

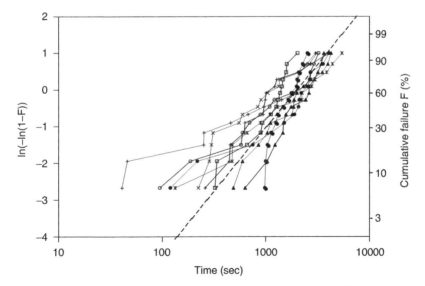

Figure 2.61. Weibull plot of a bunch of distributions generated by Monte Carlo simulation for the parameters represented by the straight, dashed line. Due to statistical variations the generated distributions can look very different from the real distribution. At high and low percentages the spread is larger than in the 50–60% range. Reprinted with permission from Ref. 202, © 2002 IEEE.

largest. Although this is not clearly seen, the spread is also larger for the highest percentages while it is the smallest in the 50–65% range, i.e., more precisely at the characteristic life of the Weibull distribution (at 63.2%). Confidence bounds around each of these distributions cover the original distribution with a certain probability defined by the confidence interval. Since a fixed sample size is used for each distribution, the frequency for the times obtained from many thousands of such experiments can be plotted for each cumulative percentage $F(t)$. The result is a frequency distribution showing the probability to get higher or lower breakdown times for a given cumulative failure percentage. The area covered by this frequency distribution represents 100% fails and by reducing this area to a lower percentage of fails by removing the low and high tails confidence bounds may be constructed. In fact a Monte Carlo simulation is one well known possibility to determine confidence limits. Comparison of different methods to determine the confidence bounds shows that generally not the same limits result and not exactly the same range is covered [248]. This is demonstrated in Figure 2.62 for three simple methods.

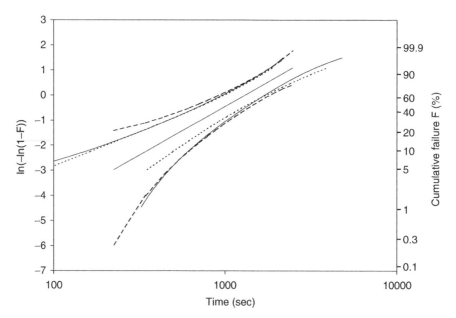

Figure 2.62. Comparison of three methods to determine confidence bounds for the distribution represented by the straight line based on 20 samples. The dashed lines represent a method that calculates the upper and lower confidence bounds for each $F(t)$, while the full [254, 255] and the dotted [248] lines represent methods calculating the upper and lower confidence bounds for $t(F)$. The full lines use a method that allows the extension of the confidence bounds beyond the range covered by the data. The methods yield different bounds, but they all get the point across that the uncertainty is considerable and significantly increases at high and low percentages.

These methods are randomly picked and do not imply any preference. The first method was described by J. Jacquelin [254, 255] and is based on the scattering of sampled data. The second method was originally proposed by Bain and Engelhardt [248]. While the third method is based on the F-distribution (Fd) with the upper limit of $F(t)$ being $F_u(t) = \frac{(r+1) \times Fd}{n-r+(r+1) \times Fd}$ and the lower limit $F_l(t) = \frac{r}{r+(n-r+1) \times Fd}$, where n is the total sample size and r is the rank of $F(t)$ for which the confidence bound is calculated. For the F-distribution, the two degrees of freedom (DF) for the upper bound are $DF_1 = 2 \times (r+1)$ and $DF_2 = 2 \times (n-r)$ and for the lower bound $DF_1 = 2 \times (n-r+1)$ and $DF_2 = 2r$. The last method determines the confidence bounds for the cumulative failure percentage, while the other two methods determine the confidence bounds for the time of each failure percentage, which can be recognized from the way the end points of the confidence bounds are aligned to the distribution (Fig. 2.62). An exception is the Jacquelin method [254, 255], which is not restricted to the sample size, but allows one to draw the confidence bounds beyond the range covered by the data. It is obvious from Figure 2.62 that the confidence bounds are somewhat different for the three methods. Nevertheless they illustrate the uncertainty of an estimate and the increasing uncertainty at high and low percentages. The true distribution can run any way completely or partially through the confidence interval, which makes extrapolation to small failure percentages undesirable, due to the large possible error. A 90% confidence interval means that there is a 90% chance that the true distribution is covered by the confidence interval. In other words, there is a 5% probability that the estimate is too optimistic and a 5% probability that the estimate is too pessimistic. Sometimes it is attempted to tighten the confidence interval by using lower confidence limits, e.g., 60% instead of 90% or higher. As shown in Figure 2.63 (a) the confidence bounds become tighter by lowering the confidence limits, but at the same time the risk of the confidence interval not covering the true distribution is already increased to 40%, which is not a solution to improve accuracy. Therefore it is necessary to adjust the sample size to get an acceptable accuracy, balanced with stress time and cost. Figure 2.63 (b) illustrates the effect of sample size on 95% confidence bounds. A high degree of accuracy requires huge sample sizes, which unfortunately is not very practical. From these examples it is obvious that the estimate for $t_{63.2}$ is more accurate than the estimate of the Weibull slope β. Hence, improving the accuracy of β is most desirable and an alternative method to accomplish this is described in Section 2.4.3. It is recommended that the reader compare this discussion to that of Section 1.4.5 where the variation of a normal distribution is compared both in terms of statistical variation and the impact of sample size. Also, the reader is reminded that in both cases we are speaking about the statistical variation in the sampled data. Both here and in Section 1.4.5, we have no choice but to assume that the sample is representative of the population of interest. Obviously repeated measurements decrease the risk of that assumption being incorrect. The variations of the estimated $t_{63.2}$ and β for small sample sizes are likely to introduce serious errors in reliability projections, as discussed by several authors in literature [154, 159, 202, 256, 257].

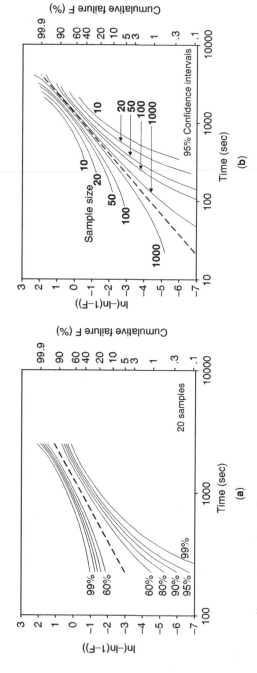

Figure 2.63. Illustration of confidence intervals and sample size relation: (a) different confidence limits for a sample size of 20 and a typical thin oxide distribution with Weibull slope of 1.7; (b) different sample sizes and their 95% confidence bounds for the same Weibull distribution as in (a). Reprinted with permission from Ref. 202, © 2002 IEEE.

The confidence bounds may also be helpful when comparing two distributions. Two distributions are most likely different if their confidence bounds—using a high confidence level—do not overlap.

2.4.5 Methods for Determination of the Weibull Shape Factor (Slope)

Obviously, the first method is to directly determine the Weibull slope from the measured distribution. A good example is already given in Figure 2.58(b) using a large sample size of ~ 900 and very tight thickness variation. A very tight and uniform distribution can be clearly seen with such a large sample number resulting in a Weibull slope of 1.65 ± 0.08. As discussed in the previous section, the accuracy of the Weibull slope continuously improves with increasing sample size even if any T_{OX} variation is excluded in the first place. Thus, the direct measurement of Weibull slopes and also the characteristic time requires thousands of samples to achieve the desired accuracy. Unfortunately, this requirement is not always practical since any reasonable reliability work involves many stress evaluations. In the direct measurement, the large uncertainty in Weibull slope determination arises from the data points at lower percentiles with wider confidence bound as seen in Figure 2.64(a). The scattering of t_{BD} data points at lower percentiles can

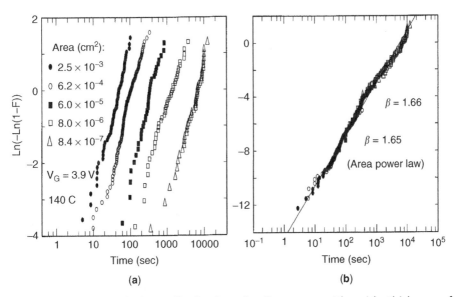

Figure 2.64. (a) Individual t_{BD} distributions for five areas with oxide thickness of 2.7 nm. (b) Normalized t_{BD} with the identical data from Figure 2.64(a) by using Equation 2.34 at a reference area of 8.4×10^{-7} cm^2. NFET structures were stressed in inversion with a stress voltage of 3.9 V applied to the gate and both diffusion and substrate grounded at 140°C. Reprinted with permission from Ref. 202, © 2002 IEEE.

significantly alter the directly measured Weibull slopes. Thus, an alternative method is to take advantage of Poisson area scaling Equation 2.34 and use all available data simultaneously. By transforming the breakdown distributions measured on different area structures to one area, the portions with wider confidence bound overlap with tighter portions of other distributions resulting in a much smoother distribution as shown in Figure 2.64(b). Additionally the range covered is significantly enlarged, which is the same as increasing the sample size. This allows a better and more accurate determination of the Weibull slope because this method tends to average out the effect of the earlier portions of intrinsic breakdown distributions with wider confidence bounds and possibly extrinsic fails. Caution must be exercised in applying this method, particularly for samples with very large area in which series resistance can yield artificially steeper Weibull slopes [232].

The third method to extract the Weibull shape factor is from the slope of T_{63} (or Q_{63}) versus area relation as shown in Figure 2.55 and Equation 2.35. The advantage of using this method lies in the fact that the confidence bounds become tighter as the failure percentiles approach 63% as indicated in Section 2.4.2. As shown in Section 2.4.2, the relative errors increase significantly with decrease in both percentiles and sample size. However, at the 63% percentile, the relative errors are the smallest and less sensitive to sample number variations. Moreover, although like Weibull slopes, the confidence intervals of T_{63} (or Q_{63}) are large on a linear scale without a large number of samples, they appear relatively small when they are expressed in a logarithmic plot. This method yields good results with a somewhat smaller sample size. In practice, oxide-thickness nonuniformity is always present in the samples used in TDDB measurements. It was found that the use of Q_{63} can provide much more accurate extraction of Weibull slopes [232].

2.4.6   Modeling Bi- or Multimodal Weibull Distributions

For dielectrics it is common that the failure distribution is not a pure, intrinsic Weibull distribution, i.e., all the collected data fall on one straight, steep line in the Weibull net, instead there is another mode before the intrinsic distribution. This means the distribution consist of at least two modes and are called bimodal. The reason for this is that the stressed population is comprised of two populations failing with two different failure mechanisms. Therefore, in this section the modeling of bi- and multimodal Weibull distributions is treated. First the two different modeling options are explained; then examples are discussed in detail where these models are applied to real cases.

2.4.6.1   Combined and Competing Distributions. In most cases the population causing a bi- or multimodal distribution consists of two or more subpopulations, which have different failure mechanisms or the same failure mechanism, but with different severities. For example, if two populations with the same dielectric thickness but different areas are (accidentally) mixed, the result is an S-shaped distribution in which the two intrinsic distributions can still be

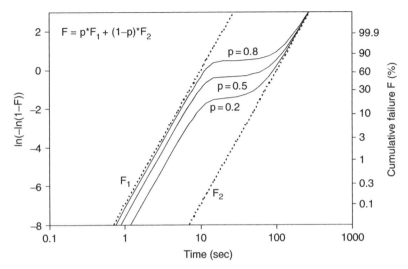

Figure 2.65. Illustration of combining or mixing two Weibull distribution with the same failure mechanism but different $t_{63.2}$ for three different contributions from the distribution failing first (F_1). The smaller the contribution of the first distribution the more the resulting mixed distribution is moved down and $t_{63.2}$ cannot be determined by just fitting the straight part at lower percentage. The same is true for the second distribution, which is even more affected and only slowly converges to the original distribution. In this case neither $t_{63.2}$ nor the slope can be obtained by fitting a single distribution to the data above the transition region directly.

recognized as illustrated in Figure 2.65. This represents the case of the same failure mechanisms, which are indicated by the same Weibull slope, but with different degrees of severity, indicated by the offset of the $t_{63.2}$ values. Since the individual distributions have the same slope a transition appears when the two distributions are combined (or mixed). This transition is shallow and contains only a few data points. It indicates the percentage contributed to the total distribution by the population that fails first. This makes it relatively easy to separate and model such populations. Combined or mixed distributions are modeled by

$$F_{mix}(t) = p \times F_1(t) + (1 - (p) \times F_2(t), \qquad (2.40)$$

where p is the percentage contributed by the population failing first, as shown for three different values in Figure 2.65 (p = 0.2, 0.5, and 0.8). It is obvious that the first distribution is shifted down by the combination of two populations due to the increasing sample size. Although the slope is the same as the original distribution, the $t_{63.2}$ is somewhat larger than for the original distribution, if the data well below the transition region would have been fitted directly. Similarly the second distribution is shifted up and slowly converges to the original distribution. As a

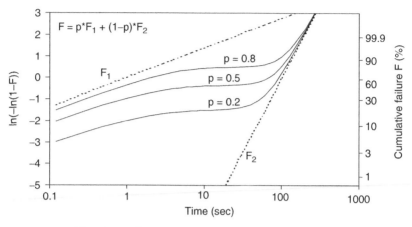

Figure 2.66. Combination of an extrinsic (F_1) and an intrinsic distribution (F_2) with the same three contributions as in Figure 2.65. The observations are the same as in Figure 2.65. However the transition region particularly for $p = 0.2$ is not as clear as before anymore. Note that the first distribution is chosen such that all samples failed before the samples of the last sample of the second distribution failed.

result the parameters of the two mixed distributions can only be determined by fitting a bimodal distribution to the data or by correctly separating the populations, which is not trivial in most cases.

Equation 2.40 is also applicable for two populations with different failure mechanisms, e.g., one population failing intrinsically only and the other population failing extrinsically due to some kind of defect. This case is illustrated in Figure 2.66, with p again assuming the same three values as in Figure 2.65. Basically the same conclusions as for Figure 2.65 are true here also. The only additional remark may be that the transition region for the case $p = 0.2$ is not as clear as before. The parameters for the two distributions were chosen in a way that the extrinsic distribution fails completely before the last fail of the intrinsic distribution. The effect of this not being the case is illustrated next in Figure 2.67.

At first glance Figure 2.67 looks pretty much like Figure 2.66. However, the transition region more or less disappears in all three cases now. Furthermore, at high failure percentages a distinct difference can be noticed: the mixed distribution deviates from the intrinsic behavior and follows again the extrinsic distribution marked by the dotted line. This happens when the intrinsically failing population fails completely and the remaining samples from the extrinsic population have a higher time to failure. Thus this is exactly the behavior expected for two independent distributions with two different failure mechanisms (one for each subpopulation). On the other hand, this behavior is not consistent with the expected behavior of dielectric structures. The reason is that for dielectrics, there is one population only; this population can fail due to two different reasons. It can fail early due to an extrinsic, defect-driven fail and if this does not happen it can

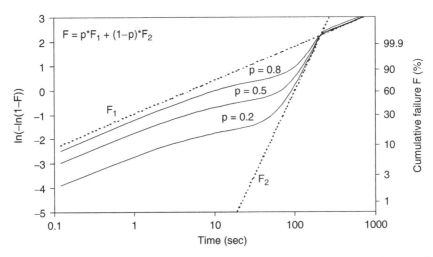

Figure 2.67. Combination of an extrinsic (F_1) and an intrinsic distribution (F_2) with the same three contributions as in Figure 2.66. While the slope of F_1 is the same as in Figure 2.66, the $t_{63.2}$ is larger and thus closer to the $t_{63.2}$ of F_2. The original distributions intersect, which means that the samples of F_2 fail completely before the ones of F_1. The surviving samples from F_1 consequently fail later, causing a tail with shallower slope than expected from F_2. This is not an artifact but the correct behavior for a mix of two independent distributions each failing with a different failure mechanism. Another observation is that the transition regions almost completely disappeared.

fail due to dielectric wearout described by the percolation model. However, once it has failed due to wearout there is nothing left that may fail due to any other reason. Hence, the description by mixed or combined distributions needs to be replaced by another model, since the fail occurs always and only due to the mechanism that happens first. Thus a model describing competing mechanisms needs to be used [258]. The probability that a sample survives a given time is described by the product of the probability to survive each of the failure mechanisms. This can be expressed using the reliability function $R(t)$.

With $R(t) = 1 - F(t)$ the probability to survive a given time is $R_1(t) \times R_2(t) = (1 - F_1(t)) \times (1 - F_2(t))$, which yields in terms of cumulative failure [259]:

$$F(t) = 1 - (1 - F_1(t)) \times (1 - F_2(t))$$

$$= 1 - \exp\left(-\left(\frac{t}{\tau_1}\right)^{\beta_1}\right) \times \exp\left(-\left(\frac{t}{\tau_2}\right)^{\beta_2}\right) \qquad (2.41)$$

where τ_1 and τ_2 are the $t_{63.2\%}$ values of the corresponding distributions. For describing a bimodal distribution that is a result of competing mechanisms, it just needs the survival probabilities and no parameter p, as in Equation 2.40 is required. In Figure 2.68 bimodal distributions based on the same two individual distributions are compared. One is calculated from Equation 2.40 with $p = 0.5$ and the other from Equation 2.41. Since in this case for the extrinsic distribution, the parameters assumed are more consistent with real observations, the resulting bimodal distribution (combined distributions) is inconsistent with the observed behavior for dielectric, while the competing distribution shows the expected behavior. It should be noted that the result for combined distributions is not wrong but the underlying assumption does not apply in this case. For the competing distribution, it can be seen that the extrinsic distribution is affected by the intrinsic distribution in a small time window only, while the intrinsic distribution is not reaching the original distribution. Thus the parameters for the intrinsic distribution can not be extracted by fitting the intrinsic part of the distribution directly, which is possible for the extrinsic part if data in the transition region and beyond are not considered. It is therefore important to fit the entire bimodal distribution in order to get the true parameters of the intrinsic distribution.

Although a typical bimodal distribution from oxide breakdown is not well fit by combined distributions, there are a lot of examples which show that the model

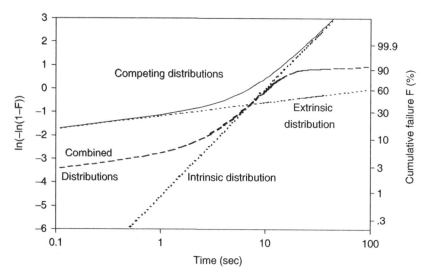

Figure 2.68. Comparison of combined distributions ($p = 0.5$) and competing distributions. The extrinsic distribution parameters are closer to distributions really observed in dielectric stresses, i.e., shallower slope than in previous figures and the $t_{63.2}$ well beyond the $t_{63.2}$ of the intrinsic distribution. The result for combined distributions is not consistent with the behavior observed in dielectric stresses, in which the intrinsic fails can only occur if the sample did not fail due to an extrinsic fail and no (extrinsic) fail is possible after wearout, meaning on the same sample two failure mechanisms compete.

of combined distributions is very useful and describes those data very well. In some cases not only two distributions are combined but several intrinsic distributions are mixed, folded, and weighted with another distribution, often a normal distribution describing the oxide thickness variation, as detailed in the following two sections. For more than two contributing distributions Equation 2.40 changes to

$$F_{mix}(t) = p_1 \times F_1(t) + p_2 \times F_2(t) + \ldots + p_n \times F_n(t) \qquad (2.42)$$

with $p_1 + p_2 + \ldots + p_n = 1$. The more distributions are combined the less recognizable are individual distributions, especially if the $t_{63.2}$ values are close together. The resulting distribution may not even look like a Weibull distribution anymore, especially if the sample size is small; the $t_{63.2}$ values are close to each other and the tails of the distribution are not clearly visible (due to the limited sample size). An example for three combined distributions is shown in Figure 2.69. The resulting distributions do not reveal that they are combined from three intrinsic distributions. For huge sample sizes, the slope of the first failing distribution may still be estimated directly from the data. But all the other parameters can only be estimated by fitting the entire distribution and some knowledge about the underlying distributions, e.g., if different oxide thicknesses are involved as

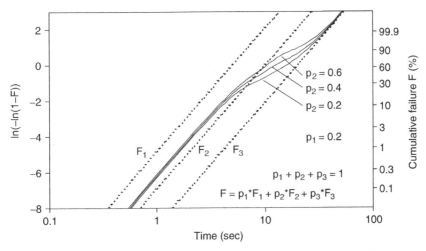

Figure 2.69. Combination or mixing of three Weibull distributions with the same failure mechanism but different $t_{63.2}$ for three different contributions (p_2) from the second distribution (F_2). The $t_{63.2}$ values are tighter together than in Figure 2.65. For small sample sizes, e.g. 20 samples, the accessible window on the cumulative failure scale ranges from 5 to 95% and the data scatter. The resulting distribution may look different than a Weibull distribution and it is not straightforward to estimate the distribution parameters. Fitting one straight line to such data will result in wrong Weibull parameters and erroneous projections.

discussed in the next section. For small sample sizes (e.g., 20 samples) the accessible range on the cumulative failure axis is reduced, starting at 5%, ending at 95%, and the data scatter tremendously, unlike the well behaved analytical calculation. In such a case, a too shallow Weibull slope would be determined if a single distribution is assumed for the data. As a result, the lifetime projection would be too pessimistic and unnecessary process optimization may be caused.

2.4.6.2 Intrinsic: Thickness Nonuniformity. For ultra-thin oxides in the direct tunneling regime, additional complications in accurate determination of the Weibull slope β can arise because of T_{OX} variations [232]. Due to the exponential dependence of direct tunneling current on T_{OX} (Fig. 2.18), any variation in T_{OX} can lead to a current modulation effect which can in turn give rise to distortions in the Weibull t_{BD} distributions and to non-Poisson area scaling effects. Figure 2.70 shows the normalized t_{BD} and q_{BD} distributions for three areas. Several observations can be made from this: 1) the T_{BD} distributions are skewed, i.e., a change of slope from higher percentiles to lower percentiles; 2) non-Poisson area scaling at higher percentiles for t_{BD} distributions; 3) q_{BD} distributions steeper than those of t_{BD} at the higher percentiles, obeying Poisson area statistics.

To understand the effect of thickness nonuniformity on t_{BD} distributions, both Monte Carlo analysis and analytic analysis are developed [232, 260]. The MC analysis treats the time-to-breakdown (t_{BD}) and oxide thickness variations as two

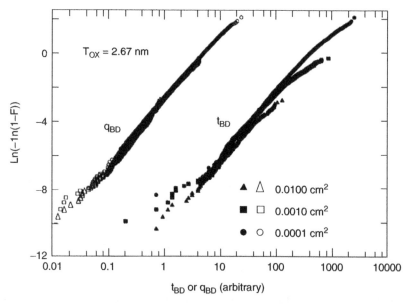

Figure 2.70. Normalized q_{BD} and t_{BD} distributions to $1 \times 10^{-4}\,cm^2$. The t_{BD} distributions are affected by current variations due to oxide thickness variations. This causes systematic deviations from Poisson area scaling that are not seen for q_{BD}. Reprinted with permission from Ref. 232, © 2000 IEEE.

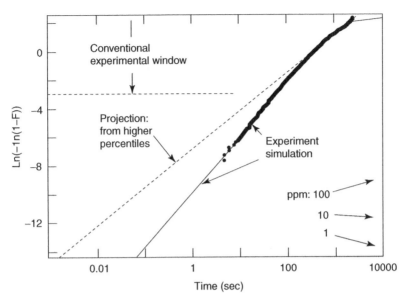

Figure 2.71. Comparison of T_{BD} distributions from experimental data and simulations shows excellent agreement. Projection from the conventional experimental window to low percentiles exhibits significant deviations leading to too pessimistic results. Reprinted with permission from Ref. 232, © 2000 IEEE.

separate random variables described by a Weibull probability function and a normal probability function, respectively [232]. The simulated t_{BD} distribution using MC analysis is compared with the experimental data as shown in Figure 2.71 with excellent agreement between the experimental results and the numerical simulation without any fitting parameters. A large sample size of up to 4000 capacitors was used to extend the measurements down to low percentiles of ~ 250 ppm. Figure 2.71 shows that the use of the slope of t_{BD} distribution at higher failure percentiles can lead to serious errors and a much more pessimistic projection [232].

2.4.6.3 Extrinsic: COP Model.
In this section the modeling also involves the folding of the Weibull distributions with a normal distribution of different oxide thicknesses similar to the previous section: however, in this case the extrinsic part of the bimodal distribution is modeled. The background for this model [160] is the observation that the slope of the extrinsic distribution is larger (i.e., steeper) at lower stress conditions of a step stress (described in Section 2.4.3) as shown in Figure 2.72 (full circles). This is not expected for defect-driven dielectric breakdown. If the voltage acceleration model for extrinsic fails were to obey the 1/E-model, it would even be expected that the slope of the extrinsic distribution would become shallower at lower voltages. Neither is the observed effect caused by the step stress, which would rather flatten the distributions, especially at short times as discussed in Section 2.4.3. Another possibility could be aggressive screening

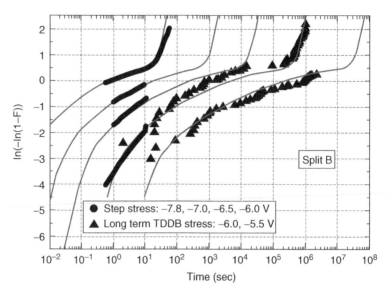

Figure 2.72. Bimodal distributions caused by COPs and observed in voltage step and constant voltage stresses at wafer and package level, respectively [160]. The solid lines are calculated using the two equations above for the different stress conditions and model the extrinsic fails very well. Reprinted with permission from Ref. 160 © 2002 IEEE.

as will be discussed in the next section in detail. However, the screening would affect the first condition of a step stress only. A final possible explanation is if the data were a result of constant voltage stresses, where a screen had been performed for every stress condition. To better understand the observed behavior, long term constant voltage stresses of packaged test structures were performed (triangles in Fig. 2.72). A large number of samples were necessary to obtain the extrinsic failure distributions with sufficient accuracy. Those stresses confirmed the step stress observations and merge nicely. The data now clearly suggests that for each stress condition at shorter times, the slope becomes as steep as an intrinsic distribution. In addition, it was known that the substrate material used in this case is the source for crystal-originated pits, the so-called COPs [160]. These are octahedral voids formed in the silicon crystal in the standard Czochralski crystal growth process. The ones that are open at the wafer surface during thermal oxidation cause locally reduced oxide thickness and are accepted as the reason for extrinsic fails [160]. During the stress, the fail occurs at the weakest spot, which is the thinnest oxide. From extensive modeling, it has been shown that the thickness ranges from the nominal thickness down to about 5 nm and is uniformly distributed [160]. This allows one to model the extrinsic fails as superposition of intrinsic Weibull distributions folded with a normal distribution of oxide thinning using [160]

$$F(t) = 1 - (1 - p_e \times F_e(t)) \times (1 - F_i(t)), \qquad (2.43)$$

with p_e being the fraction of extrinsic fails also defining the defect density and $F_i(t)$ representing the intrinsic fails due to the regular oxide thickness.

$$F_e(t) = \int_{T_{\min}}^{T_{\max}} F(t, T_{OX}) \times P(T_{OX}, \sigma) dT, \qquad (2.44)$$

where $P(T_{OX}, \sigma)$ is the probability density function of the normal distribution of the oxide thickness T_{OX} and the standard deviation σ, $T_{\min}$ is the minimum oxide thickness, and $T_{\max}$ is the maximum oxide thickness. $P(T_{OX}, \sigma)$ is used as a weighting factor for the different thicknesses involved noting that the integral of $P(T_{OX}, \sigma)$ must be equal to one. Equation 2.44 is basically the same as that used in a previous section for modeling the thickness variation. The difference is that in Equation 2.43 the parameter p_e adjusts the extrinsic part of the distribution to the observed failure level. The other difference is the competing intrinsic distribution of the regular oxide outside the COPs. This model describes the observed data excellently as shown in Figure 2.72. Other splits were also investigated and supported the model as well [160]. This has consequences for the lifetime projection, because the lifetime is not limited by extrinsic fails, but by early intrinsically behaving fails which are the result of the minimum oxide thickness. If the voltage acceleration is high enough, those fails may still pass the qualification criteria and not cause failure during the regular product lifetime. Despite this advantage there is also a potential problem, because any burn-in affects intrinsic distributions negatively, particularly if it is aggressive, as will be discussed in the next section. Thus, without the understanding developed by the modeling above, the burn-in may have been changed to more aggressive conditions, which usually helps to weed out more defects that cause early fails, but in this case without any success, and for too aggressive burn-in conditions the effect would actually be the opposite of what had been desired.

2.4.6.4 Effect of Screening and Burn-in.
Screening is an important procedure to ensure that the remaining population meets one of two requirements. The first requirement is that it is free of members that do not belong to the population to begin with, e.g., for dielectrics clean intrinsic distributions without early fails are desired for acceleration assessments. Therefore a short stress at moderate (voltage) conditions is applied to be able to sort out weak parts that do not fail intrinsically. The functional test applied to products has a similar purpose. In these cases the stress is kept small enough to not degrade the reliability of the remaining population noticeably, if at all. The second requirement is that the failure rate of the extrinsic population be improved. This requirement goes beyond the first one, which it naturally includes. A well known screen of this kind is the burn-in applied to products, where somewhat accelerated conditions are used to pick out systematically early failing parts. The difference is that those parts belong to the main population. The consequence is that the applied stress affects the remaining parts. This type of screen will be modeled in this section. Although such a screen is usually not applied during dielectric stressing, it may unintentionally be used when dealing with extrinsic distributions, as will be discussed later in this section. Also as

a product screen it needs to be considered for dielectric lifetime projection, because the dielectric is affected by burn-in and it is necessary to know to what extent.

The motivation for burn-in is given by the bathtub curve (Section 1.3.3), i.e., at the beginning of the lifetime the failure rate is very high and decreases with time. If the product is artificially aged to reach a lower failure rate before being shipped, then the customer will see less fails, because the failure rate will drop further until it bottoms out. This is the useful lifetime as explained earlier. The product is not supposed to ever reach the other end of the bathtub curve where wear out causes a steeply increasing failure rate. Burn-in is performed at somewhat increased voltage and temperature to accomplish the artificial aging in the shortest possible time. The conditions are determined by the failure rate defined for the product lifetime and by the cost for the burn-in itself. In an early stage of production, the limited maturity and a consequently higher defect density forces a more aggressive burn-in to meet the failure rate requirements. With increasing process maturity and a consequently lower defect density, the cost tradeoff between burn-in and defect density leads to a reduction of burn-in time. Modeling the effect of burn-in is simplified by the fact stated above (that no intrinsic breakdown is expected to occur during the product lifetime). Thus the modeling needs to deal with the extrinsic distribution only, although it can be applied to the entire bimodal distributions as the examples show. The percentage of screened parts can be calculated in a straightforward manner from the Weibull equation using the parameters of the extrinsic distribution at the appropriate conditions, e.g., the use or operation conditions. For this purpose, the burn-in time is converted to an equivalent burn-in time $t_{BI\ equiv}$ at those conditions using the appropriate acceleration models. The burn-in yield Y_{BI} is then $Y_{BI} = 1 - F(t_{BI\ equiv})$. The effect of burn-in and any equivalent screen is twofold. By taking the failing parts out of the population the sample size is changed, hence the population needs to be renormalized. The surviving parts did experience a stress that reduced their lifetime by a certain amount. Although all parts suffer the same reduction it affects the parts with the shortest lifetimes the most and they fail considerably earlier than without the screen. Both effects are considered in the following equation that describes the distribution a customer would observe for parts that have been burned in [163]:

$$F_{\text{post BI}}(t) = \frac{F(t + t_{BI\ equiv}) - F(t_{BI\ equiv})}{1 - F(t_{BI\ equiv})} \tag{2.45}$$

where t represents the operation time after burn-in, starting at zero. Inserting the Weibull equation and simplifying yields [163]

$$F_{\text{post BI}}(t) = 1 - \exp\left(-\left(\frac{t + t_{BI\ equiv}}{\tau}\right)^{\beta} + \left(\frac{t_{BI\ equiv}}{\tau}\right)^{\beta}\right), \tag{2.46}$$

with τ being the time at 63.2% cumulative failure and β the slope of the Weibull distribution. For modeling a bimodal distribution, Equation 2.46 is used with the Weibull parameters for both the intrinsic and extrinsic parts and the competing

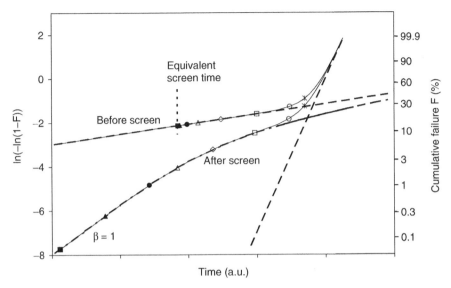

Figure 2.73. Illustration of the effect of a screen, e.g., a burn-in, on a bimodal distribution. The bold solid lines represent the bimodal distribution before and after screen. The dashed straight lines are the individual intrinsic and extrinsic distributions, and the third dashed line is the screened extrinsic distribution. The pairs of equal symbols indicate how the screen affects sample parts. Well beyond the equivalent screen time the screened distribution approaches the original distribution.

distribution model from the corresponding section above. An example [163] is shown in Figure 2.73 and shall be briefly reviewed. The bold lines represent the bimodal distribution before and after screen, where the screen refers to a short stress or a burn-in. The straight, dashed lines are the distributions of the two individual populations, while the third dashed line represents the extrinsic distribution after the screen. The pairs of equal symbols mark the positions of individual parts before and after the screen to illustrate the screen effect. The equivalent screen time is marked on the original distribution labeled "Before screen". All parts with shorter failure time than the equivalent screen time are taken out by the screen. The open triangle marks the point of twice the equivalent screen time. The parts between the equivalent screen time and twice the equivalent screen time are spread out considerably after the screen, marked by the full symbols. The slope in this range is one, which means the parts with slightly longer lifetime than the equivalent screen time exhibit a random failure rate after the screen is applied. From the remaining symbols it is obvious that the effect of the screen is smaller for parts with failure times well beyond the equivalent screen time. At long enough times after the equivalent screen time, the screened distributions slowly approaches the original distributions.

A special but realistic case is that for which the equivalent screen time represents one product lifetime. In this case the product fails all come from the period between the equivalent screen time and twice that screen time. As stated above those fails occur randomly, i.e., the bottom of the bathtub curve. Only the level of the failure rate determines whether the achieved failure rate is low enough or not. If the failure rate is so low that it exceeds the targeted requirements, the screen could be relaxed by shortening the screen time, which saves cost and improves the yield. However, if the screen is not efficient enough, the time needs to be extended or the conditions need to be more aggressive. In any case the effect of aggressive screen times or conditions on the intrinsic reliability of the dielectric must be checked. This is because the intrinsic population is weakened as the equivalent screen time approaches the intrinsic failure times. As for extrinsic fails, intrinsic distributions are also screened. However, the effect is opposite of the effect on the extrinsic distributions. As shown in Figure 2.74 the slope of the intrinsic distribution is reduced overall and reaches unity for times below the equivalent screen time. The result is an increased failure rate without any benefit.

A similar effect of nonuseful screening may be observed when dealing with extrinsic distributions [163]. This can affect the acceleration model and parameters that are to be extracted from the data. A constructed example is given in Figure 2.75. The solid lines represent three true distributions with a simple acceleration model that produces a constant acceleration factor of 1000. The screen condition is 1 s at the lowest condition. This causes the expected change of the lowest distribution as discussed above. It is unlikely that somebody would use this screen condition. However, the same screen condition still affects the distributions at higher accelerated conditions as the dashed lines show. Depending on the exact

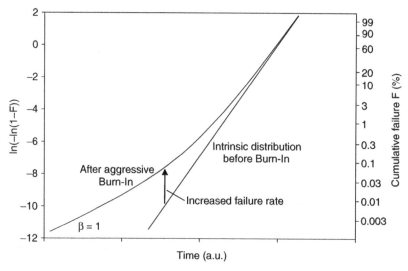

Figure 2.74. Illustration of the effect of aggressive screen conditions on intrinsic distributions. The screen reduces the slope for the entire distribution. Below the equivalent screen time the slope becomes unity. The result is an increased failure rate with no beneficial effect.

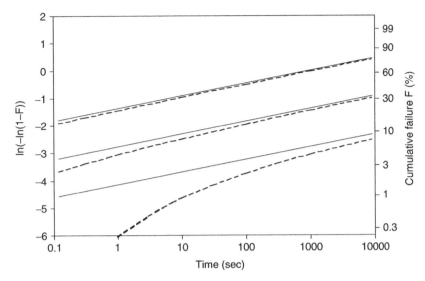

Figure 2.75. Constructed example to illustrate the effect of a screen when trying to determine the acceleration model or acceleration parameters for extrinsic distributions. The solid lines represent the true distributions and the dashed lines the result of the same screen applied in all three cases. Without the knowledge of the true behavior the resulting distributions suggest systematically steeper distributions at lower stress conditions. For a lower acceleration factor the effect becomes worse.

time window and stress conditions the resulting distributions are more or less affected and the desired true distributions are likely not to be obtained. It is obvious that a lower acceleration factor affects the results even more. For determining the acceleration model or extracting the acceleration parameters from extrinsic distributions, the screen must be considered appropriately. It may be necessary to omit the screen or base it on different criteria, because an electrical screen even at moderate conditions will take out parts that belong to the extrinsic distribution and thus skew the final result.

Another effect occurs when for some reason only parts of the population experiences the intended screen. This is a frequently encountered situation in burn-in, where due to contact problems, human error, burn-in bypass or no functionality at burn-in conditions, part of the population is not stressed [261]. Those parts are called burn-in escapes. Furthermore the stress may only reach part of the product, i.e., some of the dielectrics experience burn-in while another part is not stressed or is stressed less efficiently. In both cases the total population consists of two groups that are mixed [261]. One group is composed of the original distribution adjusted to the fraction that was not exposed to the screen and the other group is composed of parts that were screened. The latter is usually the larger fraction. Since both groups belong to the extrinsic part and both are mixed, the modeling is performed with combined distributions as described in the

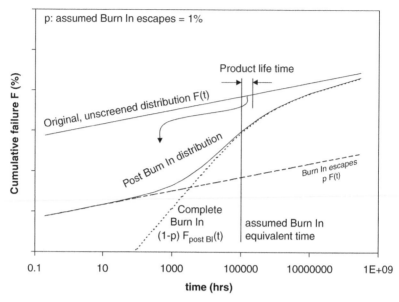

Figure 2.76. Modelling of burn-in escapes by combining the distributions of escaped parts and burned-in parts. This illustrates that a low escape rate is desired in order to not lose the beneficial effect of burn-in. The example assumes that the equivalent burn-in time corresponds to one product lifetime, which causes all fails during product operation to origin from the second lifetime period. Reprinted with permission from Ref. 261, © 1999 IEEE.

beginning of this section. The equation is given [261] by

$$F_{\text{post BI+esc}}(t) = p \times F(t) + (1 - p) \times F_{\text{postBI}}(t), \qquad (2.47)$$

where p is the fraction of escaped parts or the portion of unscreened dielectrics on the chip and $F_{\text{post BI}}$ is given by either Equation 2.45 or 2.46. The result is shown for an arbitrary example in Figure 2.76. Again the equivalent burn-in time is chosen to correspond to approximately one product lifetime and it is illustrated that the fails during product use come from the next lifetime interval. In this period the failure rate is first dominated by the burn-in escapes with a decreasing failure rate which merges with the failure rate of the burned-in parts. This illustrates that a low escape rate is desirable to meet the reliability requirements and not render the burn-in useless.

2.5   SUMMARY AND FUTURE TRENDS

This chapter reviewed fundamental characterization and measurement methods involved in silicon dioxide-based gate dielectric reliability assessment. Most of these are applicable to other dielectrics as well. The variety of addressed topics is a result of the thickness shrinking. For thinner oxides, the tunneling mechanism changes and also the oxide breakdown shows different facets. The consequences for reliability assessment are discussed throughout the chapter and range from reducing the test structure area to introducing complex and sophisticated measurement and evaluation procedures. Although silicon dioxide thickness scaling reached the point at which further reduction appears to be impossible due to the high oxide leakage, the knowledge accumulated over many decades is extremely helpful for future development and assessment of improved silicon dioxides and new dielectric materials. In new materials, other conduction mechanisms contribute and hence the degradation behavior changes. Thus, further development of the characterization methods is inevitable. On the other hand, a thin interfacial silicon dioxide layer drops the highest field in advanced dielectric stacks and is still important and beneficial for these new dielectric stacks. But development does not only move towards thinner oxides. Also, thicker oxides are increasingly important, developing better quality and reliability for use in power CMOS for automotive applications that require not only meeting more aggressive operation specs but also being delivered with zero defects in the first place. At the same time cost reduction is an important issue. In consequence, improved and faster characterization methods are required that support quantitative reliability assessment.

REFERENCES

1. Semiconductor Industry Association. The Int. Technology Roadmap for Semiconductors. Int. SEMATECH: Austin, Texas, 2004.
2. P. D. Agnello. Process requirements for continued scaling of CMOS: The need and prospects for atomic-level manipulation. IBM J. Res. and Dev., 46: 2002 pp 317–338.
3. D. Van Marum. Ann. Phys. 1: pp 68, 1799.
4. J. J. O'Dwyer. *The Theory of Electrical Conduction and Breakdown in Solid Dielectrics.* Clarendon, Oxford: 1973.
5. S. I. Raider. Time-dependent breakdown of silicon oxide films. Appl. Phys. Lett., 23: 1973, pp 34–36.
6. C. M. Osburn, S. I. Raider. The effect of mobile sodium ions on field enhancement dielectric breakdown in SiO2 films on silicon. J. Electrochem. Soc., 120: 1973, pp 1369–1376.
7. S. P. Li, E. T. Bates, J. Maserjian. Time-dependent MOS breakdown. Solid State Electron., 19: 1975, pp 235–239.
8. S. P. Li, S. Prussin, J. Maserjian. Model for MOS field-time-dependent breakdown. 1978 Int. Reliability Physics Sym. Proc, 1978, pp 132–136.
9. E. S. Anolick, G. Nelson. Low field time dependent dielectric integrity. 1979 Int. Reliability Physics Sym. Proc, 1979, pp 8–12.
10. D. Crook. Method of determining reliability screens for time dependent dielectric breakdown. 1979 Int. Reliability Physics Sym. Proc., 1979, pp 1–7.

11. Y. Hokari. Reliability of thin SiO$_2$ films showing intrinsic dielectric integrity. Digest of the 1982 Int. Electron Device Meeting, 1982, pp 46–49.

12. D. Baglee. Characteristics and reliability of 100 Å oxides. 1984 Int. Reliability Physics Sym. Proc, 1984 pp 152–155.

13. J. McPherson, D. Baglee. Acceleration factors for thin gate oxide stressing. 1985 Int. Reliability Physics Sym. Proc, 1985, pp 1–5.

14. A. Berman. Time-Zero dielectric reliability test by a ramp method. 1981 Int. Reliability Physics Sym. Proc, 1981, pp 204–209.

15. O. Hallberg. NMOS Voltage breakdown characteristics compared with accelerated life tests and field use data. 1981 Int. Reliability Physics Sym. Proc, 1981, pp 28–33.

16. E. S. Anolick, L-Y. Chen. Application of step stresses to time dependent breakdown. 1981 Int. Reliability Physics Sym. Proc, 1981, pp 23–27.

17. C. M. Osburn, E. J. Weitzman, Electrical conduction and dielectric breakdown silicon dioxide films on silicon. J. Electrochem. Soc., 119: 1972, pp 603–609.

18. Y. Hokari. Dielectric Breakdown in Electrically Stressed Thin Films of Thermal SiO$_2$. J. Appl. Phys., 49: 1978, pp 2478–2489.

19. N. Klein. Electrical breakdown of insulators by one-carrier impact–ionization. J. Appl. Phys., 53: 1982, pp 5828–5839.

20. T. H. DiStefano, M. Shatzkes. Dielectric instability and breakdown in SiO2 thin films. J. Vac. Sci. Technol., 13: 1976, pp 50–54.

21. I. C. Chen, S. E Holland, C. Hu. Electrical breakdown in thin gate and tunneling oxides. IEEE Tran. Electron Devices, 32: 1985, pp 413–422.

22. D. J. DiMaria et al. Impact ionization, trap creation, degradation, and breakdown in silicon dioxide films on silicon. J. Appl. Phys., 73(7): 1993, pp 3367–3384.

23. D. J. DiMaria. Correlation of trap creation with electron heating in silicon dioxide. Appl. Phys. Lett., 51: 1987, pp 655–657.

24. R. A. Metzler. Theoretical justification for the lognormal distribution of time-dependent breakdown in MOS Oxides. Digest of the 1979 Int. Electron Device Meeting, 1979, pp 233–235.

25. Y. Yaw, R. S. Muller. A theoretical derivation of the lognormal distribution of time-dependent dielectric breakdown in thin oxides. Solid-State Electron., 32: 1989, pp 541–546.

26. D. R. Wolters. Breakdown and wearout phenomena in SiO2. In: *Insulating Films on Semiconductors*, edited by Schulz and Pensl. Springer Verlag, Berlin: 1981, pp 180–194.

27. R.-P. Vollertsen, W. G. Kleppmann. Dependence of dielectric time-to-breakdown distributions on test structures. Proc. of IEEE 1991 Int. Conf. on Microelectronic Test Structures, 4: 1991, pp 75–80.

28. E. Y. Wu, J. H. Stathis, L.-K. Han. Ultra-thin oxide reliability for ULSI application. Semiconductor Science and Technology, 15: 2000, pp 425–435.

29. J. Suñé, I. Placencia, N. Barniol, E. Farrés, F. Martín, X. Aymerich. On the breakdown statistics of very thin SiO2 films. Thin Solid Films, 185: 1990, pp 347–362.

30. R. Degraeve, G. Groeseneken, R. Bellens, J. L. Ogier, M. Depas, P. J. Roussel, H. E. Maes. A consistent model for the thickness dependence of intrinsic breakdown in ultra–thin oxides. Digest of the 1995 Int. Electron Device Meeting: 1995, pp 866–869.

31. J. H. Stathis. Percolation models for gate oxide breakdown. J. Appl. Phys., 86: 1999, pp 5757–5766.

32. J. Suñé. New physics-based analytic approach to the thin oxide breakdown statistics. IEEE Electron Device Lett., 22: 2001, pp 296–298.

33. J. H. Stathis, D.J. DiMaria. Reliability projections for ultra-thin oxides at low voltage. Digest of the 1998 Int. Electron Device Meeting, 1998, pp 167–170.

34. U. Schwalke, M. Pölzl, U.T. Sekinger, M. Kerber. Ultra-thick gate oxides: charge generation and its impact on reliability. Microelectron. Reliab., 41: 2001, pp 1007–1010.

35. E. Cartier, D. J. DiMaria, D. A. Buchanan, J. Stathis, W. W. Abadeer, R.-P. Vollertsen. Degradation of thin SiO2 gate oxides by atomic hydrogen. Proc. of 52nd Ann. Device Research Conf., IVA–2: 1994.

36. D. J. Dumin, J. R. Maddux, R. Subramoniam. The use of low-level, pre-tunneling currents to characterize thin oxide wearout and breakdown. 1993 Int. Conf. on Microelectronic Test Structures, 6: 1993, pp 189–194.

37. E.Y. Wu, J. Suñé, W. Lai, A. Vayshenker, E. Nowak, D. Harmon. Critical reliability challenges in scaling SiO_2-based dielectric to its limit. Microelec. Reliabil., 43: 2003, pp 1175–1184.

38. E. Y. Wu, A. Vayshenker, E. Nowak, J. Suñé, R.-P. Vollertsen, W. Lai, D. Harmon. Experimental evidence of T_{BD} power-law for voltage dependence of oxide breakdown in ultra-thin gate oxides. IEEE Trans. Electron Devices, 49: 2002, pp 2244–2253.

39. K. Okada, S. Kawasaki, Hirofuji, New experimental findings on stress induced leakage current of ultra-thin oxide silicon oxides. Extended abstracts of the 1994 Int. Conf. on Solid State Devices and Materials: 1994, pp 565–567.

40. E. Miranda, J. Suñé, R. Rodríguez, M. Nafría, X. Aymerich, L. Fonseca, F. Campabadal. Soft-breakdown conduction in ultrathin (3–5 nm) gate dielectrics. IEEE Trans. Electron Devices, 47: pp 82–89.

41. M. Depas, T. Nigam, M.M. Heyns. Soft breakdown of ultra-thin gate oxide layers. IEEE Trans. Electron Devices, 43: 1996, pp 1499–1504.

42. B. E. Weir, P. J. Silverman, D. Monroe, K.S. Krisch, M. A. Alam, G. B. Alers, T. W. Sorsch, G. L. Timp, F. Baumann, C.T. Liu, Y. Ma, D. Hwang. Ultra thin gate dielectrics: they break down, but do they fail? IEEE IEDM Tech Dig, 1997, pp 73–76.

43. F. Monsieur, E. Vincent, D. Roy, S. Bruyere, J. C. Vildeuil, G. Pananakakis, G. Ghibaudo. A thorough investigation of progressive breakdown in ultra-thin oxides. Physical understanding and application for industrial reliability assessment. 2002 Int. Reliability Physics Sym. Proc., 2002, pp 45–54.

44. B. P. Linder, S. Lombardo, J. H. Stathis, A. Vayshenker, D. Frank. Voltage dependence of hard breakdown growth and the reliability implication in thin dielectric. IEEE Electron Device Lett., 23: 2002, pp 661–663.

45. E. Y. Wu, E. Nowak, J. Aitken, W. Abadeer, L. K. Han, S. Lo. Structural dependence of dielectric breakdown in ultra-thin gate oxides and its relationship to soft break-down modes and device failure. Digest of the 1998 Int. Electron Device Meeting, 1998, pp 187–190.

46. T. Pompl, H. Wurzer, M. Kerber, R. C. W. Wilkins, I. Eisele. Influence of soft breakdown on nMOSFET device characteristics. 1999 Int. Reliability Physics Sym. Proc., 1999, pp 82–87.

47. R. Degraeve, B. Kaczer, A. Keersgieter, G. Groeseneken. Relation between breakdown mode and location in short-channel nMOSFETs and its impact on reliability specifications. IEEE Trans. Device and Materials Reliability, 1: 2001, pp 163–169.

48. A. Cester, S. Cimino, A. Paccagnella, G. Ghidini, G. Guegan. Collapse of MOSFET drain current after soft breakdown and its dependence on the transistor aspect ratio W/L. 2003 Int. Reliability Physics Sym. Proc.: 2003, pp 189–195.

49. B. Kaczer, A. De Keersgieter, S. Mahmood, R. Degraeve, G. Groeseneken. Impact of gate–oxide breakdown of varying hardness on narrow and wide NFET's. 2004 Int. Reliability Physics Sym. Proc.: 2004, pp 79–83.

50. B. Kaczer, R. Degraeve, M. Rasras, K. Van de Mieroop, P. J. Roussel, G. Groeseneken. Impact of MOSFET gate oxide breakdown on digital circuit operation and reliability. IEEE Trans. Electron Devices, 49: 2002, pp 500–506.

51. B. Kaczer, R. Degraeve, M. Rasras, A. De Keersgieter, K. Van de Mieroop, G. Groeseneken. Analysis and modeling of a digital CMOS circuit operation and reliability after gate oxide breakdown: A case study. Microelec. Reliabil., 42: 2002, pp 555–564.

52. J. H. Stathis, R. Rodriguez, B. P. Linder. Circuit implications of gate oxide breakdown. Microelec. Reliabil., 43: 2003, pp 1193–1197.

53. H. Yang, J. S. Yuan, E. Xiao. Effect of gate oxide breakdown on RF device and circuit performance. 2003 Int. Reliability Physics Sym. Proc.: 2003, pp 1–4.

54. R. Rodríguez, J. H. Stathis, B. P. Linder. Modeling and experimental verification of the effect of gate oxide breakdown on CMOS inverters. 2003 Int. Reliabil. Phys. Sym. Proc.: 2003, pp 11–16.

55. N. Stutzke, B. J. Cheek, S. Kumar, R. J. Baker, A. J. Moll, W. B. Knowlton. Effects of circuit-level stress on inverter performance and MOSFET characteristics. IEEE Int. Integrated Reliability Workshop, Final Report: 2003, pp 71–79.

56. B. Kaczer, R. Degraeve, E. Augendre, M. Jurczak, G. Groeseneken. Experimental verification of SRAM cell functionality after hard and soft gate oxide breakdowns. ESSDERC Proc.: 2003, pp 75–78.

57. R. Rodriguez, J. H. Stathis, B. P. Linder, S. Kowalczyk, C. T. Chuang, R. V. Joshi, G. Northrop, K. Bernstein, A. J. Bhavnagarwala, S. Lombardo. The impact of gate-oxide breakdown on SRAM stability. IEEE Electron Dev. Lett., 23: 2002, pp 559–561.

58. K. Mueller, S. Gupta, S. Pae, M. Agostimelli, P. Aminzadeh. 6-T cell circuit dependent GOX SBD model for accurate prediction for observed VCCMIN test voltage dependence. 2004 Int. Reliability Physics Sym. Proc.: 2004, pp 426–429.

59. J. B. Bernstein, M. Gurfinkel, X. Li, J. Walters, Y. Shapira, M. Talmor. Electronic circuit reliability modeling. Microelec. Reliabil., 46: 2006, pp 1957–1979.

60. S. M. Sze. *Physics of Semiconductor Devices*. Wiley, New York: 1981.

61. E. H. Nicollian, J. R. Brews. *MOS (Metal Oxide Semiconductor) Physics and Technology*. Wiley, New York: 1982.

62. Y. Taur, T. H. Ning. *Fundamentals of Modern VLSI Devices*. Cambridge, UK: 1998.

63. A. S. Grove, E. H. Snow, B. E. Deal, C. T. Sah. Investigation of thermally oxidized silicon surfaces using metal–oxide–semiconductor structures. Solid State Elec., 8: 1965, pp 145–151.

64. Vadasz, A. S. Grove, T.A. Rowe, G. E. Moore. Silicon-gate technology. IEEE Spectrum, 6: 1969, pp 28–35.

65. C. Wong, J. Sun, Y. Taur, C. Oh, R. Angelucci, B. Davari. Doping of N+ and P+ polysilicon in a dual-gate CMOS process. Digest of the 1988 Int. Electron Device Meeting: 1988, pp 238–241.

66. M. Depas, B. Vermeire, P. W. Mertens, R. L. Van Meirhaeghe, M. Heyns. Determination of tunneling parameters in ultra-thin oxide layer poly–Si/SiO2/Si structures. Solid State Elec., 38: 1995, pp 1465–1471.

67. F. Stern, W. E. Howard. Properties of semiconductor surface inversion layers in the electric quantum limit. Phys. Rev., 163: 1967 p. 861–835.

68. Arora N. Rios, C. Huang, N. Khalil, J. Faricelli, L. Gruber. A physical compact MOSFET model, including quantum mechanical effects, for statistical circuit design applications. Digest of the 1995 Int. Electron Device Meeting: 1995, pp 937–940.

69. Arora N. Rios. Determination of ultra-thin gate oxide thickness for structures using quantum effects. Digest of the 1994 Int. Electron Device Meeting: 1994, pp 613–616.

70. C. Lu, J. Sung, H. Kirsch, S. Hillenius, T. Smith, L. Manchanda. Anomalous C–V characteristics of implanted poly MOS Structure in n+/p+ dual-gate CMOS technology. IEEE Electron Dev. Lett., 10: 1989, pp 192–194.

71. S. Lee, C. Liang, C. Pan, W. Lin, J. Mark. A study on the physical mechanism in the recovery of gate capacitance to C_{OX} in implanted polysilicon MOS structures. IEEE Electron Dev. Lett., 13: 1992, pp 2–4.

72. C. Choi, Y. Wu, J. Goo, Z. Yu, R. Dutton. Capacitance reconstruction from measured C–V in high leakage, nitride/oxide MOS. IEEE Trans. Electron Devices, 47: 2000, pp 1843–1849.

73. R. H. Fowler, L. Nordheim. Electron emission in intense electric fields. Proc. R. Soc. London, Ser. A., 119: 1928, 173.

74. C. Zener. A theory of the electrical breakdown of solid dielectrics. Proc. R. Soc. London, Ser. A, 145: 1934, pp 523–529.

75. L. Esaki. New phenomenon in narrow germanium p–n junctions. Phys. Rev., 109: 1957, pp 603–605.

76. M. Lenzinger, E. Snow. Fowler–Nordheim tunneling into thermally grown SiO2. J. Appl. Phys., 40: 1969, pp 278–283.

77. K. Nissan-Cohen, J. Shappir, D. Frohman-Bentchkowsky. Measurement of Fowler–Nordheim tunneling current MOS structures under charge trapping conditions. Solid State Elec., 28: 1985, pp 717–720.

78. K. Gundlach. Zur Berechnung des Tunnelstroms durch eine Trapezförmige Potentialstufe. Solid State Elec., 9: 1966, pp 949–957.

79. J. Maserjian. Tunneling in thin MOS structures. J. Vac. Sci. Technol., 11: 1974, pp 996–1003.

80. G. Lewicki, J. Maserjian. Oscillations in MOS tunneling. J. App Phys., 46: 1975, pp 3032–3039.

81. S. Zafar, K. Conrad, Q. Liu, E. Irene, G. Hames, R. Kuehn, J. J. Wortman. Thickness and effective electron mass measurements for thin silicon dioxide films using tunneling current oscillations. Appl. Phys. Lett., 67: 1995, pp 1031–1033.

82. W. Lai, E. A. Irene. Si/SiO2 interface roughness study using Fowler–Nordheim tunneling current oscillation. J. Appl. Phys., 87: 2000, pp 1159–1164.

83. K. J. Herbert, S. Zafar, E. A. Irene. Measurement of the refractive index of thin SiO$_2$ films using tunneling current oscillations and ellipsometry. Appl. Phys. Lett., 68: 1996, pp 266–268.

84. F. V. Fischetti, D.J. DiMaria, L. Dori, J. Batey, E. Tierney, J. Stasiak. Ballistic electron transport in thin silicon dioxide films. Phys. Rev. B, 35: 1987, pp 4404–4415.

85. J. Simmons. Generalized formula for the electron tunnel effect between similar electrodes separated by a thin insulating films. J. Appl. Phys., 34: 1963, pp 1793–1803.

86. R. Stratton, G. Lewicki, C. A. Mead. The effect of nonparabolic energy bands on tunneling through thin insulating films. J. Phys. Chem. Solids, 27: 1966 p. 1599.

87. K. F. Schuegraf, C. C. King, C. Hu. Ultra-thin silicon dioxide leakage current and scaling limit. Digest of the 1992 Sym. VLSI Technology: 1992, pp 18–19.

88. M. Hiroshima, T. Yasaka, S. Miyazaki, M. Hirose. Electron tunneling through ultrathin gate oxide formed on hydrogen-terminated Si(100) surfaces. Jpn. J. Appl. Phys., 33: 1994, pp 395–398.

89. B. Brar, G. D. Wilk, A. C. Seabaugh. Direct extraction of the electron tunneling effective mass in ultrathin SiO2. Appl. Phys. Lett., 69: 1996, pp 2728–2730.

90. J. Suñé, P. Olivio, B. Riccó. Self–consistent solution of the Poisson and Schrödinger equations in accumulated semiconductor–insulator interfaces. J. Appl. Phys., 70: 1991, pp 337–345.

91. J. Suñé, P. Olivio, B. Riccó. Quantum-mechanical modeling of accumulation layers in MOS structure. IEEE Trans. Electron Devices, 39: 1992, pp 1732–1738.

92. F. Rana, S. Tiwari, D. A. Buchanan. Self–consistent modeling of accumulation layers and tunneling currents through very thin oxides. Appl. Phys. Lett., 69: 1996, pp 1104–1106.

93. S.-H. Lo, D. A. Buchanan, Y. Taur, W. Wang. Quantum–mechanical modeling of electron tunneling current from the inversion layer of ultra–thin–oxide nMOSFETs. IEEE Electron Device Lett., 18: 1997, pp 209–211.

94. S.-H. Lo, D. A. Buchanan, Y. Taur. Modeling and characterization of quantization, polysilicon depletion, and direct tunneling effects in MOSFETs with ultrathin oxides. IBM J. of Res. and Dev., 43: 1999, pp 327–337.

95. L. F. Register, E. Rosenbaum, K. Yang. Analytic model for direct tunneling current in polycrystalline silicon–gate metal–oxide–semiconductor devices. Appl. Phys. Lett., 74: 1999, pp 457–459.

96. W. Lee, C. Hu. Modeling CMOS Tunneling Currents Through Ultrathin Gate Oxide Due to Conduction– and Valence–Band Electron and Hole Tunneling. IEEE Trans. Electron Devices, 48: 2001, pp 1366–1373.

97. N. Yang, W. K. Hesnson, J. R. Hauser, J. Wortman. Modeling study of ultra–thin gate oxides using direct tunneling current and capacitance–voltage measurements in MOS devices. IEEE Trans. Electron Devices, 46: 1999, pp 1464–1471.

98. H. Y. Yang, H. Niimi, G. Lucovsky. Tunneling currents through ultra–thin oxide/nitride dual layer gate dielectrics for advanced microelectronic devices. J. Appl. Phys., 83: 1998, pp 2327–2337.

99. D. A. Muller, T. Sorsch, S. Moccio, F. H. Baumann, K. Evans-Lutterodt, G. Timp. The electronic structure at the atomic scale of ultra-thin gate oxides. Nature, 399: 1999, pp 758–761.

100. Y. Taur, E. Nowak. CMOS devices below 0.1 µm: How high will performance go?. Digest of the 1997 Int. Electron Device Meeting: 1997, pp 215–218.

101. M. Av-Ron, M. Shatzkes, T. H. DiStefano, R. A. Gdula. Electron tunneling at Al-SiO_2 interfaces. J. Appl. Phys., 52: 1981, pp 2897–2908.

102. B. Ricco, M. V. Fischetti. Temperature dependence of the current in SiO2 in the high field tunneling regime. J. Appl. Phys., 55: 1984, pp 4322–4329.

103. J. Suñè, M. Lanzoni, P. Olivo. Temperature dependence of Fowler–Nordheim injection from accumulated n–type silicon in silicon dioxide. IEEE Trans. Electron Devices, 40: 1993, pp 1017–1019.

104. G. Pananakakis, G. Ghibaudo, R. Kies, C. Papadas. Temperature dependence of the Fowler–Nordheim current in metal–oxide–degenerate semiconductor structures. J. Appl. Phys., 78: 1995, pp 2635–2641.

105. G. Salace, A. Hadjadj, C. Petit, M. Jourdain. Temperature dependence of the electron affinity difference between Si and SiO2 in polysilicon (n +)–oxide–silicon(p) structures: Effect of the oxide thickness. J. Appl. Phys., 85: 1999, pp 7768–7773.

106. Z. A. Weinberg. On the tunneling in metal–oxide–silicon structures. J. Appl. Phys., 53: 1982, pp 5052–5056.

107. P. Olivo, J. Suñé, B. Riccó. Determination of the Si–SiO2 barrier height from the Fowler–Nordheim plot. IEEE Electron Device Lett., 12: 1991, pp 620–622.

108. M. V. Fischetti, S. Laux, E. Crabbè. Understanding hot electron transport in silicon devices: Is there a shortcut? J. Appl Phys., 78: 1995, pp 1058–1087.

109. A. Schenk, G. Heiser. Modeling and simulation of tunneling through ultra-thin gate dielectrics. J. App Phys., 81: 1997, pp 7900–7908.

110. V. W. Schottky. Über den Einfluß von Strukturwirkungen, besonders der Thomsonschen Bildkraft, auf die Elektronenemission der Metalle. Physik Zeitschrift, XV: 1914, pp 872–878.

111. P. Hesto. The nature of electronic conduction in thin insulating layers. In Instabilities in Silicon Devices, edited by Barbotttin and Vapaille. Elsevier Science, North Holland: 1986.

112. T.E. Hartman, J.C. Blair, R. Bauer. Electrical conduction through SiO films. J. Appl. Phys., 37: 1966, pp 2468–2474.

113. J. C. Schug, A. C. Lilly, D. A. Lowitz. Schottky currents in dielectric films. Phys. Rev. B: 1970, pp 4811–4818.

114. J. Frenkel. On pre-breakdown phenomena in insulators and electronic semiconductors. Phys. Rev., 54: 1938, pp 647–648.

115. S. M. Sze. Current Transport and Maximum Dielectric Strength of Silicon Nitride Films. J. Appl. Phys, 38: 1967, pp 2951–2956.

116. Y. Shi, X. Wang, T-P Ma. Electrical properties of ultra-quality ultra-thin nitride/oxide stack dielectrics. IEEE Trans. Electron Devices, 47: 2000, pp 1349–1354.

117. E. M. Vogel, J. S. Suehle, M. D. Edelstein, B. Wang, Y. Chen, J. B. Bernstein. Reliability of ultrathin silicon dioxide under combined substrate hot-electron and constant voltage tunneling stress. IEEE Trans. Electron Devices, 47: 2000, pp 1183–1199.

118. K. Umeda, K. Taniguchi. Hot-electron-induced quasibreakdown of thin gate oxides. J. Appl. Phys., 82: 1997, pp 297–302.

119. T. H. Ning. Hot electron emission from silicon into silicon dioxide. Solid State Elec., 21: 1978, pp 273–282.

120. E. Y. Wu, S.-H. Lo, W. Abadeer, A. Acovic, D. Buchanan, T. Furukawa, D. Brochu, R. Dufresne. Determination of ultra-thin oxide voltages and thickness and the impact on reliability projection. 1997 Int. Reliability Physics Sym. Proc.: 1997, pp 184–189.

121. C. Berglund. Surface States at Steam-Grown Silicon-Silicon Dioxide Interfaces. IEEE Trans. Electron Devices, 13: 1996, pp 701–705.

122. D. K. Schroder. *Semiconductor Material and Device Characterization*. John Wiley and Sons, NY: 1998.

123. C. Chang, C. Hu, R. Brodersen. Quantum yield of electron impact ionization in silicon. J. Appl. Phys., 57: 1985, pp 302–309.

124. R. B. Calligaro. Iterative determination of oxide thickness in MOS structures from one DC current/voltage pair. Electronics Lett., 20: 1984, pp 70–73.

125. R. A. Ashton. Gate oxide thickness measurement ysing Fowler–Nordheim tunneling. Proc. IEEE 1997 Int. Conf. on Microelectronic Test Structure, 4: 1991, pp 57–60.

126. J. Maserjian, G. Petersson, C. Svensson. Saturation capacitance of thin oxide MOS structures and the effective surface density of states of silicon. Solid State Elec., 17: 1974, pp 335–339.

127. M. J. McNutt, C.T. Sah. Determination of the MOS oxide capacitance. J. Appl. Phys., 46: 1975, pp 3909–3913.

128. K. Lehovec, S. Lin. Analysis of C-V data in the accumulation regime of MIS structures. Solid State Elec., 19: 1976, pp 993–996.

129. B. Ricco, P. Olivo, T. N. Nguyen, T. Kuan, G. Ferriani. Oxide-thickness determination in thin-insulator MOS structures. IEEE Trans. Electron Devices, 35: 1988, pp 432–438.

130. H. Reisinger, H. Oppolzer, W. Hönlein. Thickness determination of thin SiO2 on silicon. Solid State Elec., 35: 1992, pp 797–803.

131. E. Vincent, G. Ghibaudo, G. Morin, C. Papadas. On the oxide thickness extraction in deep-submicron technologies. Proc. IEEE 1997 Int. Conf. on Microelectronic Test Structures, 10: 1997, pp 105–110.

132. G. Ghibaudo, S. Bruyère, T. Devoivre, B. DeSalvo, E. Vincent. Improved method for the oxide thickness extraction in MOS Structures with ultrathin gate dielectric. IEEE Trans. Semiconductor Manufacturing, 13: 2000, pp 152–158.

133. C. Leroux, G. Ghibaudo, G. Reimbold, R. Clerc, S. Mathieu. Oxide thickness extraction methods in the nanometer range for statistical measurements. Solid State Elec., 46: 2002, pp 1849–1854.

134. B. Majkusiak, A. Jakubowki. A technical formula for determining the insulator capacitance in MOS structure. Solid State Elec., 35: 1992, pp 223–224.

135. S. V. Walstra, C.-T. Sah. Extension of the McNutt–Sah method for measuring thin oxide thicknesses of MOS device. Solid State Elec., 42: 1988, pp 671–673.

136. R. M. A. Azzam, N. M. Bashara. *Ellipsometry and Polarized Light*. North–Holland: 1977.

137. A. C. Diebold, D. Venables, Y. Chabal, D. Muller, M. Weldon, E. Garfunkel. Characterization and production metrology of thin transistor gate oxide films. Materials Science in Semiconductor Processing, 2: 1999, pp 103–147.

138. S. Chongsawangvirod, E. A. Irene, A. Kalnitsky, S. P. Tay, J. P. Ellul. Refractive index profiles of thermally grown and chemically vapor deposited films on silicon. J. Electrochem. Soc., 137: 1990, pp 3536–3541.

139. K. Herbert, T. Labayen, E. Irene. A measurement of the refractive index for ultra thin SiO2 films an a re-evaluation of the thermal Si oxidation kinetics in the thin film regime. In: *The Physics and Chemistry of SiO2 and the Si–SiO2 Interface, 3*, edited by Massoud, Poindexter, Heims. The Electrochemical Society, Pennington, NJ: 1996.

140. D. A. Buchanan. Scaling the gate dielectric: materials, integration, and reliability. IBM J. of Res. And Dev., 43: 1999, pp 245–264.

141. S. I. Raider, R. Flitsch. X–ray photoelectron spectroscopy of SiO2–Si interfacial regions: ultrathin oxide films. IBM J. of Res. And Dev., 22: 1978, pp 294–303.

142. M. F. Hochella Jr., A. H. Carim. A Reassessment of Electron Escape Depths in Silicon and Thermally Grown Silicon Dioxide Thin Films. Surface Science Lett., 197: 1988, pp L260–L268.

143. D. F. Mitchell, K. B. Clark, J. A. Bardwell, W. N. Lennard, G. R. Massoumi, L. V. Mitchell. Film Thickness Measurements of SiO2 by XPS. Surface and Interface Analysis, 21: 1994, pp 44–50.

144. M. Reiche. TEM Investigations of the oxidation kinetics of amorphous silicon films. In: *The Physics and Chemistry of SiO2 and the Si–SiO2 Interface, 2*, edited by Helmes, Deal. Plenum Press, New York: 1993, pp 109–116.

145. Z. H. Lu, J. P. McCaffrey, B. Brar, G. D. Wilk, R. M. Wallace, L. C. Feldman, S. P. Tay. SiO2 film thickness metrology by X–ray photoelectron spectroscopy. Appl. Phys. Lett., 71: 1997, pp 2764–2766.

146. C. J. Powell, A. Jablonski, S. Tanuma, D. R. Penn. Effects of elastic and inelastic electron scattering on quantitative surface analyses by AES and XPS. J. Electron Spectroscopy and Related Phenomena, 68: 1994, pp 605–616.

147. L. Larcher, P. Pavan, F. Pellizzer, G. Ghidini. A new model PF gate capacitance as a simple tool to extract MOS parameters. IEEE Trans. Electron Devices, 48: 2001, pp 935–945.

148. A. Pacelli, A. S. Spinelli, L. M. Perron. Carrier quantization at flat bands in MOS devices. IEEE Trans. Electron Devices, 46: 1999, pp 383–388.

149. K. Ahmed, E. Ibok, G. Bains, D. Chi, Bob Ogle, J. J. Wortman, J. R. Hauser. Comparative physical and electrical metrology of ultra–thin oxides in the 6 to 1.5 nm regime. IEEE Trans. Electron Devices, 47: 2000, pp 1349–1354.

150. J. Hauser. CVC ©1996 NCSU Software, Version 3.0. Dept. Elect. Comput. Eng., North Carolina State University, Raleigh, NC.

151. J. F. Verweij, J. H. Klootwijk. Dielectric breakdown I: A review of oxide breakdown. Microelectronics J., 27: 1996, pp 611–622.

152. D. R. Wolters, van der J.J. Schoot. Dielectric breakdown in MOS devices, Part I: Defect–related and intrinsic breakdown. Philips J. Res., 40: 1985, pp 115–136.

153. J. C. Jackson, T. Robinson, O. Oralkan, D. J. Dumin, G. A. Brown. Nonuniqueness of time-dependent dielectric-breakdown distributions. Appl. Phys. Lett., 71: 1997, pp 3682–3684.

154. B. P. Linder, D. J. Frank, J. H. Stathis, S.A. Cohen. Transistor-limited constant voltage stress of gate dielectrics. Digest of the 2001 Sym. VLSI Technology: 2001, pp 93–94.

155. A. Toriumi, S. Tagaki, H. Satake. Study of soft breakdown in thin SiO2 films by carrier-separation technique and breakdown-transient modulation. In: *The Physics and Chemistry of SiO2 and the Si–SiO2 Interface, 4, 2000–2*, edited by H. Z. Massoud, I. J. R. Baumvol, M. Hirose, E. H. Poindexter. The Electrochemical Society, Pennington, NJ: 2000, pp 399–407.

156. E. Rosenbaum, J. C. King, C. Hu. Accelerated testing of SiO2 reliability. IEEE Trans. Electron Devices, 43(1): 1996, pp 70–80.

157. A. Martin, P. O'Sullivan, A. Mathewson. Dielectric reliability measurement methods: a review. Microelec. and Reliabil., 38(1): 1998, pp 37–72.

158. T. Nigam, R. Degraeve, G. Groeseneken, M. M. Heyns, H. E. Maes. Constant current charge-to-breakdown: Still a valid tool to study the reliability of MOS structures? 1998 Int. Reliability Physics Sym.: 1998, pp 62–69.

159. E. Y. Wu, W. W. Abadeer, L.-H. Han, S.-H. Lo, G. R. Hueckel. Challenges for accurate reliability projections in the ultra-thin oxide regime. 1999 Int. Reliability Physics Sym. Proc.: 1999, pp 57–65.

160. T. Pompl, M. Kerber, G. Innertsberger, K.-H. Allers, M. Obry, A. Krasemann, D. Temmler. Modeling of substrate related extrinsic oxide failure distributions. 2002 Int. Reliability Physics Sym. Proc.: 2002, pp 393–403.

161. A. Martin, J. S. Suehle, P. Chaparala, P. O'Sullivan, A. Mathewson. A new oxide degradation mechanism for stress in the Fowler–Nordheim tunneling regime. 1996 Int. Reliability Physics Sym.: 1996, pp 67–76.

162. A. Martin, P. O'Sullivan, A. Mathewson, J. S. Suehle, P. Chaparala. Investigation of the influence of ramped voltage stress on intrinsic T_{BD} of MOS gate oxide. Solid State Elec., 41(7): 1997, pp 1013–1020.

163. R.-P. Vollertsen, W. W. Abadeer. Influence of stress design on the parameters evaluated from breakdown distributions of thin oxide. Proc. of ESREF 1993: pp 195–200.

164. A. Yassine, K. Wieczorek, K. Olasupo, V. Heinig. A novel electrical test to differentiate gate-to-source/drain silicide short from gate oxide short. 2000 IEEE Internal Integrated Reliability Workshop, Final Report: 2000, pp 90–94.

165. R. Degraeve, B. Kaczer, A. De Keersgieter, G. Groeseneken. Relation between breakdown mode and breakdown location in short channel NMOSFETs and its impact on reliability specifications. 2001 Int. Reliability Physics Sym.: 2001, pp 360–366.

166. R.-P. Vollertsen. Procedure for quantitative fWLR monitoring of gate dielectric reliability. 2004 IEEE Internal Integrated Reliability Workshop, Final Report: 2004, pp 182–185.

167. A. Kerber, M. Kerber. Fast wafer level data acquisition for reliability characterization of sub-100 nm CMOS technologies. 2004 IEEE Internal Integrated Reliability Workshop, Final Report: 2004, pp 41–45.

168. P. Hiergeist, A. Spitzer, S. Röhl. Lifetime of thin oxide and oxide–nitride–oxide dielectrics within trench capacitors for DRAMs. IEEE Trans. Electron Devices, 36: 1989, pp 913–919.

169. A. Strong, E. Y. Wu, R. Bolam. Dielectric step stress and life stress comparison. 1995 IEEE Internal Integrated Reliability Workshop, Final Report: 1995, p 165.

170. P. A. Heimann. An operational definition for breakdown of thin thermal oxides of silicon. IEEE Trans. Electron Devices, 30(10): 1983, pp 1366–1368.

171. A. Martin, P. O'Sullivan, A. Mathewson. Reliability measurements with ramped and constant voltage stress. Proc. of ESREF 1995: pp 95–100.

172. E. S. Snyder, J. Suehle. Detecting breakdown in ultra-thin dielectrics using a fast voltage Ramp. 1999 IEEE Internal Integrated Reliability Workshop, Final Report: 1999, pp 118–123.

173. F. Chen, B. Li, R. A. Dufresne, R. J. Jammy. Abrupt current increase due to space–charge–limited conduction in thin nitride-oxide stacked dielectric system. Appl. Phys. Lett., 90(4): 2001, pp 1898–1902.

174. JEDEC Standard, JESD35–A: Procedure for the Wafer–Level Testing of Thin Dielectrics. 2001.

175. J. C. Lee, I.-C. Chen, C. Hu. Modeling and characterization of gate oxide reliability. IEEE Trans. Electron Devices., 35(12): 1988, pp 2268–2278.

176. R. Moazzami, C. Hu. Projecting gate oxide reliability and optimizing reliability screens. IEEE Trans. Electron Devices, 37(7): 1990, pp 1643–1650.

177. C. Monserie, C. Papadas, G. Ghibaudo, C. Gounelle, P. Mortini, G. Pananakakis. Correlation between negative bulk oxide charge and breakdown, modeling, and new criteria for dielectric quality evaluation. 1993 Int. Reliability Physics Sym. Proc.: 1993, pp 280–284.

178. W. W. Abadeer, R.-P. Vollertsen. Physical mechanisms of dielectric breakdown in SiO2 for the range of -150°C to 150°C. Proc. of 8th European Sym. on Reliability of Electron Devices, Failure Physics and Analysis: 1995, pp 107–109.

179. G. B. Alers, B. E. Weir, M. R. Frei, D. Monroe. J-ramp on sub-3 nm dielectrics: Noise as a breakdown criterion. 1999 Int. Reliability Physics Sym. Proc.: 1999, pp 410–413.

180. P. Roussel, R. Degraeve, G. Van den Bosch, B. Kaczer, B. Groeseneken. Accurate and robust noise-based trigger algorithm for soft breakdown detection in ultra thin oxides. 2001 Int. Reliability Physics Sym. Proc.: 2001, pp 386–392.

181. A. Martin, J. v. Hagen, G. B. Alers. Ramped current stress for fast and reliable WLR monitoring of thin oxide reliability. Microelec. Reliabil., 43(8): 2003, pp 1215–1220.

182. K. Eriguchi, Y. Uraoka. Correlating charge-to-breakdown with constant–current injection to gate oxide lifetime under constant–voltage stress. Jpn. J. Appl. Phys., 35(2B): 1996, pp 1535–1539.

183. J. F. Verwey, D. R. Wolters. Breakdown Fields in thin oxide layers. In: Insulating Films on Semiconductors, Simonne, Buxo. edited by Elsevier Science. North Holland: 1986, pp 125–132.

184. G. Ghidini, D. Brazzelli. Evaluation methodology of thin dielectrics for non-volatile memory application. Microelec. Reliabil., 42: 2002, pp 1473–1480.

185. H. Wang, C. Michael, S. Geha, R. S. Guo, C. Messick, R. Lahri. An optimized gate oxide breakdown test by activating oxide traps at low fields. Digest of the 2000 Int. Electron Device Meeting: 1992, pp 143–146.

186. P. Cappelletti, P. Ghezzi, F. Pio, C. Riva. Accelerated current test for fast tunnel oxide evaluation. Proc. IEEE 1991 Int. Conf. on Microelectronic Test Structures: 1991, pp 81–85.

187. N. Dumin. A new algorithm for transforming exponential current ramp breakdown distributions into constant current TDDB space, and the implications for gate oxide Q_{BD} measurement methods. 1998 Int. Reliability Physics Sym. Proc.: 1998, pp 80–86.

188. Y. Chen, J. Suehle, C.-C. Shen, J. Bernstein, C. Messick, P. Chaparala. The correlation of highly accelerated Q_{BD} test to TDDB life tests for ultra–thin gate oxides. 1998 Int. Reliability Physics Sym. Proc.: 1998, pp 87–91.

189. Various authors and papers in Fast wafer level reliability: Methods and experiences. Special Section in Microelec. Reliabil., 44(8): 2004.

190. M. Kamolz. CSQ-test: A special J-ramp method approved for fast routine testing of thin dielectric films. Int. Wafer Level Reliability Workshop: 1992, pp 137–141.

191. G. Diestel, A. Martin, M. Kerber, A. Schlemm, H. Erlenmaier, B. Murr, A. Preussger. Quality assessment of thin oxides using constant and ramped stress measurements. Microelec. Reliabil., 41(7): 2001, pp 1019–1022.

192. R. Hijab. Product Reliability and maximum voltage limits from extrinsic gate oxide voltage ramp data. 1999 Int. Integrated Reliability Workshop Final Report: 1999, pp 98–101.

193. M.-S. Liang, Y. Choi. Thickness dependence of oxide breakdown under high field and current stress. Appl. Phys. Lett., 50: 1987, pp 104–106.

194. D. J. DiMaria. Explanation for the polarity dependence of breakdown in ultra-thin silicon dioxide films. Appl. Phys. Lett., 68: 1996, pp 3004–3006.

195. J. Bude, B. E. Weir, P. J. Silvermann. Explanation of stress-induced damage in thin oxides. Digest of the 1998 Int. Electron Device Meeting: 1998, pp 179–182.

196. C. Jiang, D. Pramanik, C. Gabriel. Effect of layout on gate oxide defects at gate edges. 1993 Int. Integrated Reliability Workshop, Final Report: 1993, pp 113–116.

197. D. J. Dumin, N. B. Heilemann, N. Husain. Test structures to investigate thin insulator dielectric wearout and breakdown. Proc. IEEE 1991 Int. Conf. on Microelectronic Test Structures: 1991, pp 61–68.

198. Y. Uraoka, H. Yoshikawa, N. Tsutsu, S. Akiyama. Evaluation of Gate Oxide Reliability using Luminescence Method. Proc. IEEE 1991 Int. Conf. on Microelectronic Test Structures, pp 69–74, 1991.

199. K. Shiga, J. Komori, M. Katsumata, A. Teramoto, M. Sekine. A New test structure for evaluation of extrinsic oxide breakdown. Proc. IEEE 1998 Int. Conf. on Microelectronic Test Structures: 1998, pp 197–200.

200. A. Martin, M. Kerber, G. Diestel. WLR monitoring stresses and suitable test structures for future product reliability targets. 1999 Int. Integrated Reliability Workshop, Final Report: 1999, pp 124–128.

201. T. Pompl, M Kerber. Failure distributions of successive breakdown events. IEEE Trans. On Device and Materials Reliability, 4, 2, pp 263–267, 2004.

202. E. Y. Wu, R.-P. Vollertsen. On the Weibull shape factor of intrinsic breakdown of dielectric films and its accurate experimental determination. Part I: Theory, methodology, experimental techniques. IEEE Trans. Electron Devices, 49(12): 2002, pp 2131–2140.

203. S. Nakajima, T. Takeda. Failure analysis in halfmicron and quartermicron eras. Proc. ESREF: 1995, pp 273–280.

204. F. Pio. Sheet resistance and layout effects in accelerated tests for dielectric reliability evaluation. Microelectronics J., 27: 1996, pp 675–685.

205. T. Shigenobu, H. Uchida, N. Hirashita. Projecting oxide lifetime by a step voltage method using electric field correction. Proc. IEEE 1993 Int. Conf. on Microelectronic Test Structures: 1993, pp 125–130.

206. A. Martin, von J. von Hagen, J. Fazekas, K.-H. Allers. Fast and reliable WLR monitoring methodology for assessing thick dielectrics test structures integrated in the Kerf of product wafers. 2002 Int. Integrated Reliability Workshop, Final Report: 2002, pp 83–87.

207. A. K. Stamper, J. B. Lasky, J. W. Adkisson. Plasma-induced gate-oxide charging issues for sub-0.5 m complementary metal–oxide–semiconductor technologies. J. Vac. Sci. Technol. A, 13: 1995, pp 905–911.

208. K. Noguchi, T. Horiuchi. Reduced Oxide reliability due to multilevel metalization process. 1996 Int. Integrated Reliability Workshop, Final Report: 1996, pp 98–103.

209. Y. Uraoka, K. Eriguchi, T. Tamaki, K. Tsuji. Evaluation technique of gate oxide damage. 1993 Int. Conf. on Microelectronic Test Structures, pp 149–154, 1993.

210. R.-P. Vollertsen, W. W. Abadeer. Comprehensive Gate–Oxide Reliability Evaluation for DRAM Processes. Proc. ESREF 1996. In: Microelec. and Reliabil., 36: 1996, pp 1631–1638.

211. S. Lombardo, A. La Magna, C. Spinella, C. Geradi, F. Crupi. Degradation and hard breakdown transient of thin gate oxides in metal–SiO2–Si capacitors: Dependence on oxide thickness. J. Appl. Phys., 86: 1999, pp 6382–6391.

212. M. Ruprecht, G. Benstetter, D. Hunt. A review of ULSI failure analysis techniques for DRAMs. Part II: Defect isolation and visualization. Microelec. Reliabil., 43: 2003, pp 17–41.

213. J. L. Karl, W. S. Berry, R. J. Finch, W. K. Tice. The V_{sub} Problem: The characterization of an unusual dielectric breakdown phenomenon. In: Proc. Int. Sym. for Testing and Failure Analysis (ISTFA): 1985, pp 72–77.

214. R. Gonella, P. Motte, J. Torres. Assessment of copper contamination impact on inter-level dielectric reliability performed with time–dependent–dielectric–breakdown tests. Microelec. Reliabil., 40: 2000, pp 1305–1309.

215. R.-P. Vollertsen. Thin dielectric reliability assessment for DRAM technology with deep trench storage node. Microelec. Reliabil., 43: 2003, pp 865–878.

216. S. Bruyere, E. Vincent, G. Ghibaudo. Quasi-breakdown in ultra-thin SiO2 films: occurrence characterization and reliability assessment methodology. 2000 Int. Reliability Physics Sym. Proc.: 2000, pp 48–54.

217. T. Pompl, C. Engel, H. Wurzer, M. Kerber. Soft breakdown in ultra-thin oxides, Microelec. Reliabil., 41: 2001, pp 543–551.

218. Q. Xiang, G. Yeap, D. Bang, M. Song, K. Ahmed, E. Ibok, M-R Lin. Performance and reliability of sub–100 nm MOSFETs with ultra thin direct tunneling gate oxides. Digest of the 1997 Sym. VLSI Technology: 1997, pp 160–162.

219. Y. Wu, Q. Xiang, D. Bang, G. Lucovsky, M. Lin. Time dependent dielectric wearout (TDDW) technique for reliability of ultrathin gate oxides. IEEE Electron Dev. Lett., 20: 1999, pp 262–264.

220. B. Neri, P. Olivo, B. Riccó. Low Frequency noise in silicon–gate metal–oxide–silicon capacitors before oxide breakdown. Appl. Phys. Lett., 51: 1987, pp 2167–2169.

221. M. Depas, T. Nigam, M. M. Heyns. Definition of dielectric breakdown for ultra-thin (<2 nm) gate oxides. Solid State Elec.: 1997, pp 725–728.

222. R. Degraeve, B. Kaczer, F. Schuler, M. Lorenzini, D. Wellekens, P. Hendrickx, J. Van Houdt, L. Haspeslagh, G. Tempel, G. Groeseneken. Statistical model for stress-induced leakage current and pre-breakdown current jumps in ultra-thin oxide layers. Digest of the 2001 Int. Electron Device Meeting: 2001, pp 121–124.

223. F. Crupi, B. Neri, S. Lombardo. Pre-breakdown in thin oxides. IEEE Electron Device Lett., 21: 2000, pp 319–321.

224. G. Alers, B. Weir, M. R. Frei, D. Monroe. Trap-assisted tunneling as a mechanism of degradation and noise in 2–5 nm oxides. 1998 Int. Reliability Physics Sym. Proc.: 1998, pp 76–79.

225. J. Schmitz, H. J. Kretschmann, H. P. Tuinhout, P. H. Woerlee. Soft breakdown triggers for large area capacitors under constant voltage stress. 2001 Int. Reliability Physics Symp. Proc.: 2001, pp 393–398.

226. E. Y. Wu, J. Suñé, W. Lai, A. Vayshenker, D. Harmon. A comprehensive investigation of gate oxide breakdown of p+–poly/PFETs under inversion mode. Digest of the 2005 Int. Electron Device Meeting: 2005, pp 407–410.

227. B. Kaczer, R. Degraeve, R. O'Connor, Ph. Roussel, G. Groeseneken. Implications of progressive wear-out for lifetime extrapolation of ultra-thin (EOT ~ 1 nm) SiON films. Digest of the 2004 Int. Electron Device Meeting: 2004, pp 713–716.

228. W. Weibull. A statistical distribution function of wide applicability. J. Appl. Mechanics: 1951, pp 293–297.

229. D. R. Wolters, J. F. Verwey. Breakdown and wearout phenomena in SiO2. In: *Instabilities in Silicon Devices*, edited by G. Barbottin, A. Vapaille. North-Holland: 1986, pp 315–362.

230. E. J. Gumbel. Statistical theory of extreme values and some practical applications. National Bureau of Standards. Applied Mathematics Series, 33: 1954, pp 13–15.

231. D. R. Wolters, A. T. A. Zegers, van Duynhoven. Dielectric reliability and breakdown of thin dielectrics. In: *Reliability Technology: Theory and Application*, edited by J. Moltoft and F. Jensen. Elsevier, Amsterdam: 1986, pp 315–323.

232. E. Y. Wu, E. J. Nowak, R.-P. Vollertsen, L.-K. Han. Weibull breakdown characteristics and oxide thickness uniformity. IEEE Trans. Electron Devices, 47: 2000, pp 2301–2309.

233. C. H. Stapper. Fact and fiction in yield modeling. Microelectronics J., 20: 1989, pp 129–151.

234. M. A. Mitchell. Defect test structures for characterization of VLSI technologies. Solid State Technology: 1985, pp 207–213.

235. S. M. Sze. *VLSI Technology*. McGraw-Hill, Singapore: 1983.

236. M. Raghavachari, A. Srinivasan, P. Sullo. Poisson mixture yield models for integrated circuits: A critical review. Microelec. Reliabil., 37(4): 1996, pp 565–580.

237. S. R. Hofstein, F. P. Heimann. The silicon insulated-gate field effect transistor. Proc. of IEEE, 51(9): 1963, pp 1190–1202.

238. W. G. Kleppmann, R.-P. Vollertsen. Variation of defect density and its influence on yield extrapolation for integrated circuits. Quality and Reliability Engineering Int., 6: 1990, pp 133–143.

239. B. T. Murphy. Cost-size optima of monolithic integrated circuits. Proc. of IEEE, 52(12): 1964, pp 1537–1545.

240. R. B. Seeds. Yield, economic, and logistic models for complex digital arrays. 1967 IEEE Int. Convention Record, Part 6: 1967, pp 61–66.

241. C. H. Stapper. Defect density distribution for LSI yield calculations. IEEE Trans. Electr. Devices, 20: 1973, pp 655–657.

242. J. Lee, I. C. Chen, C. Hu. Statistical modeling of silicon dioxide reliability. 1988 Int. Reliability Physics Sym. Proc.: 1988, pp 131–138.

243. R. Degraeve, G. Groesenken, R. Bellens, J. Ogier, M. Depas, P. Roussel, H. Maes. New insights on the relation between electron between electron trap generation and the statistical properties of oxide breakdown. IEEE Trans. Electron. Devices, 45: 1998, pp 904–910.

244. D. J. Dumin, R. S. Scott, R. Subramoniam. A Model Relating Wearout induced Physical Changes in Thin Oxides to the Statistical Description of Breakdown. 1993 Int. Reliability Physics Sym. Proc.: 1993, pp 285–292.

245. Y. Lee, N. Mielke, M. Agostinelli, S. Gupta, R. Lu. W. McMahon. Prediction of logic product failure due to thin oxide breakdown. 2006 Int. Reliability Physics Sym. Proceeding: 2006, pp 18–28.

246. W. G. Kleppmann. Reliability predictions for integrated circuits from tests structure data. Quality and Reliability Engineering Int., 5: 1989, pp 155–163.

247. W. R. Hunter. The Analysis of Oxide Reliability Data. 1998 IEEE Int. Integrated Reliability Workshop, Final Report: 1998, pp 114–134.

248. M. Cacciari, G. C. Montanari. A method to estimate the Weibull parameters for progressively censored tests. IEEE Trans. Reliability, R36: 1987, pp 87–93.

249. G. C. Montanari, G. Mazzanti, M. Cacciari, J. C. Fothergill. Optimum estimators for the Weibull distribution of censored data. Singly censored tests. IEEE Trans. Dielectric and Electrical Insulation, 4: 1997, pp 462–469.

250. A. C. Cohen. Maximum likelihood estimation in the Weibull distribution based on complete and on censored samples. Technometrics, 7: 1965, pp 579–588.

251. E. F. Chace. Right-censored, grouped life test data analysis assuming a two-parameter Weibull distribution function. Microelec. and Reliabil., 15: 1976, pp 497–499.

252. D. I. Gibbons, L. C. Vance. A simulation study of estimators for the 2-parameter Weibull distribution. IEEE Trans. Reliability, R-30: 1981, pp 61–66.

253. J.-H. Heo, J. D. Salas, K.-D. Kim. Estimation of confidence intervals of quantiles for the Weibull distribution. Stochastic Environmental Research and Risk Assessment, 15: 2001, pp 284–309.

254. J. Jacquelin. Inference of sampling on Weibull representation. IEEE Trans. Dielectrics and Electrical Insulation, 3: 1996, pp 806–808.

255. J. Jacquelin. Inference of sampling on Weibull parameter estimation. IEEE Trans. Dielectrics and Electric Insulation, 3: 1996, pp 809–816.

256. R.-P. Vollertsen, A. Strong. Analysis and optimization of stress conditions for gate oxide wearout using Monte Carlo simulations. 2000 Int. Integrated Reliability Workshop, Final Report: 2000, pp 9–13.

257. F. Monsieur, E. Vincent, D. Roy, S. Bruyere, G. Pananakakis, G. Ghibaudo. Determination of dielectric breakdown Weibull distribution parameters confidence bounds for accurate ultra-thin oxide reliability predictions. Microelec. Reliabil., 41: 2001, pp 1295–1300.

258. K. V. Sichart, R.-P. Vollertsen. Bimodal lifetime distributions of dielectrics for integrated circuits. Quality and Reliability Engineering Int., 7: 1991, pp 299–305.

259. R.-P. Vollertsen. Pragmatic procedure for the estimate of thin dielectric reliability. Quality and Reliability Engineering Int., 8: 1992, pp 557–564.

260. Ph. Roussel, R. Degraeve, A. Kerber, L. Pantisano, G. Groeseneken. Accurate reliability evaluation of nonuniform ultrathin oxynitride and high-k layers. 2003 Int. Reliability Physics Sym. Proc.: 2003, pp 29–33.

261. Rolf-P. Vollertsen. Burn-In. 1999 Int. Integrated Reliability Workshop Final Report: 1999, pp 167–173.

3

DIELECTRIC BREAKDOWN OF GATE OXIDES: PHYSICS AND EXPERIMENTS

Ernest Y. Wu, Rolf-Peter Vollertsen, and Jordi Suñé

3.1 INTRODUCTION

The main properties of SiO_2 as a gate insulator and the basics of dielectric reliability characterization and related statistics were reviewed in the Chapter 2; we focus on the description of the properties of SiO_2 breakdown (BD) in this chapter. First, we deal with the relation between wearout and breakdown because of compelling evidence that relates the generation of structural and electrical defects during stress to oxide breakdown. We use experimental evidence as a base for discussion of the role of electric field and carrier energy in the process of defect generation. The relation between the generated defects and the triggering of the breakdown will then be established in the framework of the percolation model of oxide breakdown, which provides support to the use of the Weibull statistics for oxide reliability assessment. In Section 3.3, we present the basics of the three most important physics-based defect generation models: the thermochemical model, the anode hole injection model, and the anode hydrogen release model. However, there is still disagreement about the validity of these models so that we can not claim to present a closed and complete picture of this subject. Hopefully, our discussion and comparison of the advantages and drawbacks of these models will help the reader to acquire a critical general view of this important issue.

In Section 3.4, we review the main experimental results concerning temperature activation, voltage acceleration, and their interplay. We also discuss some relavant issues such as polarity dependence and the relation between stresses

Reliability Wearout Mechanisms in Advanced CMOS Technologies. By Strong, Wu, Vollertsen, Suñé, LaRosa, Rauch, and Sullivan
Copyright © 2009 the Institute of Electrical and Electronics Engineers, Inc.

performed under DC and AC conditions. Finally, we discuss how all these results are used for reliability projection methodology. In Section 3.5, we deal with some important issues related to the BD of ultra-thin insulators. In these oxides, the device/circuit failure does not always coincide with the breakdown occurrence; hence, it is worth studying the oxide degradation after the breakdown. After discussing relevant issues such as the separation of soft and hard breakdown modes, the statistics of multiple successive BD events, and the progressive evolution of the post-breakdown current, we discuss how to include the post-breakdown phase into the reliability assessment procedures.

3.2 PHYSICS OF DEGRADATION AND BREAKDOWN

It is currently widely accepted that oxide breakdown is related to defects generated in the oxide during electrical stress. Thus, chip reliability predictions require a thorough understanding of the process of defect generation and how it finally leads to the oxide breakdown. In this framework, there are two sets of questions that require a detailed consideration: 1) How does defect generation evolve during the stress time? How it depends on the stress conditions? What are the involved physics and what is the nature of the generated defects? 2) How do the generated defects trigger the breakdown? In this section, we focus on giving state-of-the-art answers to these questions. However, while question 2 has been successfully answered and the conclusions are quite firmly established, the former is still a subject of vivid debate and controversy. First we will deal with the process of oxide degradation, jointly reviewing the main experimental results and the methods used to monitor the evolution of the density of generated defects. Second, we will discuss the roles of the oxide electric field as well as the energy of the injected carriers in the process of defect generation and breakdown. The detailed discussion of several proposed theoretical models on degradation and breakdown in terms of physical mechanisms is left for Section 3.3. Third, we discuss how very simple geometrical models are able to relate the generation of defects to the triggering of oxide breakdown, thus providing a consistent explanation for the breakdown statistics.

3.2.1 Oxide Degradation

When an oxide is subjected to electrical stress, structural defects are continuously generated in the oxide bulk and its interfaces at a rate that strongly depends on the stress conditions (mainly voltage and temperature). As a consequence, the oxide's electrical properties gradually change and, finally, the dielectric breakdown is triggered. The molecular nature of the defects that cause the final oxide break-down is not well known yet, but the effects on the electrical properties of the oxide have been studied in great detail. A variety of phenomena related to the oxide degradation have been observed: positive charge trapping; generation of neutral electron traps (and the related trapping of electrons); generation of Si/SiO_2

interface states, increase of the low-field leakage current, etc. All of these effects witness the overall degradation of the oxide microstructure and, consequently, they have been used to monitor this degradation under different stress conditions.

The generation of interface states at the Si/SiO_2 has been measured by means of CV characterization of MOS structures and also by the charge pumping technique in transistors [1–4]. Figure 3.1 shows the interface state generation probability (the number of states generated per injected electron) and the density of interface states at breakdown as measured by DiMaria and Stathis on oxides subjected to constant voltage stresses [5]. Two important results are revealed in these figures. Figure 3.1 (a) shows that the average density of defects (N_{BD}) required to trigger the breakdown decreases with the oxide thickness and that this density is independent of the applied voltage (at least to the first order). On the other hand, it is also evident that the thinner the oxide is, the larger the statistical variations of the breakdown critical density of defects. These results will be better understood in Section 3.2.4, where we consider how the breakdown is triggered by electrically active defects and how this determines the breakdown statistical distribution. Figure 3.1 (b) shows another crucial result, namely, that the defect generation rate decreases very strongly with the applied voltage. The dependence of the defect generation rate on the stress conditions is of paramount importance for reliability extrapolations and will be considered in greater detail in Section 3.4. Although Figure 3.1 demonstrates that there is a correlation between oxide breakdown and interface state generation, the formation of a local conduction path through the oxide is inherent to the dielectric breakdown and requires the involvement of bulk defects rather than that of interface states. Thus, although the generation of interface states might be used as an indicator of the overall oxide degradation, the relation between interface states and breakdown is probably not causal. Other well known effects of oxide degradation are an initial increase of

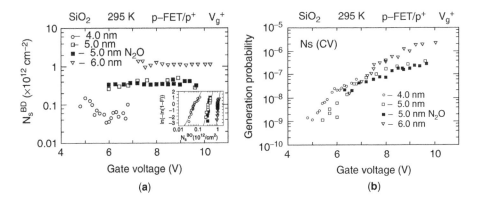

Figure 3.1. (a) Total number of interface states (per unit of area) at breakdown as a function of stress gate voltage [5] and (b) interface-state generation probability as a function of stress gate voltage [5]. Reprinted with permission from Ref. 5, Copyright 1997, American Institute of Physics.

positive charge trapping [6–8] and a long-lasting process of generation of neutral electron traps [9–12].

In relatively thick oxides (thicker than ~ 5 nm), the tunnel current evolves during stress due to the partial occupation of previously existing (native) and newly generated bulk traps. Under constant voltage stress conditions, the current increases in the initial stages of the experiments and then, after a turnaround, it continuously decreases until the breakdown, as illustrated in Figure 3.2 [13].

The initial current increase has been attributed to positive charge trapping in the oxide, while the long term decrease has been linked to electron trapping in newly generated trapping centers. Initially, some authors claimed that positive charge trapping was responsible for the triggering of the breakdown through a positive feedback mechanism involving local trapping and current increase in weak areas [6, 7, 14]. For completeness, it is worthwhile to note that in ultra-thin oxides (which do not allow the build-up of negative charge due to the easy tunneling of any trapped electron towards the electrodes) a weak positive charge transient is observed at the initial stages of stress [15]. However, in slightly thicker oxides that do allow electron trapping, the positive charge trapping is only dominant during the very early stages of the stress and, when the breakdown occurs, the net charge trapping is clearly negative. Moreover, while positive charge trapping saturates well before the breakdown occurrence (indicating that trapping occurs only in native traps), there is convincing experimental evidence of a nonsaturating negative charge build-up related to the generation of new electron trapping sites (neutral electron traps) [9–11, 13, 16, 17]. As in the case of interface states, the breakdown is observed to occur when the density of neutral electron traps reaches (on the average) a critical (thickness-dependent) value [11]. These and other experimental results suggest that electron traps are most likely the defects that are causally related to the triggering of oxide breakdown.

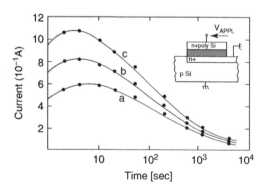

Figure 3.2. Turnaround effect in the tunnel current during constant voltage stresses on a 14-nm thick SiO_2 gate oxide. Curve (a) $V_G = 11.9$ V; (b) 12.05 V; and (c) 12.15 V. Reprinted with permission from Ref. 13, Copyright 1983, American Institute of Physics.

A trap filling step is required before sensing the trapped charge to quantitatively measure the density of generated electron traps and to compare the defect generation rate under different stress conditions. This is because most of the generated traps are empty under the high oxide electric fields inherent to the usual stress conditions of breakdown experiments. Moreover, the occupancy of the generated defects strongly depends on the oxide field [9] so that a comparison of electron-trap densities based on measurements of trapped charge performed under different stress conditions could lead to completely misleading conclusions. The substrate hot electron technique is ideally suited for filling the traps [3, 18] while the change of the flatband voltage shift as determined from the shift of the CV characteristic is a usual way to measure the trapped charge. As an example, Figure 3.3 shows the evolution of the density of (partially filled) electron traps as reported by De Blauwe et al. for 7.1 nm oxides subjected to constant current stresses [3].

In oxides thinner than about 5 nm, electron trapping is less important because electrons can easily tunnel out from the traps towards the electrodes. In these oxides, however, the generation of electron traps has the effect of remarkably increasing the leakage current at low fields, as shown in Figure 3.4.

The first experimental observation of this stress-induced leakage current (SILC) is attributed to Maserjian and Zamani [19]; the first systematic study is attributed to Olivo and colleagues [20]. The SILC has been related to the generation of electron traps [3–22] and has been explained in terms of trap-assisted tunneling mechanisms (the traps being assumed to act as stepping stones for electrons) [23–30]. The evolution of SILC with the stress time (see how the SILC increases up to the breakdown in the examples shown in Fig. 3.5) has also been used to monitor the oxide degradation [3, 22, 31, 32] and has been found to

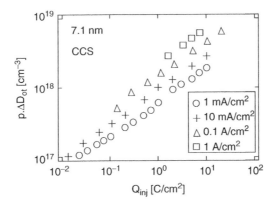

Figure 3.3. The generated oxide trap density $p \cdot N_t$ as measured on an NFET with a 7.1 nm gate oxide, versus the injected charge fluence Q_{inj}. (p represents the fraction of filled traps.) Data reported by De Blauwe et al. using substrate hot electron injection to partially fill the traps before detecting the trapped charge by measuring the flatband voltage shift. Reprinted with permission from Ref. 3, © 1998 IEEE.

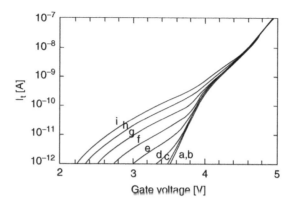

Figure 3.4. Change of the low field IV due to SILC. This figure is taken from Olivo et al. Reprinted with permission from Ref. 20, © 1988 IEEE.

be in good correlation with the results obtained by other methods. In this regard, the results of De Blauwe and colleagues [3] are particularly revealing because they reported linear relationships between the SILC, the interface state density, and the neutral electron trap density (see Fig. 3.6), thus justifying the alternative use of all these measurements to monitor the degradation of the oxide.

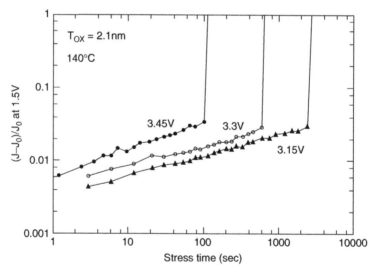

Figure 3.5. Evolution of the relative SILC current increase $\Delta J/J_o$ as a function of stress time in MOS devices with 2.1 nm oxide. J_o is the current density at the beginning of the experiment, i.e., $J(t=0)$, and $\Delta J \equiv J(t)-J_o$ is its increment at time t. The final current increase is due to breakdown. The linear relation in the log–log representation indicates that the evolution follows a power law $\Delta J/J_o = at^b$. Data from Lai et al. Reprinted with permission from Ref. 33, © 2004 IEEE.

As we will see in Section 3.2.4, the triggering of the breakdown is considered to occur when the formation of a chain of electron traps leads to a fast current runaway. The probability of formation of such a path of defects statistically depends on the average density of generated defects and increases quite abruptly (according to the percolation theory) when this density reaches a certain critical value, N_{BD}. The value of the relative SILC increase at breakdown has also been proposed as a measure of this critical defect density [34]. The fact that the statistical distribution of the SILC at breakdown is experimentally found to be relatively insensitive to the stress voltage somewhat supports this idea [34]. From the point of view of oxide reliability, however, breakdown is usually characterized in terms of the time-to-breakdown (t_{BD} or T_{BD}) or the charge-injected-to-breakdown (q_{BD} or Q_{BD}) which depend not only on the critical defect density but also on the rate at which the defects are produced during stress. The defect generation rate (which strongly depends on the stress conditions) and the critical defect density (which mainly depends on the oxide thickness and area) are the main elements required to relate the generation of defects to the breakdown.

Figure 3.7 represents the defect generation rate as a function of the stress voltage as obtained by Stathis and DiMaria from measurements performed on oxides with different thickness [35, 36]. By combining the results shown in Figure 3.7 with the measurements of the critical defect density as a function of oxide thickness, we can predict the time-to-breakdown and forecast chip reliability as a function of gate bias down to operation conditions [35, 36]. This procedure has substantially contributed to our present understanding of the relation that exists between degradation and breakdown. Nevertheless, although valuable, these predictions

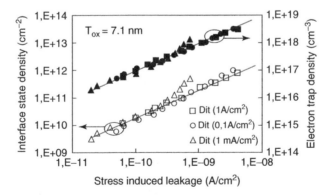

Figure 3.6. Correlations between SILC and interface state density and electron trap density generated in a 7.1 nm oxide subjected to different constant current stresses. Interface state density was measured using the charge pumping technique, electron traps by CV flatband voltage shift (after filling the traps by substrate hot electron injection) and SILC is the excess leakage measured at an oxide field of 5 MV/cm. The lines correspond to linear correlations (slope = 1). Data points are taken from De Blauwe et al. [3].

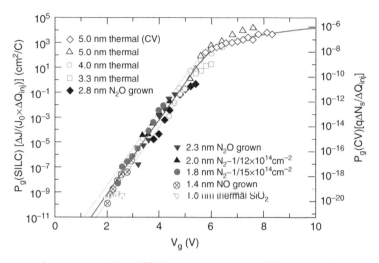

Figure 3.7. Defect generation efficiency ζ as reported by Stathis and DiMaria from SILC and CV stretch-out from measurements on different oxide thickness as a function of the stress voltage. The solid line is an exponential fit to the experimental data for $V_G < 5.6$ V. Reprinted with permission from Ref. 35, © 1998 IEEE.

rely on the assumption that the measured defects are actually those that trigger the breakdown of the oxide. Moreover, as we have seen above, all of the oxide degradation monitors are based the measurement of some electrical effects of the generated traps and not the direct measurements of the trap defect density itself. The new understanding based on the comparison of the breakdown and SILC statistics [37]; the finding of different annealing kinetics for both phenomena [38]; and on other subtleties related to the SILC definition and voltage extrapolation [39] have suggested that it is questionable to use SILC measurements to predict breakdown.

Direct breakdown measurements are required to avoid the uncertainties derived from the indirect measurement of defect generation. The problem is that these measurements require performing stresses of much longer duration (each sample must be stressed until the breakdown) in a large number of devices (typically 40–100) to obtain statistically significant values of the relevant magnitudes (mean time-to-breakdown, distribution slope, etc.). These experiments offer the ability to make reliability extrapolations based directly on breakdown data and, at the same time, provide an indirect method to obtain information about defect generation. In particular, it is possible to determine the critical breakdown defect density [40–43] from the slope of the breakdown distribution. From the mean charge-to-breakdown (or mean time-to-breakdown), the defect generation efficiency (or the generation rate) can be obtained as a function of stress conditions if we assume the validity of the percolation model for the breakdown statistics [40].

Figure 3.8 shows some results obtained with this method [44] (and the details of this "reverse engineering" procedure will be explained in a later section). The data show that at low voltages defect generation efficiency decreases rapidly with decreasing voltages while at high voltage its dependence on voltage is rather weak similar to those in Fig. 3.7. In Figure 3.7 and Figure 3.8, the two sets of data have been separately fitted to compare with the results reported from SILC and CV measurements. This comparison certainly gives further support to the idea that oxide breakdown is directly related to the generation of defects in the oxide. However, it also highlights the great difficulties when making reliable extrapolations down to operation conditions. While both SILC/CV and breakdown-based measurements of the defect generation efficiency (ζ) give results of the same order of magnitude, significant differences in the dependence on the stress voltage are revealed. While the results derived from breakdown data are nicely fitted to a power law $\zeta \propto V_G^{38}$ for $V_G < 3\,V$, the SILC measurements are better fit to an exponential law $\zeta \propto \exp(7 V_G)$ for $V_G < 5.6\,V$. As previously discussed for other degradation monitors, the "reverse engineering" approach (based on the derivation of the defect generation efficiency from breakdown data) is certainly an indirect method. However, this method has the advantage that the obtained information is certainly related to the defects that are actively involved in the process of triggering the breakdown [44].

To finish this section, it is worth referring to recent results by Degraeve et al. who have dealt with the discrete nature of SILC in small devices (oxide area down to $6.25 \times 10^{-10}\,cm^2$) with ultra-thin oxides [46]. While the breakdown is a single event (discrete in time) that is related to the random creation of a local conduction path of very small dimensions (compared to the sample area), the degradation of

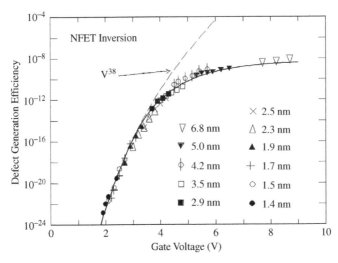

Figure 3.8. Defect generation efficiency obtained from charge-to-breakdown (the details of the procedure will be explained after discussing the percolation models for the breakdown statistics). Symbols correspond to data points and lines are empirical fits. Reprinted with permission from Ref. 45, © 2009 IEEE.

the oxide has been usually considered to be a process that is uniform across the oxide area and continuous in time. This is because the area of the test structures has usually been large enough to detect the average density of generated defects rather than the effect of each single particular trap. However, if the samples are small enough (i.e., so that few defects are generated per unit of time) and the oxides thin enough (i.e., so that each generated defect produces a large (measurable) increase of the tunnel current) it is possible to isolate the current–voltage characteristics of single-trap leakage paths and to evaluate their statistical properties [46]. Using this technique, the process of defect generation is followed trap by trap; this can eventually provide new information about the relation between defect generation and breakdown. This is particularly relevant in the case of ultra-thin oxides for which the percolation models predict that one single defect triggers the breakdown and reaches the limit of validity [41–43].

3.2.2 The Role of Electric Field and Carrier Energy in the Degradation and Breakdown of Gate Oxides

When a gate bias is applied to a MOS structure, an electric field builds up in the oxide layer while little current flows due to the insulating properties of the oxide. The fact that the oxide electric field (F_{OX}) has an important impact on the oxide degradation process has been well known to physicists and engineers from the early times of breakdown studies [47–49]. In fact, if one plots the time to breakdown versus the oxide field, an exponential $T_{BD} \propto \exp(-\gamma_F F_{OX})$ relation (or even stronger dependencies [50, 51]) is always found, as it will be explicitly shown in Section 3.4. Actually, some authors have considered the electric field is not only an important factor but the direct cause of defect generation through a process of field-assisted thermally activated bond breakage mechanisms [49, 52]. The oxide breakdown would simply be a field and time driven process within this picture. However, there is compelling experimental evidence that both electron fluence and electron energy play a crucial role [53–56] even though the current that flows through the insulator is small.

Experiments that use the injection of hot electrons from the substrate of NFETs are particularly useful to explicitly demonstrate the role of current and energy. These experiments allow the researcher the flexibility to separately change the oxide field, the stress current, and the distribution of energy of the injected carriers. To perform these experiments, special test structures like the one schematically represented in Fig. 2.25 (a) are used. These structures consist of an NFET with a separate N^+P injector. Source and drain are connected together to ground. The gate voltage is varied to change the oxide field while the transistor is kept in inversion. The injector voltage does not modify the oxide field but changes the gate current by injecting carriers to the substrate depletion region. Finally, the substrate terminal can be biased to change the energy distribution of the injected electrons, as shown schematically in Figure 2.25 (b). The breakdown results that correspond to conventional constant voltage stress conditions ($V_{inj} = 0$) and those obtained under the combination of substrate hot electron and constant voltage stress ($V_{inj} \neq 0$) are compared in Figure 3.9 [53]. The time to breakdown is drastically reduced by the injection of hot carriers (while keeping the oxide field unchanged), and this conclusively demonstrates that the breakdown is not controlled solely by the oxide electric field. On the contrary, the injection of carriers and their energy distribution at the anode are relevant variables that significantly affect the dynamics of defect generation and breakdown.

In fact, there are defect generation models that are based on the injection of carriers and the dissipation of the carrier's energy at the anode interface. This electron energy driven approach has two main versions: the anode hydrogen release (AHR) model [10, 35, 36, 44, 57–60], and the anode hole injection (AHI) model [6–8, 61–63], which will be explained in detail in Section 3.3. In these models, defect generation is described as a two-step process: 1) Electrons are injected through the oxide and lose their energy at the anode interface where they release some positively charged species. 2) The released species are reinjected into the oxide, travel towards the cathode, and generate defects by reaction with some unspecified precursor sites. This two-step process is schematically depicted in Figure 3.10. In this figure, ζ_1 represents the efficiency of the injected electrons to release species from the anode interface, ζ_2 the efficiency of the generated species to generate defects in the oxide bulk, and ζ_3 the probability to escape from the oxide bulk towards the electrodes.

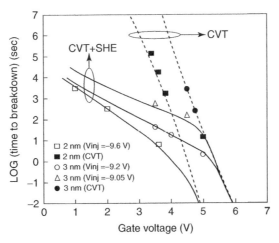

Figure 3.9. Time-to-breakdown versus gate voltage for constant voltage tunneling (CVT) stress and for the combined CVT and substrate hot electron (SHE) injection experiments as reported by Vogel and colleagues. Reprinted with permission from Ref. 53, © 2000 IEEE. Lines correspond to models as also reported in [53].

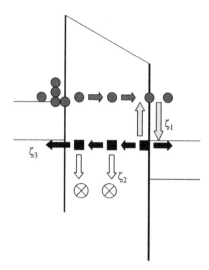

Figure 3.10. Schematic picture of the defect generation process in electron energy-based breakdown models. Electrons are injected through the oxide and partially lose their energy at the anode interface to release some positively charged species (holes or protons) with an efficiency ζ_1. On the other hand, ζ_2 represents the efficiency of the released species to create defects by reaction with bulk precursors and ζ_3 the probability of losing the generated species towards the electrodes.

Assuming that the injected carriers are responsible for the generation of defects in the oxide, the dynamics of defect generation can be generically written as

$$\frac{dN(Q)}{dQ} = \frac{\zeta}{qT_{OX}}, \tag{3.1}$$

where Q is the total charge fluence per unit area, ζ is the defect generation efficiency, and N is the density of generated defects per unit of oxide volume. To calculate ζ in terms of the efficiencies of the two processes (ζ_1 and ζ_2) involved in the two-step models of breakdown, one can start from a simple picture based on a set of two linear differential equations.

$$\frac{d\rho}{dQ} = \zeta_1 \left\{ \frac{1}{qT_{OX}} - \frac{\rho}{Q_o} \right\} - \zeta_2 \frac{\rho}{Q_o} - \zeta_3 \frac{\rho}{Q_o} \tag{3.2}$$

and

$$\frac{dN}{dQ} = \zeta_2 \frac{\rho}{Q_o}, \tag{3.3}$$

With ρ being the concentration of generated species in the oxide (implicitly assumed to be uniform in the oxide bulk) and Q_o a characteristic charge constant. The first equation determines the evolution of concentration of released species in the oxide as a balance between the three processes schematically depicted in Figure 3.10, namely, the net release of species from the anode interface; the reaction with precursors in the bulk to generate defects; and the loss of species to the silicon electrodes. The efficiency factors ζ_1, ζ_2, and ζ_3 should include the details of the physics of the corresponding processes and will strongly depend on the stress conditions. Assuming that ρ reaches a steady state and that most of the released species finally escape to the silicon electrodes or react with precursors to generate defects in the oxide bulk instead of recombining with empty sites at the interface, i.e., $\zeta_3 + \zeta_2 \gg \zeta_1$, it follows that

$$\frac{dN}{dQ} \approx \frac{\zeta_1 \zeta_2}{\zeta_3 + \zeta_2} \frac{1}{qT_{OX}} = \frac{k\zeta_1}{qT_{OX}} \tag{3.4}$$

where $k \equiv \frac{\zeta_2}{\zeta_3 + \zeta_2}$ is the probability that one released species finally generates one defect. Thus, comparison with Equation 3.1 tells us that the defect generation efficiency, ζ, can be written as the product of the efficiency of the injected electrons to release species from the anode (ζ_1) and the probability (k) that these released species react with precursors to generate a defect:

$$\zeta = k\zeta_1 . \tag{3.5}$$

This equation is basic in the formulation of the AHI and AHR models, usually assuming that k is temperature-dependent but roughly independent of the stress bias (or carrier energy), while the opposite is true for the release efficiency ζ_1. Since the actual nature of the generated defects and the generation reaction are still unknown, both the AHI and the AHR models have focused the attention on understanding the physics of hole or hydrogen release from the anode so as to obtain a reliable quantitative picture for ζ_1, while treating k as a fitting parameter.

The solution of Equation 3.1 is $N(Q)$, i.e., the evolution of the density of defects as a function of the injected charge. If ζ were independent of N and Q, the solution would be a linear relation:

$$N = \frac{\zeta}{qT_{OX}} Q \tag{3.6}$$

In our analysis of the breakdown data, we will assume that this is the case, although it must be acknowledged that some experiments suggest a power law dependence $N \propto Q^\alpha$ with $\alpha < 1$ [41, 60].

In the following subsection, we will argue that the breakdown occurs when the density of generated defects reaches a critical value (N_{BD}) that depends on the oxide thickness and oxide area in a well understood manner. Thus, the mean charge-to-breakdown can be expressed as

$$Q_{BD}(V_G, T, A_{OX}, T_{OX}) = \frac{qT_{OX}}{\zeta(V_G, T)} N_{BD}(A_{OX}, T_{OX}) \tag{3.7}$$

and, assuming that the current density remains reasonably constant during the stress experiment (an assumption that fails when electron trapping in the generated traps is important, i.e., in thick oxides), the mean time-to-breakdown is given by

$$T_{BD}(V_G, T, A_{OX}, T_{OX}) = \frac{q T_{OX} N_{BD}(A_{OX}, T_{OX})}{\zeta(V_G, T) J(V_G, T, T_{OX})}. \qquad (3.8)$$

This equation is central to our present understanding of oxide reliability since it relates the oxide lifetime, i.e., the main reliability variable, to the oxide wearout in terms of the critical density of defects required to trigger the breakdown and of the efficiency of the injected electrons to generate those defects.

3.2.3 Percolation Model for the Breakdown Statistics

In the previous sections, we have claimed that there is a direct relation between oxide breakdown and the generation of defects during stress or operation. On the other hand, we know that the breakdown is not a deterministic phenomenon, meaning that we cannot expect to find the same value of breakdown field, breakdown voltage, or time-to-breakdown (or any other breakdown variable we might choose) if we repeat the same stress experiment in nominally identical samples. On the contrary, the breakdown has an intrinsically random character and the use of statistical distributions is required to describe the experimental results. Some attempts to understand the physical origin of breakdown statistics were made in the framework of old breakdown models such as those based on multiple avalanches [47] and band-to-band impact ionization [64]. One of the most complete early treatments of this problem is due to Hill and Dissado [48], who found a justification for the Weibull distribution of the breakdown field in terms of time-dependent electric field fluctuations. However, since breakdown has been linked to defect generation, a way must be found for relating the breakdown statistics to the generation of defects. The first defect related approach to the breakdown statistics is due to Suñé and coworkers [40], who proposed a two-dimensional (2D) model based on a partition of the oxide area into small cells of area S_o ($\ll A_{OX}$) and the triggering of breakdown by the generation of a critical number of defects in one of those cells. Later, Dumin and colleagues [12, 65] revealed the potentialities of that 2D model and systematically compared it with experimental results by coupling it with their empirical picture of defect generation. They demonstrated that the model was adequate to fit the breakdown statistics [12, 65] and gave the first estimation for the breakdown spot area (10^{-14}–10^{-12} cm^2).

Degraeve made a step further and proposed the percolation breakdown model, which has been very successful [41]. This is a three-dimensional (3D) picture, in which the defects are modeled by spheres (the defect radius is the single model parameter) which are generated at random in the oxide volume, and the breakdown is triggered when a path of overlapping spheres (a percolation path of defects) connects the two electrodes. Stathis reported comparable results using a similar percolation approach though using a fixed cubic lattice [66]. This author recently reviewed the percolation models for oxide breakdown [42] and studied the case of nonhomogeneous spatial distribution in the direction perpendicular to the oxide film. The main achievement of the percolation model is the presentation of a sound explanation for the thickness dependence of the breakdown statistics. On the other hand, it is a very nice approach that allows the incorporation of the details of the degradation physics at the microscopic level, such as feedback effects at the most degraded regions of the oxide that modify the usual assumption of random defect generation [40, 67]; the geometrical dependencies of the path efficiency [42, 68]; or the microscopic effects of the temperature [69, 70].

Although some convoluted analytical version of the percolation model exists [41, 71], this approach generally requires numerical simulation and this somehow limits its application. A simpler analytical picture is highly desirable for practical reliability assessment methodologies. A first proposal in this direction was made by Chen and coworkers [72] who used the results of percolation simulations to fit the parameters of a compact analytical model based on Suñé's 2D picture [40]. Subsequently, a new physics-based 3D analytical picture was developed that provides results equivalent to those of the percolation approach [43]. We will explain this model in detail because it explicitly illustrates how the density of defects can be related to the triggering of the breakdown and because it will be used as a reference framework in the rest of this chapter to discuss important issues such as the thickness and voltage (energy) scaling of the breakdown variables.

3.2.4 Three-Dimensional Analytic Model for Oxide Breakdown Statistics

First, a cubic structure with a lattice constant (a_o) is defined in the oxide bulk. In other words, the oxide is divided in cubic cells of volume (a_o^3). In this cubic lattice, we can distinguish $N_C = A_{OX}/a_o^2$ columns of area (a_o^2) and thickness (T_{OX}), with each column being subdivided in $n_{BD} = T_{OX}/a_o$ cells. During stress, defects are considered to be generated at random in the cells of the cubic lattice. The generation of one defect is modeled as *the switching of one cell to a defective state*. Breakdown is assumed to take place when all the cells in one column are defective. Consequently, n_{BD} is the critical number of defects required to build up a path that triggers the breakdown. Figure 3.11 represents the breakdown condition and compares it schematically with the breakdown condition in the percolation model.

Under the previous assumptions, it is possible to calculate the breakdown cumulative failure distribution as a function of the average fraction of defective cells (λ), which only takes values between 0 and 1. Since λ can also be understood as the probability of each cell to be defective, the probability that all the cells in one column are defective is $\lambda^{n_{BD}}$.

Since the breakdown has an intrinsic weakest link character, i.e., since a device with N_C columns fails when one of its columns is fully defective or, to state

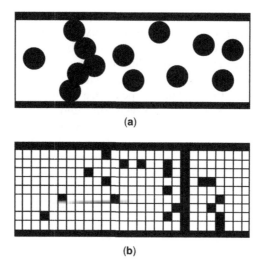

(a)

(b)

Figure 3.11. Schematic picture showing the generation of defects and break-down triggering condition in (a) the percolation approach and (b) the analytical cell-based approach. Reprinted with permission from Ref. 43, © 2001 IEEE.

it another way, when a complete defect related path has been formed, the probability of survival of the device is [43]

$$1 - F_{BD}(\lambda) = [1 - \lambda^{n_{BD}}]^{N_C}, \tag{3.9}$$

where F_{BD} is the cumulative breakdown distribution. Since F_{BD} will be conveniently approximated by a Weibull distribution, the quantity $W_{BD} \equiv Ln(-Ln(1-F_{BD}))$, known as the weibit, is convenient for graphical representation. From Equation 3.9, it follows that

$$W_{BD}(\lambda) = Ln\left[-N_C \, Ln(1 - \lambda^{n_{BD}})\right]. \tag{3.10}$$

This expression can be further simplified if the condition $\lambda \ll 1$ is satisfied at breakdown. Under this assumption, it follows that $Ln(1 - \lambda^{n_{BD}}) \approx -\lambda^{n_{BD}}$, and the weibit becomes

$$W_{BD}(\lambda) = Ln(N_C) + n_{BD} Ln(\lambda) \tag{3.11}$$

For usual devices and test structures, the error associated to this approximation is negligible at the low failure percentiles of interest for reliability and also in the range of experimentally accessible values of W_{BD}, which is limited by the number of samples used to measure the breakdown distribution. Equation 3.11 is the main equation of a single-parameter analytic model for the breakdown statistics (note that both N_C and n_{BD} are functions of a_o). Since the density of defects per unit of volume is $N = \lambda/a_o^3$, the breakdown distribution can be written as a function of N as:

$$W_{BD}(N) = Ln(A_{OX}/a_o^2) + \frac{T_{OX}}{a_o} Ln(a_o^3 N) \tag{3.12}$$

In this equation, the well-known area effect (Section 2.4.2) associated with the weakest link character of the breakdown and with the random distribution of defects appears in a natural way through the term $Ln(A_{OX}/a_o^2)$. Since W_{BD} is a straight line when plotted versus $Ln(N)$ (i.e., when F_{BD} is plotted in the Weibull plot), we confirm that F_{BD} is a Weibull distribution. The Weibull slope is

$$\beta(T_{OX}) = T_{OX}/a_o \tag{3.13}$$

and the average density of defects at breakdown N_{BD} can be calculated by imposing $W_{BD}(N = N_{BD}) = 0$ so

$$N_{BD} = \frac{1}{a_o^3} \exp\left[-\frac{1}{\beta(T_{OX})} Ln\left(\frac{A_{OX}}{a_o^2}\right) \right]. \tag{3.14}$$

Note that this model predicts that N_{BD} strongly depends on T_{OX} but does not depend on the stress voltage. This is exactly what is usually experimentally observed, as shown in Figure 3.1 (a), although some long term stress results of ultra-thin oxides suggest (a suggestion based on SILC data) a possible increase of N_{BD} at low voltages [36, 73]. This analytic model has the same predictive power as the percolation models in their standard version, i.e., considering a random distribution of defects in the oxide bulk and no local feedback effects [41, 66]. It has the advantage of providing analytical equations but the drawback of higher difficulty to introduce modifications of the basic assumptions. For example, considering a nonhomogenous distribution of the defects in the oxide bulk (with higher concentrations near the interfaces, for instance), it is a trivial issue in a numerical percolation approach [42] and very difficult (if at all possible) within this analytical framework [43]. However, the basic versions of these models are fully equivalent and we will consider them as two particular versions of the general category of percolation models.

Since the density of defects involved in oxide breakdown cannot be accurately measured, the comparison with experiment requires introducing the relation between the average density of generated defects and the stress time or the cumulative injected charge. We assume a linear relation between the density of generated defects and the injected charge for simplicity (and consistence with Eq. 3.6). Under this assumption, the weibit of the charge to breakdown statistical distribution is

$$W_{BD}(Q) = Ln(A_{OX}/a_o^2) + \beta(T_{OX})Ln\left(\frac{a_o^3 \zeta}{q\, T_{OX}} Q\right), \tag{3.15}$$

and the mean charge-to-breakdown is given by

$$Q_{BD} = \frac{q\, T_{OX}}{a_o^3 \zeta} \exp\left[-\frac{1}{\beta(T_{OX})} Ln\left(\frac{A_{OX}}{a_o^2}\right) \right] \tag{3.16}$$

The Weibull slope in Equation 3.15 is the same as that of Equation 3.12, due to the considered proportionality between injected charge and the density of generated defects. If the relation between generated defect density and injected charge were a power law $N \propto Q^{\alpha}$ (with $\alpha \neq 1$), as suggested by some experiments [3, 11, 33, 45], the Weibull slope of the q_{BD} distribution would be that of Equation 3.12 multiplied by α [11, 37]. Any other $N(Q)$ functional dependence would change the shape of this distribution so that it would no longer be a Weibull distribution. Equations 3.15 and 3.16 explicitly reveal the dependence on the oxide thickness and, assuming that ζ is independent of A_{OX} and T_{OX}, allow the comparison with the results of breakdown experiments as a function of T_{OX}. The thickness dependence of the breakdown distribution is the main successful prediction of the percolation model, as shown in the next section.

3.2.5 Thickness Dependence of Oxide Breakdown

3.2.5.1 Thickness Dependence of Weibull Slopes

The experimental Weibull slope of the cumulative distribution of t_{BD} (or q_{BD}) decreases with decreasing oxide thickness as shown in Figure 3.12a. Both the percolation model [41, 42] and the cell-based analytical model presented in the previous section [43] predict that the Weibull slope of the t_{BD} (or q_{BD}) distribution should linearly decrease with decreasing oxide thickness. As shown in Figure 3.12b, the experimental data are in agreement with this prediction and the dependence of β on T_{OX} can be fitted to a straight line. However, the extrapolation to the limit $T_{OX} \rightarrow 0$ does not yield $\beta \rightarrow 0$ and an interfacial layer thickness T_{INT} is introduced as an ad hoc concept [11, 43]:

$$\beta = (T_{INT} + T_{OX})/a_{o}. \tag{3.17}$$

Within the percolation model, the linear reduction of the Weibull slope with T_{OX} scaling is a consequence of the linear decrease of the critical number of defects (n_{BD}) required to trigger a BD path [41–43].

The results shown in Figure 3.12(b) were collected using different stress voltages that were chosen to keep t_{BD} within a reasonable experimental time window for all the values of T_{OX}. Consequently, the possibility that β is a function of the stress voltage (rather than a function of T_{OX}) must be considered [68]. Similarly, some authors have suggested that the Weibull slope could be temperature-dependent [69, 70]. This would have a significant impact on reliability projection; therefore, the voltage and temperature dependence of Weibull slopes have been studied in detail. Figure 3.13 displays the measured Weibull slopes as a function of voltage. A wide range of voltage from 5.4 V to 3.2 V was used for 2.7 nm oxides and from 5.7 V to 4.7 V for 4.2 nm oxides. The results reveal that the dependence of the Weibull slope on voltage is not significant [74]. To obtain these results within a reasonable experimental time window, structures with different oxide areas had to be used. Figure 3.14 shows the measured relationship between β and T_{OX} at various temperatures, indicating no significant temperature dependence for

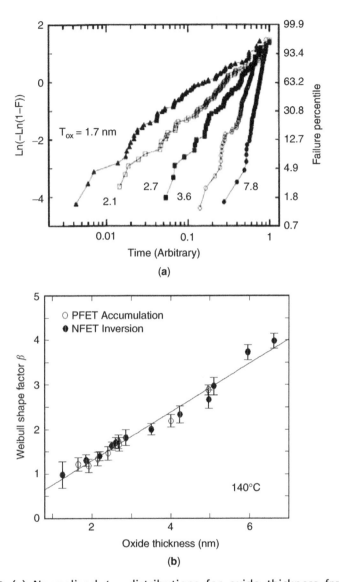

Figure 3.12. (a) Normalized t_{BD} distributions for oxide thickness from 1.7 to 7.8 nm at 140°C. The data was collected using different stress voltages. (b) Weibull shape factor (Weibull slope) as a function of oxide thickness. The linear fit of the data to Equation 3.17 yields a defect size, $a_o = 1.83$ nm and an interfacial layer thickness, $T_{INT} = 0.37$ nm. Reprinted with permission from Ref. 74, © 2002 IEEE.

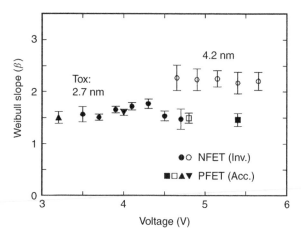

Figure 3.13. Weibull slopes versus stress voltage for two T_{OX} values of 2.7 nm and 4.2 nm oxides at 140°C. Reprinted with permission from Ref. 74, © 2002 IEEE.

Weibull slopes up to 200°C. Similar results showing voltage and temperature-independent Weibull slopes have also been reported by other authors [75].

These results show that the simple geometric picture proposed in the percolation model of breakdown [41–43] is consistent with the experimental statistics of q_{BD} or t_{BD}. Actually, the percolation picture is one of the few widely recognized solid achievements in the physical interpretation of the breakdown phenomenon. However, the case of ultra-thin oxides is particularly challenging because for $T_{OX} \sim a_o$ the validity limit of the percolation model is reached.

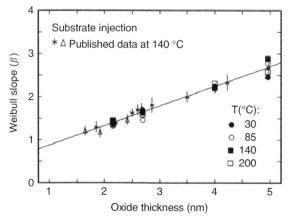

Figure 3.14. Weibull slopes versus oxide thickness measured at 30°C, 85°C, 140°C, and 200°C for 5.0, 4.0, 2.7, and 2.15 nm oxides using area-scaling method. The published data is from Figure 4 (a) in [74]. Reprinted with permission from Ref. 74, © 2002 IEEE.

At this limit, some refinements of our understanding of how electrically measured defects are related to the triggering of the breakdown are certainly required.

3.2.5.2 Thickness Scaling of Charge-to-Breakdown and Time-to-Breakdown.

While the Weibull slopes can have important impact on the extrapolations to chip area at low failure percentiles required for reliability projection, an even more critical issue is the scaling of T_{BD} or Q_{BD} with decreasing T_{OX} at a fixed stress condition. Figure 3.15 displays Q_{BD} data as a function of T_{OX} for several voltages. In order to keep a fixed voltage in a wide range of T_{OX}, the experimental data was collected by using different area structures and bonded-module stress for very long stress times up to more than half a year. Individual q_{BD} values were obtained by integrating the time dependent stress-current until time-to-breakdown is reached. Then the q_{BD} data were normalized to a reference area of 10^{-4} cm^2 (Section 2.4.2). The prediction of the percolation model for the thickness scaling of Q_{BD} is given by Equation 3.16. The experimental results show that ζ is strongly dependent on the stress conditions (voltage and temperature) but independent (to the first order) of the oxide thickness, as we will further discuss in the subsequent sections. The solid lines in Figure 3.15 correspond to the prediction of the cell-based model as given by Equation 3.16 for a single set of geometrical parameters (a_o and T_{INT}), and fitting the data using ζ as a voltage dependent parameter.

For Weibull slopes, Equation 3.17 is used with a unique set of the parameters ($a_o = 1.83$ nm and $T_{INT} = 0.37$ nm) independently derived from the β dependence

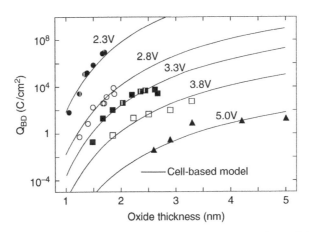

Figure 3.15. Q_{BD} thickness scaling law in comparison with the cell-based mode. Symbols represent experimental data in NFET inversion mode at 140°C. Lines for the cell-based model (see Eq. 3.17) with $a_o = 1.83$ nm, $T_{INT} = 0.37$ nm. A voltage-dependent (but thickness-independent) defect generation efficiency, $\zeta(V_G)$, was used to fit the data (i.e., the voltage independent curve resulting from Eq. 3.16 is vertically shifted for each particular value of the stress voltage). The reference area is 10^{-4} cm^2. Reprinted with permission from Ref. 161, Copyright 2005, Elsevier.

on T_{OX} (see Fig. 3.12b) and assuming voltage- and temperature-independent Weibull slopes as validated in Figures 3.14 and 3.15. Given the simplicity of the picture, the agreement between model and experiment is remarkable. While Q_{BD} is important from the viewpoint of breakdown physics, T_{BD} is more relevant for reliability assessment. For ultra-thin oxides with limited charge trapping, T_{BD} can be well approximated as the ratio of Q_{BD} to the tunneling current $J(V_G, T_{OX})$:

$$T_{BD} = Q_{BD}(A_{OX}, T_{OX}, V_G)/J(V_G, T_{OX}). \qquad (3.18)$$

For ultra-thin oxides operating in the direct tunneling regime, we can use a simple approximate expression

$$J(V_G, T_{OX}) = I_0(V_G)\exp(-\eta T_{OX}) \qquad (3.19)$$

where η is a fitting constant to describe the thickness dependence of the tunnel current at a fixed voltage. Figure 3.16 compares the thickness dependence of Q_{BD} with that of T_{BD} using $\eta = 4.5$ dec/nm as in [35]. It can be seen that for T_{OX} varying from 1.0 nm to 3.5 nm, Q_{BD} decreases by about 10 orders of magnitude, while the decrease of T_{BD} is as large as 20 orders of magnitude due to the strong increase of the direct tunneling current with decreasing T_{OX}. This demonstrates that if the tunneling current could be reduced by process improvement, this would not only alleviate the leakage current problem but significantly improve T_{BD} as well.

The experimental T_{BD} data is plotted in Figure 3.17 as a function of T_{OX} for several stress voltages. Notice that the range of T_{BD} at a fixed voltage spans over 10 orders of magnitude. As in the analysis of $Q_{BD}(T_{OX})$, an arbitrary constant (a thickness independent number which is different for each stress voltage) is used to

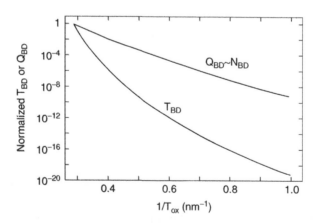

Figure 3.16. Comparison for thickness dependence of T_{BD} and Q_{BD} using Equations 3.16–3.19 with $\eta = 4.5$ dec/nm. These results emphasize the importance of the increase of the tunneling current in the degradation of the oxide reliability when oxide thickness is scaled down. Reprinted with permission from Ref. 161, Copyright 2005, Elsevier.

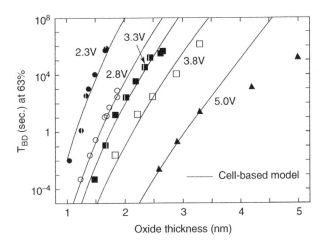

Figure 3.17. The average time-to-breakdown as a function of T_{OX} with the stress voltage as a parameter. The lines are obtained using Equations 3.16 and 3.19 with $\eta = 4.5\,dec/nm$. The reference area is $10^{-4}\,cm^2$. Reprinted with permission from Ref. 161, Copyright 2005, Elsevier.

vertically shift the curves and fit the data. This fitting parameter corresponds in this case to the product $\zeta(V_G)J_O(V_G)$ resulting from Equation 3.18. Again, the agreement between the experimental data and the modeling results is excellent for thin oxides. For thick oxides stressed at 5 V, FN tunneling currents dominate, thus the use of a direct tunnel equation overestimates the contribution of tunneling current. Figures 3.15 and 3.17 suggest that the thickness dependence of the critical defect density, N_{BD}, and that of the direct tunneling current explain the drastic degradation of T_{BD} with decreasing T_{OX}. On the contrary, the defect generation efficiency is found to be independent of oxide thickness to the first order, as we will further discuss in the following sections.

3.3 PHYSICAL MODELS FOR OXIDE DEGRADATION AND BREAKDOWN

In the previous sections we have discussed how the oxide degradation is monitored by different methods, and what the role of electric field, charge fluence, and carrier energy is. Moreover, we have related the generated defects to the breakdown statistics by means of simple geometrical models (the percolation picture). In this section, we review the present understanding of the physical mechanisms involved in the generation of the oxide defects that lead to oxide breakdown. Unfortunately, there is no unique theory to explain defect generation; there are still many open questions about the actual physics involved. Two main approaches have been considered in the most recent literature. On one hand, there is the thermochemical model of McPherson and other authors; they considered that the generation of defects is driven by the electric field [49, 52, 76–83]. On the other hand, there are several models that are based on the injection of carriers through the oxide and on the dissipation of electron energy near the anode interface [1, 8, 10, 14, 16, 35, 36, 44, 53–56, 61–63, 84, 85]. Here, we will mainly review three of these models, the thermochemical model, the anode hole injection (AHI) model, and the anode hydrogen release (AHR) model. However, an overview of older models based on hole injection and positive feedback due to positive charge trapping will also be included for completeness.

3.3.1 The Thermochemical Model

As mentioned earlier, an exponential dependence of the time-to-breakdown (T_{BD}) on the applied gate voltage or on the oxide electric field has been widely assumed for reliability extrapolation as a conservative approach. This exponential law found some theoretical support in the framework of the so-called thermochemical model [49, 52, 76–79], which describes a simple process of molecular dipole–electric field interaction as being responsible for the defect generation that ultimately causes the oxide breakdown. The involved dipole moment was first assumed to correspond to that of $O_3 \equiv Si–Si \equiv O_3$ defects that arise wherever there is an oxygen vacancy. The basic idea is that when an electric field is applied on this dipole moment, the activation energy for bond breakage is reduced, thus leading to enhanced defect generation. Figure 3.18 (a) shows a schematic drawing of the local tetrahedral environment of SiO_2 network. The bond angle for the Si–O–Si bond varies from 120° to 180°, whereas the angle of the O–Si–O bond is around 109°, as shown in Figure 3.18 (b). The strength of these bonds depends on the local environment in the SiO_2 network. If an oxygen vacancy is present, a Si–Si bond will form in place of the Si–O–Si bond as shown in Figure 3.18 (c). This molecular defect ($O_3 \equiv Si–Si \equiv O_3$) is assumed to be the precursor for the defect generation process that eventually leads to the oxide breakdown. When an electric field is applied, the permanent dipole moment of one half of the molecule tends to align parallel to the field while that of the other half has the opposite tendency (antiparallel alignment) so that the bond is heavily stressed. Under these circumstances, the atomic configuration described in Figure 3.18 (c) can become highly unstable and tends to collapse to a planar sp^2 hybridization state as shown in Figure 3.18 (d). This process eventually leads to the breakage of the Si–Si bond and to the breakdown of the oxide film when a critical number of broken Si–Si bonds are locally reached. In the following, the derivation of a simple quantitative formulation of this breakdown model is presented.

In a dielectric medium with strong ionic bonding, a net polarization P can arise as a result of bond distortion of the medium when an external electric field F_{EXT} is applied.

$$P = \chi \varepsilon_0 F_{EXT} , \tag{3.20}$$

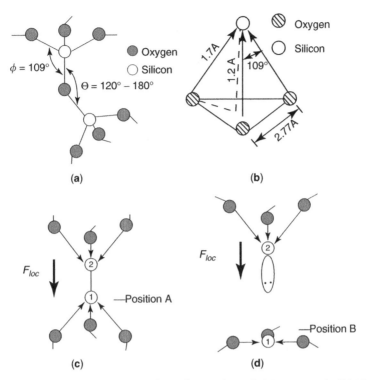

Figure 3.18. (a) The local tetrahedral configuration of SiO_2 network. (b) The bond angles for the half of molecule defect ($-Si\equiv O_3$). (c) An illustration of the Si–Si bond bridge is formed due to the oxygen vacancy. (d) An illustration of dangling bond created when the lower Si atom collapses from the sp^3 to the sp^2 hybridization state. Reprinted with permission from Ref. 52, © 1998 IEEE.

where ε_0 is the permittivity of vacuum and χ is the electric susceptibility of the dielectric ($\chi = 2.9$ for SiO_2). Due to this polarization, the local electric field acting on the dipole moment of the molecular defects of interest can be significantly larger than the external field. The relation between the local electric field and the externally applied field is given by the well known Lorentz relation:

$$F_{\text{LOC}} = F_{\text{EXT}} + L(P/\varepsilon_0) = (1 + L\chi)F_{\text{EXT}} = \left(\frac{3+\chi}{3}\right)F_{\text{EXT}}, \qquad (3.21)$$

where L is the Lorentz factor (which for solids with cubic symmetry is 1/3).

In the formulation of the thermochemical model of SiO_2 breakdown, it is first postulated that the dipole energy (which depends on the local electric field) reduces the activation energy for bond breakage [52]:

$$\Delta H = \Delta H_0 - p \cdot F_{\text{LOC}}, \qquad (3.22)$$

where ΔH_0 is the enthalpy of activation for Si–Si bond breakage in the absence of electric field, and p is the permanent dipole moment of the $Si \equiv O_3$ molecule. ΔH_0 is estimated to be 1.15 eV based on the average Si–Si bond energy of bulk samples [52]. The permanent dipole moment, p, can be calculated as

$$p = (Z^*e)r_0, \qquad (3.23)$$

where the Z^*e is the effective charge transfer from the Si atom to the O atom in the Si–O bond, and r_0 is the equilibrium interatomic bond distance. Combining Equations 3.21, 3.22 and 3.23, and given that the external field is the oxide field, $F_{EXT} = F_{OX}$, it follows that

$$\Delta H = \Delta H_0 - p_{eff}F_{OX}, \qquad (3.24)$$

where $p_{eff} \equiv p(1 + L\chi)$ is defined as an effective dipole moment. As indicated in Figure 3.18 (d), $r_0 \sim 1.2$ Å and $Z^* \sim 1$ for the Si–O bonds of our interest. Hence, p_{eff} is estimated to be around 7.2 eÅ [52], a value which is too small to account for many of the experimental observations. Further refinements of the model in the application of the Mie–Grüneisen atomic potential, which includes a repulsive and an attractive term, suggest a large value of of p_{eff} up to 13.3 eÅ in the limit of pure ionic interaction [78]. However, this assumption of ionic bond is questionable for the SiO_2 films.

In McPherson's formulation of the thermochemical model, this electric-field-enhanced process of Si–Si bond breakage is assumed to be the fundamental mechanism responsible for defect generation and for triggering the breakdown of dielectric film. Thus, a classical thermal activation process is postulated to describe the time-to-breakdown dependence on the oxide field and stress temperature:

$$T_{BD} = T_{BD0} \exp\left(\frac{\Delta H}{k_B T}\right) = T_{BD0} \exp\left(\frac{\Delta H_0 - p_{eff}F_{OX}}{k_B T}\right), \qquad (3.25)$$

where k_B is the Boltzman constant, T is the absolute temperature in Kelvin and T_{BD0} is a scale factor (the time to breakdown for $\Delta H = 0$). For such an exponential dependence, comparison with experiments is conveniently done in terms of the field acceleration factor:

$$\gamma_F \equiv -\left(\frac{\partial Ln T_{BD}}{\partial F_{OX}}\right) = \frac{p_{eff}}{k_B T}, \qquad (3.26)$$

Equations 3.24 and 3.26, which directly relate the activation energy and the field acceleration factor to the effective dipole moment and the enthalpy of average Si–Si bond, reveal the main results of the thermochemical model: 1) The activation energy (Eq. 3.24) linearly decreases with the oxide field, and 2) the field acceleration factor (Eq. 3.26) monotonically increases with decreasing temperature. Figure 3.19, (a) and (b), exhibits the predicted ΔH versus F_{OX} and γ_F versus T relations for two different sets of values of p_{eff} and ΔH_0 taken as limit cases. These predicted behaviors, however, are generally only observed within very narrow ranges of the involved variables in T_{BD} stress experiments (as we will discuss in detail in Section 3.4.1 and 3.4.2). Notice that the thermochemical model states that γ_F only depends on the effective dipole moment (p_{eff}) and temperature but remains independent of oxide field. In other words, an oxide-field dependent γ_F is incompatible with the claim of the exponential field-dependence as postulated by the thermochemical model.

The derivation presented above provides a simple description for the field enhanced bond-breakage process considered in the thermochemical model of breakdown. However, this simple picture encounters serious difficulties to explain many experimental results and also its basic assumptions are under question. Firstly, according to this model, the breakdown is a field- and time-driven process, that is, that carrier injection or tunneling current are not expected to play any fundamental role. However, as discussed in Section 3.2, there are compelling experimental evidences that both electron fluence and energy play a crucial role in the breakdown process [53–56]. Figure 3.9 in Section 3.2 shows that the time-to-breakdown (T_{BD}) decreases significantly with respect to conventional FN stress when the injected current is increased using substrate injection techniques, i.e., without changing the oxide field. Secondly, poly-depletion experiments were carried out to discriminate between the effects of oxide field and those of carrier energy. By changing the impurity concentration of the polysilicon gate, the oxide electric field can be modified without changing the gate voltage [84, 85]. In

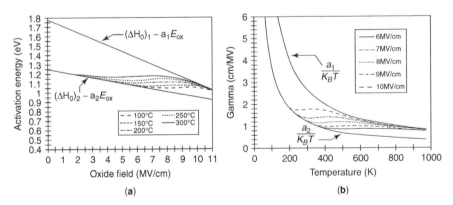

Figure 3.19. (a) Activation energy as a function of oxide field. The two solid lines represent the results of Equation 3.22 using the effective dipole moments and activation energies of $(\Delta H_0)_1 = 1.95$ eV and $a_1 = 13.0$ eÅ and $(\Delta H_0)_2 = 1.3$ eV and $a_2 = 7.4$ eÅ, respectively. (b) Field acceleration parameters as a function of temperatures. The two solid lines represent the results of Equation 3.26 with the effective dipole moments a_1 and a_2 of 6.8 eÅ and 3.0 eÅ. The results of two-state mixing are also shown in (a) and (b). Reprinted with permission from Ref. 52, © 1998 IEEE. Notice in Figure 3.19 (b) around 400 K the field acceleration parameters decreases with decreasing oxide field.

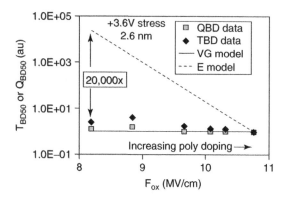

Figure 3.20. Average time-to-breakdown (T_{BD}) and charge-to-breakdown (Q_{BD}) versus oxide fields. The change of oxide field at a fixed gate voltage is achieved by varying doping concentration of polysilicon gate for NFET inversion mode. Reprinted with permission from Ref. 84, © 2000 IEEE.

Figure 3.20, the average T_{BD} and Q_{BD} are shown to remain relatively constant for a wide range of oxide field obtained by changing poly-doping concentration while keeping the same applied gate voltage. Based on the predictions of the thermo-chemical model, T_{BD} would be expected to change about four orders of magnitude in this range of electric field, while Figure 3.20 shows that it remains practically constant. This result of constant T_{BD} and Q_{BD} is perfectly consistent with those models that consider that defect generation and breakdown are driven by the energy gained by the carriers injected into the oxide. In ultra-thin oxides, the energy of the carriers at the anode interface can be identified with the gate voltage because transport in the oxide conduction band is ballistic. Thus, constant T_{BD} and Q_{BD} at fixed gate voltage are consistent with carrier-energy-driven models and in clear contradiction with the predictions of the thermochemical model.

The mechanism of bond breakage considered in the thermochemical model is quite a fundamental one, since it only involves basic mechanisms such as thermal activation and field effects on molecular defects with an intrinsic dipole moment. On the other hand, oxygen vacancies are so abundant in SiO_2 films of interest for microelectronics that they can be considered as intrinsic defects. However, in this section we have shown many contradictions between the experimental results and the predictions of the thermochemical model. This means that although the field-assisted thermal bond breakage is a likely mechanism of defect generation in SiO_2, it is not the dominant wearout mechanism responsible for oxide breakdown, at least in the usual conditions of stress experiments.

3.3.2 Hole-Induced Breakdown Models

The other two physical models, widely considered as an explanation of the defect generation that finally triggers oxide breakdown, are the so-called hole-induced

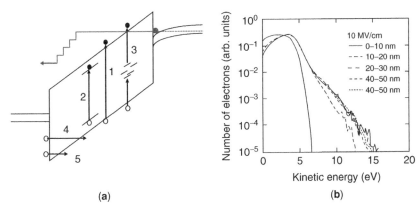

Figure 3.21. (a) Several mechanisms of hole generation process within SiO$_2$ and at Si anode. Reprinted with permission from Ref. 94, © 1986 IEEE. Process 1: intraband-to-band impact ionization; Process 2: electron–hole pair generation at the band tails; Process 3: trap-assisted impact ionization; and Processes 4 and 5 for cold and hot hole tunneling, respectively, after they are generated at the Si anode. These two process are referred anode hole injection. (b) Electron energy distribution for various thickness values obtained from MC simulation for various oxide thicknesses [93]. Notice for oxide thickness less than 10 nm, electron distribution sharply decreases below ∼7 eV. Reprinted with permission from Ref. 93, Copyright 1994, American Physical Society.

breakdown model and the anode hydrogen release model. Unlike the thermochemical model, both hole-induced BD model and hydrogen release BD model assume that breakdown is energy- and fluence-driven. In both models, electrons injected through the oxide under bias are assumed to partially dissipate the energy gained from the electric field at or near the anode interface and release some positively charged species, either holes or hydrogens (protons). These released species will diffuse across the oxide and react with some still unidentified precursors (presumably oxygen vacancies) to finally trigger the breakdown as explained in Section 3.2. Figure 3.21 (a) illustrates several mechanisms for hole generation process in oxides. Since both carrier injection (tunneling current dependence) and defect generation process play a crucial role in breakdown process, it is convenient to use the charge-to-breakdown (q_{BD} or Q_{BD}) as the breakdown variable to investigate the details of the defect generation process rather than using the time-to-breakdown (t_{BD} or T_{BD}), which jointly depends on tunneling current and defect generation efficiency.

3.3.2.1 Impact Ionization in SiO$_2$ and Early Hole-Induced Breakdown Models.
The work on hole-induced breakdown can be traced back to the time frame of the early 1970s [86–88]. Much work during these early times was based on the concept of avalanche carrier multiplication process that lead to a positive feedback process, similar to that occurring in the (reversible) breakdown of

semiconductors. In this process, electrons are injected by FN tunneling (or by other means such as avalanche injection techniques) into the conduction band of SiO_2 where they gain energy from the oxide field as they travel towards the anode. Their energy is partially lost when the electrons experience scattering processes with phonons. However, a small fraction of electrons eventually gains an energy larger than the energy gap of SiO_2 (Eg ~ 9 eV) and create electron-hole pairs by impact ionization in the oxide. The generated holes can travel towards the cathode under the influence of the field and some of them get trapped in defect sites. As a consequence, a positive charge is built up near the cathode that causes an increase of the cathode field and a subsequent enhancement of the current. Eventually, this process would continue in a positive feedback loop and finally cause a current runaway that leads to oxide breakdown.

Although the positive feedback process initiated by impact ionization appears to be a reasonable possibility, a detailed investigation of electron transport in SiO_2 was required to further exploring its feasibility. In fact, a solid understanding of impact ionization processes in SiO_2 was only achieved in the early 1990s. Early Monte Carlo simulations [89] and experimental results [90] suggested that the sole inclusion of longitudinal-optical phonon scattering can not explain the main properties of electron transport and would lead to breakdown fields much smaller than those measured in stress experiments. Subsequently, more complete Monte Carlo simulations [91] suggested that electron scattering with acoustic phonons is responsible for the stabilization of the energy of the electron population. In the mid of 1990s, a soft-X-ray-induced electron transmission technique was proposed to directly measure electron-phonon scattering rates [92]. This zero-field technique has an important advantage over other techniques because the scattering rates can be directly measured as a function of electron energy. By incorporating these directly measured scattering rates into Monte Carlo simulators, the simulated electron energy distributions showed excellent agreement with those obtained in experiments [93] and it was demonstrated that the kinetic electron energy in the tail of the distributions can exceed the 9 eV (V_G ~ 12 V) required for impact ionization as shown in Figure 3.21 (b). Subsequent experimental work revealed that some of the holes that move towards the cathode as result of impact ionization become trapped in energetically deep sites [1]. At the same time, it was argued that some of the trapped holes recombine with the injected electrons at the cathode to produce interface states and traps near the cathode [1].

The first analytic formulation of a hole-induced breakdown model, including both tunneling current and impact ionization, was introduced in the mid 1980s [14]. In this formulation, it was assumed that a critical trapped hole density (Q_{ot}^+) is required to trigger oxide breakdown. The trapped hole density is reasonably assumed to be proportional to the hole fluence, which can be calculated as a function of the injected electron density as [94]

$$Q_{ot}^+ \propto \int_0^t J_p dt = \int_0^t dt \ J_G(t) \int_{S_{ox}}^{T_{ox}} \alpha_{SiO_2}(x) dx, \qquad (3.27)$$

with J_p being the hole current density, J_G the FN tunneling electron current density, α_{SiO_2} the hole generation probability, and T_{OX} and S_{OX} the oxide thickness and tunneling distance [Fig. 2.28 (a)], respectively. Since the trapped hole density is proportional to the hole fluence, the breakdown condition can be expressed in terms of a constant hole fluence at breakdown:

$$Q_p = \int_0^{t_{BD}} J_p(t)dt. \tag{3.28}$$

On the other hand, since the charge to breakdown is given by the integral of the electron current,

$$Q_{BD} = \int_0^{t_{BD}} J_G(t)\, dt. \tag{3.29}$$

Equations 3.27 and 3.28 yield

$$Q_{BD} = \int_0^{t_{BD}} J_G(t)\, dt = \frac{Q_p}{\int_{S_{OX}}^{T_{OX}} \alpha_{SiO_2}(x)dx}. \tag{3.30}$$

In a first order approximation, the impact ionization coefficient dependence on the oxide electric field can be modeled [94] as

$$\alpha_{SiO_2} = \alpha_0 \exp(-H/F_{OX}), \tag{3.31}$$

so that the field dependence of Q_{BD} is given by

$$Q_{BD} \propto Q_p \alpha_0 \exp(H/F_{OX}), \tag{3.32}$$

Under constant current stress conditions, which were the most usual test conditions in TDDB measurements at the time that this model was developed, Q_{BD} is given by the product of J_G and T_{BD}, so that the field dependence of T_{BD} is

$$T_{BD} = \frac{Q_{BD}}{J_G} \propto T_{BD0} \exp\left\{ \frac{B+H}{F_{OX}} \right\} = T_{BD0} \exp\left\{ \frac{G}{F_{OX}} \right\} \tag{3.33}$$

where

$$G = B + H \tag{3.34}$$

and the FN gate tunneling current (Eq. 2.8) has been approximated by $J_G = J_{G0} \exp(-B/F_{OX})$. The values of B and H are quite different, B being

$\sim 240 \, \text{MV/cm}$ while the value of H for the impact ionization in SiO_2 is $\sim 78 \, \text{MV/cm}$ [94]. This analytical model, which is valid provided that the injection of electrons takes place through a triangular barrier (i.e., under the FN injection conditions) had a very strong impact in the oxide reliability community since it gave a physics-based justification for a $T_{BD} \propto \exp(G/F_{OX})$ functional dependence, as opposed to the $T_{BD} \propto \exp(-\gamma_F F_{OX})$ dependence derived from the thermochemical model. Both types of functional laws can be used to fit the experimental data in the usually quite narrow voltage window that allows performing BD stress experiments within a reasonable time frame. As discussed in Section 3.4.6, however, extrapolation with these models to operation voltage conditions yields lifetime predictions orders of magnitude different, with the $\exp(1/F_{OX})$ model providing a much more optimistic forecast.

One of the key aspects of the hole-induced BD models is the trapping of holes near the cathode, i.e., the build-up of positive oxide charge during stress [7, 14]. However, the effects of positive charge trapping on the evolution of the tunneling current were only observed during the very initial phase of the experiments ($Q_{inj} < 0.01 °\text{C/cm}^2$), while the effects of electron trapping in existing and newly generated electron traps were dominant all the way up to the BD occurrence ($Q_{inj} \sim 1\text{--}100 °\text{C/cm}^2$ under the usual conditions of stress experiments). Moreover, while a constant positive charge trapping density should be expected in the framework of the hole-induced breakdown model, several authors had reported a relatively constant trapped-electron charge at breakdown, thus suggesting that oxide breakdown occurred when a certain critical density of electron traps had been generated [9, 95]. The Berkeley group was aware of the role of electron traps in generating the breakdown so that they also suggested that defects (mainly neutral electron traps) were generated by recombination of injected electrons with the trapped holes [96].

Another crucial concept of the hole-induced breakdown model is the critical hole fluence at breakdown. Figure 3.22 represents experimental results in which the total hole fluence to breakdown Q_p, as measured by integrating the substrate current in carrier separation measurements, is rather constant ($Q_p \sim 0.1 \, \text{C/cm}^2$) in comparison with the much more voltage dependent Q_{BD}. The experimental values of Q_p, however, were reported to be strongly dependent on the oxide thickness below $10 \, \text{nm}$ [94]. This constancy of Q_p, which was considered to be a solid demonstration of the feasibility of the hole-induced breakdown concept, is now under question, as we will discuss later on in the framework of the anode hole injection model.

3.3.2.2 Anode Hole Injection Model. When the oxide thickness or the oxide field are lowered, the interband impact ionization process for hole generation in the SiO_2 film becomes ineffective. On one hand, Figure 3.21 (b) shows that the number of electrons that reach energies above $9 \, \text{eV}$ is drastically reduced when oxide field is lowered. This $9 \, \text{eV}$ threshold for interband impact ionization roughly corresponds to a gate voltage of 12V (ignoring any voltage drops in the Si substrate and gate electrode) because of the $3 \, \text{eV}$ barrier height associated with the

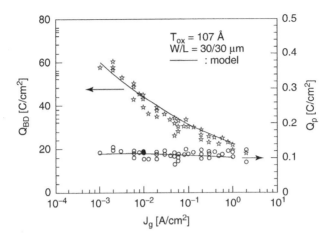

Figure 3.22. Charge-to-breakdown, Q_{BD}, and hole fluence–to-breakdown, Q_p, versus stress current for a 10.7 nm oxide stressed to breakdown under constant current conditions. Q_{BD} is found to increase significantly when the stress current decreases while the average hole fluence remains roughly constant, in agreement with the prediction of the hole-induced breakdown model as reported by Chen and colleagues. Reprinted with permission from Ref. 7, © 1986 IEEE.

Si/SiO$_2$ system due to FN tunneling current. Figure 3.21 (b) illustrates that the probability of electrons exceeding 9 eV becomes negligible at even higher electric fields when the oxide thickness is reduced. In particular, for oxide thickness below 10 nm the number of electrons above the impact ionization threshold is negligible at 10 MV/cm. On the other hand, the appreciable multiplication impact ionization factors are only reached for oxides (> 10 nm) if the oxide field exceeds 14 MV/cm [93], a condition which is rarely accessed in conventional BD experiments and which is far from the conditions of operation in MOS circuits.

Due to the difficulty of application of the impact ionization model for oxide breakdown in thinner oxides (< 10 nm), the so-called anode hole injection was proposed by the Berkeley group as alternative mechanism to explain the defect generation process and oxide breakdown [8, 97]. This model was originally suggested to be responsible for positive charge generation in thick oxides [98, 99]. Fischetti et al. first reported a theoretical formulation for anode–hole injection mechanisms in both aluminum and polysilicon electrodes based on surface plasmon excitation by incoming electrons [99]. In this scenario, a significant fraction of electrons exiting SiO$_2$ into the metal gate (anode) dissipates their energy by exciting surface plasmons. The surface plasmons are collective motions of valence band electrons localized at the anode/oxide interface. The emitted surface plasmons will rapidly decay via the creation of electron–hole pairs. The generated hot holes can be injected into the gate oxide at the anode interface. The predictions of this model were found in agreement with the experimental measurements of positive charge generation as a function of anode electric field, temperature, gate materials, and oxide thickness [99]. The gate voltage threshold for this surface-plasmon-induced hole generation was calculated to be about 7–8 V (again, ignoring voltage drops in the gate and substrate), thus explaining the generation of holes in thinner oxides and lower fields. Figure 3.23 (a) illustrates this anode-hole injection process in detail. The electrons gain sufficient energy (E_{GAIN}) with respect to the conduction band of silicon anode and, owing to the small band gap of silicon (1.12 eV), impact ionization by these electrons can effectively generate electron–hole pairs at the anode.

This process is known as conventional majority, carrier-impact ionization (as compared to the minority carrier-impact ionization that we shall discuss later). We will now proceed to provide a simple derivation of the anode injection model first developed by Schuegraf and Hu [8, 62, 97]. In their formulation, the hole tunneling current density was assumed to be proportional to the incident electron current density J_G flowing through the oxide:

$$J_p = \alpha_p \Theta_p J_G , \qquad (3.35)$$

with α_p being the hole generation probability in the anode (the probability that one injected electron generates one hole–electron pair) and Θ_p as the probability

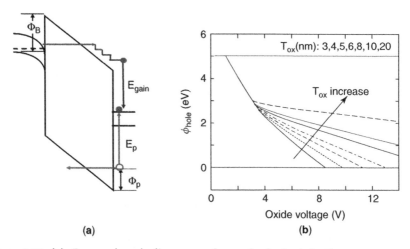

(a)   (b)

Figure 3.23. (a) Energy-band diagram of anode hole injection process after electrons tunneling through the barrier arrive at the anode with the energy, E_{GAIN}, gained from the oxide field. This energy is transferred to excite an electron deep in the valence band to the bottom of conduction band, and thus creates a hole, which tunnels back into the oxide. Reprinted with permission from Ref. 8, © 1994 IEEE. (b) The barrier height (Φ_p) for holes versus oxide voltages for several oxide thickness values.

for the generated holes to tunnel back to the oxide through the silicon/oxide hole barrier. The hole tunneling probability under FN injection conditions can be written as

$$\Theta_p \propto \exp\left(-\frac{B_p}{F_{OX}}\left[\Phi_p(V_{OX})\right]^{3/2}\right), \quad (3.36)$$

where $B_p = 4\sqrt{2m_{p,OX}}/3q\hbar$ and $m_{p,OX} = 0.2\,m_0$ [8]. In this equation, it is assumed that V_{OX} is larger than the barrier height for hole tunneling (Φ_p) so that it takes place through a triangular barrier analogous to the formulation of FN electron tunneling. The barrier height for hole tunneling is

$$\Phi_p = \left\{E_{gap}^{SiO_2} - q\Phi_B - E_{GAIN}(V_{OX}, T_{OX})\right\}/q, \quad (3.37)$$

with E_{GAIN} being the energy that the electrons gain from the oxide field while crossing the oxide layer under bias. They are given in Equations 2.20 and 2.21 for $V_{OX} > \Phi_B/q$ and $V_{OX} < \Phi_B/q$, respectively.

The barrier height for hole transport is represented in Figure 3.23 (b) as a function of V_{OX} for several values of T_{OX}. At extremely large values of V_{OX}, the effective barrier height approaches zero because holes are very hot so that they can be injected into the oxide over the barrier. On the contrary, at lower V_{OX} values, the barrier height becomes thickness-dependent and finally converges towards the valence-band hole tunneling barrier-height ($\sim 5\,eV$). The injected hole fluence can be obtained by integrating the hole tunneling current (Eq. 3.35) over time and the critical hole fluence at breakdown is

$$Q_p = \int_0^{T_{BD}} J_p(t)\,dt = \alpha_p\Theta_p\int_0^{T_{BD}} J_G(t)\,dt = \alpha_p\Theta_pQ_{BD}. \quad (3.38)$$

By inverting this equation, Q_{BD} is obtained:

$$Q_{BD} = \frac{Q_p}{\alpha_p\Theta_p} = \frac{Q_p}{\alpha_p}\exp\left(\frac{B_p}{F_{OX}}\left[\Phi_p(V_{OX})\right]^{3/2}\right), \quad (3.39)$$

and the voltage acceleration factor of Q_{BD} is

$$\gamma_{V_{OX}}^Q = -\left(\frac{\partial LnQ_{BD}}{\partial V_{OX}}\right) = -\frac{\partial LnQ_p}{\partial V_{OX}} + \frac{\partial Ln\alpha_p}{\partial V_{OX}} + \frac{\partial Ln\Theta_p}{\partial V_{OX}}. \quad (3.40)$$

The first term on the RHS should be zero because the critical hole density is assumed to be independent of the stress voltage. The contribution of the second term is also negligible because in this formulation the hole generation probability (α_p) is treated as a fitting parameter that is generally considered to be relative constant at high field [8, 62, 97]. Thus, the voltage acceleration factor is

$$\gamma_{V_{OX}}^{Q} \approx \frac{\partial \Theta_p}{\partial V_{OX}}$$

$$= \frac{T_{ox}^3 B_p [\Phi_p(V_{OX})]^{1/2}}{V_{OX}^2} \left\{ \Phi_B \left[2 - \exp\left(\frac{T_{OX}}{\lambda} \left(\frac{\Phi_B}{V_{OX}} - 1 \right) \right) \right] - E_{gap}^{SiO_2} \right\}. \quad (3.41)$$

The equation can be used to compare with experimental data as a rigorous test for anode hole injection model since it removes the dependence of tunneling current. Figure 3.24 (a) compares the Q_{BD} voltage dependence prediction of the AHI model with experimental data, indicating that the AHI model can roughly capture the general trend in Q_{BD} versus V_{OX} characteristics, consistent with the original publication [8, 62, 97]. However, the disagreement in the voltage acceleration factor, $\gamma_{V_{OX}}^{Q}$, between model prediction and experimental data is pronounced as shown in Figure 3.24 (b), thus suggesting that the AHI model has some limitations to explain the experimental data. Although in this analysis we have ignored the contribution of the voltage dependence of the impact ionization rate, it must be noticed that any contribution from α_p would only magnify the difference between experimental data and AHI prediction.

In the case of constant current stress or when the oxide is thin enough to guarantee that the initial current does not change due to charge trapping effects during a constant voltage stress experiment, the relation between charge and time

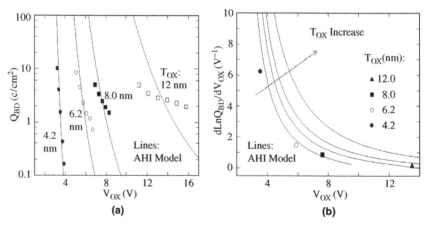

Figure 3.24. (a) Comparison of AHI with experimental results. (b) Comparison of voltage acceleration factor of AHI model with the experimental data at 140°C. The lines represent the AHI calculation of Equation 3.41 for (a) and (b). The symbols represent the experimental data. The open symbols represent the results of thermal SiO₂ when the closed symbols stand for nitrided SiO₂ [141].

to breakdown is $Q_{BD} = J_G T_{BD}$ so that T_{BD} is readily obtained as given below:

$$T_{BD} = \frac{Q_{BD}}{J_G} = \frac{Q_p}{J_G \alpha_p \Theta_p} = \frac{Q_p}{J_G \alpha_p} \exp\left(\frac{B_p}{F_{OX}} [\Phi_p(V_{OX})]^{3/2}\right)$$

$$= \frac{\Theta_p}{\alpha_p} \exp\left(\frac{G}{F_{OX}}\right), \qquad (3.42)$$

where

$$G = B + B_p \Phi_p^{3/2}(V_{OX}) \qquad (3.43)$$

is introduced to combine the field dependencies of the FN electron tunneling current and of the hole tunneling probability. The first term (B) comes from the FN electron current with $B \sim 240\,MV/cm$ while the second terms ranges from $115\,MV/cm$ to $55\,MV/cm$ for oxide field varying from $7\,MV/cm$ to $12\,MV/cm$. It is common practice in the literature to only investigate the T_{BD} field dependence and its field acceleration factor such as the linearity of T_{BD} versus $1/F_{OX}$ relation. This practice is inadequate because of the dominant contribution of FN tunneling current and its strong field dependence. A more rigorous demonstration for the validity of AHI model should be carried out by examing the Q_{BD} field dependence, for example, the comparison of experimental data with Equation 3.41.

The analytical formulation of the AHI model presented up to this point is conveniently simple but has some limitations mainly related to an overly simplistic treatment of impact ionization processes in the silicon anode. The hole generation probability can be calculated much more accurately using Monte Carlo (MC) simulation of electron transport in the substrate and a detailed consideration of the different types of impact ionization events [61, 63]. Figure 3.25 (a–c) illustrates three different impact ionization processes. The first one is the conventional majority carrier impact ionization process, i.e., the one that was already included in the analytical formulation of AHI. In this case, both the incoming (primary) electron and the secondary electron reside near the bottom of the Si conduction band after the impact ionization event. The maximum hole energy with respect to the top of the valence-band is $E_{MAX} = E_{GAIN} - E_{gap}^{Si}$ for this type of impact ionization events.

The other two impact ionization processes involve minority carriers in one way or another. In the process described in Figure 3.25 (b), the primary electron remains at the bottom of the Si conduction band after the impact ionization event, while the generated electron stays near the top of valence band. Finally, as shown in Figure 3.25 (c), another type of impact ionization process is possible in which both electrons end up in the valence band. Both types of minority impact ionization processes require the availability of empty electron states (holes) near the top of the valence band so one expects (based on these minority carrier effects) to find some impact of substrate doping and bias polarity on oxide breakdown. The maximum energies for the generated holes in the two minority carrier impact ionization processes are $E_{MAX} = E_{GAIN} + (E_V - E_F)$ and $E_{MAX} = E_{GAIN} + E_{gap}^{Si} + 2(E_V - E_F)$ [61]. MC simulations including all these types of impact ionization

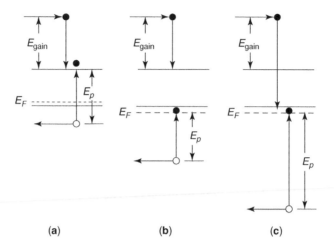

Figure 3.25. Different impact ionization channels in silicon anode: (a) Conventional impact ionization involving majority carriers; (b) and (c) impact ionization involving minority carriers. Reprinted with permission from Ref. 63, © 2000 IEEE. In (b), the incident electron ends up at the bottom of the conduction band while in (c), it ends up at the top of valence band. In contrast to conventional impact ionization, the excited valence band electron resides at the top of valence band.

processes reveals a high density of holes generated by the two minority carrier processes except when the anode is an n^+ – type polysilicon gate with $V_G > 0$, as in the case of stress experiments with NFETs biased in inversion. Therefore, for NFET inversion case, we should still expect majority impact ionization to play a dominant role while the two minority impact ionization processes provide additional channels for energetic electrons to lose the energy at the anode. The fact that minority carrier impact ionization mechanisms are only possible when there are empty states in the valence band of the anode explains the bias polarity dependence (the so-called polarity gap) of the charge to breakdown versus voltage relation, as it will be discussed in Section 3.4. In particular, it explains why this polarity gap can be cancelled if Q_{BD} is plotted versus the electron energy, provided that the electron energy is referred to the anode Fermi level and not to the anode silicon conduction band as it might seem more reasonable at first sight [100].

One of the key points of the AHI model is the postulate of a critical hole fluence at breakdown which should be rather independent of the stress conditions. Experimentally, substrate currents with the carrier separation method have been historically used as a measure of hole fluence and the existence of a critical hole fluence to breakdown has been considered as a strong evidence supporting the hole-induced breakdown models, as shown in Figure 3.22. In the case of ultra-thin oxides at low voltages, however, the measure of anode hole generation by this method becomes problematic because the valence-band electron tunneling dominates the gate current. Figure 3.26 (a) illustrates how the valence-band electron can tunnel through the oxide barrier towards the gate and leave holes behind in

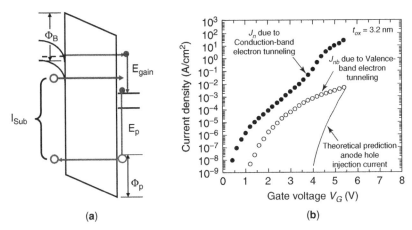

(a)   **(b)**

Figure 3.26. (a) Anode hole injection process in direct tunneling regime. In addition to the conduction band electron tunneling, electrons in the valence band at the cathode can directly tunnel through the oxide barrier and leave holes being collected as substrate currents. (b) Electron valence-band tunneling current in comparison with theoretical prediction of AHI model for substrate current in NFET inversion mode [97]. The contribution due to AHI injection is significantly smaller than tunneling currents due to valence band electrons. Reprinted with permission from Ref. 93, Copyright 1994, American Institute of Physics.

the substrate. These holes will be collected by the substrate electrode as substrate currents and have nothing to do with the anode hole generation process considered in the AHI model. Detailed studies demonstrate that for oxide thickness less than 4.5 nm, the substrate current on NFETs biased in inversion mode is completely dominated by the valence-band electron tunneling with the AHI component being negligible as shown in Figure 3.26 (b) [8]. Moreover, even in the situation in which the valence band electron tunneling component is not important, the actual physical mechanisms responsible for the substrate current are now under question. In this regard, alternative mechanisms for the injection of holes created in the anode by hot electrons coming from the gate have been proposed to explain the substrate current such as generation–recombination processes in the substrate [55] or the creation of hole–electron pairs by photons generated by energetic electrons in the gate [101]. Nevertheless, the issue is not easy to deal with and the latter interpretation has also been recently challenged [102].

The most complete description of the AHI process is obtained by Monte Carlo simulation, which includes a detailed consideration of electron transport and impact ionization in the substrate [63]. These simulation results are compared with the experimental data in Figure 3.27, thus showing that the AHI model is capable of reproducing experimental voltage-dependent voltage acceleration factors down to a gate voltage of about 4 V [63]. No further results have been reported for lower voltages within the framework of the AHI model. Based on Monte Carlo simulation, it was first suggested that at low voltages, an exponential law for T_{BD} voltage-dependence is predicted (similar to that found in the framework of the thermochemical model) while at high voltages an $\exp(1/F_{OX})$ dependence is expected [63]. A later publication suggests that for voltages below 8–9V $Ln(T_{BD})$ depends neither linearly nor inversely on F_{OX}, revealing the complications of modeling TDDB data [103].

Despite the early success of the AHI model, serious doubts have recently been raised after the publication of results concerning the efficiency of defect generation by holes injected through thin oxides [104–106]. For PFETs stressed under low voltage inversion conditions, the direct hole tunneling current (injected from the inversion channel into the oxide) becomes comparable to the electron current (injected from the gate). Using these devices, it is demonstrated that only when the hole current reaches at least the magnitude of the electron current, the effects of hole injection on additional defect build-up become observable [104]. The conclusion is quite straightforward: Under conventional conditions such as NFET inversion, the AHI mechanism that relies on a tiny anode hole current (orders of magnitude smaller than the injected electron current) cannot be the dominant defect generation mechanism. In addition, using the substrate hot hole (SHH) injection technique, it was conclusively shown that different amounts of injected holes have no impact on subsequent TDDB measurements [106]. Recently, a detailed experimental study has ruled out AHI mechanism for ultra-thin oxides because the measured defects are acceptor-type as opposed to donor-type defects required for the presence of holes [60].

Surprisingly enough, while many efforts during the last 25 years have been dedicated to developing an improved model for the generation and injection of

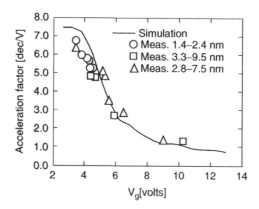

Figure 3.27. T_{BD} voltage acceleration factors as a function of gate voltage. The line represents the MC simulation results of AHI model. Reprinted with permission from Ref. 63, © 2000 IEEE. As pointed out earlier, the examamination of T_{BD} voltage dependence is inadqequate because it does not separate the voltage dependence of tunneling from hole generation probability.

holes from the anode, now the main criticisms are based on results concerning the second step of the model, i.e., the efficiency of the holes to react with precursors and generate defects (ζ_2 as defined in Section 3.2) in the oxide, a theoretical process that has not been studied in detail.

3.3.3 Anode Hydrogen Release Model

As already anticipated, another well known model for oxide breakdown is the so-called anode hydrogen release (AHR) model. Both AHR and AHI models share a common basic framework and they both assume that breakdown is energy- and carrier-fluence-driven. In both models, the electrons injected through the oxide are assumed to partially dissipate the energy gained from the electric field at or near the anode interface where they release some positively charged species, either holes or hydrogen (protons). These released species are assumed to diffuse/drift across the oxide and react with some precursors (presumably oxygen vacancies) to create breakdown defects. This process was discussed in general terms in Section 3.2 and was schematically shown in Figure 3.10.

Hydrogen is known to form a variety of defects in the silica network and one of them, the hydrogen bridge, has been recently identified as a trap with properties compatible with those required to be responsible for the stress-induced leakage current (SILC), a forerunner of the breakdown [107]. On the other hand, the influence of hydrogen on defect generation and oxide breakdown has been experimentally demonstrated by different means. The introduction of hydrogen during the fabrication process, either by a high temperature anneal or by the deposition of LPCVD Si_3N_4, was shown to strongly reduce Q_{BD} during high voltage stress [108]. On the other hand, detailed experiments using remote hydrogen plasma exposure undoubtedly showed that the trap generation causing SILC correlates well with hydrogen dose [22]. More recently, it was also shown that excessive hydrogen can lead to degraded T_{BD} and Q_{BD} and a change of the voltage acceleration factors [109, 110]. On the other hand, it was demonstrated that although the introduction of deuterium can significantly improve the lifetime of hot carrier effect [111], its impact on SILC measurements is not conclusive [112, 113]. Moreover, no improvement in Q_{BD} or T_{BD} has been reported at relatively high voltages [112, 114]. Subsequently, it was suggested that high rates of hydrogen exchange ($>80\%$) are required to observe any measurable effect on oxide breakdown data [114]. In summary, many different sources of experimental evidence suggest that hydrogen has a strong impact on defect generation and oxide breakdown. These experimental results inspired a picture of defect generation based on the release of hydrogen from the anode interface [22]. However, contrary to what happened with the AHI model, this picture has long remained as a qualitative proposal rather than as a well developed oxide breakdown model.

A quantitative formulation of AHR has been recently launched by different groups and is a model based on the inelastic coupling of the tunneling electrons with the local vibrational modes of Si–H bonds (H-passivated dangling bonds) at the anode interface [44, 57–60, 115]. The first attempt was made to connect the

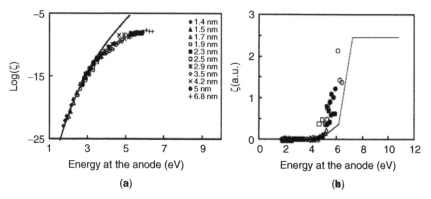

Figure 3.28. Defect generation efficiency versus maximum energy at the anode plotted in the log scale (a); reprinted with permission from Ref. 44, Copyright 2004, American Physical Society, and the linear scale (b); reprinted with permission from Ref. 115, © 2005 IEEE. The solid line in (a) represents a fit of the data with a power law $\zeta \propto E_{MAX}^n$ with $n \sim 38$. The solid line in (b) is a schematic representation of the results obtained in STM desorption experiments [116].

results of scanning tunneling microscope (STM) to the breakdown measurements [57]. The proposal is still in its infancy but promising results have been obtained by comparing oxide breakdown results with those obtained in STM experiments of hydrogen desorption from H-passivated silicon surfaces [116, 117]. In particular, the defect generation efficiency measured in breakdown experiments nicely compares with the STM desorption yield data attributed to the excitation of local vibration modes of the Si–H bond at low energies [44, 57, 115, 118, 119] and to direct electronic excitation [44, 115] at high electron energies (Figure 3.28).

The coupling of electronic and vibrational degrees of freedom at low voltages is particularly relevant to reliability because this is thought to be the mechanism which governs the generation of defects at very low operating voltages (~ 1 V) in ultra-thin oxide devices. Here is a description of this coupling: The Si–H bond stretch vibrations are usually described using a truncated harmonic potential with a ground state and a number of excited levels. The electrons that reach the bond region by means of tunneling are transiently trapped in a nonbonding state of the Si–H bond that becomes negatively charged. Then, the electron can relax to any empty state available in the anode, thus losing all or part of its energy, and can leave the Si–H bond in an excited vibration state. The barrier for bond breakage by vibrational excitation is of the order of 2.5 eV for Si–H bonds at the Si/SiO_2 interface [120–122] and it can be overcome either by the interaction with a single electron (coherent excitation) [44, 119] or with several successive electrons (thermal heating mechanism) [58, 118]. Both mechanisms explain a power law dependence of the defect generation efficiency as a function of the electron energy (applied voltage), consistent with recent extensive breakdown experiments [50, 51, 123]. Although there has been some controversy in this regard, several groups are converging to agree that a combination of coherent excitation and thermal heating mechanisms [124] provides the most likely explanation for low-voltage hydrogen release through the interaction of electrons and local vibration modes [44, 57–59, 115].

The following discussion of the physics of hydrogen release is based on the voltage/energy dependence of the defect generation efficiency ζ (see Figures 3.7) obtained through a reverse engineering approach from breakdown data [44, 115]. Instead of using a technique for directly measuring defects (see Section 3.2.1 for a discussion of the limitations of these methods) as they are generated during the stress experiment, this approach is based on assuming the validity of the percolation model for the breakdown statistics and uses the relation of the defect generation efficiency to charge-to-breakdown established in Equation 3.16, which can be inverted so as to calculate ζ directly from Q_{BD} data:

$$\zeta = \frac{qT_{OX}}{a_o^3 Q_{BD}} \exp\left[-\frac{1}{\beta(T_{OX})} Ln\left(\frac{A_{ox}}{a_o^2}\right)\right]. \tag{3.44}$$

This approach relies on two main assumptions: the validity of the percolation approach and the assumed proportionality between density of generated defects and injected charge. Its main advantage over other methods of measuring is that it ensures that the considered defects are those which are directly related to the triggering of the breakdown. Figure 3.28 (a) represents the defect generation efficiency obtained with this reverse engineering approach as a function of electron maximum energy at the anode. The raw experimental data is a very large Q_{BD} database constructed by subjecting oxides of different thickness to constant voltage stresses with V_G chosen to keep T_{BD} within a reasonable time frame. The area of the stressed devices also changed from set to set but no area normalization is required since Equation 3.44 properly takes the weakest link effect associated to A_{OX} into account for the calculation of ζ. Another important observation about Figure 3.28 (a) is that the data corresponding to structures with different T_{OX} fall onto a single curve, revealing that Equation 3.44 also properly accounts for the T_{OX} dependence of breakdown. At low energies below ~ 3 eV, the defect generation efficiency decreases rapidly with decreasing electron energy. A power law $\zeta \propto E_{MAX}^n$ with $n \sim 38$ nicely captures the energy dependence of the defect generation efficiency over this range of energies. However, at high energies above ~ 3 eV, the defect generation efficiency increases more gradually with increasing electron energy and somewhat saturate at very high energies in the log scale. The power law exponent becomes energy-dependent over this range of energies and is shown to decrease with energy in this range as we will explain with a model that involves the cooperation between vibrational excitation and electron excitation as discussed later. The defect generation efficiency [the same data of Figure 3.28 (a)] is plotted in Figure 3.28 (b) in a linear scale that allows the comparison with the yield of H desorption from an H-passivated silicon surface as obtained in STM experiments. An energy threshold of ~ 6–7 eV can be clearly identified for both BD data and STM experiments. Above this threshold of ~ 6–7 eV, the desorption yield is found to be independent of electron energy

[116, 117]. This is consistent with direct electronic excitation (EE) mechanism in which an incident carrier causes the transition of one electron from the bonding 6σ state to the $6\sigma^*$ anti-bonding state [117]. In BD experiments, the saturation of ζ should occur at higher values of V_G because electron transport in the oxide is dispersive in this regime and the relation between V_G and E_{MAX} becomes sublinear (Section 2.2.3). Therefore, the derived $\zeta(E_{MAX})$ data does not reach saturation at high E_{MAX}.

Below the threshold of direct electron excitation (EE), the excitation of the Si–H vibration degrees of freedom has been suggested to play a relevant role [44, 57–59, 115]. This involves the transient resonant trapping of the tunneling electron in the $6\sigma^*$ orbital until it decays into the available anode states and leaves the Si–H bond in an excited vibration state. Two mechanisms have been theoretically considered to explain hydrogen release by vibration excitation: the multiple carrier (incoherent) thermal heating process [116, 117] and the single carrier-induced (coherent) multiple vibration excitation [119]. A truncated harmonic potential with N eigenstates has been considered to model the Si–H bond vibrations and to study the voltage and current dependences of the involved transitions [118, 119, 124]. A schematic representation of the incoherent [118] and coherent mechanisms [119] is shown in Figure 3.29.

Under some simplifying approximations, the dissociation rate for hydrogen release via multiple incoherent excitations in the truncated oscillator model is conveniently expressed by [119]

$$R_{des}^{incoh} \approx (N - 1)\gamma \left|\frac{I_0}{e\gamma}\right|^N \left(\frac{I_1}{I_0}\right)^N , \tag{3.45}$$

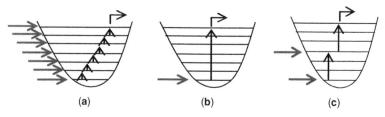

(a)   (b)   (c)

Figure 3.29. Schematic representation of HR by excitation of the local vibration modes of the Si–H bond. Reprinted with permission from Ref. 115, © 2005 IEEE. (a) The incoherent thermal heating mechanism involves the action of several electrons (red arrows) which induce one-level transitions between adjacent vibration states [118]. (b) The coherent excitation mechanism involves a single carrier to excite the bond vibration to the highest energy state [119] and (c) More general situations involve several electrons and also multiple level transitions [124]. The dependence of the release efficiency on the tunneling current is crucial to determine which is the mechanism responsible for HR. The release rate is proportional to $I^{(n-1)}$, n being the number of involved electrons.

where I_0 is the elastic current flowing through the resonance state, I_1 is the inelastic current due to excitation from the ground state to the first excited state, and γ is the relaxation rate of the first excited state. The ratio of I_1 to I_0 is known as the inelastic tunneling fraction, $f_{in} \sim I_1/I_0$[116]. On the other hand, the desorption rate for the coherent excitation mechanism can be written as [119]

$$R_{des}^{incoh} \approx N! \left| \frac{I_0}{e\gamma} \right| \left(\frac{I_1}{I_0} \right)^N \qquad (3.46)$$

Notice that both the incoherent and coherent excitation mechanisms show the same power law dependence on the inelastic tunneling fraction. However, the incoherent excitation mode shows a much stronger power law dependence on the current, I_0, whereas the coherent excitation mode only depends linearly on I_0. The inelastic tunneling fraction is experimentally found to depend on the electron energy (bias voltage) as shown in Figure 3.30, a dependence which can be approximately modelled as

$$f_{in} = \frac{I_1}{I_0} \propto E^4 \qquad (3.47)$$

Therefore, if we assume that the release of hydrogen from the anode interface is due to the coupling of the tunneling electrons to vibronic degrees of freedom of Si–H bonds, the energy dependence of defect generation efficiency for either the incoherent or the coherent excitation mechanisms is

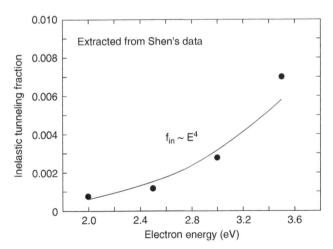

Figure 3.30. Inelastic tunneling fraction of electrons as a function of electron energy, extracted from the hydrogen desorption data of the work of Shen et al. Reprinted with permission from Ref. 116, AAAS.

given by a power law:

$$\zeta(E_{\text{MAX}}) \propto \frac{R_{des}}{I_0} \propto f_{in}^N \propto E_{\text{MAX}}^{4N} \tag{3.48}$$

The number of bound states in the Si–H bond potential can be estimated using the truncated harmonic oscillator model

$$N \approx \frac{E_{\text{D}}}{\hbar w}, \tag{3.49}$$

with E_{D} being the desorption barrier of the Si–H atomic potential, $\hbar$ is Planck's constant, and ω is the frequency of the involved vibrations. Thus, $\hbar w$ represents the vibrational energy, which is known to be very close to 0.25 eV for the stretching mode of the Si–H bond. This frequency is found to be quite insensitive to the local atomic environment [125]. However, the desorption barrier (E_{D}) can indeed be affected by the local environment. Typical values for E_{D} vary from 2.5 eV to 3 eV, based on various experimental and theoretical investigations [120, 125, 126]. Thus, based on the values for E_{D} and $\hbar w$, the number of bound states in the Si-H bond potential well would be around 10 to 12; the power law exponent of the energy dependence of $\zeta(E_{\text{MAX}})$ would be $4N \sim 40$–48, which is in reasonable agreement with $n \sim 38$ found for the fitting of the experimental breakdown-related defect generation efficiency data. This quantitative analysis suggests that power law voltage dependence of Q_{BD} or T_{BD} with a very large exponent is perfectly consistent with anode hydrogen release mechanism.

We have related defect generation to hydrogen release by direct electronic excitation at high energies (above ~ 6–7 eV) and to the coupling with Si–H bond vibrations at low energies below 2.5 eV–3 eV. In between these two extremes, the $\zeta(E_{\text{MAX}})$ power law exponent is found to decrease with E_{MAX} converging towards zero at $E_{\text{MAX}} \sim 6.5$ eV. For this intermediate energy regime, a cooperation between VE and EE processes is proposed as shown in Figure 3.31. In this process, hydrogen release takes place when one electron excites the bond vibration to the nth state and then another one causes the ulterior bond breakage by EE. To understand how this process takes place, we must take into account two facts. First, the E($6\sigma \rightarrow 6\sigma^*$) threshold decreases $\sim$ linearly with the Si–H bond distance as demonstrated by Avouris' theoretical results shown in Figure 3.32 [117]. Second, the excitation of local vibrational modes leads to an increase of the average bond distance although this effect is not captured by the truncated harmonic potential and requires a more realistic model such as the Morse potential shown in Figure 3.31. In this asymmetric potential, the average Si–H distance increases linearly with the eigenvalue n ($1 < n < N$). In summary, when an electron excites the bond vibration to the nth state, the bond distance increases and hence the energy required to excite an electron from the 6σ orbital to the $6\sigma^*$ anti-bonding orbital (thus possible triggering bond breakage) decreases. When the energy of the electrons injected by FN tunneling increases (due to the increased applied gate voltage), the required vibrational excitation is shifted to a lower

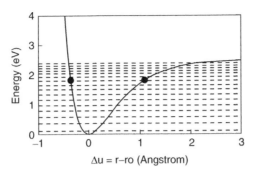

Figure 3.31. An illustration of the proposed cooperation process of vibration excitation mode and electron excitation mode for the intermediate range of electron energies between ~2.5 eV and ~6.5 eV below the threshold of direct electron excitation. The vibrational excitation leads the Si–H to an excited level, n_{th}. Then the Si–H bond is broken by direct electron excitation, leading to hydrogen release. Reprinted with permission from Ref. 115, © 2005 IEEE.

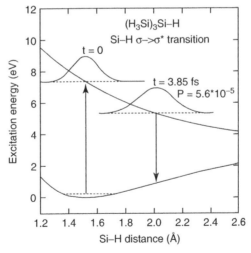

Figure 3.32. Computed potential energy curves of the ground (6σ) and excited ($6\sigma^*$) states of the Si–H bond as a function of the Si–H bond distance, according to Avouris et al. [117]. After the electronic excitation ($t = 0$), the electron wavepacket evolves in time and, even if the excitation is quenched (at $t = 3.85$ fs in this particular example), the rupture of the bond is quite probable. Reprinted with permission from Ref. 117, Copyright 1996, Elsevier.

energy state $n < N$ so that the power law exponent of Equation 3.48 is reduced from $4N$ to $4n$. Local fitting of the experimental $\zeta(E_{MAX})$ data to a power law gives a linear reduction of the exponent from ~ 40 at 3 V to ~ 0 at 6.5 V. Dividing this exponent by 4 allows calculating the eigennumber $n(E_{MAX})$ of the vibrational state required to assist the process of hydrogen release by electronic excitation. Comparison of the experimental $n(E_{MAX})$ data to the prediction of this simple model of hydrogen release by VE/EE cooperation quantitatively supports the feasibility of this picture in the intermediate E_{MAX} range, as shown in Figure 3.33.

Although both the incoherent and coherent hydrogen release mechanisms previously discussed can explain a power law dependence of the release efficiency on E_{MAX}, the purely incoherent process involves multiple (N) electrons and hence, this mechanism shows a strong dependence on the current ($\zeta \propto J^{N-1}$) while the coherent VE process does not depend on J because only one electron is involved. In general, processes involving K transitions ($1 < K < N$) are also possible (Figure 3.29) and would yield a power law dependence on the current with a smaller exponent, $\zeta \propto J^{K-1}$[124]. In the case of STM experiments, current and voltage are varied independently and the results are found to be compatible with the incoherent model [116, 118]. In breakdown experiments, the current depends on V_G so that it is more difficult to decouple the dependencies on these two variables. However, since the current also depends on T_{OX} at fixed V_G, we can examine the T_{OX} dependence of ζ as a way to indirectly check the current dependence. In this regards, based on the results shown in Figure 3.34, we can conclude that ζ does not significantly depend on T_{OX} and hence not on the tunneling current. Thus, the coherent single-electron vibrational excitation process is more likely to explain the defect generation

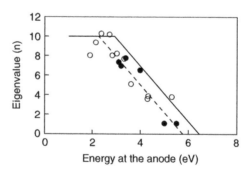

Figure 3.33. The change of power law exponent with increasing electron energy. Symbols are calculated by dividing the experimental $\zeta(F_{MAX})$ power law exponent by 4 (to calculate the eigenvalue of the vibrational state involved in the HR process). Solid line is obtained by combining the change of the $E(6\sigma$ to $6\sigma^*)$ with Si–H distance with the average Si–H bond distance when it is excited to the nth eigenstate. The dashed line better fits the data and it is just the result of an arbitrary shift of the EE threshold by 0.7 eV. Reprinted with permission from Ref. 115, © 2005 IEEE.

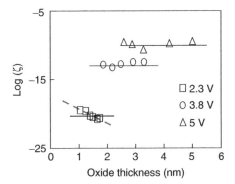

Figure 3.34. Dependence of defect generation rate on T_{OX} at fixed V_G (44, 115). The dashed line emphasizes a possible T_{OX} dependence at low V_G. Reprinted with permission from Ref. 44, Copyright 2004, American Physical Society.

efficiency data. Actually, this was already evident in Figure 3.35 since the data corresponding to devices with different T_{OX} fall onto a single curve [115].

However, the slight change of ζ with T_{OX} at the lowest stress voltage ($V_G = 2.3$ V) considered in Figure 3.34 suggests a possible transition towards multiple electron excitation mechanisms at low voltages. To further explore this possibility, we can examine the dependence of ζ on E_{MAX} at the lowest voltages. If ζ does not depend on the current, we expect that the $\zeta(E_{MAX})$ data corresponding to different T_{OX} should fall on a single "universal" curve. If, on the contrary, the relevant HR mechanism involves K electrons ($1 < K < N$), a dependence $\zeta \propto J^{K-1}$ should be found, and the magnitude giving a "universal" T_{OX}-independent curve

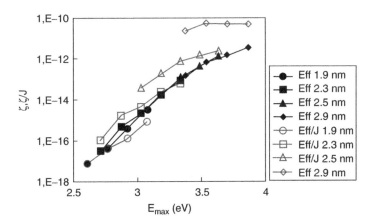

Figure 3.35. Representation of ζ and ζ/J versus E_{MAX} for devices with different T_{OX} such that the E_{MAX} range is above 2.5 eV. While $\zeta(E_{MAX})$ appears as T_{OX} independent, $\zeta/J(E_{MAX})$ shows a clear dependence on T_{OX}. This indicates that the HR mechanism involves a single electron. Reprinted with permission from Ref. 115, © 2005 IEEE.

should instead be ζ/J^{K-1}. Figure 3.35 shows that above 2.5 eV, $\zeta(E_{MAX})$ is thickness independent while $\zeta/J(E_{MAX})$ changes with T_{OX}. On the contrary, Figure 3.36b reveals that the situation is reversed for E_{MAX} below ~ 2.5 eV since $\zeta(E_{MAX})$ becomes dependent on T_{OX} while $\zeta/J(E_{MAX})$ is T_{OX}-independent. This suggests that below $E_{MAX} \sim 2.5$ eV two electrons cooperate in the HR process. This is not an unexpected result since the low-energy limit of the coherent mechanism of hydrogen release is the barrier height (E_D). When the incident electrons have an energy smaller than the barrier height, they cannot excite the Si–H bond vibration so as to cause bond breakage and at least two electrons are required. The finding of a change of the mechanism at ~ 2.5 eV as suggested by the results of Figure 3.35 indicates that this is the barrier height for Si–H bond-rupture, in perfect agreement with annealing experiments [120, 121].

Examination of the experimental dependencies of the defect generation efficiency (derived from breakdown data) on carrier energy and oxide thickness has allowed to suggest a rather detailed picture of the possible hydrogen release mechanisms involved. In summary: 1) At high electron energies (above ~ 6.5 eV), direct electronic excitation by one single electron can cause hydrogen release; 2) at very low energies (<2.5 eV) hydrogen release is controlled by vibrational excitation and a transition from single-electron coherent mechanism towards multiple-electron incoherent mechanisms is suggested by the data; and 3) in between these two extremes, the results are consistent with a cooperation of

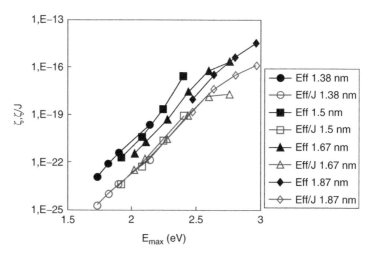

Figure 3.36. Representation of ζ and ζ/J versus E_{MAX} for ultra-thin oxides in the E_{MAX} range below 3 eV. While $\zeta(E_{MAX})$ appears as T_{OX} dependent, $\zeta/J(E_{MAX})$ becomes independent on T_{OX}, contrary to what is observed at higher voltages (higher E_{MAX}). This result suggests a transition from the coherent single-electron HR mechanism towards a combination of coherent and incoherent mechanisms at low energies. Reprinted with permission from Ref. 115, © 2005 IEEE.

vibrational excitation and electronic excitation mechanisms. The results suggest that a transition towards multiple electron hydrogen release processes at low voltages are particularly significant since they are relevant to the reliability of circuits under operating conditions.

3.4 EXPERIMENTAL RESULTS OF OXIDE BREAKDOWN

In Section 3.2, we introduced the defect generation process that leads to oxide breakdown. Moreover, we have related the generation of defects with the breakdown statistics and we have discussed its scaling properties with oxide thickness and area. In Section 3.3, we have presented three physical models that have been widely discussed in the gate dielectric research community. In this section, we will discuss some of the key experimental findings on voltage, temperature, and polarity dependence of oxide breakdown as well as the AC effect on the lifetimes of oxide breakdown. We will examine these experimental results in detail and compare them with the three oxide breakdown models discussed in Section 3.3 in an attempt to construct a self-consistent and coherent picture for gate oxide reliability. At the end of this section, we will attempt to answer the crucial questions raised in Section 2.1.2 regarding gate oxide reliability projection.

It is a daunting task to select the most significant results from an enormously large range of excellent publications. However, some considerations have helped us in this regard. First, it is interesting to note that much of the experimental TDDB data obtained prior to the early 1980s for thick oxides [127–130] largely corresponds to the extrinsic breakdown mode with the Weibull shape factors (β) being much less than one, a defining property of extrinsic breakdown [131]. The situation did not really improve until the mid 1980s, when publications showing much steeper cumulative failure distributions that corresponded to intrinsic mode of oxide breakdown became available. Since our goal is to deal with intrinsic breakdown, we will avoid including old data corresponding to the extrinsic breakdown mode. While it is important to have a historical perspective of both theoretical and experimental work, we will base our discussion on the work published since 1990s because of the improved oxide fabrication process.

First, we must note that many previous publications adopted the method of constant current stress, which has been shown to be inadequate for the evaluation of thin oxides (Fig. 2.45 and Section 2.3.1). Second, the importance of an accurate determination of oxide voltages and fields was not appreciated because the effects of poly-depletion and Si surface quantization were not fully understood within the dielectric reliability community. Finally, although the statistical nature of dielectric breakdown was well known, as shown in the work of Wolters [132], its impact on the accuracy of breakdown parameters only became particularly important when the researchers had to deal with ultra-thin gate oxides. In the following discussion, we will take the above considerations into account as the main criteria for selecting the relevant data from the available publications.

3.4.1   Voltage Dependence

We will first focus the discussion on relatively thick oxides stressed in the FN regime ($V_{OX} > \Phi_B/q$), including the experimental results mainly reported in the 1990s [133–140]. Figure 3.37 displays some typical time-to-breakdown cumulative distributions corresponding to 9.9 nm oxides stressed under constant voltage condition. These results correspond to NFET structures stressed in accumulation mode (gate injection) at different applied fields [135]. The cumulative failure distributions are presented in the Weibull plot in which $Ln\{-Ln[1-F(t)]\}$ is plotted versus $Log(t)$. The raw cumulative percentiles, $F(t)$, are specified on the left side vertical axis. It can be seen that time-to-breakdown distributions are strongly affected by the applied bias. Notice also that at the lowest bias of -8 MV/cm, the stress is terminated around 2×10^6 seconds due to the practical time limitation. Consequently, the extrapolation from these low percentile data points to high percentiles (eg., 63%) would suffer from a much increased uncertainty because the lower-percentile data are known to have much wider confidence bounds, as discussed in Section 2.4.4.

Figure 3.38 displays the time-to-breakdown values versus oxide-field (F_{OX}) in the log-linear scale for relatively thick oxides of 6.2 nm to very thick oxides of 15 nm [141]. This data was obtained from several publications [133–135, 138–140] in NFET accumulation mode at 125°C and 150°C. Although the areas of data points are not precisely the same (maximum change by 6X), there is only a small variation in T_{BD} because of the larger value of Weibull slopes for these thick oxides (Section 3.2). The lines in this figure by the least-square fit of each data set appear to suggest an exponential dependence of T_{BD} on F_{OX}. Although at 8 MV/cm, the

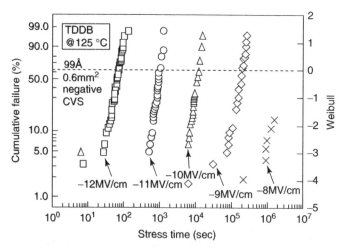

Figure 3.37. The Weibull distributions of time-to-breakdown for $T_{ox} = 9.9$ nm in NFET accumulation mode. The applied fields in this case are simply taken to be gate voltage divided by the oxide thickness. Reprinted with permission from Ref. 135, © 1997 IEEE.

time-to-breakdown appears to be roughly the same as the claim of the thermo-chemical model [76], the T_{BD} values do not overlay each other for different T_{OX} values at high F_{OX} values. Moreover, at low F_{OX} values the T_{BD} data of thinner oxides exceeds those of thicker oxides. None of these experimental observations are predicted by the thermochemical model (Eq. 3.25). The field acceleration factors, γ_F, derived from the slopes of $Ln(T_{BD})-F_{OX}$ relation are larger for thin oxides than those for thick oxides, suggesting either a field-dependent or a thickness-dependent acceleration factor. Note that the T_{BD} exponential field-dependence as proposed by the thermochemical model rules out any possibility that field acceleration factors can be field dependent. Thus, we can at least conclude that a simple force-fit is not sufficient to determine which model provides a better description of the oxide breakdown process due to the limited experimental time window available. Because of this difficulty, the constant voltage stress is extended to much longer times, up to two or three years [133, 138, 139].

Figure 3.39 (a) shows the time-to-breakdown versus oxide field while the reciprocal field dependence of T_{BD} is given in Figure 3.39 (b) for 6.2 nm oxides stressed at 140°C [139], 9 nm oxides stressed at 175°C [138], and 11 nm oxides stressed at 150°C [133]. The T_{BD} data points at the lowest field are extrapolated from the low percentile data and hence suffers from a much larger uncertainty similar to the situation found in Figure 3.37 for $F_{OX} = 8$ MV/cm. Because of

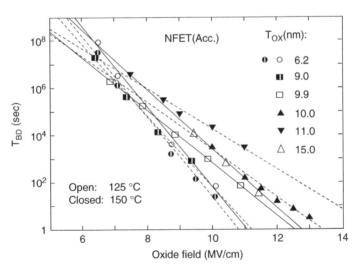

Figure 3.38. The T_{BD} values (either at 63.2% or 50% depending on the references used) plotted against oxide fields for NFET accumulation mode [45]. These data points of 15 nm, 11 nm, 10 nm, 9.9 nm, 9.0 nm, and 6.2 nm oxides are taken from five different references [133–135], respectively. The T_{BD} data of 6.2 nm oxides at 140°C [139] were properly corrected to 125°C and 150°C by the activation energies. Reprinted with permission from Ref. 45, © 2009 IEEE. The areas of these data are about 0.001 cm^2 except for 9.9 nm oxides with an area of 0.006 cm^2. For the data of 9.0 nm oxide, the area is not specified in [138].

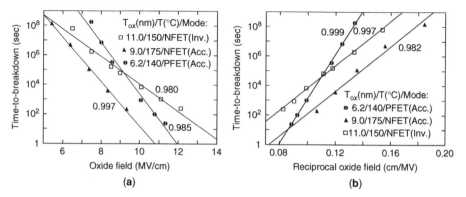

Figure 3.39. Time-to-breakdown plotted against oxide field (a) and the reciprocal oxide field (b) [141]. Three different modes are PFET accumulation, NFET accumulation, and NFET inversion modes for oxide thickness of 6.2 nm [139], 9.0 nm [138], and 11 nm [133], respectively. The value adjacent each fitting line represents its correlation coefficient.

different stress modes used in these three data sets, the absolute comparison of T_{BD} values can not be done. Also notice that this extended time window spans about six to seven orders of magnitude in time depending on the inclusion of the last data point. The experimental data reported for the 9.0 nm oxides suggest that the E-model is the best choice to describe the field-dependence of oxide breakdown [138]. On the contrary, the results of 6.2 nm and 11 nm oxides suggest that the T_{BD} data follow the $1/E$-model or $1/V$ model [139]. While the long term stress method is certainly useful to provide more information at lower stress voltages, the contradictory experimental results of Figure 3.39 suggest that the long term stress method alone cannot conclusively determine which is the best field or voltage acceleration model unless the time window of accelerated stress continues to expand (as we discuss later in the case of ultra-thin oxides). If the oxide thickness is further reduced, a change of field acceleration factor is clearly observed when stress conditions change from FN tunneling regime to DT regime as shown in Figure 3.40 [142]. In this study, a much higher stress temperature of 342°C is used to expand the experimental time window. As can be seen from Figures 3.37–3.40, there is quite a wide variety of experimental results regarding the field or voltage dependence of time-to-breakdown even in the range of relatively thick oxides.

At this point, it is worthwhile to discuss the reproducibility of experimental TDDB measurements. In usual laboratory practice, the reproducibility of experimental data can be ensured by repeating the measurements many times with appropriate data analysis. In TDDB measurements involving long term module stress, this practice is often considered economically not practical. Although the oxide fabrication process from various institutions can vary, the intrinsic properties of TDDB are usually demonstrated to be very similar [143]. For ultra-thin oxides with very large voltage acceleration factors, it becomes even more difficult

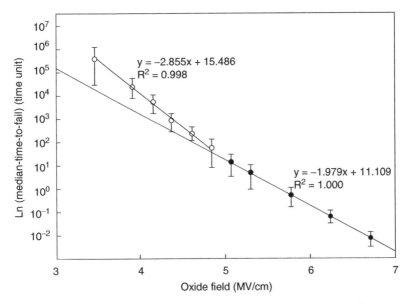

Figure 3.40. Field dependence of T_{BD} for 3.9 nm oxides stressed in direct tunneling regime for a device area of 10^{-4} cm^2 at 342°C. Reprinted with permission from Ref. 142, © 2000 IEEE.

to separate one physical model from another [144]. Facing these many limitations and uncertainties in the experimental work, it is essential to examine the experimental data with new approaches, judging the consistency of the whole picture within the available physical BD models. This is what we try to do in the following discussion.

Alternatively, the time-to-breakdown data can be expressed as a function of gate voltages as shown in Figure 3.41 for several oxide thicknesses. This practice is based on the fact that the energy of the carriers at the anode (roughly equal to the gate voltage) has been found to be the fundamental variable that controls the dielectric breakdown (rather than the electric field) [100]. This key finding was confirmed by means of poly depletion experiments that demonstrated that T_{BD}/Q_{BD} is strongly dependent on the gate voltage but rather insensitive to the oxide field (provided that V_G does not change), as discussed in Section 3.3 [84, 85]. Figure 3.41 (a) displays the T_{BD} data as a function of gate voltage for NFET inversion stress experiments performed on ultra-thin oxides at room temperature [145]. Due to the low temperature used in this study, the stress voltages remain relative high for a reasonable stress time window. Figure 3.41 (b) shows the T_{BD} versus V_G for both thin and thick oxides at 140°C [50, 51]. It is convenient to define the local voltage acceleration factor as

$$\gamma_V = -\left(\frac{\partial Ln T_{BD}}{\partial V_G}\right), \tag{3.50}$$

consistent with the widely used exponential extrapolation law, $T_{BD} \sim exp(-\gamma_V V_G)$. The gate voltage acceleration factor can be readily related to the field acceleration parameter, which is defined below:

$$\gamma_V = -\left(\frac{\partial LnT_{BD}}{\partial V_G}\right) = -\frac{\partial LnT_{BD}}{\partial F_{OX}}\frac{\partial F_{OX}}{\partial V_G} = \frac{b\gamma_F}{T_{OX}}. \tag{3.51}$$

Since $V_{OX} = F_{OX} \cdot T_{OX} = a + bV_G$ is a reasonable approximation for the V_G versus F_{OX} relation, as discussed in Section 2.2.3. The typical value of the coefficient, b, varies from 0.6 to 0.9 depending on oxide thickness and surface potential related band bending in both polysilicon gate and Si substrate. The Equation 3.51 indicates that the gate voltage acceleration factor increases with decreasing oxide thickness if the field acceleration factor is independent of oxide thickness and field.

It can be seen in Figure 3.41 that the local voltage acceleration factors extracted over the limited time window ($1-10^4$ seconds) are larger for thinner oxides than those for thicker oxides. This apparent thickness dependence of voltage acceleration factors seem to agree with the thermochemical model with constant field acceleration. However, it should be pointed out that the voltage acceleration factors are obtained over a different range of voltages. Hence, the interpretation that thinner oxides have a higher voltage acceleration factor than thick oxides could be misleading. The interpretation of this apparent thickness dependence of the voltage acceleration factor has a serious consequence in reliability projection (see Section 3.4.6). In Figure 3.41 (b), we show the

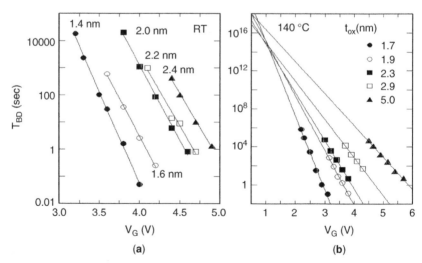

Figure 3.41. Time-to-breakdown versus gate voltage in NFET inversion mode (a) at room temperature (RT); reprinted with permission from Ref. 145, Copyright 2000, Institute of Physics, and (b) at 140°C; reprinted with permission from Ref. 50, © 2002 IEEE.

predictions of T_{BD} at low voltages based on the exponential law using their respective voltage acceleration factors, γ. This analysis suggests that if the exponential law of T_{BD} voltage dependence remains true over the entire voltage range down to the operation conditions, then the lifetime of thinner oxides would exceed that of thick oxides at lower voltage. Notice that for the extrapolation using the exponential law, the use of gate voltage or the oxide field would yield the same result as long as a linear relation between gate voltage and oxide field holds as shown in Figure 2.27. Since the critical defect density, N_{BD}, required to trigger the breakdown strongly decreases with T_{OX} (particularly in the ultra-thin oxide range), the previous interpretation seems to be unphysical because it would require a defect generation rate strongly decreasing with the oxide thickness, and such a decrease is not supported by the experimental data.

Figure 3.42 (a) summarizes a large volume of oxide field acceleration factors obtained over a wide range of oxide thicknesses (from ~ 1.0 to $10.0\,nm$) at a fixed temperature of 140°C. In obtaining oxide field acceleration factors, we adopt the method discussed in Section 2.23 to determine oxide fields accurately. Oxides fabricated with different processes such as conventional thermal SiO_2, lightly nitrided oxides, nitrogen implanted oxides, and remote plasma nitrided oxides were used. Long term module stress data (solid symbols) were also included, which generally yield higher γ_F as compared to the wafer level stress data. The experimental data reveals an apparent continuous increase in oxide field acceleration factors as oxide thickness decreases. In the framework of thermochemical model, this requires the effective dipole moment to be either thickness- or field-dependent. No independent experimental findings on thickness dependence of the effective dipole moment have been reported. A field dependence of the effective dipole moment is in contradiction with the concept of constant field acceleration factor in the thermochemical approach. For the effective dipole moment varying from 7.2 to 13.3 e$\mathring{A}$, the values of γ_F change from 2.02 to 3.74 cm/MV at 140°C by equating Equation 3.29 to Equation 3.51. These values are much smaller than the experimental values commonly found in ultra-thin oxides, as shown in the figure. Even this largest p_{eff} with the ionic bond is far too small in comparison with the results of long term module stress, which reach values up to 5.0 cm/MV.

The same data of Figure 3.42 (a) are plotted against the average oxide field, (F_{OX}) in Figure 3.42 (b). The average oxide field is taken to be the average value of individual oxide fields considered for the determination of γ_F in each data set (for each T_{OX}, the calculation of γ_F requires stressing the data at different V_G and hence different F_{OX} values). The data corresponding to the long-term module stress experiments are also plotted. This figure reveals that oxide field acceleration γ_F generally shows a field dependence. This field dependence is obtained with different oxide thickness but it is consistent with the results of long term module stress using different area structures to access the different range of oxide fields for fixed T_{OX} values. These results demonstrate that the apparent thickness-dependence of γ_F shown in Figure 3.42 (a) is caused by the fact that for thinner oxides, lower voltages are used for stress within a fixed time window. Moreover, the results in Figure 3.42 (b) reveal a polarity dependence, i.e., a different γ_F is

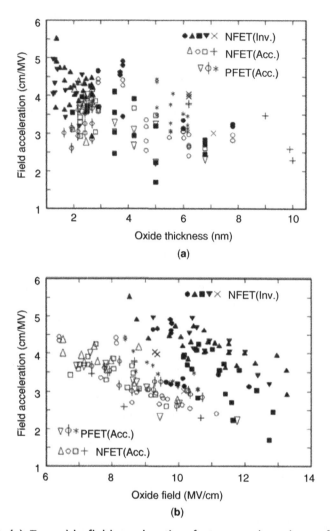

Figure 3.42. (a) T_{BD} oxide-field acceleration factors γ_F plotted as a function of oxide thickness from $\sim 1\,nm$ to $10\,nm$ at $140°C$. Reprinted with permission from Ref. 45, © 2009 IEEE. For relatively thicker oxides $>\sim 2\,nm$, the oxide fields are determined by using the CV integration technique for each thickness. For ultra-thin oxides $<2\,nm$, they are derived from the theoretical calculation which is calibrated with DT current calculation. Five different oxide processes are included: thermal SiO_2, nitrided SiO_2, SiO_2 grown on nitrogen implanted Si substrates, SiO_2 with remote-plasma nitridation process. (b) The same data is plotted as a function of oxide field (F_{OX}). For a given data set (T_{BD} versus F_{OX}), a local γ_F is derived and F_{OX} is taken to be the average of all F_{OX} values within this dataset. This represents one data point in this figure. The plus (+) and cross ($\times$) symbols represent the data from literature for 6.2 nm, 7.1 nm, 9.0 nm, 9.9 nm oxides at $140°$, $125°C$, $150°C$ and $170°C$, and $125°C$, respectively [135, 137–140].

reported for oxides stressed under the same electric field but with different polarity of the gate bias. Once again, these results are not compatible with the predictions of the thermochemical model.

In the framework of the energy- and fluence-driven breakdown, as discussed in Sections 3.2 and 3.3, the reduction in T_{OX} would only cause an increase in the tunneling current, but this change should not significantly modify the voltage acceleration factor, γ_V, except for a small contribution from the tunneling current–voltage relation which is thickness dependent. This consideration suggests that the apparent thickness dependence of voltage acceleration may just be an artifact of the data being collected at low voltages for the thinner oxides within the fixed time-window used in stress experiments due to experiment duration constraints. Therefore, the results shown in Figure 3.41, (a) and (b), may just indicate that the voltage acceleration factors increase with decreasing voltages, regardless of thickness. As already discussed in Section 2.4 and 3.2, it is universally accepted that the statistical behavior of intrinsic oxide breakdown is that of a random process with the weakest link character [132]. It has been shown both experimentally and theoretically that intrinsic breakdown defects are homogeneously distributed across the oxide area and follow the Poisson random statistics [132]. This unique behavior predicts that T_{BD} depends on area as given in Equation 2.37. Figure 3.43, (a) and (b), displays the T_{BD} data measured on four areas for a fixed T_{OX} of 2.7 nm in log-linear scale and log–log scale, respectively. It is evident that a simple forced fit to the data in Figure 3.43 (a) shows nonparallel slopes over the much wider voltage range. It can be seen the extrapolation of T_{BD} data from stress voltages would eventually lead to a crossover at lower voltages. This is another unphysical result caused by the unjustified assumption of an exponential law for voltage acceleration.

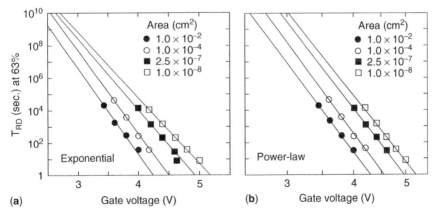

Figure 3.43. T_{BD} voltage dependence measured p+-poly/SiO$_2$/n–Si structures (i.e., PFET accumulation) using different area structures for 2.67 nm at 140°C. (a) in log-linear scale, (b) in log-log scale. Reprinted with permission from Ref. 50, © 2002 IEEE.

Secondly, the nonparallel lines in Figure 3.43 (a) are inconsistent with the voltage-independent Weibull slope (Fig. 3.13). Thirdly, the extrapolated T_{BD} values at 3V using the exponential acceleration law for four different areas would yield a Weibull slope of $\beta \sim 3$ for 2.7 nm oxides, which is totally in disagreement with the experimental findings shown in Figure 3.12 (b). These considerations demonstrate that the exponential law for T_{BD} voltage acceleration is inconsistent because it does not preserve the fundamental properties associated to the Poisson statistics of oxide breakdown. On the other hand, the plot in the power law form for T_{BD} versus V_G in Figure 3.43 (b) yields parallel lines. This result preserves the Poisson random statistics and is consistent with the experimental finding of a voltage-independent Weibull slope.

Therefore, the results in Figure 3.43 (b) may suggest that T_{BD} voltage dependence follows a power law behavior ($T_{BD} \sim V_G^{-n}$) rather than the exponential law as commonly accepted. In summary, the consistency requirement between the T_{BD} versus V_G (or F_{OX}) relationships with Poisson area scaling can provide an independent experimental methodology for checking the validity of the mathematical function used to fit the data for T_{BD} versus V_G as discussed in detail in [146].

Using this approach, we will now reconsider the voltage- or field-acceleration of time-to-breakdown in the FN regime in an attempt to resolve the discrepancy in the experimental observations of Figure 3.39. In Figure 3.44, (a) and (b), the relations of $Ln(T_{BD})$ versus V_G or $Ln(T_{BD})$ versus $1/V_G$ are examined using four different area structures for 5.0 nm oxide. Figure 3.44 (a) shows a cross-over at low voltages in violation of Poisson random statistics. Thus, the exponential law for T_{BD} voltage (or field dependence) can be reasonably ruled out. On the contrary, the $Ln(T_{BD})$ versus $1/V_G$ curves remain reasonably parallel but with some deviations as shown in Figure 3.44 (b). These deviations will be discussed in detail in [146]. This finding is

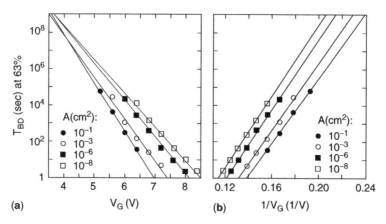

Figure 3.44. T_{BD} voltage dependence measured on n+-poly/SiO$_2$/p-Si structures (i.e., NFET accumulation) using different area structures for 5.0 nm oxides at 140°C. (a) linear voltage scale, (b) inverse voltage scale. Reprinted with permission from Ref. 146, © 2009 IEEE.

consistent with the experimental observation of T_{BD}-Exp($1/V_G$) dependence through long term module stress for 6.2 nm oxides (Fig. 3.39) [139].

As we already discussed in Section 3.2, in the formulation of energy- and fluence-driven BD, the time-to-breakdown is related to both tunneling current and defect generation efficiency. At the high voltages or electron energies considered in Figure 3.44, the voltage (or field) dependence of FN tunneling current, $(J \sim \exp(-C/F_{OX}))$, dominates over the voltage dependence of defect generation efficiency. This explains the reported linear relation of $Ln(T_{BD})$ versus $1/V_G$. Figure 3.45 (a) depicts the power law exponents as a function of T_{OX} from 1.0 to 10 nm, while the corresponding acceleration factors are also given as a reference as previously discussed. Changes in the voltage acceleration factors are as large as 200%. The power law exponents have an average value around 43 ± 3 and remain unchanged for thin oxides stressed at lower voltages in DT regime, while at high voltages in FN regime the T_{BD} power law exponents approach an average value of 31 ± 5 for thick oxides as recently reported [45, 139, 144, 146, 147]. In DT regime, the defect generation efficiency shows a strong power law voltage dependence, $\zeta(V_G) \sim V_G^{38}$, which is reflected by the larger exponent of the T_{BD} power law since the voltage dependence of direct tunneling current is weak. On the other hand, as discussed in Section 3.2 a reduction of the dependence of ζ on V_G at high voltages in FN regime asymptotically converges to an almost voltage independent $\zeta(V_G)$ (Figures 3.7 and 3.8). Therefore, the T_{BD} voltage dependence in this regime is proposed to arise from strong voltage dependence of FN tunneling current [148]. A force-fit of the J_G^{FN} versus V_G in the log–log plot yields an exponent of 31.0 ± 5.0 [146]. This value is consistent with the value obtained from the values of T_{BD} power law. In other words, a T_{BD} power law dependence in FN regime with its exponent of ~ 30 is equivalent to the linear T_{BD} dependence of $\exp(1/F_{OX})$ or $\exp(1/V_G)$, both consistent with the concept of fluence-driven BD process.

We remarked that the thickness dependence of voltage acceleration factors as seen in Figure 3.45 (b) is only an apparent effect due to the different voltage regimes used for stress. The same results can be expressed as a function of reciprocal of gate voltages as shown in Figure 3.45 (b). We deduce the experimental γ_V's from semi-log plots such as Figure 3.41. Note that the values of γ_V derived from raw experimental results are only discrete approximations for the true differential voltage acceleration. For a given γ_V, the voltage in the above equation is taken to be the average of the applied stress voltages for a set of T_{BD} data with at least three stress voltages.

To further expand the experimental time window, significant efforts to develop new experimental techniques for TDDB measurements have been made [149, 150]. In addition to a conventional source measurement units (SMU) setup, a fast measurement setup based on a peripheral computer interconnect (PCI) card [149] is used to record the breakdown times in millisecond regime. To extend the measurement times down to a few microseconds, a combined measurement setup of a pulse generator unit (PGU) [150] for an SMU for sensing was used. Furthermore, to increase the time window to 10^6 seconds, the samples are prepared at the package

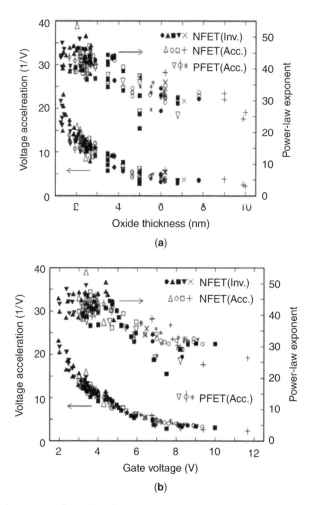

Figure 3.45. The same data in Figure 3.42 are potted by considering the gate voltage acceleration and power law exponents as a function of oxide thickness (a) and gate voltage (b). For a given dataset (T_{BD} versus V_G), a local γ_V is derived, V_G is taken to be the average of all V_G values within this dataset. Reprinted with permission from Ref. 45, © 2009 IEEE.

level (PL) so that the electrodes could be permanently connected to the module sockets. Using these four measurement techniques, the experimental time window can be over twelve orders of magnitude as seen in Figure 3.46 (a) for the T_{BD} distributions with excellent statistical basis for V_G from 2.4 V to 4.4 V for 1.5 nm oxides [123]. Figure 3.46 (b) depicts the corresponding T_{BD} at 63% as a function of voltage for 1.5 nm, 2.2 nm, and 5.2 nm oxides in the log–log scale with the power law exponents of 42, 43, and 32, respectively. For 5.2 nm oxides, the stress is performed in the FN regime, resulting in a reduced power law exponent as already discussed

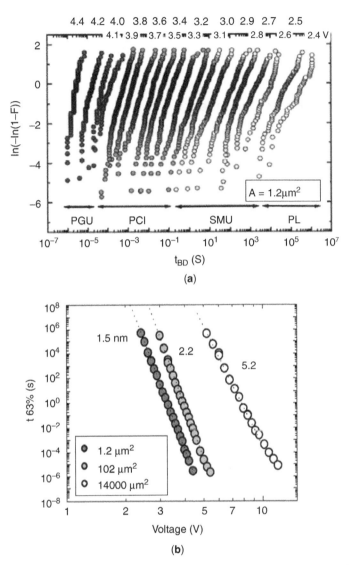

Figure 3.46. (a). The Weibull distributions for 1.5 nm oxides stressed in NFET inversion mode. (b) Voltage dependences of T_{BD} for 1.5 nm, 2.2 nm, and 5.2 nm oxides. Notice the data spans over 12 orders of magnitude. Reprinted with permission from Ref. 123, © 2006 IEEE.

above. It is worthwhile to note that T_{BD} measurements over microsecond range was also performed for oxides stressed in FN regime [151, 152] follows a linear dependence of $\exp(1/F_{OX})$, consistent with the results of 6.2 nm [139]. At lower voltages or for thinner oxides, the T_{BD} power law dependence remains valid over 12 orders of magnitude as seen in Figure 3.46 (b). Recently, the empirical T_{BD} power law was also confirmed by different groups [59, 123, 153–155]. These independent studies from different research groups unambiguously demonstrate that an exponential law is invalid to characterize the T_{BD} voltage dependence for both thin and thick oxides.

3.4.2 Temperature Dependence

It is well known that the dielectric breakdown (including that of silicon dioxide) is a thermally activated process and its temperature dependence has been the focus of numerous investigations. As early as the beginning of 1980s, a large volume of experimental data was collected for silicon dioxide and thermal activation energies ranging from as high as $\sim 1.0\,\mathrm{eV}$ to as low as $0.1\,\mathrm{eV}$. Unfortunately, as already discussed above, these data were mainly obtained without the knowledge that they actually represented the extrinsic character of SiO_2 breakdown [131]. Figure 3.47 (a) displays the time-to-breakdown as a function of temperature in the 65–400°C range reported for 15 nm oxides while in Figure 3.47 (b) the same data is plotted against oxide field [156]. It is evident from this figure that time-to-breakdown (T_{BD}) increases with decreasing temperature at a given stress voltage. The slope of LnT_{BD} versus the inverse of temperature $(1/T)$ is defined as the local activation energy at a given field or a gate voltage:

$$\Delta H = K_B \left(\frac{\partial LnT_{BD}}{\partial(1/T)} \right) \qquad (3.52)$$

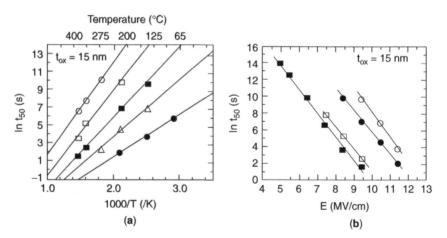

Figure 3.47. (a) Time-to-breakdown as a function of temperature. The open circles, the open squares, the filled squares, the open triangles, and the filled circles represent 7.42, 8.42, 9.42, 10.42, 11.42 MV/cm, respectively. (b) The same data of Figure 3.47 (a) plotted as a function of oxide field for 15.0 nm oxides. Reprinted with permission from Ref. 156, © 1997 IEEE.

As shown in 3.47 (a), the activation energy (the slopes of T_{BD} versus $1/T$) remains constant at lower fields (8–9 MV/cm) but decreases with increasing oxide fields (9 MV/cm–12 MV/cm). The decreasing trend of activation energy with increasing oxide field seems to be consistent with the prediction of the thermochemical model discussed in Section 3.3 with the mix of two bonding states [52]. The results in Figure 3.47 (a) show that a single value of the activation energies can not even characterize the slope only for a limited range of temperatures. Figure 3.48 summarizes the temperature-dependence of time-to-breakdown at several oxide fields for 10 nm oxides [140]. A change of slopes occurs around $T_C \sim 125°C$, suggesting two different values for ΔH are required to fully characterize the T_{BD} temperature dependence. Figure 3.49 displays the activation energies as a function of oxide fields for thick oxides, indicating that at lower temperatures ($<125°C$) activation energy can be above 1.0 eV. It is interesting to note that temperature-dependent activation energy has already been reported for extrinsic oxide failure mode [157]. These results illustrate that a conventional Arrhenius thermal activation model with a single value of ΔH has serious difficulties to explain the whole range of experimental observations. Thus, an extended Arrhenius model with the two activation energies for their respective oxide field ranges was proposed for thick oxides [140, 157].

The concept of a non-Arrhenius thermal activation was first proposed for dealing with ultra-thin oxides below 3.5 nm for which a much stronger temperature dependence of oxide breakdown was found in comparison with thick oxides [69]. Figure 3.50 reveals that for thin oxides of 2.2 nm and 3.1 nm, change by more than

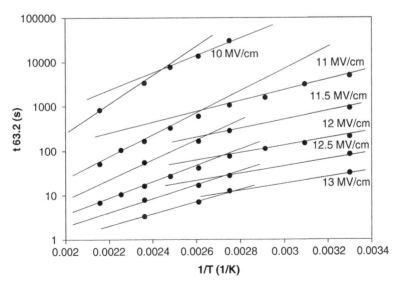

Figure 3.48. Temperature dependence of T_{BD} for 10 nm oxides stressed reveals different activation energies. Reprinted with permission from Ref. 140, Copyright 1995, John Wiley & Sons, Inc.

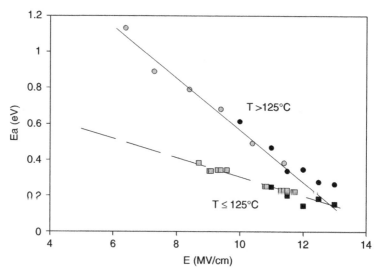

Figure 3.49. Thermal activation energies versus oxide field, revealing that activation energy varies with temperature. Reprinted with permission from Ref. 140, Copyright 1995, John Wiley & Sons, Inc.

four orders of magnitude in T_{BD} is found while for thicker oxides of 7.3 to 13.8 nm, T_{BD} changes only by one order of magnitude over the same range of temperatures (25–200°C) [70]. This stronger temperature dependence of T_{BD} found in ultra-thin oxides makes the observation of non-Arrhenius temperature dependence feasible and the nonlinearity in T_{BD} versus $1/T$ characteristics can be revealed in the conventional range of stress temperatures. It was proposed that both the temperature dependence of the defect generation rate and that of the critical breakdown defect density might be the cause for the strong temperature dependence found in ultra-thin oxides. These results regarding non-Arrhenius activation and strong temperature dependence were subsequently confirmed by many researchers [70, 158]. The nonuniqueness of activation energies already seen for thick oxides in Figure 3.49 is consistent with the concept of non-Arrhenius temperature dependence.

The stronger temperature dependence found in ultra-thin oxides was initially attributed to a thickness effect [69]. However, the t_{BD} data for various oxide thicknesses were measured over different stress voltages (as shown in Figure 3.50) due to the constraint of experimental time window and the strong thickness-dependence of t_{BD} and q_{BD}. To resolve whether this is a thickness or a stress voltage effect, the same thickness but different stress voltages should be used for t_{BD} or q_{BD} measurements. This is accomplished by using different area structures with a fixed thickness to bypass the limitation of the experimental time-window. Figure 3.51 shows the normalized Q_{BD} data at 63% as a function of the inverse of temperature. In this case, the Q_{BD} rather than T_{BD} is used to eliminate the thickness dependence of tunneling current. Several T_{OX} values were used for stress

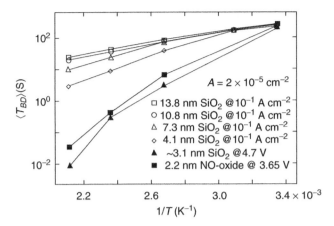

Figure 3.50. Temperature dependence of T_{BD} for several oxide thicknesses. Reprinted with permission from Ref. 70, © 2000 IEEE.

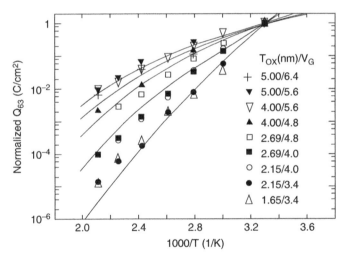

Figure 3.51. Temperature dependence of charge-to-breakdown for several T_{OX} values under different stress voltages. Reprinted with permission from Ref. 161, Copyright 2005, Elsevier.

voltages from 3.4 V to 6.4 V, with different area capacitors and much longer stress times [159]. For each voltage, two T_{OX} values were used (from 1.65 to 5.0 nm) while for a fixed T_{OX} value, two voltages were used. Note that the entire temperature range from 30 to 200°C is covered in all cases except for 1.65 nm oxide at 30°C. For four voltages, no thickness effect on temperature activation was observed as shown in Figure 3.51. Thus, it is clear that the strong temperature activation observed on ultra-thin oxides is simply a consequence of applying lower stress voltages (rather than an intrinsic thickness effect). However, the non-Arrhenius temperature dependence remains valid for both thick and thin oxides. In thick oxides, this effect is less pronounced due to the high voltages applied; but it becomes evident in thinner oxides stressed at lower stress voltages.

3.4.3   Interrelationship of Voltage and Temperature Dependence

Until now we have discussed the voltage (or where applicable, field) dependence and the temperature dependence of T_{BD} or Q_{BD}, separately. In Figure 3.47 (b), the T_{BD} data corresponding to the 15 nm oxide of Figure 3.47 (a) are plotted against oxide field with the stress temperature as a parameter. It is evident that field acceleration factors, γ_F, remain temperature independent while Figure 3.47 (a) shows that the thermal activation of T_{BD} increases with decreasing field. This important result is also confirmed for ultra-thin oxides with thickness ranging from 2.1 to 4.0 nm (regardless of whether oxide field or the gate voltage is considered as the breakdown variable [160]) and indicates that this is a general observation for both thick and thin oxides. However, it can be easily shown that voltage-dependent thermal activation is mathematically inconsistent with temperature-independent voltage acceleration, given that the voltage dependence is described by an exponential law with a constant acceleration factor and the temperature dependence by an Arrhenius law with single activation energy. Nevertheless, as already pointed out previously, such results were obtained in a limited time window ($1–10^4$ seconds) so that different ranges of voltages and temperatures were used to reach these conclusions. Again, to gain better understanding, different area capacitors and long stress times close to 10^6 seconds have to be used. The T_{BD} data as a function of V_G are plotted in Figure 3.52 for a fixed T_{OX} value of 2.67 nm and two

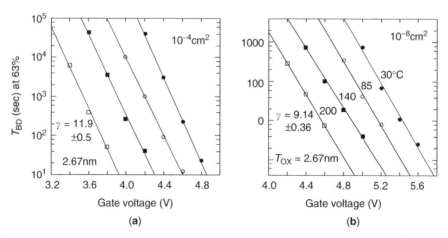

Figure 3.52. T_{BD} versus gate voltage for (a) large area capacitors and (b) small area capacitors from 30°C to 200°C [161]. The same structures of p⁺poly/n–Si capacitors as used in Figure 3.43 and stressed under substrate injection.

different oxide areas stressed at 30°C, 85°C, 140°C, and 200°C. These figures reveal that voltage acceleration factors remain relatively temperature-independent within a fixed time window, but their values at lower voltages are higher than those at higher voltages. For even thinner oxides of 1.65 nm, similar results of temperature-independent voltage acceleration factors were also found with an average value γ_V of $14.8 \pm 0.4\,V^{-1}$ for stress temperatures between 30°C and 140°C [159].

The previous observations indicate that the interpretation of experimental data in a limited observation time window is always very difficult and can easily lead to misleading and contradictory conclusions. Pursuing a more complete and clear picture for both voltage and temperature dependence of oxide breakdown, T_{BD} measurements were performed for various thicknesses, $T_{OX} = 2.67, 2.41, 2.15, 1.92$, and 1.65 nm, at four different temperatures of 30°C, 85°C, 140°C and 200°C and using capacitor areas varying from $10^{-2}\,cm^2$ to $10^{-8}\,cm^2$. Then, all the obtained T_{BD} data were normalized using area scaling and thickness scaling procedures [161]. The area scaling method is discussed in Section 2.4.2 and the relevant Weibull slopes for these thicknesses have been used. The Weibull slopes are assumed to be temperature-independent, consistent with the previously shown results. The thickness scaling is done with respect to an arbitrarily chosen reference value of $T_{OX} = 2.67$ nm and uses the thickness scaling law of N_{BD} (Eq. 3.14) and the thickness dependence of direct tunneling current (Eq. 2.13). The parameters used in this normalization are exactly the same as discussed earlier; that is, $a_o = 1.83$ nm, $T_{INT} = 0.37$ nm, and $\eta = 4.5$ dec/nm. Corresponding to four different temperatures, four universal curves can be clearly identified in Figure 3.53 (a) with

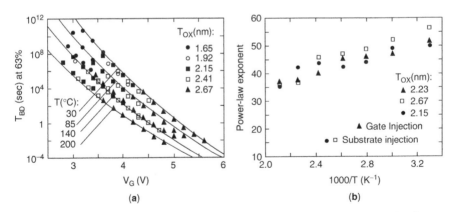

Figure 3.53. (a) Normalized T_{BD} versus V_G for 30°C, 85°C, 140°C, and 200°C for five T_{OX} values with p⁺poly/n-Si capacitors (+V_G). The two normalizations were performed using area scaling law and thickness-scaling law (see the text for discussion) [161]. The reference oxide thickness is 2.67 nm. The direct tunneling regime corresponds to $V_G < 4.6$ V where FN tunneling regime refers to above 4.6 V [161]. The lines are from the empirical T_{BD} power law using the temperature dependence of power law exponents (b). Reprinted with permission from Ref. 161, Copyright 2005, Elsevier.

distinctive features of voltage-dependent voltage acceleration. The lines represent the results of empirical T_{BD} power law as we discuss below.

The normalized T_{BD} in Figure 3.53 (a) spans over 12 orders of magnitude in time contrary to the small time-window usually considered. Figure 3.53 (a) reveals some very important findings: 1) At a fixed temperature, the local voltage acceleration factors increase with decreasing voltage; 2) at a fixed voltage, the voltage acceleration factor is larger at lower temperature than that at higher temperature; 3) at a fixed T_{BD}, local voltage acceleration factors remain roughly temperature-independent; and 4) at lower voltages, a more pronounced temperature dependence (as compared to that at higher voltages) is observed. Furthermore, this analysis demonstrates that the T_{BD} thickness dependence at a fixed stress condition can be entirely captured along with the thickness dependence of the tunneling current by the cell-based model. These results suggest that voltage and temperature dependence of T_{BD} mainly stems from both defect generation efficiency and tunneling current, and this is consistent with our initial assumptions. As shown in Figure 3.53 (a), it is evident that the overall experimental behavior can be well described the T_{BD} power law with its exponent being temperature-dependent. This temperature dependence of the power law exponent is explicitly provided in Figure 3.53 (b), although it was already implicitly revealed in [159–162]. The figure reveals this effect is independent of injection polarities and oxide thickness. These results suggest that power-law voltage dependence can well capture many experimental observations in terms of interrelation of T_{BD} voltage and temperature dependence as discussed above.

Given these results, we can say that from the experimental viewpoint, a self-consistent picture regarding temperature dependence of oxide breakdown has emerged for both thick and thin oxides. However, our present knowledge about the physical processes involved in the thermal activation of oxide breakdown is rather poor to say the least. Nevertheless, many attempts have been made to provide this type of insight based on various physical mechanisms as we will now briefly discuss. Since the very idea of the thermochemical model is based on the free-energy formulation of the bond breakage process [52], this seems to be a natural description for a thermally activated process such as oxide breakdown. First, the activation energy as discussed above for both thick and thin oxides [134, 159] is found to be temperature-independent in disagreement with the prediction of Equation 3.26. To account for this experimental observation, the model has been modified to incorporate a distribution of different precursor bonds, presumably oxygen vacancies, distorted Si–Si bonds, Si–H bonds, and many others [52]. By mixing two types of the bonding states, a constant field acceleration factor is produced over a limited range of temperatures as shown in Figures 3.19 (a). While invoking the mixing of distribution of the bond states might be plausible [52], this requires a detailed knowledge of the involved microscopic defects to provide a quantitative explanation of the experimental observations. Moreover, notice that in the range of temperature independent activation energy, γ_F is predicted to decrease with decreasing oxide field (Figure 3.19 (b)). This prediction is in contradiction with the widely reported experimental observation of the opposite trend. At a more fundamental level, the observed temperature dependence of the activation energy, sometimes referred to as the nonuniqueness of the activation energy, also poses a serious challenge to the thermochemical model. To account for this difficulty, the mixing of field-induced (thermochemical model) and current-induced degradation mechanisms has been proposed to explain the non-Arrhenius temperature dependence [80]. This approach suggests that the usual field dependence of activation energy should reappear in the limits of low temperature (30°C) and high temperature (300°C). Figure 3.54 compares the experimental data with the prediction of this combined approach of field-induced (E-model) and current-induced model (AHI model). It is seen that this approach is not adequate to explain the experimental results for both thick and thin oxides. Furthermore, by including a current-induced mechanism, it was also implicitly recognized that the thermochemical model is inadequate for both thick and thin oxides.

On the other hand, some attempts have been made to explain the temperature dependence within the framework of the hole-induced breakdown models. In the case of the AHI model, the majority of generated holes injected from the anode to the SiO$_2$ film would undergo a tunneling process rather than injection over the barrier. As already discussed in Section 2.2.2, the tunneling process is known to be strongly dependent on voltage (or field) but only weakly depends on temperature. Thus, the effort to explain the temperature dependence of oxide breakdown has involved a parameterization of the critical hole fluence to fit the experimental data [62]. However, the measured substrate current remains insensitive to temperature variation for 25–125°C, with a slight increase at 175°C by 3X at low voltages. The parameterization of Q_p seems to be inconsistent with the experimental results. Finally, it must be pointed out that while the hydrogen release model as discussed in Section 3.3 enjoys much success in explaining the voltage dependence of T_{BD}

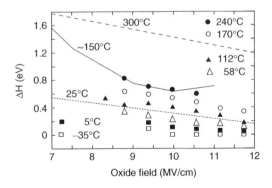

Figure 3.54. Nonuniqueness of activation energy as a function of F_{OX} measured on 6.2 nm oxides [163]. The solid line represents the results of a theoretical study considering a combination of field-driven (E-model) and current-induced (1/E-model) mechanisms [80] whereas the short and long dashed lines represent the results at the extreme cases for 1/E-model (25°C) and E-model (300°C), respectively.

and Q_{BD} for both thick and thin oxides, no serious attempt has been made yet to explain the temperature dependence of oxide breakdown within this picture.

Facing these difficulties in using a BD model to explain BD temperature dependence, many researchers adopt the independent experimental measurements of defect generation as an alternative approach. It is well known that both interface states and bulk traps are generated as a result of stress. By measuring the charge build-up for 12 nm and 8.5 nm oxides under CCS in FN regime, it was shown that bulk-trapped charge at BD is temperature-independent [164, 165]. The trapping efficiency was found to be correlated to the activation energy of Q_{BD}. The interface state density under constant current stress for 9.0–20 nm oxides has been shown to be independent of temperature while Q_{BD} changes by ∼10X [165]. The finding of non-Arrhenius temperature dependence has prompted the investigation of the temperature dependency of the critical defect density, N_{BD}, similar to the work on Q_p temperature dependence in the AHI model. The results based on SILC [69, 159] and charge-pumping measurements [70] shows that N_{BD} varies at most by 5X, which is not sufficient to explain the particularly large activation energies often found at low stress voltages [159]. In summary, our current understanding of temperature dependence of oxide breakdown is quite limited in comparison with the progress we have made on voltage acceleration of oxide breakdown over the last 40 years.

3.4.4 Polarity Dependence

Polarity dependence of oxide breakdown has been a subject of intense study for many years by several groups [100, 160, 166–168]. Hokari first reported this phenomenon in the case of very thick oxides with thickness ranging from 5.7 to 19 nm [166], and it is also been found for very thin oxides down to 1.9 nm in recent years [160]. In general, Q_{BD} or T_{BD} results measured under gate injection stress (NFET accumulation mode) were found to be worse (i.e., lower T_{BD} or Q_{BD}) than those obtained under substrate injection (NFET inversion mode) [100, 160, 166–168] either using CCS or CVS methods for the same stress current density or stress voltage, respectively. The thickness dependence of the polarity dependence of Q_{BD} found in constant current stress experiments is shown in Figure 3.55. For positive bias, the Q_{BD} values increase with decreasing oxide thickness while the opposite is found for negative bias. Thus the polarity gap appears to become larger as oxide thickness is reduced. However, this effect is partly caused by the use of CCS since for the same current density, a much reduced stress voltage is required for thinner oxides as compared to thick oxides in Fig. 2.45 in Section 2.3.1. This is especially difficult to understand in the framework of field-driven breakdown models because in NFET accumulation mode the oxide fields are much lower than those of NFET inversion mode, mainly due to the flatband voltage correction for gate injection. Thus, only qualitative reasons such as surface roughness at poly Si/SiO$_2$ interface and the existence of a transition region at the SiO$_2$/Si interface could be given in an attempt to explain the polarity dependence of oxide breakdown [166–168]. This situation did not change until the concept that the energy of an electron exiting the oxide layer at the anode is responsible for the defect generation rate was launched [100]. In

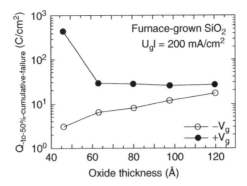

Figure 3.55. Charge-to-breakdown versus oxide thickness with n⁺poly gate and p-type substrate. Reprinted with permission from Ref. 167, © 1994 IEEE.

Figure 3.56, it was first shown that when Q_{BD} data is plotted against gate voltage, $q|V_G|$, which is approximately equal to the maximum electron energy with respect to the Fermi level at the anode, this strong polarity dependence disappears to the first order although the difference is relatively large (~10X) [100].

Reaching an understanding of the polarity gap of oxide breakdown is further complicated by the fact that different stress and evaluation methodologies often lead to different interpretations. For example, different methodologies such as constant current stress (CCS) versus constant voltage stress (CVS) and the choice of breakdown variable (either T_{BD} or Q_{BD}) can affect the interpretation. In any case, it

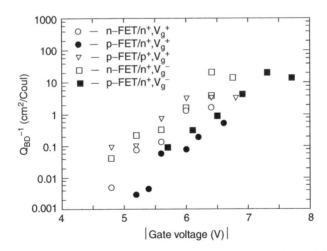

Figure 3.56. The reciprocal of charge-to-breakdown as a function of the absolute magnitude of the applied gate voltage [100]. Oxide thickness is between 3.7 and 3.9 nm at 22°C. Reprinted with permission from Ref. 100. Copyright 1996, American Institute of Physics.

has been clearly demonstrated that constant voltage stress is the relevant method for the evaluation of oxide breakdown in ultra-thin oxides as discussed in Section 2.3. A typical comparison of Q_{BD} and T_{BD} data for NFET stressed in inversion and accumulation modes is shown in Figure 3.57 (a) and (b), respectively. It is shown that Q_{BD} is larger for NFET biased in inversion than for NFET in accumulation by about 20X at a fixed $|V_G|$. On the other hand, the T_{BD} polarity dependence shows a crossover effect: T_{BD} is longer when thick-oxide NFETs are biased in accumulation rather than in inversion and the contrary is true for thin oxides. However, it was shown that this crossover can be understood considering $T_{BD} = Q_{BD}/J(V_G,T_{OX})$, since the difference in tunneling current between inversion and accumulation modes is much smaller in the DT regime than it is in the FN tunneling regime [160]. Because thick oxides are often stressed in the FN regime and thin oxides in the DT regime, a crossover effect would occur for a roughly constant factor of 20X in the Q_{BD} data measured under both polarities.

As we have already said, some authors had argued that the asymmetry of the SiO_2 interfaces could be the reason behind the polarity dependence (with gate injection being considered as more deleterious than substrate injection). However, this explanation turned out to be unlikely when a similar gap in T_{BD} and Q_{BD} was found for PFET accumulation (substrate injection) and NFET inversion (substrate injection also) data as shown Figure 3.58 (a) [169]. For both n^+poly/NFET and p^+poly/PFET stressed in accumulation mode, the anode material is p-type and the cathode is n-type so that the SiO_2/poly Si and the SiO_2/Si interfaces are accumulated in both cases the anode and cathode interfaces are opposite. In n^+poly/NFET accumulation, injection takes place from the gate towards the SiO_2/Si interface, while in p^+poly/PFET accumulation, we have substrate injection with the anode interface being the poly Si/SiO_2 interface. The similar values of Q_{BD} results reported for these two cases in Figures 3.57 (b) and Figure 3.58 (a) further indicate that the quality of the interfaces do not play a major role in determining the Q_{BD} polarity

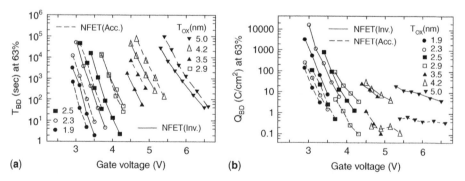

Figure 3.57. T_{BD} versus gate voltage (a) and Q_{BD} versus gate voltage (b) for n+poly/NFET devices under inversion and accumulation modes at 140°C. Reprinted with permission from Ref. 160, © 2002 IEEE.

gap. On the contrary, the whole set of results suggests that it is the type of doping of the anode that plays an important role in the breakdown process.

The results shown in Figures 3.57 suggest that the gate voltage (V_G) itself can not simply cancel the Q_{BD} polarity gap as previously claimed [100]. Nevertheless, the electron energy with respect to the Fermi level at the anode remains a valid concept. If electrons are injected into a p-type silicon anode, they can fall to empty states in the valence band at or near the anode interface. On the contrary, when they are injected into an n-type Si-electrode (either n-type poly Si gate or n-type substrate), they can only end up in states of the anode conduction band. These two cases are illustrated in Figure 3.59 for nonballistic FN tunneling regime as an example. Since the degradation of the oxide is believed to be strongly dependent on the electron energy dissipation at the anode interface, the difference of about 1V in electron energy causes a large difference in defect generation efficiency and, consequently, in Q_{BD}. Therefore, the position of the Fermi level in the anode electrode has an important impact on Q_{BD} degradation [61, 100, 169]. To resolve the Q_{BD} polarity gap, a new definition of of electron energy is required [169]. This maximum available energy is defined as the energy of electrons at the oxide/anode interface as measured with respect to either the anode silicon conduction band or to the anode valence band, depending on the availability of empty states in the valence band. Using this new definition for E_{MAX}, we plot the Q_{BD} data in Figure 3.58 (b) for NFET stressed in inversion and accumulation modes and for PFET biased in accumulation with T_{OX} ranging from 1.9 to 7.0 nm. Moreover, the Q_{BD} data are also normalized using the thickness scaling law of N_{BD} (Eq. 3.14) with the same parameters $a_o = 1.83$ nm and $T_{INT} = 0.37$ nm used previously. The

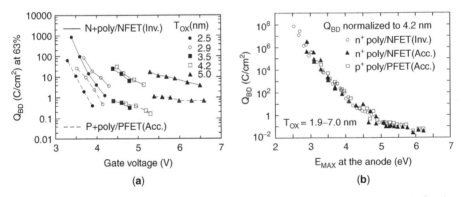

Figure 3.58. (a) Q_{BD} versus gate voltage for n⁺poly/NFET in inversion and p⁺poly/PFET in accumulation at 140°C. Substrate injection mode is used in both cases by definition. (b) The normalized Q_{BD} for oxides from 1.9 to 7.0 nm with three sets of Q_{BD} data as function of maximum energy discussed in the text. The temperature is 140°C. The concentration of n⁺poly doping is assumed to be 10^{20} cm³. The electron mean free path is 1.5 nm. Reprinted with permission from Ref. 169, © 2002 IEEE.

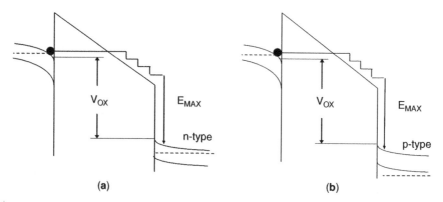

Figure 3.59. Energy band diagrams to illustrate the maximum energy dissipation of injected electrons for thick oxide in nonballistic FN tunneling region at an n-type anode (a) and at a p-type anode (b). The gate voltage is the difference between the Fermi-levels of the cathode and the anode depending on the n-type or p-type anode. Similar diagrams with the same positions of available empty states at either anode for direct tunneling regime.

results of Figure 3.58 (b) show a single universal curve for the three independent sets of Q_{BD} data. Thus, the experimental results from both thin and thick oxides can be unified, revealing a fundamental relation between Q_{BD} and the maximum electron energy that can be dissipated at the anode interface. The results shown in Figure 3.58 (b) suggest that Q_{BD} (i.e., the total electron fluence) and the electron energy with respect to the bottom of the empty states at the anode are the fundamental factors controlling the oxide BD process. Within this framework, a wide variety of T_{BD}/Q_{BD} data under different injection polarities and various anode types can be reconciled. This gives a sound physical explanation of polarity dependent breakdown and provides strong support to the idea that the release of species at the anode interface is the most relevant step in the creation of the oxide defects that finally trigger the breakdown.

3.4.5 Degradation and Breakdown Under AC Stress Conditions

As explained in previous sections, oxide reliability studies are usually performed under ramped or DC stress conditions. However, when operated in digital circuits, the MOSFET gate oxide experiences time-varying bias, usually being switched between two voltage levels at high frequencies. It is therefore reasonable to try to emulate the real situation in circuits and check how the oxide degrades to breakdown under the application of pulsed stress conditions. Several authors have considered this issue and have compared the time-to-breakdown (or the charge-to-breakdown) under DC, unipolar and bipolar pulsed stress conditions [170–175].

Pulse generators are usually used to stress the devices with voltage stress waveforms as those shown in Figure 3.60. The current waveforms can be

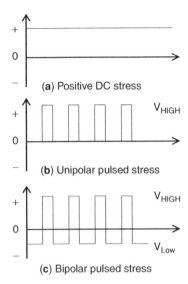

(a) Positive DC stress

(b) Unipolar pulsed stress

(c) Bipolar pulsed stress

Figure 3.60. Voltage waveforms used for DC and dynamic pulsed stresses. The ratio between the time at the high voltage and the period of the signal is the duty cycle (usually 0.5). The amplitude of the positive and negative semicycles is usually the same. However, since the time to breakdown is different under negative and positive bias for the same absolute value of the stress voltage, some authors have used asymmetric voltage waveforms so as to have similar values of T_{BD} in the corresponding positive/negative DC stresses.

visualized by measuring the voltage drop in a small series resistance by means of a digital oscilloscope which can also be used to periodically store trains of pulses. When this is done, trapping/detrapping transients are observed at the times corresponding to the transitions between the two voltage levels [174]. Moreover, if the oxide is thick enough to allow bulk trapping, the amplitude of the current pulses is found to decrease with time due to the generation of neutral electron traps (as in DC stress experiments) which become partially filled with electrons [174]. The effects of the reverse bias voltage amplitude, stress frequency, and duty cycle have been considered [170, 172, 173]. Usually, the stress experiment is periodically stopped at exponentially increasing time intervals to characterize the evolution of the oxide degradation by measuring the trapped charge distribution [134, 173, 176]; the density of interface states [134, 172, 175, 177], or the evolution of SILC [175]. The time-to-breakdown and the evolution of the oxide degradation under unipolar pulse stress conditions is found to be very similar to that obtained under pure DC conditions, provided that the total stress time is multiplied by the duty cycle. On the contrary, a significant frequency-dependent enhancement of the time to breakdown is consistently found under bipolar stress conditions. Figure 3.61 shows the frequency dependence of the time-to-breakdown measured

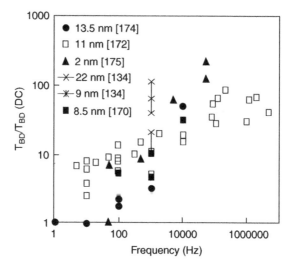

Figure 3.61. Mean time-to-breakdown obtained under bipolar pulsed stress conditions normalized to the time to breakdown measured under DC stress (with the same stress voltage). The data has been obtained from the literature as indicated in the legend and cover a wide range of stress oxide fields and oxide thickness (2 to 22 nm) from references [134, 170, 172, 174, 175].

under bipolar pulsed stress normalized to the time-to-breakdown measured under DC stress conditions. The data reported by different authors shows a clear trend of T_{BD} increasing with frequency, in spite of having included data from a wide range of oxide thickness (2 nm to 22 nm) and stress oxide electric field. Some authors have claimed that the difference between oxide lifetime under pulsed and DC bias becomes less significant as stress voltage (stress oxide field) and oxide thickness are reduced [134, 171]. However, recent results reported for gate oxides as thin as 2 nm somehow contradict this conclusion and indicate that the time-to-breakdown increases more than two orders of magnitude under bipolar stress conditions [175].

The results on oxide degradation under dynamic stress conditions and how they are correlated to the breakdown results are less conclusive. Several authors have reported that the generation of interface states under bipolar stress conditions is significantly enhanced in respect to those measured under DC stress conditions. This means that although the breakdown lifetime can be enhanced under high frequency pulsed conditions, the device lifetime can be reduced due to other phenomena such as the degradation of drain current and transconductance, which are directly related to the density of interface states [172]. The results reported about the frequency dependence of interface state generation are somehow inconsistent. While some authors have reported that the generation rate increases with frequency up to 10 MHz [177], others have found an initial increase up to $f \sim 20$ KHz, followed by a subsequent decrease at higher frequencies [172]. In any case, the generation of interface states at breakdown is much larger under bipolar stress conditions; this demonstrates that there is not a causal relation between the triggering of the breakdown and the generation of interface states. On the contrary, the generation of bulk electron traps and the corresponding negative charge trapping has been found to be much better correlated to the breakdown data [173, 176]. By measuring the evolution of the negative charge trapping, it was concluded that the trap generation rate decreases and the density of generated defects at breakdown increases with frequency under bipolar stress conditions [174, 176]. The combination of both phenomena can qualitatively explain the breakdown lifetime frequency dependence.

Using SILC measurements in ultra-thin oxides, Wang and coworkers found that the electron trap generation rate under bipolar stress was similar to that measured under DC stress conditions. However, the density of defects at breakdown as measured by the relative SILC increase was greatly increased under high-frequency bipolar pulse stress [175]. Rosenbaum and coworkers also attributed the increase of the time-to-breakdown to the reduced generation of neutral electron traps [172]. Moreover, they tried to give an explanation for the reduction of the trap generation rate in the framework of the anode hole injection model. In particular, they argued that bipolar stress conditions favor the detrapping of holes from sites that are close to the interfaces and limit the transport of the generated holes towards the oxide bulk. Hence, since fewer holes are trapped, the rate of neutral trap generation by recombination of electrons with trapped holes is reduced [172]. A similar explanation also seems reasonable for the hydrogen release model of breakdown: Protons are released at the anode interface by the incoming electrons but they fail to travel towards the cathode due to field reversal, thus reducing the rate of defect generation. The increased density of protons at the anode interface can also lead to an ulterior increase of the interface state generation by reaction with other H-passivated dangling bonds. However, the validity of these speculations is still to be determined. The larger densities of defects measured at breakdown under bipolar stress conditions are more difficult to explain in terms of our present understanding of the breakdown triggering process. Wang suggested that the electrically active defects produced during bipolar pulsed stress are less effective in causing dielectric breakdown, while noting the need for a better understanding of which electrically active defects are involved and how they trigger the dielectric breakdown [175] (Fig. 3.61).

3.4.6 Gate Oxide Reliability Projection

We have reviewed the experimental breakdown data in comparison with the three different oxide breakdown models; we now discuss the perspectives of gate oxide reliability projection. As previously stated, three empirical models have been widely considered for reliability projection:

1) The exponential field (voltage) dependence model:

$$T_{BD} \sim \exp(-\gamma_F F_{OX}) \sim \exp(-\gamma_v V_G). \tag{3.53}$$

2) The exponential dependence with reciprocal of the electric field or gate voltage:

$$T_{BD} \sim \exp(G/F_{OX}) \quad \text{or} \quad T_{BD} \sim \exp(C/V_G). \tag{3.54}$$

3) The power law voltage dependence model:

$$T_{BD} \sim V^{-n}. \tag{3.55}$$

Figure 3.62 displays the projected times-to-breakdown from high voltage data to the low voltage regime for a typical set of stress experiments considering three different models above. Although at high voltages, all three models can equally fit the data; the projected times-to-breakdown at lower voltages can vary by several orders of magnitude, depending on the model choice. The remarkable differences in the final projected results demonstrate the importance to understand the validity of an extrapolation model in real applications of reliability qualification.

Historically, the exponential law of Equation 3.53 for voltage acceleration has been routinely used for oxide reliability projection, and it has theoretical support of the thermochemical model (the so-called E-model). While the exponential law

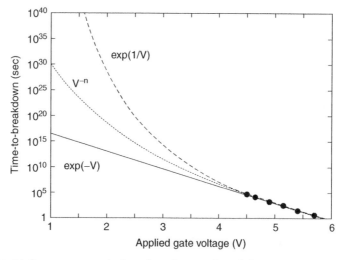

Figure 3.62. Voltage extrapolation for time-to-breakdown at 63% using three different models in Equations 3.54, 3.55, and 3.56. The symbols represent the stress data of 4.2 nm oxides obtained at a fixed area and at 140°C.

can always be used as a worst case approach for projection, its validity remains questionable both on physical and practical grounds as discussed in this section and also in Section 3.3.1. For many high performance applications of the CMOS technologies, the projection results can be unrealistic and may not be always acceptable. Rather than starting as an empirical approach for projection, the exponential law for the reciprocal field dependence or the reciprocal voltage dependence of Equation 3.54 was originally introduced on the basis of the results obtained from physics-based models such as the impact ionization model or the anode hole injection model. By its own definition, this model only remains valid in the FN regime ($V_{OX} > \Phi_B/q \sim 3\,V$) as discussed in Section 3.3.2. In the DT regime, the numerical approach using Monte Carlo simulation based on AHI model has been proposed for reliability projection [63, 103]. While the applicability of the AHI model to oxide breakdown is still under debate, a numerical approach involving a heavy consumption of computational time is not practical in the industry environment.

Prior to the introduction of ultra-thin gate oxides ($<4\,nm$) in the later 1990s, the limitations of both extrapolation methods [Eqs. 3.53 and 3.54] for a realistic projection had been recognized. Some researchers proposed to use the $1/E$-model at high voltages and the E-model at low voltages to keep a balance between a conservative exponential law (E-model) and a more optimistic model ($1/E$-model) and, at the same time, to provide a realistic projection. In other words, this approach entails the use of the local voltage acceleration parameter measured at the lowest voltages for lifetime projection but ignores the data points at high voltages [178]. However, this method is arbitrary and difficult to implement because it requires that this local voltage acceleration parameter is higher than those obtained from the entire voltage range. This cannot be true due to statistical variation. Moreover, this approach implicitly presumes that voltage acceleration increases with decreasing voltage. Similar to this approach, another empirical formulation based on combining the E-model at low fields and the $1/E$-model at higher fields was also proposed [179].

Although the voltage dependence of the T_{BD} power law model of Equation 3.55 was also originally proposed as an empirical approach without any theoretical justification [50], it fulfills the requirement of preserving the area scaling properties of the Poisson statistics, a fundamental character of oxide breakdown [132]. As shown in Figure 3.62, the projected lifetimes are more optimistic than those of the E-model but more conservative than those of $1/E$-model. The applicability of T_{BD} power law model was originally limited to low voltages in the DT regime while, recently, it has also been proposed for the FN regime [180, 181].

The empirical models defined by Equations 3.53, 3.54, and 3.55 share a common approach in that they do not explicitly deal with the separate roles of tunneling current and defect generation in the breakdown process. In the case of exponential law of Equation 3.53, the current is not expected to play any role according to the thermochemical model. In the case of power law (Eq. 3.55), it was recognized that defect generation has a much more important impact than the tunneling current in the T_{BD} voltage acceleration for the DT regime [50, 51]. The empirical nature of these approaches limits their applicability to different ranges of voltage and thickness since experimental data cannot be collected over the same voltage range for different thickness values or vice versa due to the constrained time window. A transition to a smaller power law exponent from the DT regime to the FN regime shows the situation is even more complicated [146].

In the framework of energy- and fluence-driven breakdown, both tunneling current and defect generation can contribute to the T_{BD} reduction. From the practical point of view of projection methodology it is convenient to separately consider the roles of tunneling current and defect generation efficiency. In this regard, the formulation of the time-to-breakdown given in Equation 3.8 explicitly provides a comprehensive and realistic methodology for gate oxide reliability projection over a wide range of voltages and oxide thickness. In this formulation, the time-to-breakdown is directly shown to be inversely proportion to the tunneling current, $J(V_G, T_{OX})$ and to the defect generation efficiency, $\xi(V_G)$ [45]. The voltage and thickness dependencies of tunneling currents $J(V_G, T_{OX})$ in either FN or DT regimes are well understood as already described in Section 2.2. The energy (voltage) dependence of defect generation efficiency, $\xi(V_G)$ [45] shown in Figure 3.8 is revealed to be thickness independent to the first order. As already discussed in Section 3.3.3, rather than relying on an assumed physical mechanism, this defect generation efficiency is directly obtained from the breakdown experimental data with a reverse-engineering approach based on two assumptions: the validity of the percolation model for the breakdown statistics [44, 45] and a linear dependence of the density of generated defects on the total fluence. Therefore, this methodology is a physics-based model and will be discussed in detail in [45]. It has the advantage of incorporating well established knowledge such as the voltage and thickness dependence of the tunneling currents, as well as the area and thickness scaling properties of the breakdown statistics, as we have discussed throughout Sections 2.4 and 3.2.

Putting all the pieces together, we can now compare the projected results of all these models, as shown in Figure 3.63 (a). Three different thickness values of 1.2 nm, 3.5 nm, and 7.0 nm are included for comparison. The voltage acceleration parameters of three models (Eqs. 3.53, 3.54, and 3.55) are directly derived from the corresponding stress data from wafer-level stress, but not included for clarity. To implement the defect generation breakdown model described in Section 3.2 (see Eq. 3.8), the tunneling currents are calculated using compact DT and FN tunneling expressions (as described in Sections 2.2.3 and 2.24). The defect generation efficiency is considered to be that derived from Figure 3.2.8. It is evident that the E-model gives the most conservative projections. For thin oxides in the DT regime, the projection of the empirical T_{BD} power law (Eq. 3.55) is identical to that of the defect generation model (Eq. 3.8). Since the voltage dependence of the DT tunneling current is relatively weak, the T_{BD} power law essentially captures the dominant role of defect generation efficiency, $\xi(V_G) \sim V^{38}$. In the other limit of 7.0 nm, the projection using the defect generation model is comparable to that of the reciprocal voltage dependence, a simplified version of the $1/E$-model (Eq. 3.54). In this case, the T_{BD} voltage dependence is dominated by the FN tunneling current because the defect generation efficiency is nearly

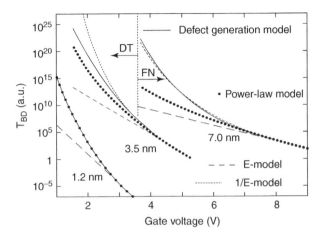

Figure 3.63. Comparison of various projection models for oxides of 1.2, 3.5, and 7.0 nm. The stress data at highest voltages was not included for clarity. The vertical dashed line separates the DT and FN regimes for 3.5 nm and 7.0 nm oxides, only. For 1.2 nm oxides, the entire range of stress voltages is in DT regime for V_{OX} below $\Phi_B/q \sim 3$ V. Reprinted with permission from Ref. 45, © 2009 IEEE.

voltage independent in the FN regime. In the case of 3.5 nm oxides, the stress data is usually available at high voltages in the FN regime, but the projection must be extended to low voltages in the DT regime. The projection of defect generation model agrees with that of the $1/E$-model in the FN regime, but disagrees in the DT regime because the voltage dependence of the defect generation efficiency becomes progressively more dominant and the DT current becoming more weakly dependent on voltage. For 3.5 nm and 7.0 nm oxides, the projected lifetime using the empirical T_{BD} power law is consistently lower than that obtained from the defect generation model, thus representing a conservative projection approach. The disagreement stems from the fact that different contributions of tunneling currents and defect generation efficiency are lumped together in the T_{BD} power law model.

To deal with the projection from FN regime to DT regime, a combined version of T_{BD} power law with two power-law exponents has been proposed [180] as follows:

$$T_{BD} \sim (V/V_t)^{-n} + (V/V_t)^{-m}, \qquad (3.56)$$

where n and m are the power exponents derived from the stress data in the DT and FN regimes, respectively, with a voltage fitting parameter, V_t ($V_t = 5.3$ V). Figure 3.64 compares the results of the defect generation model, the combined T_{BD} power law model, and the power law with single exponent. For 1.2 nm oxides, an excellent agreement is obtained as expected. However, the differences between the two models become larger for the thicker oxides of 7.0 nm operated in the FN

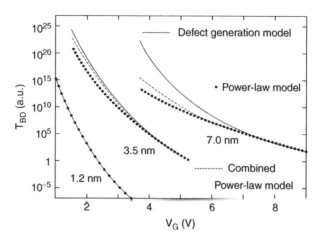

Figure 3.64. Comparison of various projections for oxides of 1.2, 3.5, and 7.0 nm using the defect generation model and the combined T_{BD} power law model (Eq. 3.56). The symbols represent the projection results of T_{BD} power law with single power. Reprinted with permission from Ref. 45, © 2009 IEEE.

regime. As previously remarked, the FN tunneling currents plays a major role in T_{BD} voltage acceleration at high voltages where the defect generation efficiency saturates. Thus, the differences are caused by the fact that the FN tunneling currents are approximated by a power law form, rather than using the more accurate physical description based on the results presented in Section 2.2.3. For the 3.5 nm oxides, the differences are much reduced because the first term on the RHS of Equation 3.56 dominates at lower voltages so that the combined power law model becomes equivalent to the single exponent power law. On the other hand, the results are also closer to those of the defect generation breakdown model because the defect generation efficiency dominates the T_{BD} voltage dependence. Thus, for oxide thickness below 4.0 nm, this model can be used with reasonable accuracy, judging from the agreement with the physics-based defect generation breakdown model.

In addition to the voltage acceleration models, a T_{BD} lifetime extrapolation from stress temperature to use temperature is also required for reliability projection. As already discussed previously, non-Arrhenius temperature dependence can cause an erroneous lifetime projection, which is usually over optimistic if an Arrhenius model is used for extrapolation outside of experimental data range. Furthermore, as already discussed in Section 3.4.3, the power law exponent is temperature-dependent (as shown in Fig. 3.53b) [163] and adds complexity to the projection. However, the T_{BD} temperature dependence becomes weaker at higher voltages; therefore, the differences in T_{BD} from stress temperature to use temperature are relatively small at high voltages for most applications. In practice, it is recommended to perform the stress experiments over the range of temperatures for real applications but at high voltages and thereby avoid the necessity of temperature extrapolation. If temperature extrapolation needs to be included in the reliability projection procedure, it is important to first construct an empirical non-Arrhenius temperature model based on the stress data at a fixed high voltage. Then, a voltage extrapolation using power law model can be used to project down to the use voltage with the exponent corresponding to this temperature. This procedure is illustrated in Figure 3.65. This means, in principle, that a full set of experimental T_{BD} data at various voltages and temperatures (as shown in Fig. 3.53) is required. This exercise can be time consuming and impractical in routine qualification. In practice, this can be simplified by performing stresses for T_{BD} voltage dependence at relevant use temperature to maximize the benefit of the temperature dependence of the power law exponent for reliability projection by trading off the measurement of T_{BD} voltage dependence at high temperatures.

Until now, we have mainly focused on reliability issues and projections of n^+-poly/NFET devices under inversion and accumulation modes. As we already discussed above, gate oxide breakdown properties of p^+-poly/PFET accumulation mode can be well understood since its electron injection and energy release is analogous to those for n^+-poly/NFET in inversion mode. For p^+-poly/PFET devices stressed under inversion mode, the situation becomes rather complicated. In this case, gate leakage currents under stress voltages can not be exclusively attributed to electron tunneling currents either from the conduction-band or from the valence-band of p^+-poly Si gate [182]. This is resolved later by an interface-state tunneling mechanism being responsible for current–voltage characteristics observed at high voltages in FN regime [183].

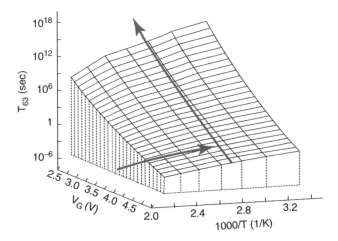

Figure 3.65. Three-dimensional plot for time-to-breakdown versus voltage and temperature including both effects of power law voltage dependence and non-Arrhenius temperature dependence in reliability projection. Reprinted with permission from Ref. 162, © 2001 IEEE. The bold-faced arrows indicate the projection procedure should be carried from high temperature to low temperature, or vise versa as discussed in the text.

For thin oxides, carrier conduction through gate oxides involves both electron and hole injection for conduction band and valence band tunneling, depending on applied gate voltages. In particular, at lower voltages around $\sim 1.0\,\text{V}$, the gate current is dominated by valence-band hole-tunneling current, rather than by valence-band electron-current above $\sim 1.0\,\text{V}$. For thin oxides ($<2.9\,\text{nm}$) in TDDB reliability of p^{+}-poly/PFET inversion mode, the Weibull slopes are comparable compared to those of n^{+}-poly/NFET inversion and accumulation modes; those for p^{+}-poly/PFET accumulation mode, the Weibull slopes are found to be smaller at the same oxide thickness for oxides thicker than 2.9nm [184]. The root cause of this difference is not understood [184]. It is found that defect generation efficiency for thick oxides ($>2.9\,\text{nm}$) remain the same as those of the other three modes using the defect size derived from the Weibull slope versus oxide thickness relation [184]. At low voltages for thinner oxides, the defect generation efficiency is shown to be larger; thus consistent with the finding that T_{BD} voltage acceleration factors (or power law exponents) are smaller than those of the other three modes [154, 184]. Additional difficulty for the detection of first breakdown events arises from the fact that gate currents become extremely noisy [184, 185]. Therefore, it can be reasonably concluded that our understanding of dielectric breakdown in p^{+}-poly/ PFET inversion mode is rather limited, especially when compared to the other modes. Much more work is required to construct a complete model. We have outlined the reliability projection procedures for SiO_2-based dielectrics ranging from medium thickness ($< \sim 100\,\text{Å}$) down to ultra-thin thickness ($\sim 10\,\text{Å}$). Due to time constraints, stress voltages are very high for wafer level reliability evaluation for ultra-thick oxides above $\sim 100\,\text{Å}$. Depending on the specific thickness under consideration, the degradation mode of band-to-band impact ionization ($E_G \sim 9\,\text{eV}$) as discussed in Section 3.3.2 can be encountered if stress voltages are too high, e.g., $V_G > 12\,\text{V}$. Contrary to the experimental TDDB observation of thin oxides, an anomalous change of field acceleration factors (γ_F) was observed for ultra-thick oxides: The measured γ_F values at higher stress voltages are found to be much larger than those measured at lower stress voltages [186]. This effect appears to be less pronounced at higher temperatures [186]. This increase in the field acceleration factors at high fields was attributed to the positive charge generation due to the onset of band-to-band impact ionization for $F_{\text{OX}} > 14.5\,\text{MV/cm}$ while the T_{BD} data at lower stress voltages ($F_{\text{OX}} < 14.5\,\text{MV/cm}$) were interpreted by trap generation model due to hydrogen release mechanism [187]. On the other hand, anode-hole-injection mechanism was later introduced to play a role in defect generation below the threshold voltage for impact ionization [8, 104, 187]. More work is required to clearly demonstrate which mechanism plays the dominant role in controlling $T_{\text{BD}}/Q_{\text{BD}}$ degradation in ultra-thick oxides since the published work for ultra-thick oxides remains scarce. For practical projection in dealing with thick oxides, using the higher field-acceleration factors obtained at high fields can certainly result in serious errors. In this case, it is recommended that stress should be performed at lower voltages in these situations.

Finally, it is worthwhile to comment on the singularity of T_{BD} reciprocal exponential law and power law models, which are discussed in detail in

Section 3.3. When the voltage approaches zero, the time-to-breakdown diverges to infinite, implying the dielectric would never break down according to these models as pointed out in [52]. It is evident that this is an unrealistic situation because other physical mechanisms would certainly cause the oxide failure at the lowest voltages. The uncertainty about the voltage range of applicability of breakdown models is a concern that affects not only those models showing a singularity at zero voltage (field); it is unavoidable because breakdown experiments at the lowest voltages of interest (even the operation voltage) are not possible because times to breakdown are too long. In any case, based on all extensive experimental data from many groups down to 2 V, we can conclude that T_{BD} exponential law is not applicable, even if this model has no a zero-field singularity.

3.5 POST-BREAKDOWN PHENOMENA

In previous sections, we have shown that the time-to-breakdown decreases many orders of magnitude when the oxide thickness is scaled down and the voltage is kept fixed. This reduction of T_{BD} is related to the smaller average density of defects required to trigger the breakdown, as well as to the increased direct tunneling current, which allows a faster generation of these defects. It is true that the gate voltage is also scaled down and that this reduces the defect generation efficiency (see Fig. 3.8), but this reduction is not enough to compensate the other factors and the result is a significant reduction of the oxide reliability margin.

The standard reliability predictions and the reliability assessment methodology are based on the statistics of the time to the first breakdown. In the previous sections, we have seen that this methodology is based on the combination of two elements: a detailed knowledge of the dynamics of defect generation (how the defect generation efficiency depends on the stress conditions) and a physics-based weakest link model for the breakdown statistics, which gives the correct dependencies on the oxide thickness and area. However, the first breakdown might not always be the best definition of device failure because many MOS digital circuits remain functional after the first breakdown, provided that the post-breakdown gate leakage is low enough [188]. In other words, some circuits might tolerate even several breakdown events, and this should provide the much required additional reliability margin. For this reason, significant research efforts in the last years have been dedicated to the study of the post-breakdown phenomena. Some researchers have focused on studying the impact of breakdown on device performance, with the final goal of establishing a relation between oxide breakdown and device failure [189–193]. Since the criterion of device failure inevitably depends on the application, others have studied the impact of breakdown on different circuits [193–195]. And last but not least, other authors have tried to understand the physics of the post-breakdown phenomena with the goal of developing new statistical tools and reliability methodologies [196–199].

The post-breakdown experimental results show a very rich variety of phenomena. In relatively thick oxides ($T_{OX} \sim 3$–5 nm) two apparently stable breakdown modes have been identified as hard breakdown (HBD) and soft breakdown (SBD) [200–202]. Since the post-breakdown conduction through the oxide is orders of magnitude higher for HBD than for SBD, several approaches to post-breakdown reliability are based on the idea that SBD is tolerated while HBD causes the circuit failure. Two examples of such approaches are the HBD prevalence ratio model, which is based on the assumption that the circuit failure is only caused by HBD [203–205], and the successive breakdown approach, which considers that all the breakdown events are soft at use conditions so that the superposition of several SBD events is required to cause the device or chip failure [196, 197, 206]. A completely different description is the progressive breakdown approach [185, 198, 199, 207, 208], which is based on the fact that the post-breakdown current growth is usually progressive in oxides thinner than ~ 2 nm. In these oxides, the breakdown event does not appear as a sudden abrupt jump of the current as in thicker oxides, and a certain time is required for the progressive growth of the BD current up to a value that perturbs the device performance. In all these approaches, the extra time required to reach the device failure condition after the first breakdown gives additional useful life to the devices and circuits. This extra time, which we refer to as the residual time, is the main element of the post-breakdown reliability methodologies.

3.5.1   Review of Post-Breakdown Experimental Observations

3.5.1.1   Oxide Breakdown Versus Device Failure
Experimental results show that a breakdown event does not always cause a complete loss of the device and circuit functionality. This issue was first discussed in 1997 by Weir and coworkers, who reported device functionality after oxide breakdown [189], showing that the threshold voltage and the transistor transconductance do not degrade significantly after SBD.

Another important work in this field is due to Kaczer and colleagues, who reported the reasonable performance of a ring oscillator after inducing multiple breakdown events in several transistors [188, 209]. After the breakdowns, the logical functionality of the circuit remained unaltered while the stand-by current and the oscillating frequency were somewhat affected. After these seminal papers, many groups began studying the impact of the breakdown on the device characteristics and on the performance of different circuits. An example of the effects of a soft breakdown on the transistor transfer characteristics is shown in Figure 3.66 [210]. The considered device is an NFET with $T_{OX} = 1$ nm (i.e., at the limit of thickness scaling), $W = 0.1$ μm, and $L = 0.1$ μm. This transistor was subjected to CVS at 2.4 V and the first breakdown was detected by measuring a small jump in the current at 1 V. The evolution of the gate current during the stress experiment at 2.4 V and the current measured periodically at 1 V are shown in Figure 3.66 (a). The first breakdown is seen to occur at $t_{BD} \sim 30$ sec, when the low-voltage current suddenly jumps to ~ 4 μA. However, no sign of change appears in

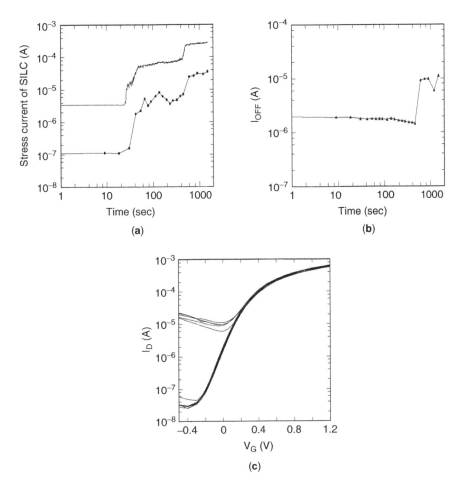

Figure 3.66. An example of the impact of oxide soft breakdown on MOS transistor characteristics. The device was an NFET with $T_{OX} = 1$ nm, $W = 0.1\,\mu$m and $L = 0.1\,\mu$m. This transistor was subjected to CVS at 2.4 V and the first breakdown is detected by a current jump measured at 1 V. The evolution of the gate current during the stress experiment at 2.4 V and the current measured periodically at 1 V are shown in (a). The evolution of the transistor OFF current, I_{OFF}, is reported in (b). In (c), the evolution of the transistor transfer characteristics is depicted. Reprinted with permission from Ref. 211, © 1999 IEEE.

the evolution of the transistor OFF-current, I_{OFF}, until ~ 500 sec of stress, as shown in Figure 3.66 (b). This change in I_{OFF} correlates with the second abrupt change of I_G, which is observed to jump to $\sim 30\,\mu$A at $V_G = 1$ V, as shown in Figure 3.66 (a). The effects of breakdown on the transfer characteristics are shown in Figure 3.66 (c), where it is evident that the quantity that is most significantly affected by the breakdown is I_{OFF}, as reported by Pompl [211].

However, this issue is not as simple as it might appear at first sight, and it has been demonstrated that breakdowns located at or near the gate-to-drain or gate-to-source overlaps have more pronounced effects on the device characteristics [190, 191]. Consequently, the operation of short channel devices tends to be much more affected by breakdown than their long channel counterparts. Since the location of the breakdown path along the transistor channel appears as a parameter relevant for device failure, Degraeve and colleagues developed an electrical technique to measure the breakdown position based on comparing the magnitude of the gate-to-drain and gate-to-source leakage currents [212]. The results obtained with this simple method have been found to be in good correlation with the locations derived from high-resolution transmission electron microscopy images [213]. Recent works have also suggested that narrow channel devices can also be more deeply affected by soft breakdown than wider devices. In this regard, Cester et al. reported a collapse of the drain current and the transistor transconductance caused by SBD in narrow transistors [192]. Again, a simple geometrical picture provides a successful explanation. The breakdown spot is considered to block the transistor channel in a region of diameter D around the breakdown conduction path. According to this picture, the transistor drain current is expected to drop according to the ratio $(W-D)/W$, in nice agreement with experiments. These results have been recently reexamined and mostly confirmed by Kaczer [193].

Since the final goal is the assessment of the reliability of integrated circuits, the study of the impact of breakdown on the performance of circuits is essential to define device failure criteria. In this regard, several groups have begun studying the performance of simple circuit blocks as a function of the breakdown current and location. Rodríguez and colleagues have studied the influence of SBD events on SRAM-cell stability [194] and on the performance of CMOS inverters [195]. Their results clearly reveal that, although the most deeply affected quantity might be I_{OFF}, the impact on circuit functionality primarily depends on how much the transistor ON current, I_{ON}, is affected. An example of the influence of breakdown on inverter performance is shown in Figure 3.67. In this figure, a SPICE simulation of the transfer characteristic of a 5 V inverter is shown for various SBD events of increasing severity. In all the cases, breakdown is assumed to introduce a conduction path between input and output with the power law IV typical of SBD, i.e., $I = aV^b$. The results are shown with the effective resistance of the breakdown path, $R_{EFF} \equiv (1/a)$ as a parameter. Notice that while a severe breakdown event ($R_{EFF} = 10\,\Omega$) causes the inverter failure, its functionality remains almost unaltered in the case of typical SBD resistance ($R_{EFF} = 10\,K\Omega$).

The potential impact of the breakdown on the performance of dynamic logic circuits has also been examined. In this regard, Kaczer and Groeseneken showed that the retention time of soft nodes decreases proportionally to the gate-to-drain breakdown leakage [214]. In general, we can conclude that, although a breakdown always somehow affects the circuit performance, it might remain functional if the breakdown current is small enough. Since the impact of breakdown on circuit functionality depends on the type of circuit, it is strictly impossible to establish a

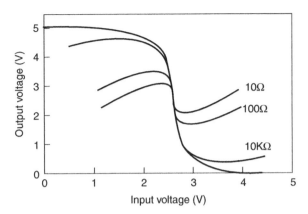

Figure 3.67. SPICE simulation of the transfer characteristic of a 5 V-CMOS inverter as a function of the hardness of a breakdown. The breakdown path is assumed to introduce a conduction path between input and output with the power law IV characteristic typical of SBD, i.e., $I = aV^b$. The parameter used to label the curves is the effective resistance, $R_{EFF} = 1/a$.

universal device failure criterion. Moreover, it has been shown that the circuit failure may be influenced by the coupled effects of the degradation of several devices. In particular, Mueller et al. have demonstrated the coupled effects of NBTI damage in a PFET and breakdown current in an NFET of an SRAM cell [215]. In conclusion, reliability assessment becomes strongly application dependent.

3.5.1.2 *Hard Breakdown and Soft Breakdown.* Gate oxides with thickness in the 3 to 5 nm range show at least two different types of breakdown events that have been classified in terms of their severity as soft breakdown (SBD) and hard breakdown (HBD). An example of the evolution of the gate voltage during a constant-current stress is shown in Figure 3.68. In this particular case, two SBD events are detected as small voltage drops of about 0.05–0.2 V while the HBD event causes a large drop of about 3 V.

Although it is arbitrary to establish a current threshold to separate the two modes, there is no ambiguity in determining whether a breakdown event is a SBD or a HBD in terms of the associated post-breakdown IV characteristic. Figure 3.69 shows that the typical post-SBD current is orders of magnitude smaller than the post-HBD current. The functional dependence of the post-breakdown current on the gate voltage is also qualitatively different. While the SBD IV characteristic has been fitted to a power law or an exponential law [216], the post-HBD IV is essentially linear. Although SBD and HBD are considered to be different breakdown modes, they share a common origin (a percolation path) and, in both cases, the post-breakdown current flows through an extremely small oxide area of nanometer-scale dimensions [218].

Although, SBD and HBD are clearly identified in oxides with thickness between 3 nm and 5 nm, the situation is more complicated in oxides thinner than

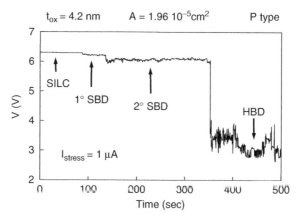

Figure 3.68. Evolution of the gate voltage during a constant-current stress test during which several breakdown events are detected as abrupt voltage drops for oxide thickness of 4.2 nm with an area of 1.96×10^{-5} cm^2. Reprinted with permission from Ref. 201, © 2000 IEEE.

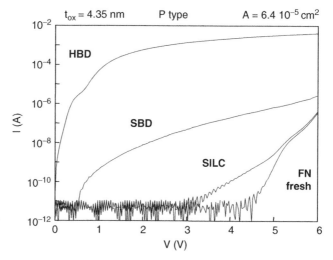

Figure 3.69. Current–voltage characteristics at different stages of oxide degradation. Fresh $I(V)$, stress induced leakage current (SILC), and typical post-SBD and post-HBD $I(V)$ curves are shown. Reprinted with permission from Ref. 217, © 2001 IEEE.

3 nm and the existence of SBD has even been questioned. As shown in Figure 3.70, examination of the post-BD IV characteristics in a 1.35 nm thick oxide suggests that two BD modes still exist, although the measured post-SBD currents are orders of magnitude larger than in thicker oxides. In the case of the 4.3 nm oxide

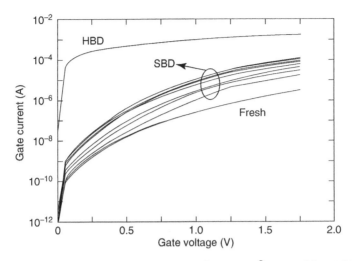

Figure 3.70. Post-BD IV characteristics measured in a $8\,\mu m^2$ NFET with a 1.35 nm-thick oxide. SBD and HBD modes are clearly distinguished. A family of SBD curves is shown. During stress the current is seen to evolve progressively (see Fig. 3.71) and the post-BD IV scans this family of curves towards higher and higher currents. Reprinted with permission from Ref. 219, © 2006 IEEE.

(Fig. 3.69), the post-SBD current at $V_G = 1.5\,V$ is below 10 nA for all the samples; in the 1.35 nm oxide considered in Figure 3.70, it ranges from $10\,\mu A$ to $100\,\mu A$ range at the same gate voltage. On the other hand, since the HBD current is comparable in both cases, the difference between HBD and SBD currents is much reduced in the case of the 1.35 nm thick oxides (less than two orders of magnitude) as compared to the results shown for 4.3 nm oxides (more than five orders of magnitude). Moreover, in ultra-thin oxides, the HBD is progressive [198, 199, 208] and the SBD is unstable [219]. For these reasons, and taking into account that the background tunneling current can often mask the softer breakdown events, the classification into SBD and HBD modes sometimes becomes extremely difficult if it is only based on the evolution of the stress current. Some examples of the evolution of the gate current are shown in Figure 3.71. In some cases, an apparently instantaneous HBD is recorded. In others, the post-breakdown current nearly stabilizes after an S-shaped transition to a current level that corresponds to the SBD mode, but it finally and suddenly jumps to a much larger value. In these samples it is impossible to decide whether these are HBD events or SBD events based on an objective criterion. However, we will see in the next section that the statistical distribution of residual time (the time elapsed from first BD to device failure) still allows an alternative way to make a meaningful classification of the breakdown events into two modes.

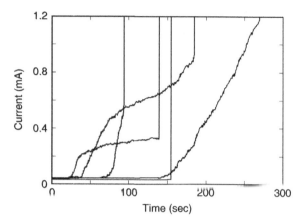

Figure 3.71. Evolution of the post-BD current in several $8\,\mu m^2$ NFETs with a 1.35 nm-thick oxide stressed under CVS at $V_G = 2.75$ V. Reprinted with permission from Ref. 219, © 2006 IEEE.

3.5.1.3 Progressive Breakdown. In ultra-thin oxides ($T_{OX} \sim 2$ nm and below) subjected to CVS, the breakdown appears as a progressive evolution of the leakage current, not as an "instantaneous" current jump. Several examples of the evolution of the breakdown current are presented in Figure 3.71.

This progressive BD behavior was first reported by two groups [198, 199] and then it has been studied in detail by a number of authors [208, 219–224]. Since a certain time is required to reach breakdown current values that seriously perturb the device performance, the progressive evolution of the breakdown current might give some extra reliability margin. Monsieur and colleagues studied the breakdown progressiveness in terms of the residual time from breakdown detection to an arbitrary wearout current value, as shown in Figure 3.72.

Since the detection of the breakdown event is difficult in ultra-thin oxides, Linder and co-workers defined the rate of BD current growth between $10\,\mu A$ and $100\,\mu A$ as an alternative way to characterize the breakdown progressiveness. Figure 3.73 shows how this rate scales with the stress voltage for different values of T_{OX} and two different temperatures. For a 1.5 nm oxide, they found that the degradation rate scales from $1\,mA/s$ at 4 V to less than $1\,nA/s$ at 2 V and extrapolates to $\sim 10\,fA/s$ at 1.2 V. At this rate, a BD path requires more than 10 years to reach a current of $10\,\mu A$, thus promising a substantial reliability relief at operation conditions [199].

Voltage, area, and percentile scaling of the residual time distribution have been subject of research interest due to their importance for reliability extrapolation. In this regard, it is remarkable that the statistical distribution of progressive BD time has been found to be log normal for large values of the failure current, while at small failure currents it tends to behave as a Weibull distribution similar to the first breakdown distribution [224]. On the other hand, recent results by Pompl et al. [223] suggest that, as originally claimed by Monsieur et al. [198], the

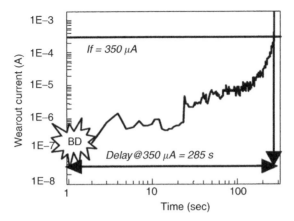

Figure 3.72. Definition of the residual time. If I_f is an arbitrary threshold current then the residual time (*delay@350 μA*) is defined as the running time between the breakdown detection and the moment where the wearout current is equal to $I_f = 350$ μA. Reprinted with permission from Ref. 221, © 2003 IEEE.

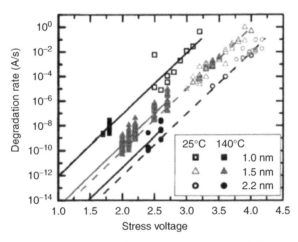

Figure 3.73. The rate of increase of stress current for a 1.5 nm oxide after the commencement of hard breakdown. An exponential dependence continues for 10 orders of magnitude over a wide range of voltages. The degradation rate is defined as the inverse of time required for the current to grow from 10 μA to 100 μA. Reprinted with permission from Ref. 199, © 2002 IEEE.

voltage acceleration of the residual time (due to breakdown progressiveness) is the same as that of the initial breakdown path formation. In fact, they showed that voltage acceleration of the residual time is compatible with the power law voltage acceleration model of oxide breakdown in a time scale covering nine orders of

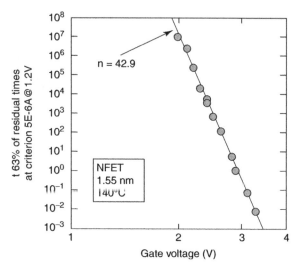

Figure 3.74. Double-logarithmic plot of voltage acceleration of residual time of progressive wearout from formation of the breakdown path until the current level at use voltage reaches 5 μA. Package and wafer level data are described by the power law model with $n = 42.9$. Reprinted with permission from Ref. 223, Copyright 2006, Elsevier.

magnitude (see Fig. 3.74). Recent experiments also show power law voltage acceleration with a failure-current-dependent power law exponent [224]. As far as area scaling is concerned, Monsieur et al. [198] were the first to report that the progressive delay time does not depend on the device area, as expected for the localized degradation of the breakdown spot. Recent results by Pompl and colleagues also support the area independence of the progressive BD time [223]. Linder and colleagues confirmed this result provided that multiple breakdown events do not occur during the progressive evolution of the breakdown current [220]. In this regard, a recent publication discusses under which conditions the residual time coincides with the degradation of a single localized path and how this impacts area scaling [207]. When the residual time is larger than the time to first BD, several BD paths progressively degrade in competition with each other, and the residual time becomes area dependent and smaller than the time required for the degradation of a single BD path [207].

The time dependence of the progressive current evolution has also been subject of study. Linder showed that the breakdown current evolves exponentially with time in its early stages of growth [220]. Hosoi et al. also reported similar results and paid attention to the reduction of the degradation rate at long stress times, relating this "quasi-saturation" to the series resistance that effectively limits the maximum breakdown current [208]. Figure 3.75 shows a typical example of the breakdown current evolution in an ultra-thin oxide, including the initial exponential growth and the final quasi-saturation. A compact description of the

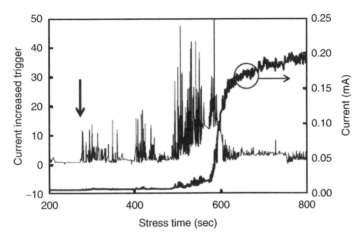

Figure 3.75. Current growth measured during the progressive breakdown of an ultra-thin PFET ($T_{OX} = 1.25$ nm) stressed under $V_G = -2.6$ V. Reduction of the current growth rate and quasi-saturation of the current (logarithmic growth as reported by Hosoi [208]) is observed at long stress times. The breakdown occurrence (indicated by the arrow) is almost unnoticeable in this linear plot of the current. The relative current increase trigger undoubtedly captures the event. Reprinted with permission from Ref. 238, © 2008 IEEE.

current growth during progressive breakdown is due to Miranda and Cester who proposed a model based on the logistic equation that captures the initial exponential growth and the roughly linear transition towards a saturated value [225]. A similar phenomenological model recently presented by Kaczer nicely captures the progressive breakdown transient and also the long time quasi-saturation limit in which the wearout evolves logarithmically [185]. Some authors have distinguished two different phases in the evolution of the current during progressive breakdown before this final quasi-saturation [226–228]. In the very early stages after the breakdown occurrence, the current remains rather constant or grows very slowly and it is very noisy, showing large multilevel random fluctuations. Then, after a certain period of time, the current suddenly evolves much faster towards quasi-saturation. While the IV is very noisy and unstable during the first phase, it becomes rather stable during the second phase. The first and second phases of the progressive breakdown current evolution have sometimes been named digital-SBD and analog-SBD, taking these terms from a work of Sakura [229]. The separation of these two phases might be of great importance for reliability assessment because only the first one remains in metal gated insulators and the second one is replaced by an abrupt transition to hard breakdown [230]; and also because some authors have recently reported a critical voltage below which the transition from digital to analog breakdown phases hardly occurs [227]. This means that below this critical voltage the reliability margin provided by progressive breakdown might be significantly enlarged and other effects (such as the generation of multiple SBD events) might pose the real lifetime limit [231].

Before ending this section, it is worth quoting recent results concerning extrinsic breakdown. Actually, it has been recently claimed that the extrinsic breakdown mode of ultra-thin oxides shares the feature of progressive current growth and that the statistical properties of the breakdown progressiveness are essentially identical for intrinsic and extrinsic breakdowns [232]. These findings might also have strong impact on product reliability methodologies.

3.5.2   Modeling the Post-Breakdown Statistics

Since the first breakdown event does not necessarily cause the device or chip failure, the statistical distribution of the time-to-first-breakdown is not the relevant distribution for reliability assessment. In this section, we focus on the formal description of the failure statistics associated to different post-breakdown failure models. First, we consider the so-called hard-breakdown prevalence ratio model [203, 204]. Then, we focus on the description of the failure distribution in the progressive breakdown model [198, 199, 208] and finally, we will deal with the statistics of successive breakdown events [196, 197, 206].

3.5.2.1   The Hard-Breakdown Prevalence Ratio Picture. The basic assumptions of the HBD prevalence ratio model are [203–205]: 1) A breakdown event can either be an HBD or a SBD with probabilities α_{HBD} and $(1-\alpha_{HBD})$, respectively; 2) SBD is tolerated by the devices in any time scale of interest for chip reliability; and 3) HBD instantaneously causes the device failure. Since it is assumed that only HBD causes the device failure, each device can suffer a number of SBD events before failure, as directly revealed by optical emission microscopy [233, 234], and the failure distribution is that of the first HBD event of each sample. In other words, the device failure distribution is the distribution of HBD times obtained by completely ignoring the occurrence of SBD events. It can be demonstrated in a straightforward manner that the relation between the first breakdown distribution and the final HBD distribution is

$$Ln(-Ln(1 - F_{HBD}(t))) = Ln(-Ln(1 - F_{BD}(t))) + Ln(\alpha_{HBD}), \qquad (3.57)$$

where α_{HBD} is the probability of each event to be a HBD, i.e., the so-called HBD prevalence ratio. Equation 3.57 reveals that these two distributions are parallel in the Weibull plot and vertically separated by $Ln(\alpha_{HBD})$. Note that this shift is always negative because $\alpha_{HBD} < 1$; this is consistent with the fact that the final HBD distribution is always displaced towards longer times. The results of some experiments (see Fig. 3.76) fully confirm that these distributions are parallel and shifted by $Ln(\alpha_{HBD})$ in the Weibull plot [203, 204].

Equation 3.57 also defines a simple reliability methodology for SBD tolerant applications. If a particular circuit can tolerate the SBD events without failure, the relevant failure distribution for this application would no longer be the first

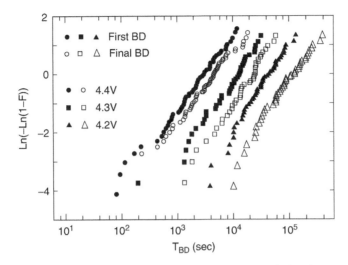

Figure 3.76. Distributions of first breakdown event and final HBD for PFET capacitors ($T_{OX} = 2.7$ nm) stressed under CVS conditions at 4.2 V, 4.3 V, and 4.4 V and temperature of 30°C. The vertical shift between each pair of SBD and HBD distributions is $\text{Ln}(\alpha_{HBD})$ as required by Equation 3.57. Reprinted with permission from Ref. 203, © 2001 IEEE.

breakdown distribution but the HBD distribution (F_{HBD}). The additional lifetime margin can be estimated as

$$T_{HBD} = \frac{T_{BD}}{\alpha_{HBD}^{1/\beta}}, \qquad (3.58)$$

where T_{HBD} and T_{BD} are the scale factors (i.e., characteristic times for $F \sim 63\%$) of F_{HBD} and F_{BD}, respectively, and β is the Weibull slope, which is common to both distributions. The value of α_{HBD} has been experimentally found to increase with the gate voltage [196, 203, 204] and the stress temperature [235], and to decrease with the oxide thickness [236]. Thus, projection to operation conditions should consider not only the voltage and temperature extrapolation of the first BD distribution but also the voltage, thickness and temperature dependencies of α_{HBD}. Figure 3.77 reports the measured voltage dependence of α_{HBD} with the oxide thickness as a parameter for PFETs stressed at 30°C and NFETs stressed at 140°C. It is found that the HBD prevalence changes quite steeply from ~ 1 to ~ 0 at a threshold voltage V_{G0} which decreases linearly with T_{OX}. At the same stress voltage, the HBD probability is much larger for thinner oxides. However, this is likely to be compensated by operation at scaled voltages.

Figure 3.77 reveals that α_{HBD} depends on voltage, temperature, and oxide thickness. However, for extrapolation to operation conditions, the voltage dependence of α_{HBD} in the vicinity of 0 should be studied in detail for each

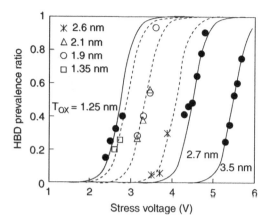

Figure 3.77. Hard breakdown prevalence ratio as a function of the stress voltage and with T_{OX} as a parameter. Solid symbols and lines correspond to PFETs stressed at 30°C. Open symbols and dashed lines to NFETs stressed in inversion at 140°C. Reprinted with permission from Ref. 236, © 2003 IEEE.

T_{OX}. However, the required experiments are strongly limited by sample size and long stress times and, as a consequence, large uncertainty remains about the values of α_{HBD} extrapolated to operation conditions. Alam and colleagues claimed that, based on their model of energy controlled BD runaway, the HBD prevalence ratio is about 10^{-15} for a 1.5 nm oxide at the operating gate voltage of 1 V [237]. If this optimistic prediction were true, one should conclude that oxide breakdown is not an obstacle to continued scaling of integrated circuits, since they would never fail due to this mechanism [237]. However, purely empirical extrapolation based on the results shown in Figure 3.77 yields much larger values of $\alpha_{HBD} \sim 10^{-4}$ to 10^{-3} for the same oxides at identical operation conditions.

3.5.2.2 Progressive Breakdown Distribution. If one ignores recent results claiming that the current of some BD events would hardly evolve if the stress voltage is kept below a critical value [227], all the breakdown events within the progressive breakdown (PBD) approach are assumed to cause the device failure and no distinction is made between SBD and HBD. However, the breakdown does not cause the immediate failure of the device which is delayed by the progressive growth of the breakdown current. In Section 3.5.1.3 we discussed the progressive breakdown in terms of the mean residual time [198] or the mean degradation rate [199]. However, a reliability methodology needs the complete failure distribution to be able to extrapolate to the low failure percentiles required for a chip reliability forecast. Linder and Stathis studied this distribution by means of Monte Carlo (MC) simulations [220], arbitrarily assuming a uniform distribution for the current growth rate in the range of ± 10X the median rate. Their results showed that the increase of lifetime at low percentiles can be significant even if the mean residual time is much shorter than the mean time to the first breakdown

($T_{RES} \ll T_{BD}$). In this framework, the relation between the failure distribution F_{FAIL} and the first BD distribution depends on the statistical distribution of residual time, and the device failure probability density is given by

$$f_{FAIL} = \int_0^t f_{BD}(t - t_{RES}) f_{RES}(t_{RES}) dt_{RES} \qquad (3.59)$$

where f_{BD} and f_{RES} are the probability density functions for first BD and residual time, respectively. In some particularly simple cases, this integral has an analytical solution. For example, if, $f_{RES} = \delta(t - T_{RES})$, the distribution of device failure is equal to F_{BD} with a delay, i.e., $F_{FAIL} = 0$ for $t < T_{RES}$ and $F_{FAIL} = F_{BD}(t - T_{RES})$ for $t \geq T_{RES}$. This is the simplest case and the one that gives the biggest reliability margin within the PBD framework.

However, this approach is somewhat unrealistic and overestimates the chip lifetime because the residual time distribution is expected to have a finite width due to various reasons. Among these reasons, we must emphasize the distribution of the percolation path resistance of the as-formed BD paths [205]. To deal with more realistic distributions, other simple cases can be analyzed. If both f_{BD} and f_{RES} are exponential distributions with scale factors T_{BD} and T_{RES}, respectively, Equation 3.59 also has an analytical solution

$$F_{FAIL}(t) = \frac{1}{(T_{RES} - T_{BD})}$$
$$\times \left\{ T_{RES} \left[1 - \exp\left(-\frac{t}{T_{RES}}\right) \right] - T_{BD} \left[1 - \exp\left(-\frac{t}{T_{BD}}\right) \right] \right\}. \qquad (3.60)$$

Figure 3.78 compares (in a Weibull plot) the cumulative device failure distributions corresponding to different distributions of residual time. These failure distributions converge to F_{BD} at high failure percentiles while a larger Weibull slope is obtained at low percentiles. For the delta distribution, the distribution is vertical at $t = T_{RES}$. For the exponential distributions, the slope converges to $\beta = 2$ at $t \ll T_{BD}$, T_{RES}. Other intermediate cases correspond to f_{BD} and f_{RES} being Weibull distributions with arbitrary slopes β_{BD} and β_{RES}. Figure 3.78 reveals the importance of the shape of the distribution of the residual time in determining the improvement of chip lifetime: the wider f_{RES} (for the same mean value, T_{RES}), the smaller the lifetime margin available at a fixed failure percentile. It is also worthwhile noting that even if the value of T_{RES} is much smaller than T_{BD}, a very important lifetime margin can be obtained at low percentiles (as pointed out by Linder and Stathis [220]). Recently, there has been much interest in the development of new reliability assessment methodologies which take into account the breakdown progressiveness [226, 238]. These methodologies are based on F_{FAIL}, which has been demonstrated to scale in area according to the Poisson scaling law (as the first BD distribution) even if the PBD time distribution is area independent. To facilitate the application of these methodologies, a compact model has recently been developed which is valid both when a single BD spot evolves

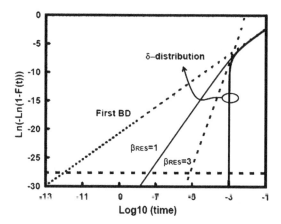

Figure 3.78. Cumulative distributions of first BD and final device failure in the PBD model. The first BD distribution is assumed to be the exponential distribution ($T_{BD} = 1$; $\beta_{BD} = 1$) while several distributions of residual time are considered (keeping $T_{RES} = 10^{-3}$ s): the delta distribution, the exponential distribution ($\beta_{RES} = 1$), and a Weibull distribution with $\beta_{RES} = 3$. In the first two cases, the solution is analytical as discussed in the text. For $\beta_{RES} = 3$, the corresponding line (dashed) represents the low percentile asymptotic behavior. The horizontal line corresponds to $F = 10^{-12}$ and represents the device failure percentile required to ensure a 100 ppm ($F_{CHIP} = 10^{-4}$) lifetime specification for a chip of 10^8 devices. Reprinted with permission from Ref. 219, © 2006 IEEE.

towards causing device failure and when several spots degrade in competition with each other [239]. This model provides an analytical expression for F_{FAIL}, provided that F_{BD} and F_{PBD} are Weibull distributions.

3.5.2.3 Successive Breakdown Statistics. It is well known that when a stress experiment is not stopped after the first BD event, it is possible to observe the occurrence of subsequent BD events. This is related to the fact that the breakdown is an extremely localized phenomenon that occurs in an area which is usually much smaller than the device area. Thus, if the stress continues, it is possible to generate many defect-related percolation paths at different locations of the oxide area. The mathematical description of the statistics of successive BD events is already well established [196, 197, 206, 240, 241]. Although successive BD events are found to be somewhat correlated in space and time [206, 240], we will assume for simplicity that the generation of successive BD paths is uncorrelated and uniform across the oxide area. This assumption allows the use of the Poisson model for the distribution of BD paths. For the sake of generality, it is convenient to formulate the problem in terms of the average number of BD paths per device, μ, without explicit reference to the time dependence of the generation of defects in the oxide. According to the Poisson distribution, the probability of finding K

uncorrelated BD events in one device when the average is μ is

$$P_K(\mu) = \frac{\exp(-\mu)\mu^K}{K!} \qquad (3.61)$$

The first BD cumulative failure distribution is $F_{BD} = 1 - P_o$. However, if the devices can tolerate $K-1$ events without failure, the relevant cumulative failure function would be F_K, which is given by

$$F_K(\mu) = 1 - \sum_{i=0}^{K-1} P_i(\mu). \qquad (3.62)$$

If P_K is given by Equation 3.61, $F_K(\mu)$ is the Gamma distribution, which is widely used to deal with the reliability of systems with redundancy [242]. In the Weibull plot, only the first breakdown distribution is a straight line with both slope and scale factor equal to 1 as shown in Figure 3.79.

For $K \neq 1$, the distributions converge towards the first BD distribution at high percentiles and a Weibull approximation $W_K(\mu) \approx K\,Ln(\mu) - Ln(K!)$ is adequate at low percentiles. To deal with a measurable variable such as t_{BD}, the change of variable $\mu = (t_{BD}/T_{BD})^\beta$ is required, where β is the Weibull slope and T_{BD} is the

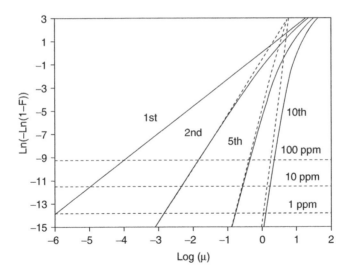

Figure 3.79. Some examples of the Kth successive breakdown event failure distributions ($K = 1, 2, 5$ and 10) as a function of the normalised variable $\mu = (t_{BD}/T_{BD\%})^\beta$. The solid lines are the exact distributions as calculated from Equation 3.61 and the dashed lines correspond to the low percentile Weibull approximations considered in the text. The horizontal dashed lines correspond to the fixed percentile of failure at 1, 10, and 100 ppm. Reprinted with permission from Ref. 197, © 2003 IEEE.

scale factor. The mean time to the Kth breakdown event was previously shown to be $\langle T_{BDK} \rangle \approx K^{\frac{1}{\beta}} T_{BD}$ [197]. This basic theory of successive BD events has been compared to Monte Carlo simulations [243] and to direct measurements of the successive BD distributions [196, 206, 240, 244]. Figure 3.80 reveals the occurrence of successive BD events in the evolution of the gate current measured during the stress of NFETs with 2.6 nm-thick oxide. Figure 3.81 successfully compares theory with experiments performed on 2.7 nm PFETs stressed until the HBD under 4.2 V at 140°C. The main assumption required for the application of the successive BD theory to chip reliability assessment is that a fixed number of BD events can be tolerated by all the chips (or by all the devices in each chip) without causing their failure. If this assumption were adequate, the successive BD theory would provide very simple and convenient reliability methodologies. In particular, two possible methodologies associated with two failure criteria have already been proposed. The first one (chip level model) assumes that the chip fails at the Kth BD event in the chip, no matter in which device [197, 241]. The second one (device level model) assumes that the chip fails when K breakdown events have occurred in the same single device of the chip [196]. The cartoons of Figure 3.82 illustrate the application of these two methods to chip reliability for the particular case of $K = 4$.

The successive BD model at the chip level is the adequate statistical description for the chip leakage (i.e., power consumption) limited reliability, as considered by Okada [245]. On the other hand, the results obtained by its application at the device level are largely overoptimistic according to a recent analysis of the limits of the successive BD models [206]. The application at the device level is limited by total power consumption, HBD probability and lack of

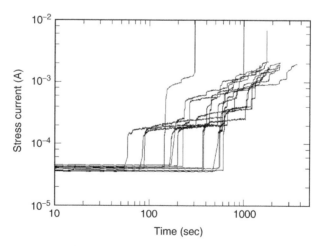

Figure 3.80. Evolution of the stress current of many samples recorded during a CVS experiment performed at 140°C and $V_G = 3.7$ V on NFET capacitors with $T_{ox} = 2.6$ nm and $A_{ox} = 6.2 \times 10^{-4}$ cm². Successive SBD events are evident. Final HBD events are observed before the end of the stress for only three of the samples.

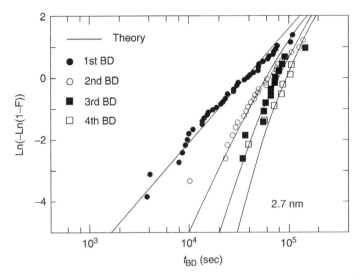

Figure 3.81. Statistical distributions of the first, the second, the third and the fourth breakdown events as measured (marks) in 2.7 nm PFET samples stressed at 4.2 V and 30°C. The solid lines correspond to the distributions given by the theory of uncorrelated breakdown events (Eqs. 3.61 and 3.62). Reprinted with permission from Ref. 244, © 2003 IEEE.

sufficient SBD stability [206]. Moreover, since the number of SBD events required for device failure is not always exactly the same, the cumulative distribution of failure would not be given by F_K but by a weighted combination of such distributions, $F_{FAIL} = \sum_K p_K F_K$ with $\sum_K p_K = 1$, as recently considered by Pompl and Kerber to deal with the failure of PFET arrays [243].

3.5.2.4 Conclusions.
Section 3.5 presented different (apparently contradictory) models which aim to consider the extra reliability margin associated with the tolerance (or delayed failure) of MOS devices and circuits to the breakdown of the gate oxide. All these models are based on experimental observations and capture different aspects of the reality. The reason why different authors have not coincided in the analysis of the experiments is related to the fact that there are different possible failure modes which are in competition [236]. The mode that most probably causes the device failure is the one that determines the shape of the device failure distribution. Moreover, the dominating mode changes with the sample characteristics (mainly area and thickness) and the stress conditions (voltage and temperature) so that the conclusions of different studies can easily appear to contradict each other. In particular, the HBD prevalence ratio methodology and the successive BD approach were developed for oxides with T_{OX} ranging from 3 to 5 nm, in which SBD and HBD modes could be clearly separated, the SBD appeared as a stable mode and the HBD seemed to cause the device

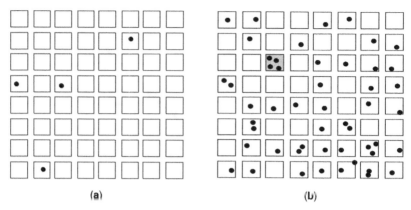

(a)   (b)

Figure 3.82. Schematic representation of the chip failure criteria in the application of the successive BD theory to chip reliability. Small squares represent the
devices of the chip and black dots are BD spots. The figure on the left represents
the chip failure for $K = 4$ for the chip-level application of the successive BD model.
In this case, the chip fails when four breakdown events accumulate in the chip
(most probably in different devices) The figure on the right also corresponds to
the chip failure for $K = 4$ but for device-level application of the successive BD
methodology. Chip fails when one single device (small square with grey background) accumulates four breakdown events. Reprinted with permission from
Ref. 206, © 2004 IEEE.

failure instantaneously. However, when the considered oxides are thinner than
about 2 nm, the relevant post-BD phenomenology seems to change considerably.
First, the SBD current appears to be much larger than in thicker oxides (as shown
in Fig. 3.70). Second, it takes a finite and measurable time (the residual time
associated to progressive breakdown) for the HBD current to grow up to the level
that seriously perturbs the device functionality. Third, the SBD does not appear as
a stable mode but it causes the device failure after a certain time. In general, we
can say that much work is still to be done for the complete characterization of the
post-breakdown failure modes in the ultra-thin oxides of interest for the present
and future technologies.

REFERENCES

1. D. J. DiMaria et al. Impact ionization, trap creation, degradation, and breakdown in
   silicon dioxide films on silicon. J. Appl. Phys., 73(7): 199, pp 3367–3384.
2. D. J. DiMaria, D. A. Buchanan, J. H. Stathis, R. E. Stahlbush. Interface
   states induced by the presence of trapped holes near the silicon–silicon-dioxide
   interface. J. Appl. Phys., 77:1995, p 2032.

3. J. Blauwe, J. van Houdt, D. Wellkens, G. Groeseneken, H. E. Maes. SILC-related effects in flash E^2PROM's-Part I: A quantitative model for steady-state SILC. IEEE Trans. Electron Devices, 45: 1998, p 1745.

4. A. Ghetti, E. Sangiorgi, J. Bude, T. W. Sorch, G. Weber. Low voltage tunneling in ultra-thin oxides: a monitor for interface states and degradation. Digest of the 1999 Int. Electron Device Meeting: 1999, pp 731–734.

5. D. J. DiMaria, J. Stathis. Explanation for the oxide thickness dependence of breakdown characteristics of metal-oxide-semiconductor structures. Appl. Phys. Lett., 70: 1997, pp 2708–2710.

6. I. C. Chen, S. Holland, C. Hu. Hole trapping and breakdown in thin SiO_2. IEEE Electron Device Lett., 7: 1986, pp 164–167.

7. I. C. Chen, S. Holland, K. K. Young, C. Chang, C. Hu. Substrate hole current and oxide breakdown. Appl. Phys. Lett., 49: 1986, pp 669–671.

8. K. F. Schuegraf, C. Hu. Hole injection SiO_2 breakdown model for very low voltage lifetime extrapolation. IEEE Trans. Electron Devices, 41: 1994, pp 761–767.

9. Y. Nissan-Cohen, J. Shappir, D. Frohman-Bentchkowsky. Trap generation and occupation dynamics in SiO_2 under charge injection stress. J. Appl. Phys., 60: 1986, p 2024.

10. D. J. DiMaria, J. W. Stasiak. Trap creation in silicon dioxide produced by hot electrons. J. Appl. Phys., 65: 1989, pp 2342–2356.

11. R. Degraeve, G. Groesenken, R. Bellens, J. Ogier, M. Depas, P. Roussel, H. Maes. New insights on the relation between electron trap generation and the statistical properties of oxide breakdown. IEEE Trans, Electron Devices, 45: 1998, pp 904–910.

12. D. J. Dumin, R. S. Scott, R. Subramoniam. A model relating wearout induced physical changes in thin oxides to the statistical description of breakdown. 1993 Int. Reliability Physics Sym. proc.: 1993, pp 285–292.

13. P. Olivo, B. Riccò, E. Sangiorgi. Electron trapping/detrapping within thin SiO_2 films in the high field tunneling regime. J. Appl. Phys., 54: 1983, pp 5267–5276.

14. I. C. Chen, S. E. Holland, C. Hu. Electrical breakdown in thin gate and tunneling oxides. IEEE Tran. Electron Device, 32: 1985, pp 413–422.

15. M. Depas, B. Vermeire, P. W. Mertens, van Meirhaeghe R. L., M. Heyns. Determination of tunneling parameters in ultra-thin oxide layer poly $Si/SiO_2/Si$ structures. Solid State Electronics, 38: 1995, pp 1465–1471.

16. J. Suñé, I. Placencia, N. Barniol, E. Farrés, X. Aymerich. Degradation. breakdown of gate oxides in VLSI Devices. Phys. Stat. Sol. A, 111: 1989, pp 675–685.

17. W. D. Zhang, J. F. Zhang, M. Lalor, D. Burton, G. Groeseneken, R. Degraeve. Two types of neutral electron traps generated in the gate silicon dioxide. IEEE Trans. Electron Devices, 49: 2002, pp 1868–1874.

18. R. Degraeve, G. Groeseneken, I. D. Wolf, H. E. Maes. Oxide and interface degradation and breakdown under medium and high field injection conditions: A correlation study. Microelectronic Engineering, 28: 1995, p 313.

19. J. Maserjian, N. Zamani. Behavior of the Si/SiO_2 interface observed by Fowler–Nordheim tunneling. J. Appl. Phys., 53: 1982, pp 559–567.

20. P. Olivo, T. N. Nguyen, B. Riccó. High-field induced degradation in ultra-thin SiO_2. IEEE Trans. Electron Devices, 20: 1988, pp 2259–2267.

21. D. J. Dumin, J. R. Maddux. Correlation of stress-induced leakage current in thin oxides with trap generation inside the oxides. IEEE Trans. Electron Devices, 40: 1993, pp 986–993.

22. D. J. DiMaria, E. Cartier. Mechanisms for stress-induced leakage currents in silicon dioxide films. J. Appl. Phys., 78: 1995, pp 3883–3894.

23. R. Moazzami, C. Hu. Stress-induced current in thin silicon dioxide films. Digest of the 1992 Int. Electron Device meeting: 1992, pp 139–142.

24. E. Rosenbaum, L. F. Register. Mechanism of stress-induced leakage current in MOS capacitors. IEEE Trans. Electron Devices, 44: 1997, pp 317–323.

25. B. Riccó, G. Gozzi, M. Lanzoni. Modeling and simulation of stress-induced leakage current in ultra-thin SiO_2 films. IEEE Trans. Electron Devices, 45: 1998, p 1554.

26. S. Takagi, N. Yasuda, A. Toriumi. Experimental evidence of inelastic tunneling in stress-induced leakage current. IEEE Trans. Electron Devices, 46: 1999, pp 335–341.

27. J. Wu, L. F. Register, E. Rosenbaum. Trap-assisted tunneling current through ultra-thin oxide. 1999 Int. Reliability Physics Sym. Proc.: 1999, pp 389–395.

28. D. Ielmini, A. S. Spinelli, A. L. Lacaita, D. J. DiMaria, G. Ghiaini. A detailed investigation of the quantum yield experiment. IEEE Trans. Electron Devices, 48: 2001, pp 1696–1702.

29. F. Schuler, R. Degraeve, P. Hendrickx, D. Wellkens. Physical description of anomalous charge loss in floating gate based NVMs and identification of its dominant parameter. 2002 Int. Reliability Physics Sym. Proc.: 2002, pp 26–33.

30. R. Degraeve, F. Schuler, B. Kaczer, M. Lorenzini, D. Wellkens, P. Hendrickx, M. van Duuren, G. J. M Dormans, J. van Houdt, L. Haspeslagh, G. Groeseneken, G. Tempel. Analytical percolation model for predicting anomalous charge loss in flash memories. IEEE Trans. Electron Devices, 51: 2004, pp 1392–1400.

31. K. Okada, K. Yoneda. A consistent model for time-dependent dielectric breakdown in ultra-thin silicon dioxides. Digest of the 1999 Int. Electron Device Meeting: 1999, pp 445–448.

32. R. Rodríguez, E. Miranda, R. Pau, J. Suñé, M. Nafría, X. Aymerich. Monitoring the degradation that causes the breakdown of ultra-thin (< 5 nm) SiO_2 gate oxides. IEEE Electron Device Lett., 21: 2000, pp 251–253.

33. W. L. Lai, E. Wu, J. Suñé. Impact of stress induced-leakage current on power consumption in ultra-thin gate oxides. 2004 Int. Reliability Physics Sym. Proc.: 2004, pp 102–109.

34. D. A. Buchanan, J. H. Stathis, E. Cartier, D. J. DiMaria. On the relationship between stress induced leakage currents and catastophic breakdown in ultra-thin SiO_2 based-dielectrics. Microelectronic Engineering, 36: 1997, pp 329–332.

35. J. H. Stathis, D. J. DiMaria. Reliability projections for ultra-thin oxides at low voltage. Digest of the 1998 Int. Electron Device meeting: 1998, pp 167–170.

36. J. H. Stathis. Physical and predictive models of ultra thin oxide reliability in CMOS devices and circuits. 2001 Int. Reliability Physics Sym. Proc.: 2001, pp 132–149.

37. E. Y. Wu, J. Suñé, E. Nowak, W. Lai, J. McKenna. Weibull slopes, critical defect density, and the validity of stress-induced-leakage current (SILC) measurements. Digest of the 2001 Int. Electron Device meeting: 2001, pp 125–128.

38. L. Pantisano, K. P. Cheung. Stress-induced leakage current (SILC) and oxide breakdown: Are they from the same oxide traps? IEEE Trans. Dev. Mat. Reliab, 1: 2001, pp 109–112.

39. M. A. Alam. SILC as a measure of trap generation and predictor of T_{BD} in ultathin oxides. IEEE Trans. Electron. Devices, 49: 2002, pp 226–231.

40. J. Suñé, I. Placencia, N Barniol, E Farrés, F Martín, X. Aymerich. On the breakdown statistics of very thin SiO_2 films. Thin Solid Films, 185: 1990, pp 347–362.

41. R. Degraeve, G. Groeseneken, R. Bellens, J. L. Ogier, M. Depas, P. J. Roussel, H. E. Maes. A consistent model for the thickness dependence of intrinsic breakdown in ultra-thin oxides. Digest of the 1995 Int. Electron Device Meeting: 1995, pp 866–869.

42. J. H. Stathis. Percolation models for gate oxide breakdown. J. Appl. Phys., 86: 1999, pp 5757–5766.

43. J. Suñé. New physics-based analytic approach to the thin oxide breakdown statistics. IEEE Electron Device Lett, 22: 2001, pp 296–298.

44. J. Suñé, E. Wu. Hydrogen release mechanisms in the breakdown of thin SiO_2 films. Phys. Rev. Lett., 92: 2004, p 87601.

45. E. Y. Wu, J. Suñé, R.-P. Vollertsen. Comprehensive physics-based breakdown model for reliability assessment of oxides with thickness ranging from 1nm up to 12nm. 2009 Int. Reliability Physics Symp., Proc.: 2009, pp 708–717.

46. R. Degraeve, B. Govoreanu, B. Kaczer, J. van Houdt, G. Groeseneken. Measurement and statistical analysis of single-trap current-voltage characteristics in ultra-thin SiON. 2005 Int. Reliability Physics Sym. Proc.: 2005, pp 360–365.

47. N. Klein. A theory of localized electronic breakdown in insulating films. Advances in Physics, 21: 1972, pp 605–625.

48. R. M. Hill, L. A. Dissado. Examination of the statistics of dielectric breakdown. J. Phys. C: Solid State Physics, 16:1983, pp 4447–4468.

49. J. McPherson. Stress dependent activation energy. 1986 Int. Reliability Physics Sym. Proc.: 1986, pp 12–18.

50. E. Y. Wu, A. Vayshenker, E. Nowak, J. Suñé, R.-P. Vollertsen, W. Lai, D. Harmon. Experimental evidence of T_{BD} power-law for voltage dependence of oxide breakdown in ultra-thin gate oxides. IEEE Trans. Electron Devices, 49: 2002, pp 2244–2253.

51. E. Y. Wu, J. Aitken, E. Nowak, A. Vayshenker, P. Varekamp, G. Hueckel, J. McKenna, D. Harmon, L. K. Han, C. Mostrose, R. Dufresne, R.-P. Vollertsen. Voltage-dependent voltage acceleration of oxide breakdown for ultra-thin oxides. Digest of the 2000 Int. Electron Device Meeting: 2000, pp 54–57.

52. J. W. McPherson, H. C. Mogul. Disturbed bonding states in SiO_2 thin-films and their impact on time-dependent dielectric breakdown, 1998 Int. Reliability Physics Sym. Proc.: 1998, pp. 47–56.

53. E. M. Vogel, J. S. Suehle, M. D. Edelstein, B. Wang, Y. Chen, J. B. Bernstein. Reliability of ultra-thin silicon dioxide under vombined substrate hot-electron and constant voltage tunneling stress. IEEE Trans. Electron Devices, 47: 2000, pp 1183–1191.

54. K. Umeda, K. Taniguchi. Hot-electron-induced quasibreakdown of thin gate oxides. J. Appl. Phys., 82: 1997, pp 297–302.

55. D. J. DiMaria. Defect generation under substrate-hot-electron injection into ultra-thin silicon dioxide layers. J. Appl. Phys., 86: 1999, pp 2100–2109.

56. L. Selmi, D. Esseni, P. Palestri. Towards microscopic understanding of MOSFET reliability: The role of carrier energy and transport simulations. Digest of the 2003 Int. Electron Device Meeting: 2003, pp 333–336.

57. W. McMahon, A. Haggag, K. Hess. Reliability scaling issues for nanoscale devices. IEEE Tran. Nanotechnology, 2: 2003, pp 33–38.

58. G. Ribes, S. Bruyere, M. Denais, D. Roy, G. Ghibaudo. Modeling charge-to-breakdown using hydrogen multivibrational excitation (thin SiO_2 and high-K dielectrics). IEEE Int. Integrated Reliability Workshop, Final Report, 2004, pp 1–3.

59. A. Haggag, N. Liu, D. Menke, M. Moosa. Physical model for the power-law voltage and current acceleration of TDDB. Microelectronics Reliability, 45: 2005, pp 1855–1860.

60. P. E. Nicollian, A. T. Krishnan, C. A. Chancellor, R. B. Khamankar. The traps that causes breakdown in deeply scaled SiON Dielectrics. Digest of the 2006 Int. Electron Device Meeting: 2006, pp 743–746.

61. J. Bude, B. E. Weir, P. J. Silvermann. Explanation of stress-induced damage in thin oxides. Digest of the 1998 Int. Electron Device Meeting: 1998, pp 179–182.

62. K. F. Schuegraf, C. Hu. Effects of temperature and defects on breakdown lifetime of thin SiO_2 at very low voltages. 1994 Int. Reliability Physics Sym. Proc.: 1994, pp 126–135.

63. M. A. Alam, J. Bude, A. Ghetti. Field acceleration for oxide breakdown: Can an accurate anode hole injection model resolve the E versus 1/E controversy? 2000 Int. Reliability Physics Sym. Proc.: 2000, pp 21–26.

64. M. Shatzkes, M. Av-Ron. Determination of breakdown rates and defect densities in SiO_2. Thin Solid Films, 91: 1985, p 217.

65. R. Subramonian, R. S. Scott, D. J. Dumin. A statistical model of oxide breakdown based on the physical description of wearout. Digest of the 1992 Int. Electron Device Meeting: 1992, pp 285–288.

66. J. H. Stathis. Quantitative model of the thickness dependence of breakdown in ultra-thin oxides. Microelectronic Engineering, 36: 1997, pp 325–328.

67. M. A. Alam, J. Bude, B. E. Weir, P. J. Silverman, A. Ghetti, D. Monroe, K. Cheung, S. Moccio. An anode hole injection percolation model for oxide breakdown: The doom's day scenario revisited. Digest of the 1999 Int. Electron Device Meeting: 1999, pp 715–718.

68. K. P. Cheung, C. T. Liu, C.-P. Chang, J. I. Colonell, W. Y.-C. Lai, R. Liu, J. F. Miner, C. S. Pai, H. Vaidya, J. T. Clemens, E. Hasegawa. Field dependent critical trap density for thin gate oxide breakdown. 1999 Int. Reliability Physics Sym. Proc.: 1999, pp 52–56.

69. D. J. DiMaria, J. H. Stathis. Non-Arrhenius temperature dependence of reliability in ultra-thin silicon dioxide films. Appl. Phys. Lett., 74: 1999, pp 1752–1754.

70. B. Kaczer, R. Degraeve, N. Pangon, G. Groeseneken. The influence of elevated temperature on degradation and lifetime prediction of thin silicon-dioxide films. IEEE Trans. Electron Devices, 47: 2000, pp 1514–1521.

71. R. Degraeve, Ph. Roussel, G. Groeseneken, H. E. Maes. A new analytic model for the description of the intrinsic oxide breakdown statistics of ultra-thin oxides. Microelectronic Engineering, 11/121996, pp 1639–1642.

72. M.-J. Chen, H.-T. Huang, J.-H. Chen, C.-S. Hou, M.-S. Liang. Cell-based analytic statistical model with correlated parameters for intrinsic breakdown of ultra-thin oxides. IEEE Electron Device Lett, 20: 1999, pp 523–525.

73. J. H. Stathis. Reliability limits for the gate insulator in CMOS technology. IBM J. Res. and Dev., 46: 2002, pp 265–286.

74. E. Y. Wu, J. Suñé, W. Lai. On the Weibull shape factor of dielectric films and its accurate experimental determination, Part II. Experimental results and the effects of stress conditions. IEEE Tran. Electron Devices, 49: 2002, pp 2141–2150.

75. W. Mizubayashi, Y. Yoshida, S. Miyazaki, M. Hirose. Statistical analysis of soft breakdown in ultra-thin gate oxides. Digest of the 2001 Sym. on VLSI Technology, 2001, pp 95–96.

76. J. W. McPherson, R. B. Khamankar. Molecular model for intrinsic time-dependent dielectric breakdown in SiO_2 dielectrics and the reliability implications for hyper-thin gate oxide. Semicond. Sci. Technol., 15: 2000, pp 462–470.

77. J. W. McPherson, R. B. Khamankar, A. Shanware. Complementary model for intrinsic time-dependent dielectric breakdown in SiO_2 dielectrics. J. Appl. Phys., 88: 2000, pp 5351–5359.

78. J. W. McPherson. Determination of the nature of molecular bonding in silica from time-dependent dielectric breakdown data. J. Appl. Phys., 95: 2004, pp 8101–8109.

79. J. McPherson. Extended Mie-Grüneisen molecular model for time dependence dielectric breakdown in silica detailing the critical roles of $O–Si\equiv O_3$ tetragonal boding, stretched bonds, hole capture, and hydrogen release. J. Appl. Phys., 99: 2006, pp 083501.

80. A. Shanware, R. B. Khamankar, J. W. McPherson. Resolving the non-uniqueness of the activation energy associated with TDDB for SiO_2 thin films. Digest of the 2000 Int. Electron Device Meeting: 2000, pp 549–553.

81. K. P. Cheung. A physics-based, unified gate-oxide breakdown model. Digest of the 1999 Int. Electron Device Meeting: 2000, pp 719–722.

82. K. P. Cheung. Unifying the thermal-chemical and anode-hole-injection gate-oxide breakdown models. Microelectronics Reliability, 41: 2001, pp 193–199.

83. K. P. Cheung. Temperature effect on ultra-thin SiO_2 time-dependent dielectric breakdown. Appl. Phys. Lett., 83: 2003, p 2399.

84. P. Nicollian, W. R. Hunter, J. C. Hu. Experimental evidence for voltage driven breakdown models in ultra-thin gate oxides. 2000 Int. Reliability Physics Sym. Proc.: 2000, pp 7–15.

85. J. McKenna, E. Wu, S.-H. Lo. Tunneling current characteristics and oxide breakdown in P+Poly gate PFET capacitors. 2000 Int. Reliability Physics Sym. Proc.: 2000, pp 16–20.

86. J. J. O'Dwyer. *The Theory of Electrical Conduction and Breakdown in Solid Dielectrics* Clarendon: Oxford. 1973.

87. N. Klein. Electrical breakdown of insulators by one-carrier impact-ionization. J. Appl. Phys., 53: 1982, pp 5828–5839.

88. T. H. DiStefano, M. Shatzkes. Impact ionization model for dielectric instability and breakdown. Appl. Phys. Lett., 25: 1974, pp 685–687.

89. H. Fitting, J. Friemann. Monte Carlo studies of the electron mobility in SiO_2. Phys. Stat. Sol., 69: 1982, pp 349–358.

90. T. N. Theis, D. J. DiMaria, J. R. Kirtley, D. W. Dong. Strong electric field heating of conducting-band electrons in SiO_2. Phys. Rev. Lett., 52: 1984, pp 1445–1448.

91. M. V. Fischetti, D. J. DiMaria, S. D. Brorson, T. N. Theis, J. R. Kirtley. Theory of high-field electron transport in silicon dioxide. Phys. Rev. B, 31: 1985, pp 8124–8142.

92. F. R. McFeely, E. Cartier, L. J. Terminello, A. Santoni, M. V. Fischetti. Soft-X-ray-induced core-level photoemission as a probe of hot-electron dynamics in SiO_2. Phys. Rev. Lett., 65: 1990, pp 1937–1940.

93. D. Arnold, E. Cartier, D. J. DiMaria. Theory of high-filed electron transport and impact ionization in silicon dioxide. Phys. Rev. B, 49: 1994, pp 10278–10297.

94. I. C. Chen, S. Holland, C. Hu. Oxide breakdown dependence on thickness and hole current – enhanced reliability of ultra thin oxides. Digest of the 1986 Int. Electron Device Meeting: 1986, pp 660–663.

95. Y. Hokari. Conduction and trapping of electrons in highly stressed ultra-thin films of thermal SiO_2. Appl. Phys. Lett., 30:1977, pp 601–603.

96. I. C. Chen, S. Holland, C. Hu. Electron trap generation by recombination of electron and holes in SiO_2. J. Appl. Phys., 61: 1987, pp 4544–4548.

97. K. F. Schuegraf, C. Hu. Metal-oxide-semiconductor field-effect-transistor substrate current during Fowler-Nordheim tunneling stress and silicon dioxide reliability. J. Appl. Phys., 76: 1994, pp 3695–3700.

98. Z. A. Weinberg. Hole injection and transport in SiO_2 films on Si. Appl. Phys. Lett., 27: 1975, pp 437–439.

99. M. V. Fischetti. Model for the generation of positive charge at the $Si–SiO_2$ interface based on hot-hole injection from the anode. Phys. Rev. B, 31: 1985, pp 2099–2113.

100. D. J. DiMaria. Explanation for the polarity dependence of breakdown in ultra-thin silicon dioxide films. Appl. Phys. Lett., 68: 1996, pp 3004–3006.

101. M. Rasras, I. De Wolf, G. Groeseneken, B. Kaczer, R. Degraeve, H. E. Maes. Photo-carrier generation as the origin of Fowler–Noidheim substrate hole current in thin oxides. IEEE Tran. Electron Devices, 48: 2001, pp 231–238.

102. P. Palestri, A. Dalla Serra, L. Selmi, M. Pavesi, P. Luigi Rigolli, A. Abramo, F. Widdershoven, E. Sangiorgi. A comparative analysis of substrate current generation mechanisms in tunneling MOS capacitors. IEEE Tran. Electron Devices, 49: 2002, pp 1427–1435.

103. M. A. Alam, B. Weir, J. Bude, P. Silverman, A. Ghetti. A comprehensive model for oxide breakdown: theory and experiments. Microelectronic Engineering, 59: 2001, pp 137–147.

104. D. J. DiMaria, J. H. Stathis. Anode hole injection, defect generation, and breakdown in ultra-thin silicon dioxide films. J. Appl. Phys., 89: 2001, pp 5015–5024.

105. E. M. Vogel, M. D. Edelstein, J. S. Suehle. Defect generation and breakdown of ultra-thin silicon dioxide induced by substrate hot-hole injection. J. Appl. Phys., 90: 2001, pp 2338–2346.

106. D. Heh, E. M. Vogel, J. B. Bernstein. Impact of substrate hot hole injection on ultra-thin silicon dioxide breakdown. Appl. Phys. Lett., 82: 2003, pp 3242–3244.

107. P. E. Blöchl, J. Stathis. Hydrogen electrochemistry and stress-induced leakage current in silica. Phys. Rev. Lett., 83: 1999, pp 372–375.

108. Y. Nissan-Cohen, T. Gorczyca. The effect of hydrogen on trap generation, positive charge trapping, and time-dependent dielectric breakdown of gate oxides. IEEE Electron Dev. Lett., 9: 1988, pp 287–289.

109. C. Gelatos, H. H-. Tseng, S. Filipiak, D. Sielofff, J. Grabt, P. Tobin, R. Cotton. The effects of passivation and post-passivation anneal on the integrity of thin oxides. Digest of the 1997 Sym. VLSI Technology, 1997, pp 188–192.

110. T. Pompl, K.-H. Allers, R. Schwab, K. Hofmann, M. Roehner. Change of accelera-tion behavior of time-dependent dielectric breakdown by the BEOL process: Indica-tions for hydrogen induced transition in dominant degradation mechanism. 2005 Int. Reliability Physics Sym. Proc.: 2005, pp 393–402.

111. K. Hess, I. C. Kizilyalli, J. W. Lyding. Giant isotope effect in hot electron degrada-tion of metal oxide silicon devices. IEEE Trans. Electron Devices, 45: 1988, pp 406–410.

112. J. Wu, E. Rosenbaum, B. MacDonald, E. Li, J. Tao, B. Tracy, P. Fang. Anode hole injection versus hydrogen release: The mechanism for gate oxide breakdown. 2000 Int. Reliability Physics Sym. Proc.: 2000, pp 27–32.

113. Y. Mitani, H. Satake, A. Toriumi. Experimental evidence of hydrogen-related SILC generation in thin gate oxides. Digest of the 2001 Int. Electron Device Meeting: 2001, pp 29–132.

114. W. Clark, E. Cartier, E. Y. Wu. Hot carrier lifetime and dielectric breakdown in MOSFETs processed with deuterium. Proc. of the 6th Int. Sym. on Plasma- and Process-Induced Damage, 2001, pp 80–85.

115. J. Suñé, E. Wu. Mechanisms of hydrogen release in the breakdown of SiO2-based gate oxides. Digest of the 2005 Int. Electron Device Meeting: 2005, pp 399–402.

116. T.-C. Shen, C. Wang, G. C. Abeln, J. R. Tucker, J. W. Lyding, Ph. Avouris, R. E. Walkup. Atomic-scale desorption through electronic and vibrational excitation mechanisms. Science, 268: 1995, pp 1590–1592.

117. Ph. Avouris, R. E. Walkup, A. R. Rossi, T.-C. Shen, G. C. Abeln, J. R. Tucker, J. W. Lyding. STM-induced H atom desoprtion from Si(100): Isotope effects and site selectivity. Chem. Phys. Lett., 257: 1996, pp 148–154.

118. K. Stokbro, B. Y-K. Hu, C. Thirstrup, X. C. Xie. First-principles theory of inelastic currents in a scanning tunneling microscope. Phys. Rev. B, 58: 1998, pp 8038–8041.

119. G. P. Salam, M. Persson, R. E. Palmer. Possibility of coherent multiple excitation in atom transfer with a scanning tunneling microscope. Phys. Rev. B, 49: 1994, pp 10655–10662.

120. K. L. Brower. Dissociation kinetics of hydrogen-passivated (111) Si–SiO$_2$ interface defects. Phys. Rev. B, 42: 1990, p 3444.

121. A. Stesmans. Influence of interface relaxation on passivation kinetics in H2 of coordination Pb defects at the (111) Si/SiO$_2$ interface revealed by electron spin resonance. J. Appl. Phys., 92: 2002, p 1317.

122. K. Hess, B. Tuttle, F. Register, D. K. Ferry. Magnitude of the threshold energy for hot electron damage in metal–oxide–semiconductor field effect transistors by hydrogen desorption. Appl. Phys. Lett., 75: 1999, p 3147.

123. M. Röhner, A. Kerber, M. Kerber. Voltage acceleration of T_{BD} and its correlation to post breakdown conductivity of N- and P-Channel MOSFETS. 2006 Int. Reliability Physics Sym. Proc.: 2006, pp 76–81.

124. C. Stipe, M. A. Rezaei, W. Ho, S. Gao, M. Persson, B. I. Lundqvist. Single-molecule dissociation by tunneling electrons. Phys. Rev. Lett., 78: 1997, pp 4410–4413.

125. Y. J. Chabal, C. K. N. Patel. Infrared absorption in a-Si:H first observation of gaseous molecular H2 and Si–H overtone. Phys. Rev. Lett., 53: 1984, p 210.

126. B. Tuttle, C. G. van de Walle. Structure, energetics, and vibrational properties of Si–H bond dissociation in silicon. Phys. Rev. B, 59: 1999, p 12884.

127. E. S. Anolick, G. Nelson. Low field time dependent dielectric integrity. 1979 Int. Reliability Physics Sym. Proc.: 1979, pp 8–12.

128. D. L. Crook. Method of Determining reliability screens for time dependent dielectric breakdown. 1979 Int. Reliability Physics Sym. Proc.: 1979, pp 1–7.

129. D. Baglee. Characteristics and reliability of 100lÅ oxides. 1984 Int. Reliability Physics Sym. Proc.: 1984, pp 152–155.

130. J. McPherson, D. Baglee. Acceleration Factors for Thin Gate Oxide Stressing. 1985 Int. Reliability Physics Sym. Proc.: 1985, pp 1–5.

131. As discussed in Section 2.1 of Chapter 2 in earlier publications [127–130] above, the T_{BD} measurements were analyzed using the lognormal distribution. The data has been reanalyzed using the Weibull distributions. Consequently, we found the results presented in these publications actually correspond to the extrinsic mode with Weibull slopes less than one.

132. D. R. Wolters, J. F. Verwey. Breakdown and wear-out phenomena in SiO$_2$. In: *Instabilities in Silicon Devices*, edited by Vapaille. North Holland 1986, pp 315–362.

133. N. Shiono, M. Itsumi. A lifetime projection method using series model and acceleration factors for TDDB failures of thin gate oxides. 1993 Int. Reliability Physics Sym. Proc.: 1993, pp 1–6.

134. J. Suehle, P. Chaparala, C. Messick, W. M. Miller, K. C. Boyko. Field and temperature acceleration of time-dependent dielectric breakdown in intrinsic thin SiO$_2$. 1994 Int. Reliability Physics Sym. Proc.: 1994, pp 120–125.

135. M. Kimura. Oxide breakdown mechanism and quantum physical chemistry for time-dependent dielectric breakdown. 1997 Int. Reliability Physics Sym. Proc.: 1997, pp 190–200.

136. A. Martin, M. Kerber, G. Diestel. Investigation of initial charge trapping and oxide breakdown under Fowler–Nordheim Stress. IEEE IRW 1998, Final Report, 1998, pp 99–104.

137. A. Teramoto, H. Umeda, K. Kobayashi, K. Shiga, J. Komori. Study of oxide breakdown under very low electric field. 1999 Int. Reliability Physics Sym. Proc.: 1999, pp 66–71.

138. J. McPherson, V. Reddy, K. Banerjee, Hue Le. Comparison of E and 1/E TDDB models for SiO_2 under long-term/low-field test conditions. Tech. Digest Int. Electron Devices Meeting: 1998, pp 171–174.

139. R.-P. Vollertsen. Thin dielectric reliability assessment for DRAM technology with deep trench storage node. Microelectronics Reliability, 43: 2003, pp 865–878.

140. R.-P. Vollertsen, W. W. Abadeer. Upper voltage and temperature limitations of stress conditions for relevant dielectric breakdown projections. Quality and Reliability Engineering International, 11(4): 1995, pp 233–238.

141. E. Y. Wu et al. To be submitted for publication.

142. A. Yassine, H. E. Nariman, M. McBride, M. Uzer, Kola R. Olasupo. Time dependent breakdown of ultra-thin gate oxide. IEEE Trans. Electron Devices, 47: 2000, pp 1416–1420.

143. J. H. Stathis, A. Vayshenker, P. Varekamp, E. Y Wu, C. Monstrose, J. McKenna, D. J. DiMaria, L.-K Han, E. Cartier, R. Wachnik, B. P. Linder. Breakdown measurements of ultra-thin SiO_2 at low voltage. Digest of the 2000 Sym. on VLSI Technology, 2000, pp 94–95.

144. R.-P. Vollertsen, E. Y. Wu. Gate Oxide Reliability Parameters in the Range 1.6 to 10 nm. IEEE Int. Integrated Reliability Workshop, Final Report, 2003, pp 10–15.

145. B. Weir, M. Alam, J. Bude, P. Silverman, A. Ghetti, F. Baumann, P. Diodato, D. Monroe, T. Sorsch, G. Timp, Y. Ma, M. Brown, A. Hamad, D. Hwang, P. Mason. Gate oxide reliability projection to the sub-2 nm regime. Semicond. Sci. Technol., 15: 2000, pp 455–461.

146. E. Y. Wu and J. Suñé, On voltage acceleration models of time-to-breakdown, Part I: Experimental analysis and methodology, Trans on Electron Device, 56: 2009, pp 1433–1440.

147. R. Duschl, R.-P. Vollertsen. Voltage acceleration of oxide breakdown in the sub-10 nm Fowler–Nordheim and direct tunneling regime. IEEE Int. Integrated Reliability Workshop, Final Report, 2005, pp 44–48.

148. R.-P. Vollertsen, T. Pompl, R. Duschl, A. Kerber, M. Kerber, M. Röhner, R. Schwab. Long term gate dielectric stress – a timely method? Digest of the 2005 Int. Electron Device Meeting: 2006, pp 747–750.

149. A. Kerber, M. Kerber. Fast wafer level data acquisition for reliability characterization of sub-100 nm CMOS technologies. IEEE Int. Integrated Reliability Workshop, Final Report, 2004, pp 41–45.

150. A. Kerber, M. Röhner, C. Wallace, L. O'Riain, M. Kerber. From wafer-level gate-oxide reliability towards ESD failures in advanced CMOS technologies. IEEE Trans. Electron Devices, 53: 2006, pp 917–920.

151. B. Weir, C. Leung, P. J. Silverman, M. A Alam. Gate dielectric breakdown in the time-scale of ESD events. Microelectronics Reliability, 45: 2005, pp 427–436.

152. J. Wu, E. Rosenbaum. Gate Oxide Reliability Under ESD-Like Pulse Stress. IEEE Trans. Electron Devices, 51: 2004, pp 1528–1532.

153. A. Hiraiwa, D. Ishikawa. Thickness dependent power-law of dielectric breakdown for voltage dependence of oxide breakdown in ultra-thin gate oxides. Microelectronics Engineering, 80: 2005, pp 374–377.

154. K. Ohgata, M. Ogasawara, K. Shig, S. Tsujikawa, E. Murakami, H. Kato, H. Umeda, Kubota. 2005 Int. Reliability Physics Sym. Proc.: 2005, pp 372–376.

155. T. Pompl, M. Röhner. Voltage acceleration of time-dependent breakdown of ultra-thin gate dielectrics. Microelectronics Reliability, 45: 2005, pp 1835–1841.

156. J. S. Suehle, P. Chaparala. Low electric field breakdown of thin SiO$_2$ films under static and dynamic stress. IEEE Trans. Electron. Devices, 44: 1997, pp 801–808.

157. R. Moazzami, J. Lee, C. Hu. Temperature acceleration of time-dependent dielectric breakdown. IEEE Trans. Electron Devices, 36: 1989, pp 2462–2465.

158. E. Y. Wu, D. Harmon, L.-K. Han. Interrelationship of voltage and temperature dependence of oxide breakdown for ultra-thin oxides. IEEE Electron Device Lett, 21: 2000, pp 362–364.

159. E. Y. Wu, J. Suñé, W. Lai, E. Nowak, J. McKenna, A. Vayshenker, D. Harmon. Interplay of voltage and temperature acceleration of oxide breakdown for ultra-thin oxides, In special issue: 2001 Insulating Films On Semiconductors (INFOS). Solid State Electronics, 46: 2002, pp 1787–1798.

160. E. Y. Wu, W. Lai, M. Khare, J. Suñé, L.-K. Han, J. McKenna, R. Bolam, D. Harmon, A. Strong. Polarity-dependent oxide breakdown of NFET devices for ultra-thin gate oxide. 2002 Int. Reliability Physics Sym. Proc.: 2002, pp 60–72.

161. E. Y. Wu, J. Suñé. Power-law voltage dependence: A key element for ultra-thin gate oxide reliability. Microelectronics Reliability, 45: 2005, pp 1809–1834.

162. E. Wu, E. Nowak, A. Vayshenker, J. McKenna, D. Harmon, and R.-P. Vollertsen. (Invited paper), New global insight in ultra-thin oxide reliability using accurate experimental methodology and comprehensive database, IEEE Trans. Device and Material Reliability, 1: 2001, pp 69–80.

163. E. Wu et al. Journal of Applied Physics, submitted for publication.

164. E. Vincent, C. Papadas, G. Ghibaudo. Temperature dependence pf charge build-up mechanisms and breakdown phenomena in thin oxides under Fowler–Nordheim Injection. Solid State Electronics, 41: 1997, pp 1001–1004.

165. M. Kerber, U. Schwalke. Interface degradation and dielectric breakdown of thin oxides due to homogeneous injection. 1989 Int. Reliability Physics Sym. Proc.: 1989, pp 17–21.

166. Y. Hokari. Stress voltage polarity dependence of thermally grown thin gate oxide wear-out. IEEE, Trans. Electron Device, 35: 1988, pp 1299–1304.

167. L. K. Han, M. Bhat, D. Wisters, J. Fulford, D. L. Kwong. Polarity dependence of dielectric breakdown in scaled SiO$_2$. Digest of the 1994 Int. Electron Device Meeting: 1994, pp 617–620.

168. P. P. Apte, K. C. Saraswat. SiO$_2$ degradation with charge injection polarity. IEEE Electron Device Lett, 14: 1993, pp 512–514.

169. E. Y. Wu, J. Sune. New Insights in polarity dependent oxide breakdown for ultra-thin breakdown for ultra-thin gate oxides. IEEE Electron Device Letters, 23: 2002, pp 494–496.

170. M. S. Liang, S. Haddad, W. Cox, S. Cagnina. Degradation of very thin gate oxides MOS devices under dynamic high field/current stress. Digest of the 1986 Int. Electron Device Meeting: 1986, pp 394–398.

171. H. Hwang, J. Lee. Anomalous breakdown behaviour in ultra-thin oxides and oxynitrides under dynamic electrical stress. IEEE Electron. Device Lett, 13: 1992, pp 485–487.

172. E. Rosenbaum, Z. Liu, C. Hu. Silicon dioxide breakdown lifetime enhancement under bipolar bias conditions. IEEE Trans. Electron. Devices, 40: 1993, pp 2287–2295.

173. D. J. Dumin, S. Vanchinathan. Bipolar stressing, breakdown, and trap generation in thin silicon oxides. IEEE Trans. Electron. Devices, 41: 1994, pp 936–940.

174. M. Nafría, J. Suñé, D. Yélamos, X. Aymerich. Degradation and breakdown of thin silicon dioxide films under dynamic electrical stress. IEEE Trans. Electron. Devices, 43: 1996, pp 2215–2226.

175. B. Wang, J. S. Suehle, E. M. Vogel, J. B. Bernstein. Time-dependent breakdown of ultra-thin SiO_2 gate dielectrics under pulsed biased stress. IEEE Electron Device Lett, 22: 2001, pp 224–226.

176. R. Rodríguez, M. Nafría, J. Suñé, X. Aymerich. Trapped charge distributions in thin (10 nm) SiO_2 films subjected to static and dynamic stresses. IEEE Trans. Electron Devices, 45: 1998, pp 881–888.

177. S. Zhu, A. Nakajima, T. Ohashi, H. Miyake. Interface trap generation induced by charge pumping current under dynamic oxide field stresses. IEEE Electron Device Lett., 26: 2005, pp 216–218.

178. W. R. Hunter. The analysis of oxide reliability data. 1998 IEEE Int. Integrated Reliability Workshop, Final Report, 1998, pp 114–134.

179. C. Hu, Q. Lu. A unified gate oxide reliability model. 1999 Int. Reliability Physics Sym. Proc.: 1999, pp 47–51.

180. R. Duschl, R.-P. Vollertsen. Is the power-law model applicable beyond the direct tunneling regime? Microelectronics Reliability, 45: 2005, pp 1861–1867.

181. R.-P. Vollertsen, E. Y. Wu. Voltage acceleration and t63.2 of 1.6–10 nm gate oxides. Microelectronics Reliability, 44: 2004, pp 906–916.

182. V. E. Houtsma, J. Holleman, C. Salm, I. R. de Haan, J. Schmitz, F. P. Widdershoven, P. H. Woerlee. Minority carrier tunneling and stress-induced leakage current for p^+ gate MOS capacitors with poly Si and poly $Si_{0.7}Ge_{0.3}$ gate material. Digest of the 1999 Int. Electron Device Meeting: 1999, pp 457–460.

183. T. Pompl, M. Kerber, H. Wurzer, I. Eisele. Contribution of interface traps to valence-band electron tunneling in PMOS devices. Proc. European Solid State Device Research Conf. (ESSDERC), 2000, pp 292–295.

184. E. Y. Wu, J. Suñé, W. Lai, A. Vayshenker, D. Harmon. A comprehensive investigation of gate oxide breakdown of p^+-poly/PFETs under inversion mode. Digest of the 2005 Int. Electron Device Meeting: 2005, pp 407–410.

185. B. Kaczer, R. Degraeve, R. O'Connor, Ph. Roussel, G. Groeseneken. Implications of progressive wear-out for lifetime extrapolation of ultra-thin (EOT ~ 1 nm) SiON films. Digest of the 2004 Int. Electron Device Meeting: 2004, pp 713–716.

186. W. W. Abadeer, R.-P. Vollertsen, R. J. Bolam, D. J. DiMaria, and E. Cartier, Correlation between theory and data for mechanisms leading to dielectric breakdown, Digest of the 1994 Symposium on VLSI Technology, 1994, pp 43–44.

187. D. J. DiMaria, E. Cartier, D. A. Buchanan. Anode hole injection and trapping in silicon dioxide. J. Appl. Phys., 80: 1996, pp 304–317.

188. B. Kaczer, R. Degraeve, G. Goeseneken, M. Rasras, S. Kubicek, E. Vandamme, G. Badenes. Impact on MOSFET oxide breakdown on digital circuit operation and reliability. Digest of the 2000 Int. Electron Device Meeting: 2000, pp 553–556.

189. B. E. Weir, P. J. Silverman, D. Monroe, K. S. Krisch, M. A. Alam, G. B. Alers, T. W. Sorsch, G. L. Timp, F. Baumann, C. T. Liu, Y. Ma, D. Hwang. Ultra thin gate dielectrics: They break down, but do they fail? IEEE IEDM Tech. Digest, 1997, pp 73–76.

190. E. Y. Wu, E. Nowak, J. Aitken, W. Abadeer, L. K. Han, S. Lo. Structural dependence of dielectric breakdown in ultra-thin gate oxides and its relationship to soft break-down modes and device failure. Digest of the 1998 Int. Electron Device Meeting: 1998, pp 187–190.

191. W. K. Henson, N. Yang, J. J. Wortman. Observation of oxide breakdown and its effects on the characteristics of ultra-thin-oxide nMOSFETs. IEEE Electron Device Lett., 20:(12)1999, pp 605–607.

192. A. Cester, S. Cimino, A. Paccagnella, G. Ghidini, G, Guegan. Collapse of MOSFET drain current after soft breakdown and its dependence on the transistor aspect ratio W/L. 2003 Int. Reliability Physics Sym. Proc.: 2003, pp 189–195.

193. B. Kaczer, A. De Keersgieter, S. Mahmood, R. Degraeve, G. Groeseneken. Impact of gate-oxide breakdown of varying hardness on narrow and wide NFETs. 2004 Int. Reliability Physics Sym. Proc.: 2004, pp 79–83.

194. R. Rodriguez, J. H. Stathis, B. P. Linder, S. Kowalczyk, C. T. Chuang, R. V. Joshi, G. Northrop, K. Bernstein, A. J. Bhavnagarwala, S. Lombardo. The impact of gate-oxide breakdown on SRAM stability. IEEE Electron Dev. Lett., 23: 2002, pp 559–561.

195. R. Rodríguez, J. H. Stathis, B. P. Linder. A model for gate-oxide breakdown in CMOS inverters. IEEE Electron Dev. Lett., 24: 2003, pp 114–116.

196. M. A. Alam, R. K. Smith, B. E. Weir, P. J. Silverman. Statistically independent soft-breakdowns redefine oxide reliability specifications. IEDM Technical Digest, 2002, pp 151–154.

197. J. Suñé, E. Y. Wu. Statistics of successive breakdown events in gate oxides. IEEE Electron Device Lett, 24(4): 2003, pp 272–274.

198. F. Monsieur, E. Vincent, D. Roy, S. Bruyere, J. C. Vildeuil, G. Pananakakis, G. Ghibaudo. A thorough investigation of progressive breakdown in ultra-thin oxides. physical understanding and application for industrial reliability assessment. 2002 Int. Reliability Physics Sym. Proc.: 2002, pp 45–54.

199. B. P. Linder, S. Lombardo, J. H. Stathis, A. Vayshenker, D. Frank. Voltage dependence of hard breakdown growth and the reliability implication in thin di-electric. IEEE Electron Device Lett., 23: 2002, pp 661–663.

200. S. Lee, B. Cho, J. Kim, S. Choi. Quasibreakdown of ultra-thin gate oxide under high field stress. Digest of the 1994 Int. Electron Device Meeting: 1994, pp 605–608.

201. E. Miranda, J. Suñé, R. Rodríguez, M. Nafría, X. Aymerich, L. Fonseca, F. Campabadal. Soft-breakdown conduction in ultra-thin (3–5 nm) gate dielectrics. IEEE Trans. Electron Dev., 47: 2000, pp 82–89.

202. M. Depas, T. Nigam, M. M. Heyns. Soft breakdown of ultra-thin gate oxide layers. IEEE Trans. Electron Devices, 43: 1996, pp 1499–1504.

203. J. Suñé, E. Y. Wu, D. Jiménez, R. P. Vollertsen, E. Miranda. Understanding Soft and Hard breakdown statistics, prevalence ratios and energy dissipation during break-down runaway. IEEE Int. Electron Devices Meeting: 2001, pp 120–123.

204. J. Suñé, E. Y. Wu, D. Jiménez, W. L. Lai. Statistics of soft and hard breakdown in thin SiO$_2$ gate oxides. Microelectronics Reliability, 43(8): 2003, pp 1185–1192.

205. M. A. Alam, B. E. Weir, P. J. Silverman. A study of soft and hard breakdown—Part I: Analysis of statistical percolation conductance. IEEE Trans. Electron Devices, 49: 2002, pp 232–238.

206. J. Suñé, E. Y. Wu, W. L. Lai. Successive oxide breakdown statistics: Correlation effects, reliability methodologies, and their limits. IEEE Trans. Electron Devices, 51: 2004, pp 1584–1592.

207. E. Wu, J. Suñé. Statistical and Voltage scaling properties of post-breakdown for ultra-thin-oxide PFETs in inversion mode. 2006 Int. Reliability Physics Sym. Proc.: 2006, pp 54–62.

208. T. Hosoi, P. L. Ré, Y. Kamakura, K. Taniguchi. A new model of time evolution of gate leakage current after soft breakdown in ultra-thin gate oxides. Digest of Technical Papers of the 2002 Int. Electron Devices Meeting: 2002, pp 155–158.

209. B. Kaczer, R. Degraeve, M. Rasras, K. van de Mieroop, P. J. Roussel, G. Groeseneken. Impact of MOSFET gate oxide breakdown on digital circuit operation and reliability. IEEE Trans. Electron Devices, 49: 2002, pp 500–506.

210. J. Suñé, E. Y. Wu, and W. Lai, Limits of successive breakdown statistics to assess chip reliability, Microelectronic Engineering, 72: 2004, pp 39–44.

211. T. Pompl, H. Wurzer, M. Kerber, R. C. W. Wilkins, I. Eisele, Influence of soft breakdown on nMOSFET device characteristics, International Reliability Physics Symposium Proceedings, 1999: 82–87.

212. R. Degraeve, B. Kaczer, A. Keersgieter, G. Groeseneken. Relation between breakdown mode and location in short-channel nMOSFETs and its impact on reliability specifications. IEEE Trans. Device and Materials Reliability, 1: 2001, pp 163–169.

213. K. L. Pey, R. Ranjan, C. H. Tung, L. J. Tang, W. H. Lin, M. K. Radhakrishnan. Gate dielectric degradation mechanism associated with DBIE. 2004 Int. Reliability Physics Sym. Proc.: 2004, pp 117–121.

214. B. Kaczer, G. Groeseneken. Potential Vulnerabity of Dynamic CMOS Logic to Soft Gate Oxide Breakdown. IEEE Electron Device Lett., 24: 2003, pp 742–744.

215. K. Mueller, S. Gupta, S. Pae, M. Agostimelli, P. Aminzadeh. 6-T cell circuit dependent GOX SBD model for accurate prediction for observed VCCMIN test voltage dependence. 2004 Int. Reliability Physics Sym. Proc.: 2004, pp 426–429.

216. E. Miranda, J. Suñé, R. Rodríguez, M. Nafría, X. Aymerich. A function-fit model for the sofá-breakdown failure mode. IEEE Electron Device Lett, 20: 1999, pp 265–267.

217. E. Miranda and J. Suñé, Analytic modeling of leakage current through multiple breakdown paths in SiO$_2$ films, International Reliability Physics Symposium Proceedings, 2001, pp 367–379.

218. J. Suñé, G. Mura, E. Miranda. Are soft breakdown and hard breakdown of ultra-thin gate oxides actually different failure mechanisms? IEEE Electron Device Lett., 21: 2001, pp 167–169.

219. J. Suñé, E. Y. Wu, W. L. Lai. Statistics of competing post-breakdown failure modes in ultra-thin MOS devices. IEEE Trans. Electron. Devices, 53: 2006, pp 224–234.

220. B. P. Linder, J. H. Stathis. Statistics of progressive breakdown in ultra-thin oxides. Microelectronic Engineering, 72:2004, pp 24–28.

221. F. Monsieur, E.Vincent, G. Ribes, V. Huard, D. Roy, G. Pananakakis, G. Ghibaudo, Evidence for defect-generation-driven wear-out of breakdown conduction path in ultra thin oxides, International Reliability Physics Symposium Proceedings, 2003, pp 424–431.

222. L. J. Tan, K. L. Pey, C. H. Tung, R. Ranjan, W. H. Lin. Study of breakdown in ultra-thin gate dielectrics using constant voltage stress and successive constant voltage stress. Microelectronic Engineering, 80: 2005, pp 170–173.

223. T. Pompl, A. Kerber, M. Röhner, M. Kerber. Gate voltage and oxide thickness dependence of progressive wear-out of ultra-thin gate oxides. Microelec. Reliab., 9–11: 2006, pp 1603–1607.

224. E. Y. Wu, S. Tous, J. Suñé. On the progressive breakdown statistical distribution and its voltage acceleration. Digest of the 2007 Int. Electron Device Meeting: 2007, pp 493–496.

225. E. Miranda, A. Cester. Degradation dynamics of ultra-hin gate oxides subjected to electrical stress. IEEE Electron Device Lett, 24: 2003, pp 604–606.

226. S. Sahhaf, R. Degareve, Ph.J. Roussel, T. Kauerauf, B. Kaczer, G. Groeseneken. TDDB Reliability prediction based on the statistical analysis of hard breakdown including multiple soft breakdown and wear-out. IEEE Int. Electron Devices Meeting Techn. Digest, 2007, pp 501–504.

227. V. L. Lo, K. L. Pey, C. H. Tung, D. S. Ang. A critical gate voltage triggering irreversible gate dielectric degradation. Proc. of the 45th Int. Reliability Physics Sym. IEEE, Piscataway, NJ, 2007, p 576.

228. J. S. Suehle, B. Zhu, Y. Chen, J. B. Bernstein. Detailed sudy and projection of hard breakdown evolution in ultra-thin gate oxides. Microelectronics Reliability, 45: 2005, pp 419–426.

229. T. Sakura, H. Utsunomiya, Y. Kamakura, K. Taniguchi. A detailed study of soft- and pre-soft-breakdowns in small geometry MOS structures. IEEE Int. Electron Devices Meeting Techn. Digest, 1998, pp 183–186.

230. T. Kauerauf, R. Degraeve, M. B. Zahid, M. Cho, B. Kaczer, Ph. Roussel, G. Groesenken, H. Maes, S. De Gendt. Abrupt breakdown in dielectric/metal gate stacks: a potential reliability limitation? IEEE Electron Device Lett, 26: 2005, pp 773.

231. V. L. Lo, K. L. Pey, C. H. Tung, X. Li. Multiple digital breakdowns and its consequence on ultra-thin gate dielectrics reliability prediction. IEEE Int. Electron Devices Meeting Techn. Digest, 2005, pp 497–500.

232. E. Wu, J. Suñé. Post-breakdown characteristics of extrinsic failure modes for ultra-thin gate oxides. 2006 Int. Reliability Physics Sym. Proc.: 2006, pp 36–45.

233. T. Pompl, C. Engel, H. Wurzer, M. Kerber. Soft breakdown in ultra-thin oxides. Microelectronics Reliability, 41: 2001, pp 543–551.

234. J. C. Tsang, B. P. Linder. Characterization of breakdown in ultra-thin oxides by hot carrier emission. Appl. Phys. Lett., 84: 2004, pp 4641–4643.

235. J. Suñé, E. Y. Wu. Temperature-dependent transition to progressive breakdown in thin silicon dioxide based gate dielectrics. Appl. Phys. Lett., 86: 2005, p 193502(1–3).

236. E. Y. Wu, J. Suñé, B. Linder, J. Stathis, W. Lai, Critical assessment of soft breakdown stability time and the implementation of new post-breakdown methodology for ultra-thin gate oxides. Digest of the 2003 Int. Electron Device Meeting: 2003, pp 919–922.

237. M. A. Alam, B. E. Weir, P. J. Silverman. A Study of soft and hard breakdown—Part II: Principles of area, thickness and voltage scaling. IEEE Trans. Electron Devices, 49: 2002, pp 239–246.

238. J. Suñé, E. Y. Wu, S. Tous. Failure-current based oxide reliability assessment methodology. Proc. of the 46th Int. Reliability Physics Sym. IEEE, Piscataway, New Jersey, 2008, pp 230–239.

239. S. Tous, E. Y. Wu, J. Suñé. A compact model for oxide breakdown failure distribution in ultra-thin oxides showing progressive breakdown. IEEE Elec. Dev. Lett., 29: 2008, pp 949–951.

240. M. A. Alam, R. K. Smith. A phenomenological theory of correlated multiple soft-breakdown events in ultra-thin gate dielectrics. 2003 Int. Reliability Physics Sym. Proc.: 2003, pp 406–411.

241. J. Suñé, E. Y. Wu. Statistics of successive breakdown events in gate oxides. Digest of the 2002 Int. Electron Device Meeting: 2002, pp 147–150.

242. NIST/SEMATECH e-Handbook of Statistical Methods. http: //www.itl.nist.gov/ div898/handbook/

243. T. Pompl, M. Kerber. Failure distributions of successive breakdown events. IEEE Trans. on Device and Materials Reliability, 4:(2)2004, pp 263–267.

244. E. Y. Wu, J Suñé. Successive breakdown events and their relation with soft and hard breakdown modes. IEEE Electron Device Lett., 24: 2003, pp 692–694.

245. K. Okada. The gate oxide lifetime limited by B-mode stress induced leakage current and the scaling limit of silicon dioxides in the direct tunneling regime. Semiconductor Science and Technology, 15: 2000, p 478.

4

NEGATIVE BIAS TEMPERATURE INSTABILITIES IN pMOSFET DEVICES

Giuseppe LaRosa

4.1 INTRODUCTION

Negative bias temperature instability (NBTI) is a pMOSFET wearout mechanism resulting in the generation of interface states (ΔN_{it}) at the Si/SiO_2 interface and build-up of positively charged bulk defects when biased at low vertical gate oxide fields ($F_{ox} \leq 6$–$10\,MV/cm$) in inversion ($V_g < 0\,V$) and elevated temperatures. (Typical stress temperature ranges between 30 to 200°C). Because of NBTI's impact to key pMOSFET parameters, such as threshold voltage (V_{th}), linear (I_{dlin}) and saturation (I_{dsat}) drain current, and transconductance (g_m), it has become the most critical MOSFET reliability concern in current circuit design. NBTI was originally considered to belong to the same group of MOSFET phenomena as bias temperature instabilities (BTI) associated with the drift of ionic contaminants such as Na^+ Li^+, K^+, Ca^{++} and Mg^{++}. While the nature of the NBTI damage relates to an intrinsic wearout mechanism, ionic BTI phenomena can be, however, misleading in quantifying the impact the intrinsic NBTI shift or can even be the main cause of the measured device parametric shift. NBTI was observed in thermal oxides since the early phases of MOSFET development [1, 2]. Because of the use of buried channel (n^+ poly) pMOSFETs in early, single workfunction CMOS technologies, NBTI was not considered of any reliability concern. The nMOSFET channel hot carrier was typically found to be the main contributor to circuit performance degradation. The current aggressive CMOS scaling adopted

Reliability Wearout Mechanisms in Advanced CMOS Technologies. By Strong, Wu, Vollertsen, Suñé, LaRosa, Rauch, and Sullivan

to improve circuit performance (<0.13 um groundrules) has produced an enhancement of NBTI degradation. The reason is fourfold.

1. The introduction of dual workfunction technologies to replace buried channel (n^+ poly gate) with surface channel (p^+ poly) pMOSFET and to reduce short channel effects with device scaling. The difference in workfunction makes the surface channel pMOSFETs more sensitive to the NBTI damage because of the increased gate oxide electric field.

2. The implementation of nitridation processes with gate oxide scaling to improve the transistor performance. Gate oxide nitrization is used to give both a better control of the gate leakage current and reduce boron penetration in p^+ poly surface channel pMOSFETs. The introduction of nitrogen at the SiO_2/Si interface of surface channel pMOSFETs has produced an enhancement of the NBTI degradation.

3. Nonlinear reduction of the operating voltage compared to more aggressive device scaling produces both the gate oxide electric fields and the chip temperatures to increase to values large enough to accelerate the NBTI damage ($F_{ox} \approx 5$–$10\,MV/cm$, $T \approx 100°C$) at operating conditions.

4. Power supply reduction increases the effect of the same NBTI damage to device and circuit level parameters.

Despite the extensive research conducted on NBTI in recent years, there is still a significant discrepancy on the nature of the NBTI damage. In particular, no general consensus has been achieved on the relative contribution of the N_{it} generation and positively charged bulk defects to the NBTI kinetics and their impact to the NBTI time evolution, V_g or F_{ox} dependence, as well as stress temperature. Several issues contribute to this confusion in the open literature. First, the large diversity in experimental data on NBTI is certainly related to the strong NBTI sensitivity to CMOS processes. Without paying careful attention to differences in the gate oxide process such the nitrogen content, thermal treatment, as well the details of the front-end-of-line (FEOL) and back-end-of-line (BEOL) processes, large differences in the NBTI sensitivity of a given technology could be experienced. In addition, it is important to emphasize that not all the reported data is generated with the stress condition appropriate to activate only the intrinsic NBTI damage. As we will see later, NBTI should be characterized by stressing in the cold holes regime ($F_{ox} \approx 5$–$7\,MV/cm$). Aggressive V_g stress conditions can activate other forms of unwanted transistor damage, such as hot holes injection, and thus generating a lot of confusion in the NBTI picture. The stress and test NBTI procedures must also to be selected carefully so that the measured pMOSFET parameters shift reflects the intrinsic NBTI degradation and is not corrupted by any testing artifacts. The overall result is that no general physical framework has been established that allows for making lifetime projections from modeling at accelerating conditions.

In this chapter we attempt to give a consistent overview of the physics of the NBTI damage and its impact to current MOSFET design/processes and to circuit scaling, with particular focus in surface channel pMOSFETs of advanced bulk CMOS technologies.

4.2 CONSIDERATIONS ON NBTI STRESS CONFIGURATIONS

NBTI phenomena take place at high temperatures as long as the pMOSFET channel is inverted with or without carrier conduction. Figure 4.1 describes two typical configurations during which a pMOSFET does experience NBTI. In Figure 4.1a, the transistor is stressed in a uniform (capacitor-like) bias condition with the channel in inversion ($V_g < 0$) while the source, the drain, and the n-well are grounded. In this case no channel current is measured since $V_{ds} = 0$ V. The NBTI stress is symmetric under this uniform bias condition. This configuration is typical of a pMOSFET in an inverter stage with the input signal low. NBTI damage is part of pMOSFET degradation in the conductive channel condition ($V_{ds} < 0$) as well (Fig. 4.1b). In this nonuniform NBTI condition, the NBTI damage is localized under the inverted channel region. The NBTI degradation may occur simultaneously with other wearout mechanisms such as the channel hot carrier mechanism ($V_{ds} < V_{dsat}$). This configuration is typically found in analog applications [98]. Recently, NBTI damage in a nonuniform configuration has been attributed to be the main cause of pMOSFET device shift at channel hot carrier degradation configuration ($V_{gs} = V_{ds} = V_{dd}$) due to drain junction heating at stress conditions [99]. The corresponding channel heating is caused by ballistic phonons generated into the drain self-heating in saturation.

Among the nonuniform NBTI configurations that have recently been developed, a lot of interest is shown in the on-the-fly (OTF) NBTI monitoring technique adopted to characterize NBTI without the impact of recovery effects. We will discuss this methodology in Section 4.7.7.

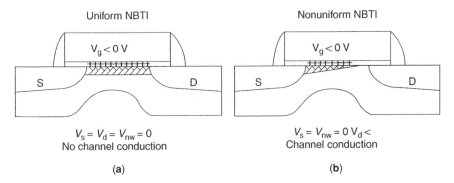

Figure 4.1. Schematic illustration of two typical NBTI configurations: (a) uniform off condition (no channel conduction); (b) Nonuniform (conducting channel) condition. In both cases the NBTI damage takes place along the SiO$_2$/Si interface above the inverted channel (dashed area).

Since the uniform configuration is used to investigate the NBTI sensitivity of pMOSFET devices during a technology reliability qualification, it will be the main topic of this chapter. In particular, we will focus on the NBTI sensitivities of surface channel PMOS of dual function technologies, typically used in advanced CMOS scaling.

A closer look at the uniform NBTI stress configuration suggests that two separate areas of the device are exposed to different NBTI bias conditions for the same applied V_g, respectively the inverted channel region and the accumulated overlap region between the gate and the source/drain junctions. Figure 4.2 shows these regions in the case of a pMOSFET device with single source and drain engineering. In the inverted channel region the oxide field is dependent on the p^+ poly to n-well workfunction difference, while the overlap region between the gate and the source/drain junction experiences a lower fringe electric field which still can produce NBTI damage. Being the single drain/source junctions p^+ type, NBTI in this area will take place with the interface at accumulated conditions. Under the inverted channel condition F_{ox} is given by

$$F_{ox} = \frac{(V_{gs} - V_{FB} - 2\Phi_B)}{T_{ox}}, \qquad (4.1)$$

where V_{FB} is the flatband voltage and Φ_B is the Fermi potential given by

$$\Phi_B = (k_B T/q) \times \ln(N_A/n_i). \qquad (4.2)$$

As seen in Figure 4.3, the gate/junction overlap region, however, has higher hole concentration [31]. In addition, the BF_2 self-aligned source and drain implant, typically used in pMOSFET source and drain engineering, may produce a process-induced defect density in the overlap region, higher than in the center of the channel, making this region more sensitive to the NBTI damage. The NBTI

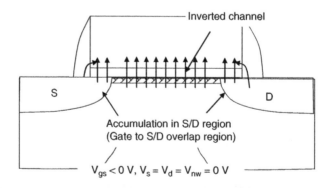

Figure 4.2. Different type of NBTI damage can be experienced in the inverted channel and in the gate–source/drain overlap region during a uniform NBTI stress.

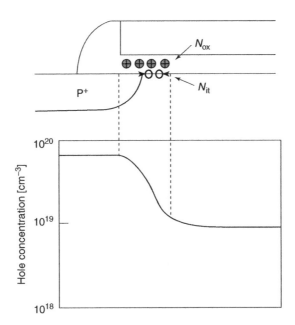

Figure 4.3. Simulated hole concentration profile at SiO_2/Si interface during an NBTI stress. Reprinted with permission from Ref. 31, © 1999 IEEE.

damage at the gate to source/drain overlap region obviously may occur simultaneously with the NBTI degradation in the inverted channel. Each of these NBTI-induced damages will, however, contribute differently to the pMOSFET parametric shift. Their relative contribution depends on the device design as well as the adopted CMOS processes.

4.3 APPROPRIATE NBTI STRESS BIAS DEPENDENCE

Before running NBTI stressing activities in a given technology qualification, it is extremely important to choose the appropriate NBTI stress bias configuration to characterize the "intrinsic" NBTI damage experienced in operating conditions. The lifetime projection of NBTI induced device parametric shifts involves accelerated stressing (typically wafer level) at sufficiently high V_g conditions to get a reasonable parametric shift within short stress duration (3000–10000 sec). (This is also done for other device wearout mechanisms.) The objective of this activity is to develop the understanding and modeling that allows the extrapolation of the impact of NBTI at application conditions. Given that the uniform NBTI stress is similar to a typical gate TDDB reliability stress configuration in inversion, it is extremely important to carefully define the appropriate stress V_g conditions representative of NBTI damage activated at the operating conditions.

At sufficiently high V_g conditions, other types of charges can be generated (for example, bulk defects) that are associated with the TDDB rather than the "intrinsic" NBTI damage. Despite the extensive research activities in the last few years, the lack of careful monitoring of the reported NBTI stress condition is probably one of the main sources of discrepancy and confusion on the physical understanding and modeling of NBTI.

To clarify this point, we will analyze the different carrier injection mechanisms that the gate oxide will experience with increasing vertical oxide electrical field under a symmetric NBTI configuration. Figure 4.4 shows the poly $Si/SiO_2/Si$ energy band diagrams for a surface channel pMOSFET in a uniform NBTI configuration and the corresponding dependence of the hole and electron currents on V_g in a thin gate oxide ($T_{ox} \leq 20$ A) experiencing gate tunneling currents at the applied V_g bias conditions. We assume, for simplicity, that the main contribution is coming from both hole- and electron-assisted direct tunneling. In this case, the gate current I_g has contributions from the electrons from the poly valence band (EVB) and holes from the Si valence band (HVB). In Figure 4.4, the condition where EVB contribution is zero is given. In this case I_g is mainly given by the HVB tunneling ($I_g = I_{h1}$). The injected holes are supplied by the source and drain junctions ($I_d = I_s = -(1/2)I_{h1} < 0$). These holes in the inverted channel are not subject to any kinetic energy-gaining process and are thermalized with the lattice ($T_{hole} = T_{lattice}$). In this situation, they are called cold holes. In Figure 4.4b, the V_g is negative enough that EVB tunneling is activated. In this situation, electrons, tunneling from the poly gate, are injected into the n-well, thus contributing to the n-well current ($I_{nw} = I_{e1}$). At the same time, cold holes in the inverted channel are injected into the gate (I_{h1}) so that $I_g = I_{e1} + I_{h1}$. As more negative voltage is applied, electrons tunneling into the gate oxide may gain enough energy in the n-well to produce impact ionization ($E_e > E_{II}$). In Figure 4.4c holes and electrons produced by the e–h generation process in the n-well contribute to the I_{h2} and I_{e2} currents respectively. A percentage of these holes will gain kinetic energy (hot holes) in the n-well as they are accelerated by the vertical electric fields due the valence band bending. In this situation $I_g = I_{e1} + I_{h1} + I_{h2}$, while $I_{nw} = I_{e1} + I_{e2}$. If the impact ionization is very effective, a given portion of the hot holes can flow toward the source and drain junctions, overcoming the flow of holes delivered to the inverted channel ($I_d = I_s = I_{h2} - I_{h1}$), and changing in this way the direction of the drain and source current respect to the cold holes regime ($I_d > 0$ V). The hot holes are not thermalized and obey the same physics as experienced under the substrate hot carrier injection regime. In this case the total oxide damage is the contribution of both cold and hot holes activated at the applied V_g conditions.

The onset of both the cold and hot hole regimes is easily observed in a CMOS bulk technology by the use of the carrier separation technique. A carrier separation technique applied to the CMOS technology, given in [27], finds the transition from cold holes ($I_d < 0$) to hot holes ($I_d > 0$) at V_g around –5 V (Fig. 4.5).

From the above observation we will assume that the intrinsic NBTI damage is activated only in the cold holes regime, which is represented by (a) and/or (b) in above figure. While the (a) is observed in thick oxides at aggressive V_g condition,

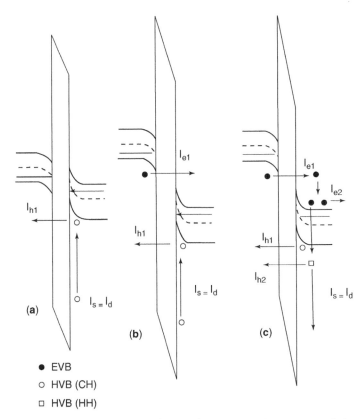

Figure 4.4. Energy band diagram of a surface channel pMOSFET (p$^+$ poly Si, n-type substrate (n-well) under channel inversion ($V_g < 0$): (a) HVB cold holes (CH) regime with no EVB contribution (b) HVB CH regime with EVB contribution (c) hot holes (HH) due to impact ionization by the EVB and HVB assisted regime. Electron and hole currents are shown.

(b) describes the NBTI configuration in pMOSFETs that is typically experienced in thin oxides (<0 A) even at use conditions. As we will see later, the direct tunneling of the electron valence band (poly side) can introduce the generation of positively charged bulk oxide defects that will contribute to the V_{th} shift. This component of the NBTI damage is only activated at stress conditions if the application bias condition in inversion is represented by (a).

It is important to carefully define the stress bias configurations to qualify a given CMOS process sensitivity to NBTI; as an example, we discuss the different nature of the interface states generation [106] depending on the degradation regime (cold versus hot holes) dominating at a given uniform NBTI stress configuration. N_{it} generation may be produced from the breaking of both $Si_3 \equiv Si-H$ and $Si_3 \equiv Si-O$ at the Si/SiO$_2$ interface. The relative contribution of the breaking of each of these bonds to the N_{it} generation in pMOSFETs in

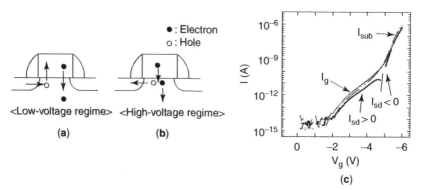

Figure 4.5. Carrier separation in Bulk CMOS pMOSFET with source and drain grounded under channel inversion conditions ($V_g < 0\,V$). Two hole regimes are observed a cold (low voltage $V_g > -5\,V$) ($I_{sd} > 0$) and a hot (high voltage $V_g < -5\,V$) hole regime with ($I_{sd} < 0$). These regimes are graphically described in (a) and (b). The different hole and electron current contributions at each regime are illustrated in (c).

inversion mainly depends on the channel holes kinetic energies which, in turn, are modulated by the gate (V_g) and n-well (V_{nw}) voltages at stress.

Figure 4.6 is a clarification of this point, shown by the authors, in the case of a uniform ($V_{ds} = 0\,V$) stress configuration. Figure 4.6a describes the time evolution of the N_{it} generation for both an NBTI ($V_{nw} = 0\,V$, V_B in the figure) and FN ($V_{nw} = 2\,V$) stress configuration with constant V_g ($-3.1\,V$) applied on pMOSFET

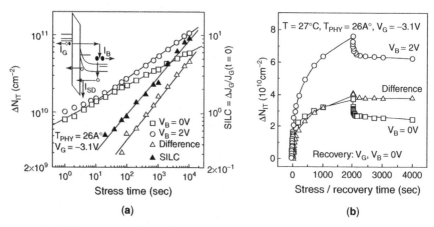

Figure 4.6. (a) Time evolution of N_{it} generation for uniform NBTI and FN stress configuration at different V_B, but fixed V_g (V_g in figure). $V_B > 0\,V$ stress induced enhanced N_{it} and SILC are also shown. SILC was measured at $V_g = -2.5\,V$; (b) time evolution of N_{it} generation and recovery during and after a uniform stress at different V_B, but fixed V_g. Reprinted with permission from Ref. 106, © 2006 IEEE.

with nonnitrided gate oxides ($T_{PHY} = 26$ A). Increasing V_{nw} at stress will generate an increased amount of hot holes (HH) in the inverted channel by impact ionization. A different power law time evolution is observed for each stress configuration. The HH regime is characterized by a power-law exponent ($n \approx 0.5$) larger than what is measured for the NBTI contribution ($n \approx 0.16$). This phenomenon is responsible for the enhanced N_{it} generation observed in respect to the NBTI configuration. The difference between the N_{it} generation under the FN and the NBTI configuration follows the same power-law time trend observed for the SILC current (I_g at $V_g = -2.5$ V). This observation suggests that the enhanced HH-induced N_{it} generation and the SILC increase are related to the same physical reason. Figure 4.6b shows the N_{it} generation during both the NBTI and HH stress followed by a recovery phase with $V_{nw} = 0$ V. It is observed that, while the N_{it} generated during the NBTI stress do recover, the additional N_{it} generated during the $V_{nw} > 0$ V phase (the difference in the figure) are permanent and do not recover during the recovery phase.

This experiment suggests that two different types of N_{it} are generated during the cold ($V_{nw} = 0$ V) and hot ($V_{nw} > 0$ V) holes driven NBTI stress configuration. From these observations it is clear that the NBTI damage generated in the cold holes regime is expected to have N_{it} contribution mainly due to breaking of the $Si_3 \equiv Si\text{--}H$ bonds, with a fraction of them being recoverable after the NBTI stress is removed. When a hot hole regime is established (for example, during aggressive NBTI-accelerated stresses activating a FN condition), the N_{it} generation has an extra contribution controlled by breaking of the $O_3 \equiv Si\text{--}O$ bonds as indicated by the power-law time equivalency with SILC. These generated N_{it} are not expected to be recoverable. With these conclusions, we assume that N_{it} contribution of the NBTI damage to be correctly predicted only under the NBTI cold holes regime. In this case N_{it} generation is assumed to be an assisted reaction–diffusion process related to the breaking of the $Si_3 \equiv Si\text{--}H$ bond with possible recovery by bond repassivation. In Section 4.8, we will describe the reaction–diffusion kinetics and the corresponding modeling of the NBTI-induced N_{it} damage associated with this physical process. We are going to give further details on the role cold holes have in the NBTI damage topic in Section 4.7.2.

4.4 NATURE OF THE NBTI DAMAGE

Probably the most direct experimental evidence of the different nature of the NBTI damage is given in [6]. In this experiment, a symmetric NBTI stress (phase A), $V_g < 0$ is followed by a relaxation phase (B) in accumulation ($V_g > 0$). The shift of both the threshold voltage, V_{th}, and interface states surface concentration, N_{it} (measured by charge pumping), were monitored during both phases. In Figure 4.7 both the relative shift of the interface states generation and V_{th} increase are normalized to 1 at the time the NBTI stress is stopped (2500 sec). During the accumulation phase, the relative N_{it} concentration does not seem to change, while the relative V_{th} decreases very rapidly to a saturation value of around 0.4 after 500

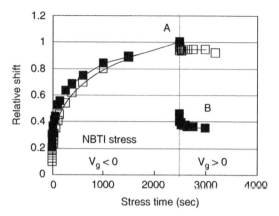

Figure 4.7. Relative shifts of interface states density (open squares) and the threshold voltage (filled squares). Reprinted with permission from Ref. 6, © 2003 IEEE.

seconds. The different relaxation behavior of the N_{it} and V_{th} suggests the dual nature of the NBTI damage. In fact, it is well known that by applying a positive bias after a negative bias stress, a possible neutralization of both trapped holes and/or positive charge slow states can take place [100]. Since N_{it} is observed to be constant during the B phase, the NBTI-induced V_{th} shift can not be completely accounted for soley by the positive charge due to the donor-type portion of the N_{it}, which is electrically activated in a pMOSFET during the V_{th} measurement. Notice that in this accumulation phase, the V_{th} recovers to a value higher that the initial (before stress) V_{th}. From this experiment we observe that the NBTI physical damage results in positive-charge build up that has two possible contributions: interface states generation (ΔN_{it}) and activation or generation of positively charged defects. In turn, these have two components: a fixed (nonrelaxable) $\left(\Delta N_f^+ \right)$ as well as recoverable, positively charged defects $\left(\Delta N_R^+ \right)$. The total charge generated or measured at given time t is

$$\Delta Q^+(t) = q\left(\Delta N_{it}^d + \Delta N_f^+ + \Delta N_R^+ \right), \tag{4.3}$$

where ΔN_{it}^d is the donor-type portion of the interface state generated during the NBTI stress and electrically activated at the time t in the pMOSFET under investigation. We will see later that the nature of the recoverable part ΔN_R^+ has been associated with both neutralization of activated hole traps and/or repassivation of the generated interface states. Both physical processes may be active during the stress phase in cold holes regime. When we consider that only donor-type interface states contribute to the flatband shift in a pMOSFET device and call $\Delta D_{it}^d(E)$ the NBTI-induced increase in donor-type interface-state energy-distribution density between the Fermi level at test E_F and $E_G/2$ (assume the donor-type N_{it} populate only the lower part of the Si bandgap), we have the following equation:

$$\Delta N_{it}^d(E_F, t) = \int_{E_F}^{E_G/2} \Delta D_{Id}^d(E, t) dE \qquad (4.4)$$

When we call $\Delta\rho_p(x, t)$ the increase in positively charged defects density at point x inside the gate oxide ($x = 0$ corresponds to the poly/SiO$_2$ interface) and at a given time during the stress, the total charge density generated or activated during the NBTI stress by positively charged defects is given by

$$\Delta N_f^+ + \Delta N_R^+ = \frac{1}{T_{ox}} \times \int_0^{T_{ox}} x\Delta\rho_p(x, t) dx \qquad (4.5)$$

during an NBTI stress ($V_g < 0$ V) followed by an accumulation ($V_g > 0$ V) phase.

4.5 IMPACT OF THE NBTI DAMAGE TO KEY pMOSFET TRANSISTOR PARAMETERS

The NBTI-induced physical damage impacts the key device parameters in several ways. Figure 4.8 illustrates an example of the typical evolution of the I_d–V_g curves in linear mode ($V_{ds} \ll V_{dsat}$) measured for a surface channel pMOSFET during a uniform NBTI stress. The linear drain current I_{dlin} is reduced. This reduction is mainly due to the degradation of both the flatband voltage (V_{FB}) and hole effective channel mobility (μ_{eff}) during the stress. The change in flatband is associated with an increase in total positive charge ($Q^+(t)$) in the gate oxide and corresponding decrease of the extrapolated threshold voltage ($V_{th,ext}$). $V_{th,ext}$ is

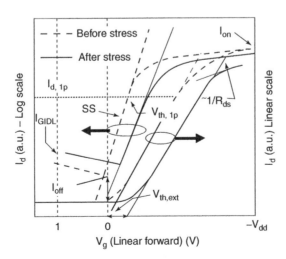

Figure 4.8. Impact of NBTI damage to the I_d–V_g characteristics of a pMOSFET.

experimentally extracted from the intercept along the V_g axis of the $I_{dlin} - V_g$ linear relation with the slope estimated at the point of maximum transconductance defined as $g_m = \partial I_{dlin}/\partial V_g\big|_{V_{ds}}$. The change in effective channel mobility is controlled by the N_{it} generation during the stress. A drain current that is typically reported is the one measured at $V_g = -V_{dd}$ conditions (I_{on}). It has recently been observed that a decrease in I_{on} can be attributed to the formation of NBTI damage along the source and drain gate overlap region, resulting in an increase in the source and drain series resistance (R_{ds}). The NBTI damage can affect the subthreshold conduction ($V_g \ll V_{th,ext}$) in two ways: the increase of the subthreshold swing (SS) and the creation of a parallel shift of the subthreshold slope related to $\Delta V_{th,ext}$. Both shifts impact the single point V_{th} ($V_{th,1p}$), defined as the change in V_g value corresponding to a constant drain current value ($I_{d,1p}$) measured in the subthreshold region during the stress. Because of the $V_{th,ext}$ decrease, the channel off current (I_{off}) is initially decreased; however, if the gate-induced drain leakage (GIDL) increases during the NBTI stress, the drain junction leakage contribution to the drain current in off conditions ($V_g = 0\,\text{V}$) may become dominant. The NBTI damage also results in a decrease of the saturated drain current (I_{dsat}) with both the forward and reverse values degrading identically as expected from the symmetry ($V_s = V_d = V_{nw} = 0\,\text{V}$) of the stress. We will give more details in the next two paragraphs on the relations between the nature of the NBTI damage and both the device physical and electrical parameters.

The degradation of these device parameters are the effect of 1) the build-up of positive charge at the Si/SO$_2$ interface or in the bulk of the gate oxide and 2) the increase of interface states density.

4.5.1   Impact of the NBTI Damage to pMOSFET Physical Parameters

As discussed in Section 4.4, the NBTI physical damage has two main charge contributions: interface states generated charge (N_{it}) and activation or generation of positively charged defects. Two key physical device parameters are directly impacted by the NBTI damage: the flatband voltage (V_{FB}) and the effective channel mobility (μ_{eff}).

The flatband voltage V_{FB} is

$$V_{FB} = \phi_{ms} - \frac{Q^+}{C_{ox}}, \qquad (4.6)$$

where C_{ox} is the gate oxide capacitance (ε_{ox}/T_{ox}), Q^+ is the equivalent oxide charge per unit area at the oxide/silicon interface, and ϕ_{ms} is the workfunction difference which, in the case of a surface channel pMOSFET device (p$^+$ poly gate on n-type silicon with doping concentration N_d), can be approximated through

$$\phi_{ms} = \frac{E_g}{2q} + \frac{kT}{q} \ln\left(\frac{N_d}{n_i}\right). \qquad (4.7)$$

From Equation 4.6, the change in flatband voltage due to the NBTI damage is given by

$$\Delta V_{FB} = -\frac{q}{C_{ox}} \times \left(\Delta N_{it}^{D} + \Delta N_{f}^{+} + \Delta N_{R}^{+} \right). \tag{4.8}$$

Following Liang et al. [33], the effective channel mobility μ_{eff} on V_g can be derived from the following empirical equation:

$$\mu_{eff} = \mu_0 / \{ 1 + (F_{eff}/F_0) \}^{\nu}, \tag{4.9}$$

where for holes of surface channel pMOSFETs at 30°C, $F_0 = 7.0 \times 10^5$ (V/cm), $\nu = 1$, and $\mu_0 \approx 160 \, \text{cm}^2/\text{V sec}$. μ_0 is the low field surface mobility that has contributions from scattering events between the channel carriers and phonons or Coulombic scattering centers, as well as the Si/SiO_2 interface roughness. F_{eff} is the effective transversal electric field defined as the electric field distribution averaged over the carrier distribution in the depletion (Q_b) and inversion (Q_i) layer. In inversion F_{eff} is related to the bulk depletion charge Q_b and the inversion charge Q_i by

$$F_{eff} = [Q_b + \varsigma Q_i]/\varepsilon_0 \varepsilon_{si}, \tag{4.10}$$

where $\varsigma = 0.25 - 0.3$ for pMOSFETs. Taking into account the above equations, the V_g dependence of μ_{eff} in circuit simulation has been approximated (SPICE MOSFET level 3 and 4) as

$$\mu_{eff} = \mu_0 / \{ 1 + \theta (V_g - V_{th}) \}, \tag{4.11}$$

where θ (V^{-1}) is the mobility degradation coefficient. We will use this formulation in the rest of this chapter. According to Equation 4.11, the induced degradation of the effective channel mobility ($\Delta \mu_{eff}/\mu_{eff}$) is directly related to the decrease of the surface channel mobility ($\Delta \mu_0/\mu_0$) and the change in V_{th}. The decrease in μ_0 during the NBTI damage is related to the increase in ΔN_{it} which acts as Coulombic scattering centers. It is well known that

$$\frac{\Delta \mu_0}{\mu_0} \approx \frac{1}{1 + \Delta N_{it}}. \tag{4.12}$$

Figure 4.9 reports the experimentally measured [59] decrease of effective channel mobility degradation as a function of F_{eff} during an NBTI stress. The channel mobility degradation is larger at low F_{eff} but decreases with increasing F_{eff}. The large degradation in channel mobility is due the increased Coulomb scattering centers by the interface states density increase, and the decrease at higher fields is related to increase charge screening by the inversion charge.

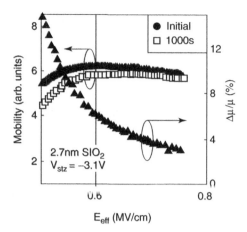

Figure 4.9. μ_{eff} versus F_{eff}, showing that mobility degradation is large at low F_{eff}, but decreases with increasing F_{eff}. The large degradation in mobility at low fields is due to Coulomb scattering by interface traps, while the observed decrease at higher fields is induced by inversion charge screening. Reprinted with permission from Ref. 59, © 2003 IEEE.

4.5.2 Relation Between Key Physical and Electrical pMOSFET Parameters

As discussed in Section 4.5, the flatband voltage (V_{FB}) and the effective channel mobility (μ_{eff}) are the two key device physical parameters impacted by the NBTI damage. In this section, we will describe their relation to the key device electrical parameters. In order to quantify this relation, we need to spend few words on drain current dependence on these physical parameters.

For simplicity, we start by considering the drain current in linear regime (I_{dlin}), assuming the gradual channel approximation. In the case of channel conduction ($V_{gs} > V_{th,ext}$) it is possible to write [30] I_{dlin} as follows:

$$I_{dlin} = \frac{C_{ox} + \left(V_{gs} + V_{th,ext}\right) \times W \times \mu_{eff} \times V_{ds}}{L + C_{ox}\left(V_{gs} - V_{th,ext}\right) \times W \times \mu_{eff} \times R_{sd}}. \qquad (4.13)$$

In this relation, R_{sd} is the total source and drain series resistance. The extrapolated V_{th} is directly related to the flatband voltage by

$$V_{th,ext} = V_{FB} + 2\Phi_B + \frac{\sqrt{4\varepsilon_s q N \Phi_B}}{C_{ox}}, \qquad (4.14)$$

where $\Phi_B = (k_{BT}/q) \times \ln(N_A/n_i 0)$. From Equation 4.14, it is easy to verify that the shift in $V_{th,extr}$ ($\Delta V_{th,extr}$) relates to the change in flatband voltage through this equation:

$$\Delta V_{\text{th,extr}} = \Delta V_{\text{FB}}, \tag{4.15}$$

with ΔV_{FB} given by Equation 4.8.

Calling $R_{\text{L}} \equiv L/(C_{\text{ox}}(V_{\text{gs}} - V_{\text{th,ext}}) \times W \times \mu_{\text{eff}}$ the channel resistance, we consider two approximations depending on the relation between R_{L} and R_{sd}.

1. $R_{\text{sd}} \ll R_{\text{L}}$. In this case Equation 4.13 reduces to

$$I_{\text{dlin}} = C_{\text{ox}}\left(V_{\text{gs}} - V_{\text{th,ext}}\right) \times W \times \mu_{\text{eff}} \times V_{\text{ds}}. \tag{4.16}$$

Under these conditions, it is easy to find out that the NBTI-induced I_{dlin} absolute fractional change at a given V_{gs} is related to the absolute values of the $V_{\text{th,ext}}$ and μ_{eff} shift by the equation

$$\frac{\Delta I_{\text{dlin}}}{I_{\text{dlin0}}} \approx \frac{\Delta V_{\text{th,ext}}}{V_{\text{gs}} - V_{\text{th,ext}}0} + \frac{\Delta \mu_{\text{eff}}}{\mu_{\text{eff0}}}, \tag{4.17}$$

with $V_{\text{th,ext0}}$, μ_{eff0}, and I_{dlin0} being the initial (before stress) values of $V_{\text{th,ext}}$ and μ_{eff}, respectively.

Notice that at the $V_{\text{gs}} = V_{\text{dd}}$ bias conditions, we can approximate the I_{on} shift as

$$\frac{\Delta I_{\text{on}}}{I_{\text{on0}}} \approx \frac{-\Delta V_{\text{th}}}{\left(V_{\text{dd}} - V_{\text{th,ext0}}\right)}. \tag{4.18}$$

In the above equation, it is assumed that the NBTI-induced mobility degradation does not contribute to the I_{dlin} shift at $V_{\text{gs}} = V_{\text{dd}}$. This approximation is justified experimentally by the observation that the mobility degradation decreases at high V_{g}. Notice that the relation between the I_{dlin} and $V_{\text{th,ext}}$ shift is controlled by the V_{gs} initial overdrive, defined as the difference $(V_{\text{g}} - V_{\text{th,ext0}})$. In particular, for the same $\Delta V_{\text{th,ext}}$ shift, a larger I_{dlin} degradation is experienced for a lower overdrive. Figure 4.10a shows the relation between I_{on} and $V_{\text{th,extr}}$ for two different gate oxides (2.7 and 7 nm) [59]. The values predicted by Equation 4.18 are also reported (dashed lines). In this case, the described thin oxide device has larger I_{dlin} shift for the same $\Delta V_{\text{th,ext}}$ due to the smaller overdrive. This observation is particularly important in MOSFET scaling, where, with power supply reduction, it is expected that the V_{g} overdrive will decrease. Under the $R_{\text{sd}} \ll R_{\text{L}}$ condition, it is easy to quantify a linear dependence between the saturation current (I_{dsat}) and I_{dlin} shift, since

$$\frac{\Delta I_{\text{dsat}}}{I_{\text{dsat0}}} \approx \theta \times \frac{\Delta V_{\text{th,ext}}}{\left(V_{\text{gs}} - V_{\text{th,ext0}}\right)}, \tag{4.19}$$

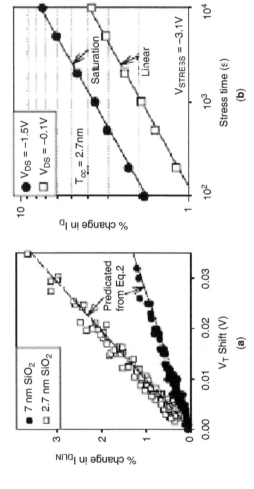

Figure 4.10. (a) I_{dlin} versus $V_{th,extr}$ shift, with the dotted lines showing the values predicted by Equation 4.18. Higher I_{dlin} degradation is measured for the same $V_{th,extr}$ shift for thinner gate dielectric ($T_{ox} = 2.7$ nm) due to the smaller V_g overdrive ($V_g - V_{th,extr}$). (b) I_{dsat} and I_{dlin} evolution as function of time during an NBTI stress. Reprinted with permission from Ref. 59, © 2003 IEEE.

where $1 \le \theta \le 2$. The above equation suggests that both I_{dlin} and I_{dsat} follow the same time behavior during the NBTI stress, but I_{dsat} has almost twice the I_{dlin} shift at the same stress time as indicated by Figure 4.10(b).

A direct measurement of the shift of the effective channel mobility degradation is the fractional change the maximum transconductance (g_m), defined as $g_m = (\partial I_{\text{dlin}} / \partial V_{\text{gs}})_{V_{\text{ds}}}$. Under the above conditions it is easy to prove

$$\left| \frac{\Delta g_m}{g_{m0}} \right| \approx \left| \frac{\Delta \mu_{\text{eff}}}{\mu_{\text{eff0}}} \right|. \tag{4.20}$$

In the example described in Figure 4.11 from [48], both the V_{th} and g_m experience the same time evolution under the same symmetric NBTI stress, indicating that in this case both parameters are associated with the shift of the same physical parameters (N_{it} increase).

The extraction of $V_{\text{th,extr}}$ is subject to an inherent statistical variation due to the estimation of the derivative ($\Delta I_{\text{dlin}} / \Delta V_{\text{gs}}$). A more direct and simpler way of characterizing the V_{th} shift is based on the constant drain current (one single point) measurement V_{th} ($V_{\text{th,1p}}$). The procedure to extract $V_{\text{th,1p}}$ is described below.

Select a drain current value ($I_{\text{d,1p}}$) in the subthreshold region of the $I_d - V_{\text{gs}}$ curve. Calling V_{g0} and V_{g1} the value of the gate voltage (V_g) associated to $I_{\text{d,1p}}$ respectively, before and after the NBTI stress, we define $\Delta V_{\text{th,1p}}$ as

$$\Delta V_{\text{th,1p}} = V_{g1} - V_{g0}. \tag{4.21}$$

Typically, $I_{\text{d,1p}}$ is a current level in the subthreshold region proportional to the $W_{\text{des}} / L_{\text{des}}$ ratio ($I_{\text{d,1p}} \approx 50\text{--}100\,\text{nA} \times W_{\text{des}} / L_{\text{des}}$). It is straightforward to define a relation between $\Delta V_{\text{th,1p}}$ and $\Delta V_{\text{th,ext}}$. It is well known that in the subthreshold

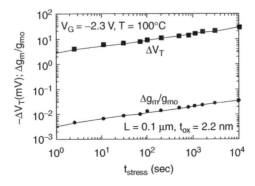

Figure 4.11. V_{th} and g_m degradation induced by NBTI during stress. Reprinted with permission from Ref. 48, © 2000 IEEE.

region [30], the following equation applies:

$$I_d \approx C_{ox} \frac{W}{L} \times \left(\frac{kT}{q}\right)^2 e^{q(V_g - V_{th,ext})/m \times kT} \left(1 - e^{-qV_{ds}/k_B T}\right). \tag{4.22}$$

In the above equation, m is the body effect coefficient given by:

$$m = 1 + \frac{C_{dm}}{C_{ox}}, \tag{4.23}$$

with C_{dm} being the bulk depletion capacitance per unit area at $\psi_s = 2\psi_B$. Since C_{dm} is contributed by the depletion (C_{Depl}) and interface states capacitance (C_{it}), we also have

$$\frac{1}{C_{dm}} = \frac{1}{C_{Depl}} = \frac{1}{C_{it}}. \tag{4.24}$$

In the case of a pMOSFET interface states capacitance, C_{it} is defined by

$$C_{it}(\psi_s) \equiv q \times \frac{dN_{it}^d(\psi_s)}{d\psi_s}, \tag{4.25}$$

where $N_{it}^d(\psi_s)$ is the donor-type N_{it} at the Si/SiO$_2$ interface. The increase of interface states during an NBTI stress will increase $C_{it}(\psi_s)$ with consequently decrease in C_{dm}.

Typically, the subthreshold I_d dependence on V_g is characterized by the subthreshold swing (SS) defined as

$$SS = \left(\frac{d(\log_{10}(I_{dlin}))}{dV_{gs}}\right). \tag{4.26}$$

Its relation to m is given by

$$SS = 2.3 \times \frac{mkT}{q} = 2.3 \times \frac{kT}{q}\left(1 + \frac{C_{dn}}{C_{ox}}\right). \tag{4.27}$$

Typically, its value is of the order of 70–100 mV/decade. In the case of a large N_{it} density at the Si/SiO$_2$ interface, it is expected that SS is controlled by the C_{Depl}/C_{ox} ratio. For a given T_{ox} and well-doping concentration, this ratio is constant and SS is mainly a function of temperature.

From Equation 4.22, it is easy to show that for a given V_{ds} and temperature T

$$\Delta V_{th,1p} - \Delta V_{th,ext} = \left(\frac{SS_1}{SS_0} - 1\right) \times \left(V_{th,1p0} - V_{th,ext0}\right), \tag{4.28}$$

where SS$_0$ and SS$_1$ are the subthreshold slope values before and after the NBTI stress, respectively.

Since SS is mainly temperature dependent, we can assume that during an uniform NBTI stress at a given stress temperature $SS_1 \approx SS_0$ so that $\Delta V_{th,1p} \approx \Delta V_{th,ext}$.

The second approximation we want to discuss is given by $R_{sd} \gg R_L$. In this case, Equation 4.13 can be approximated to

$$I_{dlin} = V_{ds}/R_{ds}. \tag{4.29}$$

Under these conditions the NBTI induced I_{dlin} change is given by

$$\frac{\Delta I_{dlin}}{Id_{lin0}} = \frac{\Delta(1/R_{ds})}{1/R_{ds0}}. \tag{4.30}$$

The NBTI-induced build-up of positive charge in the gate/source–drain overlap region will produce a depleted interface, causing an increase in the parasitic source and drain series resistance (R_{sd}) with further degradation in pMOSFET performance. This phenomenon is similar to what was described for the nMOSFET in channel hot carrier. In the case of symmetric NBTI damage, both the source and the drain overlap region will experience this resistance increases. An increase in R_{sd} degradation as a function of channel length has been demonstrated in a 0.18 μm CMOS technology [34]. Figure 4.12 suggests that the percent R_{sd} (%R_{sd}) increase, as function of time, is not dependent on the channel length. As expected, for a given process, the number of defects in the overlap region between the gate and the source and drain junction is the same, irrespective of the channel length. The effect of the R_{sd} increase on the drive current depends, however, on the channel length and on the source–drain engineering.

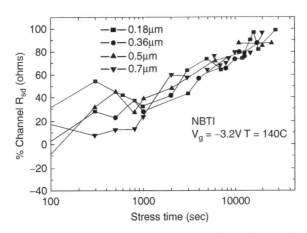

Figure 4.12. Increase in source and drain parasitic series resistance (R_{sd}) as function of stress time during a uniform NBTI stress of pMOSFET of a 0.18 um CMOS technology (stress: $V_g = -3.2$ V, $T = 140$°C). The change in series resistance is not dependent on the device channel length. Reprinted with permission from Ref. 34.

4.6 PHYSICAL MECHANISMS CONTRIBUTING TO THE NBTI DAMAGE

As discussed in Section 4.4, positive charge build-up in the gate oxide is induced during the NBTI stress ($V_g < 0$ V) and, at the same time, recovery of this charge is observed after stress during an off or accumulation phase ($V_g \geq 0$ V). The positive charge build-up has two main contributions: interface states generation and possible positively charged defects generation and/or activation in the bulk of the gate oxide (see Equation 4.3). In this chapter, we focus on the transistor damage in NBTI uniform stress configuration that is generated by cold holes in the inverted channel (case a and b in Section 4.3). Based on the large amount of experimental data collected on NBTI over the years, several basic competing physical mechanisms have been identified to contribute the NBTI damage in this regime and are graphically illustrated in Figure 4.13(a).

1. The N_{it} generation (Si*) mainly results from the breaking of Si–H bonds at the Si/SiO$_2$ interface through a reaction–diffusion process consisting of an electrochemical reaction (2) coupled with the diffusion of a hydrogenated stable (H*) species (3) in the gate oxide and eventually into to the poly-gate. The N_{it} generation is assisted by the presence of thermal holes

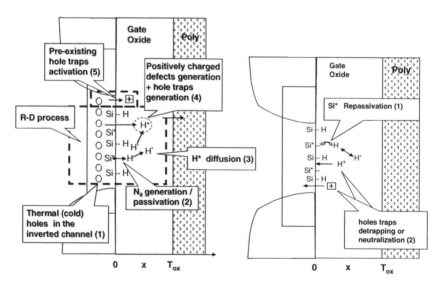

Figure 4.13. (a) Schematic representation of the key physical mechanisms contributing to NBTI-induced positive charge build-up (N_{it} assisted reaction–diffusion process and generation of positively charge bulk defect and/or activation of preexisting hole traps) during an NBTI stress ($V_g < 0$ V). It is assumed that cold holes are participating to the NBTI kinetics. (b) Physical mechanisms associated with NBTI recovery (N_{it} repassivation and hole traps neutralization).

(cold holes) in the inverted channel (1). The interface states rate is initially controlled by a balance between Si–H breaking and possible S* repassivation and sequentially by the H* diffusion limited process.

2. The generation and/or the activation of positively charged bulk defects is generally associated to three different physical mechanisms activated during the NBTI stress:

   I. Positive, fixed-charge formation as a by-product of the reaction–diffusion process responsible for the N_{it} generation (2).

   II. Bulk trap formation (4) by trapping of the diffusing species H* in process induced trapping centers (precursors) in agreement with the hydrogen release picture.

   III. Activation of preexisting hole traps (5) by hole assisted tunneling at stress conditions.

While steps I and II are sequential to the N_{it} generation through the reaction–diffusion process, the possible activation of preexisting hole traps (III) is expected to occur parallel to the N_{it} generation. As we will see later, this observation has profound implications on the characterization and modeling of the NBTI damage.

The observed NBTI recovery at $V_g \geq 0$ V conditions after the NBTI stress has being attributed to two main physical processes, illustrated in Figure 4.13 (b). The process of N_{it} reduction by Si* repassivation is controlled by the availability of H at the Si/SiO$_2$ interface (1) when the stress is stopped, which is, in turns, determined by the thermal equilibrium between the reaction by-product H and the H* species stably diffusing in the gate oxide. The role of N_{it} repassivation as part of the NBTI recovery is still subject of controversy. A possible explanation is given by the different nature of N_{it} generation activated at the NBTI stress conditions (see Section 4.3). The process of hole traps detrapping or neutralization (2) is also shown.

In the following sections we will give some more details on each of these degradation paths that may contribute to the NBTI damage. We start with describing some key features of three components of the charge build-up.

4.6.1 Interface Traps Generation

Silicon in its crystalline state has a diamond lattice structure; each Si atom is bonded in a tetrahedron configuration to four Si atoms by a covalent bond where two electrons (valence electron) are shared by a Si–Si pair. The possible bonding configurations along the Si/SiO$_2$ interface (after Si oxidation) are illustrated graphically in Figure 4.14 [115]. While most of the Si atoms are bonded to the oxygen, few Si atoms bond to hydrogen. Interface states (Si$_3 \equiv$ Si*) are Si atoms with three complete bonds to other Si atoms ($\equiv$) and a fourth unpaired valence electron in the dangling orbital (dangling bond). For clarification, the following different definitions for interface states will be used in this chapter: D_{it} (cm^{-2}eV^{-1})

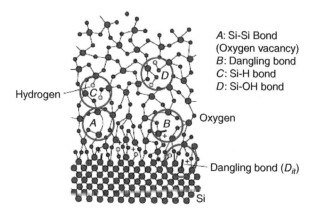

Figure 4.14. Graphical representation of the possible types of bonds created at the SiO_2/Si interface. Other possible bonds (Si–Si, Si–OH) forming in the gate oxide are also illustrated.

(interface states energy distribution density) and N_{it} (cm^{-2}) (interface states surface density).

Interface states are also called P_b centers. Along the [100] orientation two defects, named P_{b0} and P_{b1}, have been measured by electron spin resonance and shown to be related to strain relaxation at the Si/SiO_2 interface. These two interface defects have a different dangling bond axis of symmetry and differ in the defects electronic density of states. Recent work has established that both P_{b0} and P_{b1} centers are chemically identical [13]. However, while the P_{b0} are believed electrically active, a general agreement has not been achieved on the electrical nature of the P_{b1} centers. While Stesmans et al. [14] found P_{b1} centers to be electrically inactive, recent reports suggest they are electrically active at temperatures higher than room temperature [15] and have two narrow peaks close to the Si midgap [16]. Figure 4.15 gives a schematic representation of the P_{b1} and P_{b0} energy density distribution in the Si bandgap [114]. While the P_{b0} centers have two broad peaks in the lower and upper side of the Si bandgap, the P_{b1} centers have two very narrow peaks below the lower part of the bandgap.

Experimental evidence that N_{it} generation is part of the NBTI damage is given by the spin-dependent recombination (SDR) spectra [55, 56], obtained after a uniform NBTI stress (140°C, $V_g = -5.7$ V). The generation of the P_{b0} and P_{b1} interface defects (Fig. 4.16) during the stress is shown by the peaks at g = 2.0060±0.0003 and g = 2.0006±0.0003, respectively. A possible third peak (g = 2.0007±0.0003) was also measured. This peak is attributed to E′ centers, which, as we will see later, are associated with hole trapping centers. Based on this experimental evidence, we will assume (in this chapter) that the generation of P_{b0} centers has its main contribution to N_{it} generation during the NBTI stress.

The electrical nature of the P_{b0} centers is amphoteric, meaning that the dangling bonds can be occupied by zero, one, or two electrons. In addition, they

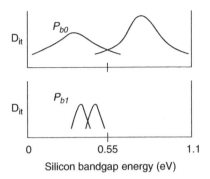

Figure 4.15. Schematic representation of the density of states of the Pb_0 and Pb_1 centers as function of the bandgap energy. The energy density of the Pb1 centers distribution is much narrower than that of the Pb0 centers and skewed towards the lower part of the SI bandgap.

are found to be distributed in the Si bandgap, having an acceptor- and donor-like nature, respectively, in the upper and lower part of the bandgap. This means that depending on the Fermi level (E_F) at the Si/SiO_2 interface, the interface states can be respectively positively charged (donor-type), neutral, or negatively charged

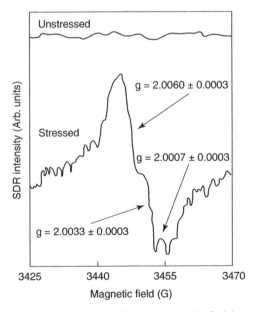

Figure 4.16. SDR traces of pMOSFET with the magnetic field vector perpendicular to the ($1\overline{0}0$) surface both before and after the application of NBTI (200°C for 20,000 sec with $V_g = -5.7$ V on the gate contact). Reprinted with permission from Ref. 55, © 2006 IEEE.

(acceptor-type). In particular, calling E_{it} a given interface state energy level in the bandgap, its electrical activation is as follows: positive $(E_{it} > E_F)$ or neutral $(E_{it} < E_F)$ for donor-type and negative $(E_{it} < E_F)$ or neutral $(E_{it} > E_F)$ for acceptor-type interface states.

Even if both acceptor and donor-type interface states are generated during the NBTI stress, only the donor-type interface states above the Fermi level contribute to the positive charge built up measured in a pMOSFET in inversion (Fig. 4.17).

A charge pumping (CP) measurement procedure is typically used to electrically monitor the N_{it} generation during stressing. Considering a two-level CP technique [35], the total interface states surface concentration $N_{it}(t)$, generated at given stress time t, is equal to $\langle D_{it} \rangle \times \Phi_{Swing}$ with $\langle D_{it} \rangle$ being the mean interface states energy density averaged over the surface potential swing Ψ_{Swing}. This parameter is estimated by pulsing the V_g between established high and low bias so that $\Phi_{Swimg} = (E_{High} - E_{Low})/q$ at the Si/SiO$_2$ interface. The N_{it} shift during the stress is linearly proportional to the total increase in maximum CP current (I_{CPmax}) according to the equation:

$$\Delta N_{it} \cong \Delta I_{CP}/q^2 \, A_{eff} F_p \Phi_{Swing}. \tag{4.31}$$

The CP current (I_{CP}) is measured by monitoring the bulk current (or source and drain current). A_{eff} is the device effective area $(A_{eff} = W_{eff} \times L_{eff})$ and F_p the pulse frequency. Figure 4.18 gives an example of NBTI induced increase in I_{CP} while keeping a constant pulse Φ_{Swing} and increasing the low level pulse (E_{Low}) from accumulation to inversion [17]. Notice that, in this case, the I_{CP} seems to increase with stress time with no evidence of a flatband shift, characteristic of fixed charge formation during the stress.

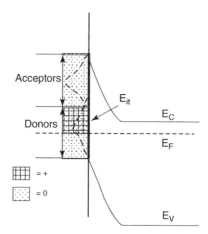

Figure 4.17. The P$_{b0}$ centers are found acceptor-type in the upper half and donor-type in the lower half region of the Si bandgap. Only the donor-type interface states (E_{it}) above the Fermi level (E_F) contribute to the positive charge build-up in a pMOSFET device in inversion.

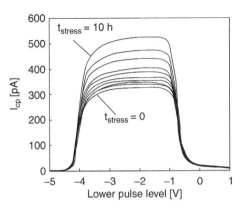

Figure 4.18. Increase in charge pumping current during an NBTI stress. Reprinted with permission from Ref. 17, © 1999 IEEE.

Following Equation 4.4, and assuming a mean increase $\left(\overline{\Delta D_{it}^d}\right)$ of donor-type energy density distribution D_{it} the NBTI stress, the fraction of the interface states between the midgap and the Fermi level (E_F) leading to a positive charge (ΔQ_{it}^+) contribution to the flatband shift in pMOSFET device is given by

$$\Delta Q_{it}^+ = q\Delta N_{it}^d = q\overline{\Delta D_{it}^d}\left(\frac{E_g}{2} - E_F\right). \qquad (4.32)$$

By measuring the increase of the NBTI induced interface state generation near flatband (by LV–SILC measurements) and midgap (by DCIV measurements) in a pMOSFET with RTO and RTO plus plasma nitridation, it was found [18] that the upper bandgap N_{it} distribution after NBTI is different between thermal and nitrided oxides. This observation suggests that either the electronic structure of the SiH dangling bond is changed or other defects appear in the presence of nitrogen in the gate oxide. Interface states act as generation/recombination centers and contribute to junction leakage currents (GIDL enhancement), 1/f noise, channel mobility reduction, flatband shift, drain current, and transconductance degradation.

4.6.2 Positive Charge Defects Generation/Activation

While it is generally agreed that the NBTI-induced interface states generation occurs by Si–H bond breaking, the origin of the observed build-up of positively charged bulk oxide defects is not well understood. A simple argument, however, rules out the possibility that NBTI is the effect of mobile charge shift. If, in fact, positive mobile ions (Na, K, Li, etc.) reach the gate oxide during the CMOS processing, they will be accumulating towards the poly/SiO$_2$ interface during the high temperature NBTI stressing ($V_g < 0$ at stress) so that, calling $V_{th,i}$ and $V_{th,f}$ the measured V_{th}, respectively, before and after the stress, $V_{th,f}/V_{th,i} < 1$ (Fig. 4.19a). The opposite behavior is observed during an NBTI stress

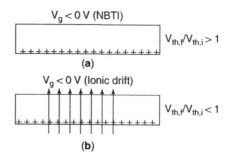

Figure 4.19. Ionic drift has an opposite behavior than NBTI.

($V_{th,f}/V_{th,i} > 1$) (Fig. 4.19b) since, in this case, positive charge is generated closer to the Si/SiO$_2$ interface.

Generally, four different mechanisms have been observed to contribute to oxide defects formation or activation during NBTI stressing in a cold holes regime.

1. Positive fixed charge as a by-product of Si–H bond breaking reaction.
2. Bulk traps formation by trapping of the H* diffusing species by-product of the Si–H bond breaking reaction.
3. Deep level hole traps generation by an electron tunneling assisted process.
4. Activation of preexisting hole traps due to the cold holes tunneling during NBTI stressing.

The first two mechanisms are considered by-products of the electrochemical reactions responsible for interface states generation. While mechanism 2 and 3 are related to the existence of neutral gate oxide defects that are made positively charged by H* trapping or loss of an electron from the interaction of a tunneling electron from the poly valence band, mechanism 4 is due to hole trapping by cold holes tunneling. The oxide defects participating to mechanisms 2, 3, and 4 are preexisting, before the NBTI stress (precursors), and can be produced by the FEOL and BEOL processes applied after the gate oxide formation step. As we will see later, the nature of the formation of the oxide defects strongly depends on the gate oxide (SiO$_2$ versus SiN gate oxide), and the following CMOS processes and their characterization strongly depends on the testing methodology used to estimate the NBTI damage. In the rest of this chapter, we will briefly discuss the experimental evidence suggesting the activation of these four mechanisms of charge defects formation during NBTI aging.

4.6.2.1 *Positive Fixed Charge Formation as a By-Product of Si–H Bond Breaking.* The formation of positive fixed charge during the NBTI stress has been well known since the initial investigations on NBTI. Blat et al. [7] observed that in SiO$_2$ dry oxides (thermally growing an oxide at 1050°C in dry oxygen to a

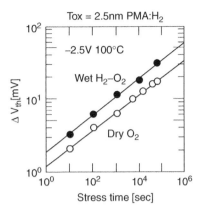

Figure 4.20. Comparison of NBTI sensitivity between wet and dry thermal oxides. Reprinted with permission from Ref. 53, © 1999 IEEE.

thickness of 560 A), NBTI can be driven by the following electrochemical reaction:

$$\{Si-H + A\} + h^+ \leftrightarrow Si^* + B^+. \tag{4.33}$$

The Si–H bond breaking is facilitated by the presence of the water-neutral-related species (A). The electrochemical reaction is activated by holes in the inverted channel and results into interface states generation (Si*) and formation of a positively charged water-related species (B^+). Evidence of the role of water species to contribute to the NBTI damage is given by the observed NBTI enhancement (Fig. 4.20) in wet respect to dry thermal gate oxides [53].

The electrochemical reaction suggested above implies that, for the same hole density, the existence of $\{Si-H + A\}$ interface defects in the gate oxide enhances both the rate of interface states generation as well as the positive fixed charge formation. In addition, the concentration of the water species by-product (B^+) is equal to the increase in interface states density during the stress.

Further evidence on the generation of positive fixed charge formation in pure SiO_2 is given by the following electrochemical reactions [64]:

$$Si_3 \equiv Si-H + h^+ \leftrightarrow Si_3 \equiv Si^* + H^+ \tag{4.34}$$

$$O_3 \equiv Si-H + h^+ \leftrightarrow O_3 \equiv Si^+ + H_0 \tag{4.35}$$

$$[H^+, H_0]_{interface} \leftrightarrow [H^+, H_0]_{bulk} \tag{4.36}$$

In these cases, interface states ($Si_3 \equiv Si^*$) and positive fixed oxide charges ($O_3 \equiv Si^+$) are generated in parallel from the cold holes-assisted dissociation of the ($Si_3 \equiv Si-H$) and ($O_3 \equiv Si-H$), respectively, during the NBTI stress. The balance between these two competing processes at a given V_g and temperature determines the net generation of the interface states and fixed positive charges. It is also observed that depending on the nature of the electrochemical reactions involved in the positive charge build-up, the increase in positive fixed charge $\left(\Delta N_f^+\right)$ is linearly proportional to N_{it} generation (ΔN_{it}), with the constant of proportionality depending on the electrochemical reactions responsible for the interface states generation. Under these conditions, the V_{th} shift will be given by $\Delta V_{th} = q(\Delta N_{it}^d + \Delta N_F) = q\gamma\Delta N_{it}$, and be linearly proportional to the interface states generation. The overall V_{th} shift can be modeled by evolution of the N_{it} generation during the NBTI stress.

4.6.2.2 Bulk Traps Formation.

The formation of bulk trap defects during an NBTI stress is still a source of active debate in the NBTI community. The main reason of this confusion is twofold. First, not all the reported studies carefully monitor if a cold holes regime is active at the applied NBTI accelerated stress conditions. In addition, even in a similar stress configuration, the formation of bulk defects is strongly dependent on the gate oxide process. In the case of cold holes regime (Fig. 4.4a,b) the following two main physical pictures have been suggested for the buildup of positively charges bulk defects in nitrided oxides.

1. Trapping of migrating H* species in oxide defects precursors (hydrogen release picture).
2. Deep level hole traps formation from electron tunneling assisted dissociation of oxide defect precursors.

In the rest of this section we are going to discuss the experimental evidence supporting both physical mechanisms. Direct experimental evidence of NBTI-induced oxide bulk defects build-up is gained from the increase in stress-induced leakage current (SILC) observed during NBTI stressing. SILC current increase has been typically correlated to the formation of bulk defects in gate oxide [109]. The SILC measurement is performed at sufficiently low gate voltage ($V_g < 0$ in pMOSFET) to reduce the electron direct tunneling contribution and monitor the gate current (I_g). The bulk defects generated in the gate oxide provide additional leakage paths for both the increased contribution of the electron and hole current to the gate current. Huard et al. [19] observed a proportionality between the hole and electron SILC current components (Fig. 4.21a) in nitrided oxides ($T_{ox} = 2\,nm$), suggesting that the primary cause of the I_g leakage increase is the electron and hole recombination in bulk oxide defects during the NBTI stress. Figure 4.21 (b) suggests that the N_{it} increase is strongly correlated to the increase in gate leakage current that suggests a common root cause of the N_{it} generation as well as the bulk trap formation. The assumed physical picture explaining this common origin is the hydrogen release model. In this case the bulk traps (X_{bulk}) are generated by trapping a diffusing hydrogenated species H* in the gate oxide defects according to the equation $H^* \to X_{bulk}$. The bulk oxide defects become positively charged by hole trapping in X_{bulk} and will contribute to the V_{th} shift. The diffusing species is the by-product of the electrochemical reaction $Si-H \to Si^* + H^*$, which results in interface state generation Si^*. An example is given by Tan [64] for nitrided oxides. In this

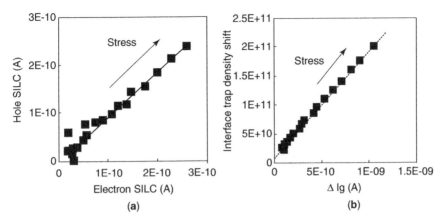

Figure 4.21. (a) Hole SILC as function of electron SILC, both measured during an NBTI uniform stress (−2 V). (b) Interface states density increase as function of SILC induced I_g increase measured during an NBTI stress V_g (−2 V). Reprinted with permission from Ref. 19, © 2003 IEEE.

case the nitrogen-enhanced NBTI sensitivity is associated with the increased trapping in oxide defects of the hydrogen ions (H^+) and neutral hydrogen atoms (H_0) released by the electrochemical reactions, described in Equations 4.34 and 4.35 near the Si/SiO$_2$ interface. Further evidence of the direct link between the N_{it} generation and increase in SILC-related bulk traps is given by a comparison of the NBTI sensitivity between the hydrogenated and deuterated gates oxides. It is well known that Si–D bonds at the Si/SiO$_2$ interface have a slower desorption rate in respect to Si–H and, according to the hydrogen release model, should yield a reduced generation of SILC related bulk traps. Experimental evidence [6] is given in Figure 4.22, which shows that the use of deuterium reduces both the generation of interface states and oxide bulk defects (monitored as increase in I_g).

Bulk defects generation due to electron direct tunneling from the polysilicon gate during NBTI stressing has been suggested by Ang et al. [52, 110]. As seen in the inset (a) of Figure 4.23 (a), during the NBTI stressing in cold holes regime, electron direct tunneling from the polysilicon electron valence band (EVB) is confined in a narrow energy window (Arrow 1). This electron injection allows a possible field and thermally assisted dissociation of oxide defect precursors. The subsequent loss of released electrons into the n-well conduction band leaves behind positively charged defects with energy above E_C (deep level hole traps) as seen in inset (b) (Arrow 2). Experimental evidence of these phenomena is given by the observed correlation between $|\Delta V_{th}|$ and $|\Delta N_{it}|$ shift for a device (A1) with lightly nitrided gate dielectric (<1% at % nitrogen; equivalent oxide thickness = 21.6 A). The dashed line represents the estimated $|\Delta V_{th}|$ due to $|\Delta N_{it}|$ ($|\Delta V_{th}| = q \, |\Delta N_{it}|/C_{ox}$). Typically, CP measurements only yield the average energy density of the donor and acceptor interface states situated $+/-$ 0.4 eV in respect to the intrinsic Fermi level. For a given $|\Delta N_{it}|$, the measured $|\Delta V_{th}|$ under a static stress (constant $V_g = -2.6$ V

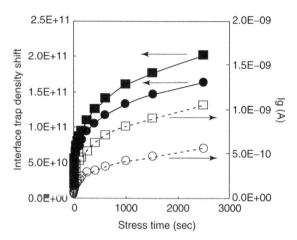

Figure 4.22. Interface states generation for H_2 (filled squares) and D_2 (filled circles) devices and the corresponding gate current leakage increase. Reprinted with permission from Ref. 6, © 2003 IEEE.

at stress) is much larger that what is estimated as contributed by $|\Delta N_{it}|$. On the other end, this difference is significantly reduced under a unipolar and bipolar AC voltage pulse at stress (50% duty cycle, 100 kHz rectangular pulse). The gate voltage during the recovery phase is 0 and 1.5 V respectively for the unipolar and bipolar stress. As it can be seen from the inset (b), during the recovery in accumulation ($V_g = 1.5$ V) electrons from the n-well conduction band can neutralize the deep hole traps. A fraction of the bulk traps, however, is not neutralized by electron injection even with $V_g > 0$. These hole traps have energy states pinned by the Si/SiO_2 conduction band offset and contribute as apparent positive fixed charged defects to the V_{th} shift. The unipolar AC stress exhibits a two-stage behavior of the $|\Delta V_{th}| - |\Delta N_{it}|$ dependence, suggesting that the generation of the deep hole traps is independent of the N_{it} generation. The generation of deep hole traps is initially dominating. In later stages $|\Delta V_{th}|$ lies between what observed for the static and bipolar stress, suggesting that during the unipolar recovery phase a larger amount of positive charge bulk defects are unrecovered. The observation that the two-stage phase (Fig. 4.23b) depends on the nitrogen concentration at the interface further confirms two effects.

1. The deep-level hole-traps precursors density is strongly related to nitrided oxides interface concentration. Higher N concentrations increase the initial oxide defects (precursors) density. In Section 4.10.2 we are going to give more evidence of these effects.

2. The deep-level traps contributions is dominant initially, but at later stages the V_{th} shift is mainly contributed by the N_{it} generation mechanism.

Further evidence of bulk traps generation is provided by Tsujikawa et al. [51] from the suggested neutralization by electron trapping during an accumulation phase

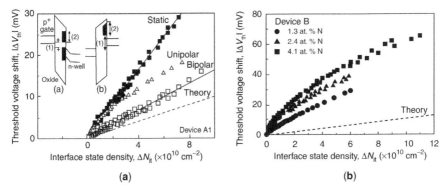

(a)   **(b)**

Figure 4.23. (a) Threshold voltage shift $|\Delta V_{th}|$ versus stress-induced interface state density ΔN_{it} (device A1) for different stress configuration (static ($V_g = -2.6$ V), unipolar ($V_g = 0$), and bipolar ($V_g = 1.5$ V)). The temperature was 125°C. Inset: Energy band diagrams showing (a) the generation of positive interfacial/oxide defects (shaded region) under negative gate biasing and (b) relaxation of the positive defects under positive gate biasing. Arrow 1 shows the energy window of electron direct tunneling. Arrow 2 depicts deep-level positive defect states outside the energy window of electron direct tunneling. (b) Threshold voltage shift $|\Delta V_{th}|$ versus stress-induced interface state density ΔN_{it} (device B). Reprinted with permission from Ref. 110, © 2006 IEEE.

following the NBTI stress. Evidence of this effect is given by monitoring the dependence of the subthreshold slope shift (Δ_{SS}) on the V_{th} decrease (ΔV_{th}) during an NBTI stress ($V_g = -2.8$ V, $T = 125°C$), immediately followed by a recovery phase both with $V_g = 0$ V and in accumulation ($V_g = 1.2$ V). In the first case (Fig. 4.24a), the same linear equation ($\Delta_{SS} = C \times \Delta V_{th}$) is found during the NBTI stress as well as the recovery ($V_g = 0$ V) phase. This result suggests that N_{it} generation and repassivation are responsible for the V_{th} shift during the NBTI stress and the recovery. The effect of applying a positive bias immediately after the NBTI stress (Fig. 4.24b) results in a ΔV_{th} recovery larger than what is expected from the above linear equation for the same Δ_{SS} recovery. This phenomenon is assumed to be due to electron injection from the accumulated channel into the gate oxide. The injected electrons are trapped by positive bulk traps (X_{bulk}^{+}) generated during the NBTI stress.

A key conclusion of the observations thus far is that positive charge bulk traps build-up is expected to be a component of the NBTI damage in cold holes regime. However, the physical picture is not clear because the dominance of the hydrogen release versus the EVB tunneling assisted damage strongly relates to the gate oxide process. Two important conclusions can be drawn at this point:

1. Both mechanisms are activated by the initial concentration oxide defects precursors.
2. At a given stress/test sequence, the deep level traps contribution to V_{th} is expected to be linearly dependent on the N_{it} increase at later stress times.

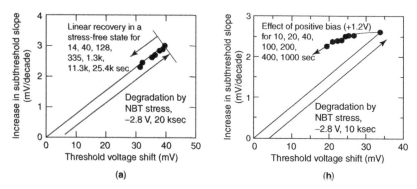

Figure 4.24. Typical behavior of threshold voltage and subthreshold slope during a $V_g = 0\,V$ (a) and ($V_g = +1.2\,V$) (b) recovery phase immediately after NBTI stress ($V_g = -2.8$). Reprinted with permission from Ref. 51, © 2006 IEEE.

Under these conditions, the total V_{th} shift due to positively charged oxide defects is easily monitored in terms of the N_{it} evolution. In addition, the electrical behavior of these bulk traps defects is equivalent to positive fixed (no recoverable) charge.

4.6.2.3 *Hole Traps Activation.* During the NBTI stress, hole traps activation is assumed to take place parallel to the N_{it} generation and the possible, eventual positively charged defect formations described previously. Its contribution is represented by Equation 4.37.

$$X_{th}(0) + h^+ \leftrightarrow X_{ht}^+, \tag{4.37}$$

In this case, preexisting (before stress) CMOS process-induced neutral traps ($X_{ht}(0)$) are activated (positively charged) by holes in the channel in inversion ($V_g < 0$), typically at the NBTI stress configuration.

The NBTI-induced activation of preexisting hole traps has been experimentally observed by the trapping [4] and detrapping phenomena observed during the NBTI stress and recovery phase. An example is given in Figure 4.25. In this case, the V_{th} shift is monitored during a long, continuous NBTI stress (open squares) at $V_g = -2.5\,V$, a pulsed (50% duty cycle) NBTI stress (filled circles) with V_g between $V_g = -2.5$ and $1.5\,V$ (electron injection phase) and a pulsed NBTI stress (open diamonds), where a positive bias ($V_g = +1.5\,V$) and a small negative bias ($V_g = -0.75\,V$) phase follow the $V_g = -2.5\,V$ NBTI stress. The total time the pMOSFET is exposed to a $V_g = -2.5\,V$ stress condition is equal for all the three stresses. No difference is measured between the two pulsed experiments. Both trapped holes and positively charged slow states can be neutralized by the electron injection phase ($V_g = 1.5\,V$), but they should exhibit a different recharging behavior under a small negative bias ($V_g = -0.75\,V$). In particular slow states should be recharge, while hole traps can not since they require larger energies [54].

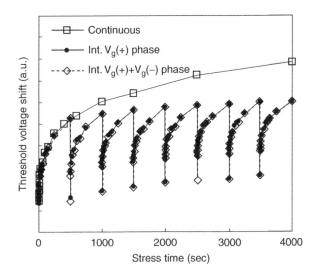

Figure 4.25. Threshold voltage shifts for a continuous stress (open squares) with a constant gate voltage stress ($V_g = -2.5$ V), for interrupted stress with an inserted positive bias phase ($V_g = +1.5$ V) (filled circles) and for an interrupted stress with an inserted positive bias phase followed by a negative bias phase ($V_g = -0.75$ V) (open diamonds). Reprinted from V. Huard, M. Denais, F. Perrier, N. Revil, A. Bravaix, E. Vincent. A thorough investigation of MOSFETs NBTI degradation. *Microelectronics Reliability*, 45: 2005, pp 83–98. Reprinted with permission from Elsevier.

These experiments lead to the conclusion that slow states do not participate to the NBTI damage. The nature of the hole traps has not been identified yet. Oxygen vacancies as E' centers have been pointed out as possible hole trapping centers. Nitrogen incorporation in the gate oxide increases the hole trapping phenomenon during an NBTI stress [4]. In Section 4.9 we will describe the kinetics of the hole trapping and detrapping during a NBTI stress and recovery phase.

4.7 KEY EXPERIMENTAL OBSERVATIONS ON THE NBTI DAMAGE

In this section we will describe some of the key experimental observations of the NBTI phenomena that must be explained by any physical picture of NBTI.

4.7.1 pMOSFET NBTI: Worst MOSFET Bias Temperature Condition

In order to have a good insight on the NBTI physics, we will explain why NBTI is worst in pMOSFET transistors in inversion. As indicated in Figure 4.26 from [4], bias temperature instabilities are experienced by both pMOSFET and nMOSFET devices at whatever negative or positive V_g (same $|V_g|$) applied to the device at stress. Exception is typically observed for the nMOSFET device in inversion

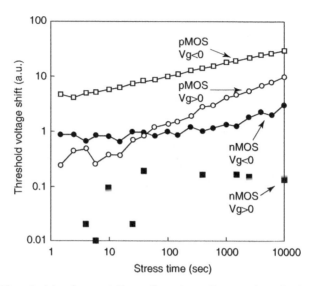

Figure 4.26. Threshold voltage shifts as function of stress time for both nMOSFET and pMOSFET transistors under inversion and accumulation bias conditions ($T = 125°C$). Reprinted from V. Huard, M. Denais, F. Perrier, N. Revil, A. Bravaix, E. Vincent. A thorough investigation of MOSFETs NBTI degradation. *Microelectronics Reliability*, 45: 2005, pp 83–98. Reprinted with permission from Elsevier.

($V_g > 0$) where the V_{th} shift is negligible. Focusing on the NBTI configuration ($V_g < 0$ V) for both devices, since it is an experimental fact that NBTI is driven by the presence of holes in the channel (see the next paragraph) even an nMOSFET device in accumulation should experience similar V_{th} shifts as the pMOSFET in inversion. In both devices, in fact, the NBTI damage should proceed with a V_{th} shift associated to positive charge build-up. The NBTI configuration for pMOSFETs is, however, the worst BT configuration. There are two main reasons why.

1. In the case of pMOSFET devices in inversion ($V_g < 0$), both the electrically activated N_{it} (donor-type) (ΔN_{it}^D) and positively charged defects (ΔNp) contribute to the total V_{th} shift. Conversely the V_{th} shift of nMOSFET, stressed in accumulation ($V_g < 0$), results from the opposite contributions from the negative charged acceptor-type interface states $((\Delta N_{it}^A))$ and the positive charged defects in the gate oxide during the stress.

2. The workfunction difference (≈ 1 V) between the pMOSFET p^+ and the nMOSFET n^+ polygate implies a reduced oxide field in an nMOSFET in accumulation for the same V_g at stress. A better quantification of these effects is given by [5].

Figure 4.27 shows the normalized I_d shift for p- and nMOSFETs ($T_{PHY} = 26$ A) stressed at room temperature in both inversion and accumulation ([5]). The

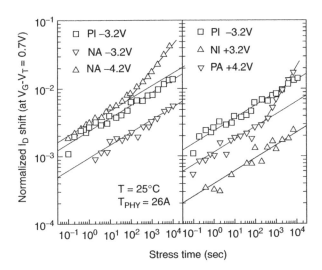

Figure 4.27. Time evolution of normalized linear drain current degradation measured at fixed gate overdrive ($V_g - V_{th} = 0.7$ V) for pMOSFETs and nMOSFETs ($T_{PHY} = 26$ A) stressed, at room temperature (25°C), both in inversion (PI, NI) and accumulation (PA, NA). Reprinted with permission from Ref. 5, © 2004 IEEE.

drain current was normalized by measuring at a fixed gate overdrive ($V_g - V_{th}$) to better quantify the total charge generated. Note that, due to differences in flatband voltage, in order to get the same gate oxide field (F_{ox}) in both cases it is necessary to apply a different V_g so that $V_{g|acc} = V_{g|inv} \pm 1$ V for inversion and accumulation bias conditions at stress.

Using this stress and test normalization procedure we can make two conclusions:

1. pMOSFET stressed in inversion and nMOSFET stressed in accumulation show similar normalized I_d shift at identical F_{ox}. Assuming that the same positively charged defects are produced in both stresses, the similarity between the normalized I_d shifts suggests that both donor-(pMOSFET sensitive when tested in inversion) and acceptor-(nMOSFET sensitive when tested in inversion) type interface states are created during the pMOSFET stress in inversion and nMOSFET stress in accumulation at the same F_{ox}.

2. For the same V_g at stress the normalized I_d shift is largest for the pMOSFET stressed in inversion and nMOSFET stressed in accumulation (holes at Si/SiO$_2$ interface), followed by pMOSFET stressed in accumulation (holes tunnel from poly gate) and nMOSFET stressed in inversion (holes generate and tunnel from poly gate). This observation suggests that holes are associated to the NBTI damage.

Notice that for sufficiently high V_g ($+4.2$ V for the pMOSFET in accumulation, -4.2 V for the nMOSFET in inversion) the I_{dlin} time evolution follows at late times a high slope associated to the setting of a hot hole injection regime. The net from this exercise is that both nMOSFET and pMOSFET devices are sensitive to hole assisted bias temperature aging and similar type of oxide damage is experienced when stressing at the same F_{ox}. Since in practical circuit level applications both the pMOSFET and nMOSFET devices are used in inversion, the pMOSFET NBTI configuration is the worst bias temperature aging mechanism expected to be experienced in the field. For this reason we will focus in this chapter to the pMOSFET NBTI degradation.

4.7.2 Role of Cold Holes

Contrary to other wearout mechanisms such as gate oxide dielectric breakdown, NBTI damage does not relate to any channel transport through the gate oxide either by direct tunneling or Fowler–Nordheim injection. The most obvious observation supporting this conclusion is that NBTI degradation has been measured in pMOSFETs with both thick ($T_{ox} \approx 6.5$ nm) and very thin ($T_{ox} \approx 2.0$ nm) gate oxides at similar oxide electric fields [24]. As seen in Figure 4.28, both gate oxides exhibit the same NBTI behavior under similar gate oxide fields. This similarity in NBTI degradation takes place despite being the 6.5 nm oxide stressed with a V_g such that a low oxide electric field is applied below the detection limit of Fowler–Nordheim injection current ($I_g < 10^{-13}$ A), while a large tunneling current ($I_g > 10^9$ A) is experienced by the thin oxide. Further evidence is reported by [20]. In this work the authors indicated that the number of bulk traps formed (I_g increase) for the same level of NBTI damage (same ΔV_{th}) seems to be

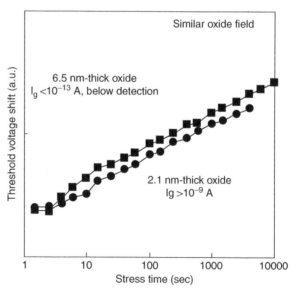

Figure 4.28. Threshold voltage shift for 2 nm- and 6.5 nm-thick gate oxide under similar gate oxide field at 125°C. Reprinted with permission from Ref. 24.

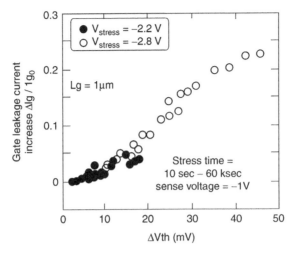

Figure 4.29. Relation between increase in gate leakage current and V_{th} shift during an NBTI stress.

independent of the stressing regime, whether the NBTI stress is taking place at the electron tunneling ($V_g = -2.8$ V) (of Fig. 4.4b) or the hole tunneling ($V_g = -2.2$ V) (Fig. 4.4a) regime (see Fig. 4.29).

All these observations suggest that the NBTI damage is not related to energetic (hot) hole injection. In addition, no carrier transport occurs along the channel since the stress conditions are symmetrical between source and drain. This means that the NBTI damage is not generated by heated holes as in the case of channel hot carrier or anode hole injection mechanisms, but by holes in thermal equilibrium in the inverted channel. These are typically called cold holes. In the case of strong inversion in the channel (typical of the NBTI stress) the cold holes concentration (p_h) at given V_g is given by:

$$p_h = C_{ox}\left(V_g - \psi_s(inv)\right) \approx C_{ox}\left(V_g - V_{th}\right) \qquad (4.38)$$

Even if the presence of holes in the inverted channel is required for the NBTI damage to occur (see Figure 4.26), the hole concentration does not seem to be the rate limiting factor [26]. Figure 4.30 shows the NBTI degradation of several NBTI stresses with V_g at a constant -2.1 V while varying the n-well bias ($V_{nw} = 0, 1.5, 3$ V) at 140°C [26]. In this case, the device V_{th} is modulated by the n-well bias (V_{SUB} in the figure). As expected from Equation 4.38, the hole population increases proportionally with increasing (V_g–V_{th}), but neither the V_{th} shift (ΔV_{th}) nor the interface state density increase (ΔD_{it}) are observed to be dependent on the different hole channel population (see Fig. 4.30). As discussed in the next paragraph, the similar NBTI shift is related to the same oxide electric field applied during the stress. The same relation between the V_{th} shift and the increase in interface states (D_{it}) is found independently of the oxide field applied during the stress as seen in Figure 4.31 [116].

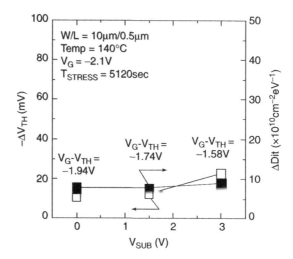

Figure 4.30. V_{th} shift (ΔV_{th}) and interface state generation (ΔD_{it}) as a function of n-well bias (V_{SUB}). NBTI stress was performed under a constant gate voltage ($V_g = -2.1 \, V$), which means that the amount of holes in the inversion layer was varied under applying constant oxide field. It should be noted that both ΔV_{th} and ΔD_{it} are independent from the hole concentration (different $V_g - V_{th}$). Reprinted with permission from Ref. 26, © 2002 IEEE.

4.7.3 Dependence on Gate Oxide Electric Field

The NBTI dependence on the gate oxide field (F_{ox}) is clearly observed in Figure 4.32 [26]. This figure describes an NBTI uniform stress ($T = 140°C$) where, for different n-well biases ($V_{sub} = 0, 1.5, 3 \, V$ in the figure), V_g at stress is

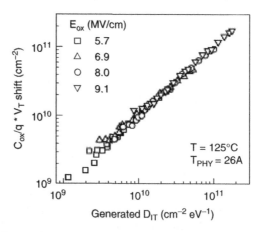

Figure 4.31. Correlation between V_{th} shift and N_{it} generation for a 26 pMOSFET stress in inversion at 125°C in cold hole regime under various gate oxide fields. Reprinted with permission from Ref. 116, © 2004 IEEE.

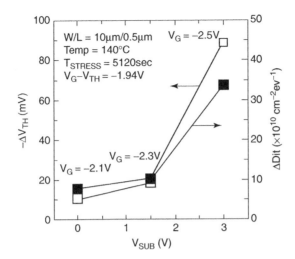

Figure 4.32. Threshold voltage shift (ΔV_{th}) and interface state generation (ΔD_{it}) as a function of substrate voltage (V_{SUB}). NBTI stress was performed under $V_g - V_{th} =$ constant, which means that the amount of holes in the inversion layer was made constant, and the oxide field was varied. Note that both ΔV_{th} and ΔD_{it} are increased with V_g and V_{SUB}. Reprinted with permission from Ref. 26, © 2002 IEEE.

modulated in a way to keep the channel hole population constant ($V_g - V_{th} = -1.94\,\text{V}$). It is observed that both the V_{th} and the D_{it} shift increase with increase in the n-well bias with a corresponding increase in F_{ox}. Further evidence that the NBTI damage is controlled by the oxide electric field (F_{ox}) is provided by the observation that both pMOS and nMOS devices equally degrade for the same electric field at the interface [105]. The actual gate oxide field dependence of the NBTI damage is still a source of very active research. At this point two empirical models have being adopted, e.g., exponential model ($\Delta V_{th} \approx \exp(\gamma F_{ox})$) and power-law model ($\Delta V_{th} \approx (F_{ox})^m$). Here we limit ourselves to provide some experimental evidence supporting these different NBTI oxide field-dependent models.

Experimental evidence of an exponential F_{ox} dependence of NBTI [84] is shown in Figure 4.33. From a wide variety of experimental data the authors established that the NBTI induced interface states generation follows an exponential dependence on F_{ox} and V_g at stress as:

$$\Delta N_{it} \propto \exp(\gamma F_{ox}) \propto \exp(\gamma_V V_g) \tag{4.39}$$

where γ and γ_V are respectively the field and voltage acceleration factor. Defining EOT the effective oxide thickness (obtained from CV simulations and physical thickness) we have $\gamma_V = \gamma/\text{EOT}$. In this work an experimental value of $\gamma \approx 0.6$ was estimated. As will see later, the physical mechanism considered to be responsible for this observed F_{ox} dependence of NBTI is the field dependence Si–H dissociation controlled by a reaction–diffusion (R–D) process. It is assumed that this

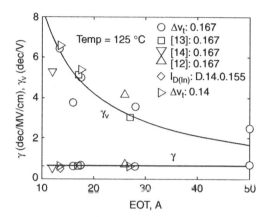

Figure 4.33. γ and γ_V (voltage acceleration factor) variation with electrical oxide thickness (EOL) (n is mentioned for each case). For field-dependent NBTI model, γ remains almost constant ($\sim 0.6 \pm 0.05$; dashed line) and $\gamma_V \sim 1/$EOT. Reprinted with permission from Ref. 84, © 2006 IEEE.

dissociation is preceded by the capture of inversion layer holes by the Si–H covalent bonds. This phenomenon weakens the existing Si–H covalent bond, which can be easily broken at higher temperatures. The NBTI field dependence has been associated to the assisted field hole tunneling current and /or the further field assisted weakening of the dipole energy associated to the bond.

An empirical power-law dependence of the NBTI damage on F_{ox} is observed in [37]. In this work the NBTI degradation is investigated for different gate oxide thicknesses (T_{ox} between 1.5 and 2.3 nm) with the same gate oxide process. The total positive charge generated after a 1000 sec NBTI stress is monitored by multiplying the V_{th} shift by the capacitance (C_{ox}) so that: $q \times \left(\Delta Np + \Delta N_{it}^D \right) = \Delta V_{th} \times C_{ox}$. The normalized V_{th} shift ($\Delta V_{th} \times C_{ox}$) is the result of the donor-type interface states and the positive oxide defects formed during the stress. The total charge generated seems to follow a universal law when plotted as a function of the gate oxide field applied during the stress (see Fig. 4.34). This result suggests a power law dependence on the F_{ox} in the oxide electric field range evaluated with a power law exponent (b_1) around 4. No clear physical argument support the experimentally power law F_{ox} behavior.

An interesting observation about the NBTI oxide field dependence is that the lifetime NBTI contribution can be strongly related to the stress and test methodology adopted to characterize NBTI as well as the end-of-life (EOL) target selected to estimate the NBTI sensitivity of a given CMOS process. Figure 4.39, [39], describes the situation of NBTI induced V_{th} shift (ΔV_{th}) experimentally observed to follow a saturating behavior with stress time at low voltage in a log-log scale (see Section 4.7.5). Following the argument by Aono et al. [39], we select an EOL target τ (dashed line in Fig. 4.35a) such that the V_{th} shift shows saturation with stress time at low stress voltage. If we assume that the ΔV_{th}

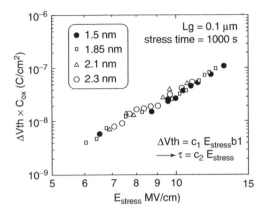

Figure 4.34. Observed power law electric field (E_{stress}) of normalized V_{th} shift ($\Delta V_{th} \times \overline{C_{ox}}$) during a DC NBTI stress. Reprinted with permission from Ref. 37, © 2003 IEEE.

saturating time behavior associated to this target can be described by the equation $\Delta V_{th} \approx C \times \exp(\gamma R_{ox}) \times (\exp \beta t)^{\alpha}$, it is easy to conclude that the lifetime τ follows a power law dependence on F_{ox} such that $\tau \approx (\log \Delta V_{th} \times F_{ox})^{1/\alpha}$. At the same time, if the lifetime target is selected such that ΔV_{th} follows the power law time behavior described by the equation $\Delta V_{th} \approx C \times \exp(\gamma F_{ox}) \times t^{n}$ τ has an exponential dependence on F_{ox} such that $\tau \approx (\Delta V_{th2} \times \exp(-\gamma F_{ox}))^{1/n}$.

These observations suggest how important it is to make sure that the observed saturation effects are not apparent since driven by the dominant role of the NBTI recovery at the initial phase of NBTI stressing at sufficiently low voltage. As we

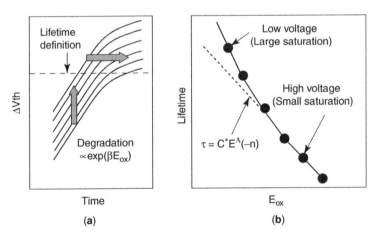

Figure 4.35. Schematic representation of lifetime (τ) extrapolation dependence on the gate oxide electric field ($E_{ox} \equiv F_{ox}$) and EOL targets. A different F_{ox} dependence of τ is observed depending on the contribution of ΔV_{th} saturation to the lifetime extrapolation. Reprinted with permission from Ref. 39, © 2004 IEEE.

will see later, by not taking this contribution into account, may yield very misleading results in terms of NBTI characterization and modeling.

4.7.4 Dependence on Stress Temperature

The NBTI degradation is a thermally activated process. As suggested by Figure 4.36 from [113], NBTI has been experimentally observed to follow an Arrhenius-like dependence on stress temperature such that, for a given device parameter Δ, $\Delta \approx \exp(-E_a/kT)$ where E_a is the temperature activation energy. In this case it is observed that the threshold voltage and the interface state density shift have different experimentally measured values of the E_a ($E_a(N_{it}) \sim 0.156$ eV, $E_a(V_{th}) \sim 0.063$ eV) giving further evidence of the dual nature of the oxide damage generated during the NBTI stress. If the threshold voltage shift would have been controlled only by interface states generation, similar E_a values should be calculated for both device parameters.

Further insights on the competing contribution of interface states generation and hole trapping to the NBTI temperature dependence is given by Ang et al. [38, 111]. In this work it is shown that two distinct degradation mechanisms are contributing independently to the NBTI damage in pMOSFETs with ultra-thin nitrided gate oxides. pMOSFETs with two different nitridation processes were investigated. Device A has a 21.6 A oxynitride gate dielectric (dry oxidation and N_2O annealing at 950°C), while device B has Si_3N_4-SiO_x gate stack (9 A SiO_x + 19 A Si_3N_4). NBTI stresses were carried out with $V_g = -2.6$ V, which yielded $F_{ox} = 8.8$ MV and 7.8 MV/cm for device A and B, respectively. Figure 4.37 (a) shows the Arrhenius plots for the NBTI induced V_{th} shift of device A and B. The

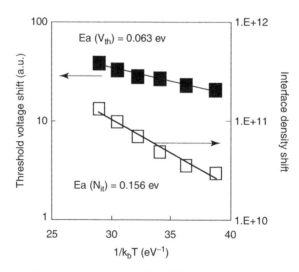

Figure 4.36. Arrhenius plot for both V_{th} (filled symbols) and interface states density shift (open symbols) during an NBTI stress with $V_g = -2.5$ V. Reprinted with permission from Ref. 113, © 2002 IEEE.

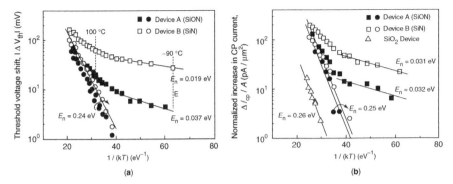

Figure 4.37. (a) Arrhenius plots for NBTI stress-induced $|\Delta V_{th}|$ shift of devices A and B. The circles denote the respective Arrhenius plots of devices A and B after eliminating the contribution of mechanism I. (b) Arrhenius plots for NBTI stress-induced interface state (N_{it}) generation in devices A and B. The circle symbols denote the respective Arrhenius plots of devices A and B after eliminating the contribution of mechanism I. Also shown for comparison is the measured Arrhenius plot of the SiO$_2$ gate devices. Reprinted with permission from Ref. 38, © 2005 IEEE.

square symbols represent the measured V_{th} shifts as function of $1/kT$. Two different temperature regimes are observed; a low temperature regime (I) ($1/kT \geq 40$ (eV^{-1})) with an activation energy $E_a \approx 0.037$ and 0.019 eV respectively for device A and B. This regime has been associated with the spontaneous hole trapping of interface Nitrogen precursor sites and not related to direct tunneling currents. The Arrhenius-like plots of the second regime (II) (circle symbols) are obtained by subtracting the extrapolated V_{th} contributions of mechanism I from the measured V_{th} shifts in the high temperature region. The nearly similar extracted activation energy ($E_a \approx 0.24$ eV) suggests that, despite of the different nitridization processes, the nature of this mechanism II is comparable for both devices and similar with what reported for SiO$_2$ thermal oxide processes (NO nitride).

The enhanced NBTI measured in device B is related to the high N concentration at the interface. In Figure 4.37 (b), the corresponding Arrhenius plots for the ΔN_{it} generation (measured from the increase in charge pumping current $\Delta I_{cp}/A$, with A being the active gate area) for both mechanism I and II are given. Again, eliminating the contribution of mechanism I, an Arrhenius plot is obtained for mechanism II, whose E_a value ($E_a \approx 0.25$ eV) is very close to that one measured for pMOSFETs with SiO$_2$ gate oxide (open triangles). This result suggests that, for what concerns the temperature activation, the N_{it} shift due to mechanism II is suggested not to be affected by the nitridation process. The enhanced N_{it} degradation in nitrided gate oxides is due to mechanism I and seems independent from the specific nitridation process ($E_a \approx 0.031$ eV). This result is different from what observed in Figure 4.37 (a). In this case the lower E_a measured for device B ($E_a \approx 0.019$ eV) may be due to trapping of positive charge at the SiN/SiOx interface and/or deep level traps activation during the stress. The small activation energy of mechanism I indicates that N_{it} generation in nitrided oxides can not be uniquely explained in terms of Si–H bond dissociation and H transport. The N incorporation seems to introduce a new degradation mechanism related to hole trapping which may be dominant at some application temperatures.

The examples given so far suggest that the temperature dependence of the NBTI induced N_{it} generation follows an Arrhenius like behavior with a measured activation energy E_a around 0.1 to 0.3 eV. As it will be described in Section 4.8, the physical explanation of this Arrhenius like temperature dependence relates to N_{it} generation by Si–H bond dissociation through a reaction–diffusion process controlled by the diffusion of a stable hydrogenated species (H*) in the gate oxide and/or poly gate.

Under these conditions we will show that the measured temperature activation energy E_a is determined by diffusivity D_{H*} of the H* species ($D_{H*} = D_{H*,0} \times \exp(-E_a(H*)/kT)$) through the equation $E_a = E_a(H*) \times n$ with n (the time power law exponent) being equal to 1/4 or 1/6, respectively, for H_0 or H^2 neutral species diffusion and $E_a(H*)$ is the activation energy associated to the diffusion of the H*. Experimental evidence of the H* diffusion controlled Errhanius like temperature dependence of N_{it} contribution has been recently given by Varghese et al. [107] on pMOSFET with DPN gate oxides and different N_2 concentration at the interface. By adopting an on-the-fly (see Section 4.7.7) stress/ test methodology to monitor the linear V_{th} shift, they conclude that the observed V_{th} shift is mostly due to N_{it} generation. On the basis of the observation that the V_{th} shift for both low and high N_2 dose (and their difference) shows similar E_a, they

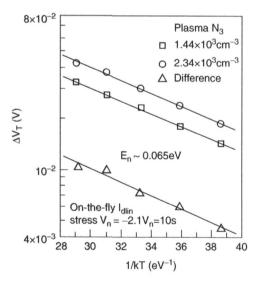

Figure 4.38. T activation of ΔV_T (V_{th} shift) data for various N_2 dose. Lower and higher N_2 dose (and their difference) show identical E_a. Reprinted with permission from Ref. 107, © 2005 IEEE.

suggest that the NBTI enhancement corresponds to an increase N_{it} generation rate with a larger N_2 dose (see Fig. 4.38). The measured temperature activation energy E_a is around 0.085 eV which supports H_2 diffusion with $E_a(H_2) = 0.5$ eV [108].

A key observation on the Arrhenius-like temperature dependence of the NBTI induced N_{it} generation is that it is implicitly assumed that the H_2 diffusion is taking place in an homogeneous oxide. This assumption leads to the conclusion of a non temperature dependence of the time power law exponent n. Recent work (see next paragraph) has suggested a possible dispersive diffusion of H* which has experimental verification in the dependence of the time power law exponent n on stress temperature.

Figure 4.37 suggests also that the hole trapping temperature dependence is associated to a very small activation energy (E_a around 0.06) as expected in the case of hole trapping mainly contributed by direct tunneling.

4.7.5 Time Evolution

The adequate characterization of the time evolution of the NBTI damage is probably the most direct evidence of the level of confusion existing today in the available literature about the physics of NBTI. To give an example let's look to the following Figure 4.39 taken from [39]. In this work symmetric NBTI stresses were applied at elevated temperature (200°C) with V_g ranging between −4.0 to −6.5 V on pMOSFETs with $T_{ox} = 6.7$ nm. The stress time was in the range from 10^1 to 10^6 sec. Figure 4.39 (a) shows the threshold voltage shifts (ΔV_{th}) versus stress time for several gate voltage at 200°C. ΔV_{th} seems to have a tendency to saturate with increasing stress time. To quantify this observed saturation effect, the stress time axis was divided into few time regions and a power-law ($\Delta V_{th} \approx At^m$) coefficient ($m$) was calculated in each region. As shown in Figure 4.39 (b), the coefficient (m) decreases with increasing stress time and reaches the value 0.16. It is also important to observe that this observed saturation effect appears stronger at lower stress voltages. The power law exponent is, in fact, higher for lower stress voltages ($|V_g| = 4.0$ V) at early stress times (region 1). This increase of the m exponent as lower $|V_g|$ values is also associated to less V_{th} shift during the initial stress phases. Several interpretations can be given of these observed phenomena:

1. The observed saturation effects are the by-product of the stressing/testing methodology used to characterize the NBTI damage. In this case the NBTI testing was carried out by using I–V curve characterization with not an accurate control of the delay time between the end of stress before testing and the beginning of the I–V measurements. This hypothesis is supported by the observation that an m value around 0.16 is measured independently of the of the stress voltage in the region 4, which corresponds to large V_{th} shift ($|\Delta V_{th}| > 100$ mV). These shifts are probably larger than the error expected from expected NBTI recovery (see next paragraph) during the switching from stressing to testing in region 1 then in region 4. NBTI saturation effects are, hence, apparent and such that NBTI modeling based

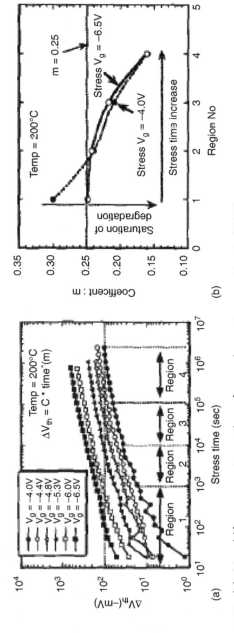

Figure 4.39. (a) V_{th} shift versus stress time for various stresses V_g at 200°C degradation. V_{th} shift seems to saturate as stress time increases. (b) Coefficient (m) of power law versus stress time region. m decreases with increasing stress time. Reprinted with permission from Ref. 39, © 2004 IEEE.

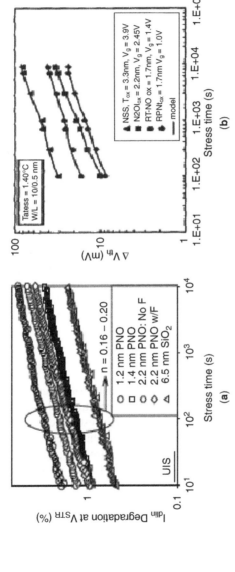

Figure 4.40. (a) I_{dlin} degradation versus stress time during uninterrupted NBTI stresses (different $V_{gstress}$ for different gate oxide processes) showing a power law time behavior with a little difference in time slope n (≈ 0.16–0.2) for different oxide thickness, nitrogen and fluorine concentration. Similar power law behavior is observed in (b) with a time n slope around 0.25. Reprinted with permission from Ref. 86, © 2001 IEEE.

on these stressing/testing procedures will give the wrong physics and end of life projections. This interpretation is confirmed by NBTI experiments where the NBTI shift was measured during the stress (uninterrupted stress). Figure 4.40 (a) shows the I_{dlin} shift measured during an uninterrupted stress (V_{ds} in linear mode) for different oxide processes [85]. The NBTI damage is observed to follow a power law time dependence with a time slope n around 0.16–0.2 and does not show any significant dependence on nitrogen, fluorine concentration, or voltage and temperature. Similar power law behavior but with $n \approx 0.25$ (Fig. 4.40b) has being observed by other researchers [86]. As we will describe later, an n value of $n \approx 0.16$ and 0.25 are explained, under the reaction–diffusion picture, as due to interface state generation respectively under a H_2 and H_0 diffusion controlled process.

2. The effects of recovery are negligible even during the early phases of the degradation (region 1–2) so that the observed V_{th} saturation effects are real. These assumptions are the baseline for NBTI models like the stretched exponential models which are described in [4, 40] and are based on the dispersive nature of the N_{it} generation mechanism. A direct evidence of is the observed increase of the n exponent with increase in temperature as discussed in Section 4.8.3 (see Fig. 4.63).

Here we want to emphasize that caution must be exercised when comparing NBTI time evolution data with different experimental setups and different stressing and testing procedures. These conclusions may be inconsistent and lead to wrong insights on the physics of the NBTI damage and possible NBTI modeling for a correct end-of-life extrapolation.

4.7.6 Recovery Phenomena

Recently, NBTI recovery has been receiving a lot of focus considering that this phenomenon may both determine, under AC operation, a less severe net MOSFET parameter shift compared to NBTI induced shift estimated at DC conditions and impact the stress/test methodologies adopted to characterize and model the NBTI damage. As shown schematically in (b) of Figure 4.13, recovery, after an NBTI stressing is been interrupted, has been attributed to two possible physical processes:

1. Hole detrapping and/or generated bulk trap neutralization by electrons injection.
2. N_{it} repassivation (supported by R–D physics).

No general agreement has been established on the relative contribution of these two physical processes to this phenomenon. The root cause of this confusion is, as usual for NBTI, attributable to the strong NBTI dependence on the details of the adopted CMOS process for the pMOSFET device under investigation as well as the methodology used for characterize this phenomenon. In this section, we will limit ourselves to describe some key experimental findings supporting the nature of this phenomenon.

By the use of charge pumping, Huard et al. [41] conclude that the observed V_{th} recovery is due to the detrapping of the hole traps from the gate oxide during the passivation phase with the interface states density (N_{it}) remaining constant. Experimental evidence of this hole detrapping contribution is shown in Figure 4.41. In this case the NBTI recovery is attributed to the positive charge neutralization by electrons when the channel is in accumulation after stress. As suggested by Figure 4.41 (a), the generated N_{it} during the stress do not seem to repassivate, while the V_{th} recovery is controlled by hole detrapping during the accumulation phase and the damage resets to the same level reached before the accumulation phase when the stress is restarted. Figure 4.41 (b) suggests that the hole trapping level (N_{ot} in the figure) basically reaches a steady state level which is mainly controlled by the gate voltage bias conditions. This particular observation suggest hole trap activation, rather than generation at given V_g at stress.

Evidence of hole detrapping is further given by the log(t) behavior (Fig. 4.42) observed during a fast monitoring of drain current increase during the recovery phase [88]. As it will described later (Section 4.9) the log(t) behavior is a sign of hole detrapping to follow a possible tunneling front kinetics. The similar time evolution of fast recovery at different stress times is mainly dependent on the V_g stress–V_g test difference. This is expected if the NBTI damage responsible for the recovery is the hole detrapping from traps filled during the NBTI stress. It also observed that deviations from log(t) behavior are observed after 1 sec stress. For t(recovery)>0.1 sec a different recovery mechanism is observed (possibly N_{it} repassivation).

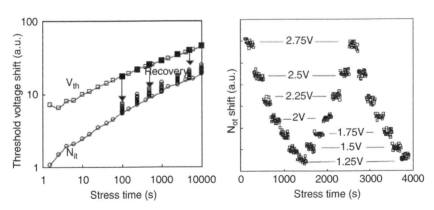

Figure 4.41. (a) V_{th} shift (squares) and N_{it}-induced V_{th} shift (circles) during an NBTI DC stress. Some devices (filled symbols) get an interrupted stress, which was followed by a recovery phase for different interruption times. The permanent threshold voltage shift component remaining after recovery corresponds to the interface traps creation. (b) Hole trap shift (N_{ot}) as function of the stress time. The gate voltage is changed during the stress. Reprinted with permission from Ref. 41, © 2004 IEEE.

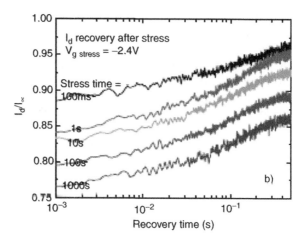

Figure 4.42. The same fast recovery in drain current is seen after stress is removed at different stress time and despite of the long stress time. Reprinted with permission from Ref. 88, © 2006 IEEE.

NBTI recovery by neutralization of generated bulk traps due to electron injection is observed by Tsujikawa et al. [87]. N_{it} generation and repassivation is suggested by the authors from the proportionality between the V_{th} and the subthreshold slope S (ΔS) shift during the NBTI stress and a recovery phase measured at $V_g = 0$ V (Fig. 4.43a). The nonproportionality between ΔS and ΔV_{th} observed during an accumulation phase $(V_g = 1.2$ V) after the NBTI stress suggests the V_{th} shift in this case is due to just a flatband shift decrease suggesting bulk traps neutralization due to electron injection. These data suggest that bulk traps generation can take place during an NBTI stress. These traps act as fixed charge unless neutralized by electron injection.

Evidence that the NBTI recovery is due to a N_{it} repassivation at the Si/SiO$_2$ interface has been found by the use of a DC current–voltage method (DCIV) ([42–44]). The suggestion that the V_{th} recovery is due by a H$_0$ diffusion controlled N_{it} repassivation process is given for the first time by Chen et al. [43]. The authors observed that during a pulsed NBTI experiment [$T = 100°C$, $V_g = -2.5$ V (1000 s), $V_g = +2.5$ V (1000 sec)] the V_{th} recovers with the same power law time evolution as observed during the NBTI stress. The time slope parameter n was estimated around 0.25 in both cases, suggesting that the N_{it} generation and repassivation is contributed by a reaction–diffusion process mainly controlled by H$_0$ diffusion in a homogeneous medium. See Figure 4.44.

A very interesting observation confirming the role of N_{it} repassivation during the NBTI recovery is given by the lock-in effect [89]. Figure 4.45 (a) shows the I_{dlin} fractional recovery ($I_{dlin}(t)/I_{dlin0}$ with I_{dlin0} measure at 1 ms delay after the stress interruption) remaining as a function of recovery time for stresses done at 125, 25 and −40°C for the range of stress fields and oxide thicknesses investigated in this work. Recovery is almost 90% after 24 hr for the 25°C data. The 25 and −40°C

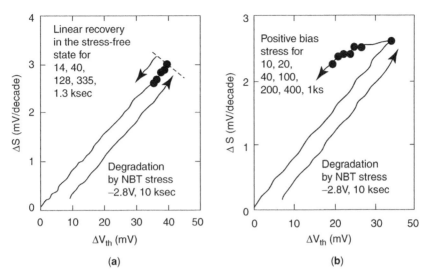

Figure 4.43. (a) Recovery of V_{th} and substhreshold slope (SS) in $V_g = 0\,V$ condition after an NBTI stress. (b) Recovery of V_{th} and SS shift during a positive bias ($V_g = 1.2\,V$) accumulation phase after an NBTI stress. Reprinted with permission from Ref. 87, © 2003 IEEE.

curves overlap independently of stress temperature and oxide field, as well as the amount of degradation. At 125°C the recovery is slower and seems to saturate before completion of the recovery phase. Physically, it is a question of whether the high temperature slows down recovery, if recovery has negative activation energy, or whether the higher temperature accelerates a competing mechanism. To answer this question, a device was stressed and recovered for 24 hr at 125°C (Fig. 4.45b). The temperature was then reduced to 25°C for 24 hr and finally the temperature was increased back to 125°C and the I_{dlin} measured. Little recovery occurred at 25°C even though 24 hr at 25°C normally results in 90% recovery. These observation suggest that high temperature (125°C) accelerates a permanent hardening of the oxide damage, locking-in the NBTI and reducing recovery rate. The main role of N_{it} repassivation in this high temperature process is suggested by the observed similarity between the fractional charge pumping and I_{dlin} increase during recovery (Fig. 4.45a). The proposed physical picture underlying this lock-in phenomenon is the following: NBTI recovery is mainly due to N_{it} repassivation. The lock-in phenomenon observed at high temperature is due to H not becoming available for repassivation due to the loss of H through diffusion in the poly or through high temperature assisted formation of stable hydrogenic species (bulk traps) which do not participate to the recovery phenomena.

Two comments can be raised on the conclusions drawn by the authors:

1. The fractional recovery seems to follow a log of time behavior at any temperature. This observation in the contest of N_{it} R–D model can be

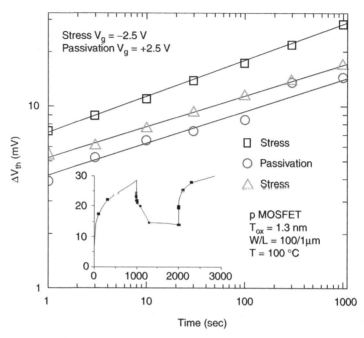

Figure 4.44. V_{th} shift versus stress time for stress–passivation–stress sequence. The similar time slope ($n \approx 0.25$) measured during the stress and passivation phases suggests a similar physical mechanism controls both phases. In this case it is assumed that H_0 diffusion controlled N_{it} generation and passivation is responsible for the recovery. Reprinted with permission from Ref. 43, © 2003 IEEE.

explained only by N_{it} repassivation by back diffusion in a dispersive medium.

2. The overlap in the observed fractional I_{dlin} recovery at -40 and 25°C seems to indicate that at least at these temperatures this recovery phenomenon is the result of a physical process not thermally activated suggesting, in this case, the possible role of hole detrapping.

The observations so far provided suggest that in trying to develop a physical understanding and modeling of NBTI (DC and AC) during the qualification of a given CMOS process four important aspects of the NBTI recovery need to be investigated:

1. Understand and quantify the role of interface states (donor) and hole traps and bulk traps formation during NBTI stress. Different recovery levels are expected depending on the role of interface states and hole traps and bulk traps during the NBTI degradation for a given gate oxide process and thickness.

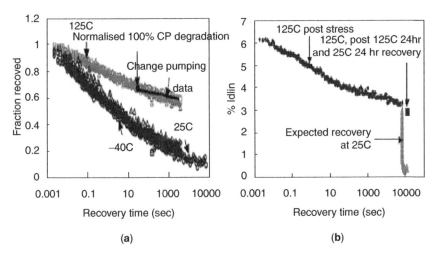

Figure 4.45. (a) Fraction remaining as function of recovery time. Recovery is independent on the amount of degradation. Lock-in effect is observed at 125°C. Similar recovery between charge pumping and I_{dlin} shift suggest the N_{it} repassivation is the main contributor. (b) Lock-in effect is permanent. The device shows no additional recovery even if the temperature is reduced to 25°C proving that there is a permanent change that is accelerated with increase in temperature. Reprinted with permission from Ref. 89, © 2003 IEEE.

2. Estimate and model the NBTI recovery rate as function of voltage (stress/ test) and temperature.
3. The effect of the recovery on the overall AC NBTI degradation. In particular the relation between NBTI recovery rate and the off state ($V_g \geq 0$) condition
4. Impact of NBTI recovery phenomena to stress and test methodology adopted to characterize NBTI. This particular topic has being the focus of a lot of investigation in recent years given the fact that without a well defined and agreed procedure for characterizing NBTI different conclusions about the physics of the NBTI damage can be drawn for a given process.

4.7.7 Impact of NBTI Recovery to NBTI Stressing/Testing Methodologies

Here we will spend few words on one aspect of the NBTI recovery phenomena that has a profound impact on the way we model NBTI: the effect of NBTI recovery to device characterization. Traditionally to characterize the effect of the NBTI damage to a given device parameter (V_{th}, I_d, G_{max}, etc.) at the assumed use bias conditions, it is necessary to interrupt the NBTI stress at a desired time intervals to make a measurement. The stress is resumed with alternate stress and test time sequence. Typically the device parameters are obtained from I_d–V_g and I_d–V_d measurements. As discussed in the previous paragraph, it is expected that the NBTI recovery phenomena will take place during the testing phase impacting the NBTI characterization. In particular, delays that are not carefully monitored between the end of the stress period and the beginning of the testing dramatically impact the monitoring of the device degradation during the NBTI stress and the corresponding NBTI modeling.

As example Figure 4.46 qualitatively illustrates, the measurement error obtained due to the fast recovery of the threshold voltage. When the stress gate voltage is removed to make the V_{th} measurement at $V_g = V_{gtest}$ part of the degradation is relaxing. Figure 4.46 (a) gives a graphical description of the evolution of the NBTI damage measured with controlled delay ($\Delta = 0.4$– 100 sec) between the stop of the stress time, t_{stress}, (at $V_g = V_{gstress}$) and beginning of the of the V_{th} measurement (at $V_g = V_{gtest}$). The corresponding change in the power law time slope (n) as function of Δ [32] is given in Figure 4.46 (b). This apparent increase of the n value with the increase measurement delay (0.2–0.24) is also strongly dependent on the allowed stress time and is characteristic of wafer level stressing procedure with uncontrolled measurements delays and traditionally used relatively short stress times (3000–10000 sec). It also important to notice that the time delay Δ is the sum of the time needed to swing the gate from $V_{gstress}$ to V_{gtest} plus the time needed by the instrumentation to make accurate

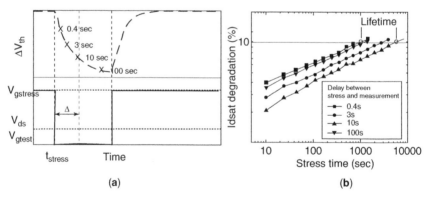

(a)   (b)

Figure 4.46. (a) Evolution of the NBTI damage (ΔV_{th}) measured with controlled delay (0.4 to 100 sec) between the stop of the stress and beginning of the measurement. (b) Apparent increase with increasing delay. (a) I_{dlin} shifts (%) measured at different measurement gate voltages (V_{gsm}) plotted against V_{th} shift. For a given V_{gsm} the same linear relation is found after various stress biases ($V_{gstress} = -1.5$ to -2.5 V) at the same temperature measurement (125°C) as long as the device is in strong inversion ($V_{gsm} \ll V_{th}$). (b) I_{dlin} changes from OTF stress at -2.0 V followed by recovery at 0 V (I_{dlin} measured at -1 and -0.5 V) converted by using Equation 4.41 into V_{th} shift (shown as DVt). Reprinted with permission from Ref. 32, © 2003 IEEE.

measurements. Limitations from the testing equipment may imply some constraints on getting accurate very short time delays ($\Delta \ll 1$ ms). In this situation a noncontrolled use of the measurement delay is expected to result into an obvious wrong estimate of the end of life NBTI induced impact at use conditions. This fast transient effect has been shown by Ershov et al. in 2003 [22, 57].

The understanding that an accurate estimate of time delays (Δ) during measurements is quite important led to the development of several measurement techniques to estimate the real NBTI shifts as accurately as possible. We will describe in the next paragraph the basic ideas of two general approaches: the on-the-fly (OTF) and the fast switching (FS) techniques.

4.7.7.1 *On-The-Fly Techniques.* The OTF stress/test methodology was introduced for the first time by Huard et al. [24]. The main objective of this approach it to capture the total device degradation activated during the NBTI stress. This is accomplished by monitoring the linear drain current ($I_{\text{dlin,st}}$) without interrupting the NBTI stress (no delays). The applied V_{ds} is small enough not to disturb the gate oxide field at a given V_{gstress} along the channel, thus the same NBTI degradation is measured respect to the NBTI symmetric configuration.

A key challenge of this methodology is the evaluation of the V_{th} shift (ΔV_{th}) from the measured $I_{\text{dlin,st}}$ shift ($\Delta I_{\text{dlin,st}}/I_{\text{dlin,st}}$). In addition of being threshold voltage shift a device parameter that directly related to the NBTI physical damage through a change in flatband shift, it can be used as a normalization parameter to calculate the I_{dlin} at a given V_{gs} conditions used in a circuit through the equation:

$$\frac{\Delta I_{\text{dlin}}}{I_{\text{dlin}}}(t) \propto \frac{\Delta V_{\text{th}(t)}}{Vgs - V_{\text{th}(0)}} \qquad (4.40)$$

Several variations of the OTF procedure has being proposed ([24, 91]) to overcome these limitations. Here we describe an OTF approach recently proposed by Parthsarathy et al. [25], which has the advantage to extract ΔV_{th} by experimental means. The main idea of this procedure is based on the observation that, as long as the pMOSFET device is in inversion ($V_{\text{gs}} \ll V_{\text{th}}$), an experimental linear relation between the I_{dlin} and V_{th} shift is found such that

$$\frac{\Delta I_{\text{dlin}}}{I_{\text{dlin}}}(t) = M_{\text{exp}}(V_{\text{gsm}}) \times \Delta V_{\text{th}} \qquad (4.41)$$

The experimentally extracted slope, M_{exp}, defines a linear relation between the V_{th} and I_{dlin} shift at a given V_{gs} measurement bias (V_{gsm}). Experimental evidence of this linearity is given in Figure 4.47 (a) (see [25] for details). Equation 4.41 is observed to hold as long as the device is kept in inversion. If the device is in subthreshold or weak inversion ($V_{\text{gsm}} \geq V_{\text{th}}$) the dependence between the I_{dlin} and V_{th} shift is not linear any longer as expected from the subthreshold characteristics. In addition it is found that Equation 4.41 holds at the same V_{gsm} for different V_{gstress}.

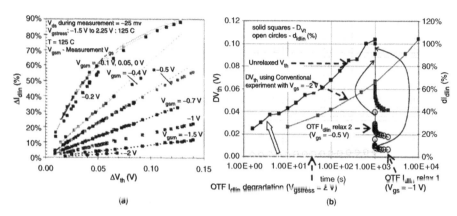

Figure 4.47. (a) I_{dlin} shift (%) measured at different V_{gsm} plotted against V_{th} shifts. (b) I_{dlin} changes from OTF stress at -2.0 V followed by recovery at 0 V. Reprinted with permission from Ref. 25, © 2005 IEEE.

Based on these findings, the following procedure is recommended:

1. Perform an interrupted OTF measurement of $I_{dlin,st}$ for a given $V_{ds}, V_{gstress}$ and temperature.

2. On different chips from the same process, perform NBTI experiments using conventional methods to measure I_{dlin} versus V_{gs} curves at the same V_{ds} conditions as stress. A long delay (> 50 sec) needs to be applied before each measurement to reduce the effect of the NBTI recovery on the measurements. The V_{th} and I_{dlin} shift at different V_{gsm} (V_{gsm} between V_{th} and $V_{gstress}$).

3. Evaluate the relation Equation 4.41 at $V_{gsm} = V_{gstress}$.

4. Calculate the ΔV_{th} shift from $\frac{\Delta I_{dlin,st}}{I_{dlin,st}}$.

Although there are no measurement delays, this OTF procedure suffers some key limitations.

1. Measurement of the first data point after the initiation of the stress requires a certain amount of time from the moment the stress is started (≈ 1 msec). This delay depends strongly on the instrumentation used for the NBTI stress. Hence this method will not be able to take into account any possible large shift experienced by the device during the very early phase of the degradation (within few msec), leading to an incorrect estimate of the $\% I_{dlin,st}$ shift. An example of this difficulty is given in Figure 4.48.

2. The $\% I_{dlin,st}$ shift measured with the OTF procedures is sensitive to the hole traps that are activated at $V_{gs} = V_{gstress}$ conditions. This means that I_d measured at stress conditions is strongly dependent on a trapping level that

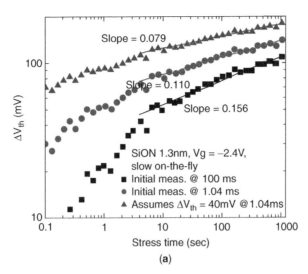

Figure 4.48. On-the-fly measurement with slow determination of the initial V_{th}. Reprinted with permission from Ref. 88, © 2006 IEEE.

is not experienced if measured at use conditions ($V_{gstress} < V_{use}$). Therefore the I_{dlin} evolution at stress and consequently any lifetime estimate based on this parameter is strongly dependent on the hole trapping density as result of the process. Figure 4.49 clearly demonstrates how generally the OTF procedure depends on the initial trapping level in the oxide [58].

4.7.7.2 Fast Switching Methodologies.

The basic idea of the fast switching (FS) methodologies originated from the need to monitor the NBTI damage at use conditions by means of a very short and controlled time delay to control and eventually minimize the effect of recovery to the testing procedure. Kaczer et al. [90] proposed a simple characterization method based on a single V_g measurement of the linear drain current at use conditions to reduce this delay.

This procedure is outlined in Figure 4.50. Instead of performing a full I_d versus V_{gs} sweep to characterize the effect of the NBTI damage to the device degradation, at a given time t_{stress} the gate voltage at stress ($V_{gstress}$) is reduced to a constant value at test (V_{gtest}) to measure I_{dlin}. V_{gtest} is constant during each readout and selected around the initially measured value of V_{th} ($V_{gtest} \approx V_{th0}$). At this bias condition only one value for the drain current, I_{dtest}, is measured after a given time delay Δ. Using the initial $I_d - V_{gs}$ dependence, an approximation for the new threshold voltage (ΔV_{th}) can be extracted by horizontally shifting the unstressed characteristics. The drain voltage is kept constant between the stress and test sequence (V_{ds} in linear mode) to reduce further switching delays. As only one drain current has to be obtained, instead of a range of drain currents for different gate biases, this approach is drastically faster and the errors due to the fast recovery are reduced.

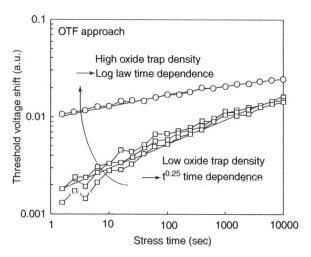

Figure 4.49. V_{th} shift for two different processes with low (squares) and high (circles) oxide trap densities as monitored by the OTF approach. When trap density is high a log of time dependence is found. When the hole trap density is low a power law time dependence is observed. Reprinted with permission from Ref. 58, © 2006 IEEE.

Because of the small dependence of the subthreshold slope shift as compared to the V_{th} shift in thin oxide, the V_{th} shift at the test time ($t_{stress} + \Delta$) can estimated by selecting V_{gtest} in the subthreshold region of the device and using the relation:

$$\Delta V_{th}(t_{stress} + \Delta) = \frac{\log_{10}\left(L_d(0)/I_d(t_{stress} + \Delta)\right)}{SS(0)}$$

where $SS(0)$ and $I_d(0)$ is respectively the initial value of the subthreshold slope and drain current.

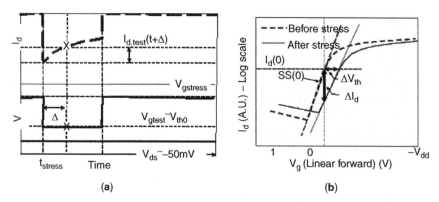

Figure 4.50. The basic experimental procedure at the basis for the fast switching methodology.

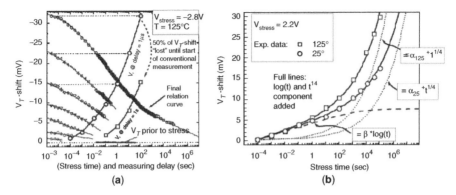

Figure 4.51. (a) Results of a typical stress measurement. Stress is interrupted at predetermined times and V_{th} is monitored for a short time. Then stress is continued. The same time axis is used for stress time and relaxation time. Also shown is the degradation curve for conventional ($\Delta = 1$ sec) and very fast switch ($\Delta = 1$ μs) measurement technique. (b) V_{th} shift as function of stress time for two different temperatures (25°C and 125°C). Reprinted with permission from Ref. 23, © 2006 IEEE.

Figure 4.51 gives an example of a fast switching technique were the V_{th} (in saturation) shift is directly measured rather than been estimated only indirectly from an analysis of the I_d shifts by the $I_d V_g$ behavior [23]. This direct V_{th} measurement during the NBTI stress and recovery is obtained by the use of an operational amplifier based circuit allowing the V_{th} to be measured with very short time delay ($\Delta \approx 1$ μs) after stress.

This work clearly describes the impact of V_{th} shift recovery free behavior ($\Delta \approx 1$ μs) as well as function of each recovery time from a single experiment. Any interruption of stress influences the degradation curve due to recovery. However, if the stress period following the readout is more than a factor of 100 longer than the readout time, the sample completely "forgets" the effect of the recovery. Figure 4.51 (a) illustrates the fast measured ($\Delta = 1$ μs) V_{th} shifts during stress as well as its recovery for each stress time. The same horizontal scale is used for both types of curves. The V_{th} degradation curve one obtains from a conventional measurement technique was taken from the relaxation curves at a delay of 1 sec. The difference between the "fast" and the conventional technique is about 50%. For the points at shorter stress times the relative "loss" gets even higher. This clearly illustrates that lifetimes taken from conventional measurement technique might be seriously misleading.

Figure 4.51 (b) shows a comparison of V_{th} shifts ($\Delta = 1$ μs) for two stress temperatures (25°C, 125°C) covering stress times between 10^{-4} to 10^6 sec. Both curves can be divided in a short term part up to about 1 sec which is common to both curves and a long term part from 1 sec to the longer stress times. The long term part of both curves is thermally activated and shows a power-law time behavior with slope $n = 1/4$. This portion of the V_{th} evolution has been attributed to N_{it} generation driven by a reaction–diffusion process (Section 4.8). The activation energy is about 0.1 eV. The short term part, which is only detected for stress times < 1 sec and needs a fast V_{th} determination, is observed to be the same for both temperatures and follows a time dependence linear in log(t). These two facts suggest this short term part is due to hole trapping in traps precursors. The de-trapping data given in Figure 4.51 (a) can also be well explained by hole traps effects. We will give more details on this phenomenon in Section 4.9.

The observed V_{th} shift (full line) is suggested to be the sum of two contributions. Hole trapping, following a log(t) behavior (---), is observed independently of the stress temperature and saturates after 1 sec stress (hole trapping in bulk traps precursors). N_{it} generation (....) with a power law time behavior ($n = 1/4$) which is dominant for stress time larger that 1 sec.

4.7.8 Dynamic NBTI

Since the initial observations of the worsening of the DC NBTI degradation with CMOS scaling, it also has recognized that NBTI may be the dominant FEOL wearout mechanism contributing to circuit aging, particularly at high temperature operations. An example of this is illustrated for the case of ring oscillator (RO) delay in [95]. ROs are often used as prototype circuits to monitor device performance during process technology development and can be also used to investigate circuit level degradation under typical operating conditions. During an RO stress, a number of different wearout mechanisms like channel hot carrier (CHC) and NBTI will be contributing to the RO delay. In Figure 4.52, a comparison between measured RO delays during RO stresses at room (30°C) and high (125°C) temperature is shown. The total contribution of both the CHC nMOSFET/pMOSFET as well NBTI pMOSFET to the RO aging was estimated

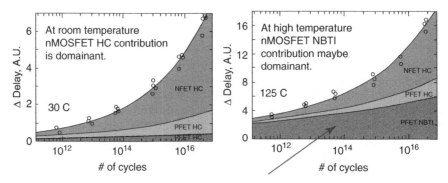

Figure 4.52. The effect of nMOSFET/pMOSFET (nfet/pfet in the figure) CHC and pMOSFET NBTI to RO delay were simulated and compared with experimentally measured values (open circles) at 30°C and 125°C. pMOSFET NBTI is expected to play a dominant role to circuit level degradation at high temperature. Reprinted with permission from Ref. 95, © 2002 IEEE.

by a simulation based on both DC generated CHC and NBTI models. It is observed that while in the technology under investigation, the nMOSFET CHC is the main contributor to the total RO delay at 30°C, the pMOSFET NBTI does play a critical role in RO aging at higher temperature (125°C). A larger weight of NBTI is expected with CMOS scaling [96].

In the simple case of an RO, the relation between the delay of an inverter stage (τ_D) and the NBTI induced V_{th} shift (ΔV_{th}) in a pMOSFET device is given by

$$\tau_D = \frac{C_{ox} \times V_{dd}}{2} \left(\frac{1}{I_{ds,n}} + \frac{1}{I_{ds,n}} \right), \tag{4.42}$$

where C_{ox} is the oxide capacitance, V_{dd} is the application bias and $I_{ds,n}$ and $I_{ds,p}$ are the drain current in saturation ($V_{gs} = V_{dd}$), respectively, for the nMOS and pMOSFET. It is easy to estimate from the above equation the effect to τ_D if only the NBTI-induced V_{th} shift (ΔV_{th}) contributes to RO aging. Since, assuming ballistic transport $I_{ds,p} \approx (V_{gs} - V_{th})$, we can write the degradation in RO delay ($\Delta \tau_D$) in terms of pMOSFET NBTI aging as follows:

$$\Delta \tau_D = \frac{C_{ox} \times V_{dd}}{2} \left(\frac{1}{V_{dd} - V_{th}(0) - \Delta V_{th}} - \frac{1}{V_{dd} - V_{th}(0)} \right), \tag{4.43}$$

with $V_{th}(0)$ being the initial pMOSFET V_{th}.

A frequency decrease (τ_D increase) is expected as V_{dd} is reduced. This trend was observed by Reddy et al. [97] by running static NBTI stresses in ROs. As discussed in the previous paragraph, it is expected that NBTI recovery does play a key role in reducing the effect of NBTI at circuit level compared to DC operation. In the case of an RO, recovery is, in fact, active 50% of the time (pMOSFET biases: $V_g = 0$ V, $V_s = V_d = V_{nw} = 0$ V) and produces an effective reduction to RO frequency degradation. A clear experimental evidence of this behavior is given by [97].

In Figure 4.53 from [96], the frequency degradation at 175°C is shown for both 100 MHz and 3 GHz as well as the I_{dsat} shift due to DC NBTI. A four-decade improvement in lifetime is observed for the RO (as compared to DC stressed devices). In addition, the NBTI frequency degradation seems to be a weak function of the RO frequency. No clear experimental evidence has been found in the past on the frequency dependence of the NBTI damage. Some works have claimed that the NBTI shift is frequency independent [43], while others have given experimental evidence that the frequency dependence, even if weak, is nonetheless important [103]. As already indicated previously, a lot of these results suffer from the stress and test methodology adopted to monitor their AC NBTI behavior. Figure 4.53.

Perhaps the most updated account of the NBTI dynamic effects has been given by Fernandez et al. [94]. In this work, some key experimental results are given about the relation between DC and AC NBTI behavior. This comparison between DC and AC NBTI was carried out by the use of the on-chip circuit illustrated in Figure 4.54.

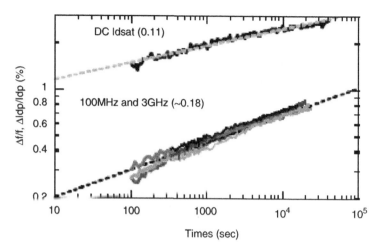

Figure 4.53. RO frequency degradation at 175°C. The I_{dsat} degradation for DC stress on a PMOS transistor at the same temperature is also shown. Reprinted with permission from Ref. 96, © 2006 IEEE.

In this case, AC NBTI experiments on a single device were carried out by using RO oscillators ($V_{select} = V_{cc}$) or external pulse generators applying the pulsed V_g ($V_{select} = 0$ V) to a single pMOSFET. The DC stresses are applied to the single pMOSFET in a similar way ($V_{select} = 0$ V). By the use of this on-chip circuit the authors were able to generate, for the first time, an experimental comparison

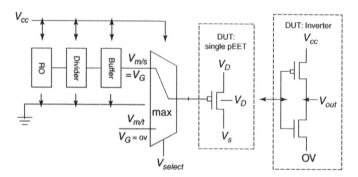

Figure 4.54. The circuit for on-chip studies of DC and AC NBTI consists of a 41-stage ring oscillator, a frequency divider, a buffer, a pass-gate-based multiplexer, and the device under test (DUT), which is either a single pMOS or an inverter. When $V_{select} = 0$ V, V_g is connected to the DUT, allowing for direct stressing (DC or pulsed waveform) and characterization. When $V_{select} = V_{cc}$, the DUT is connected to the oscillator, resulting in very high frequency AC stress (duty cycle 50%). $L = 0.13$ μm, pMOSFET $W = 2$ μm, inverter $W_p = 6$ μm, $W_n = 3$ μm. Reprinted with permission from Ref. 94, © 2006 IEEE.

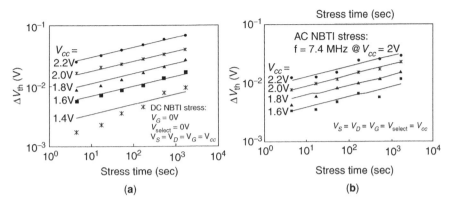

Figure 4.55. ΔV_{th} as a function of stress time for varying stress voltage for (a) DC and (b) AC NBTI stress conditions. Both data sets are fitted with a power law with the respective exponents common for all stress voltages. Typically observed exponents are 0.17–0.20. Reprinted with permission from Ref. 94, © 2006 IEEE.

between DC and AC NBTI (50% duty factor) that covers a range between 1 Hz to 2 GHz. Three important observations are provided in this work:

1. *Similar NBTI behavior is observed between DC and AC NBTI.* In this experiment the same stress and test (I_d–V_g measurement) sequence was carried out between a series DC stresses ($V_{gs} = -1.4$ to $-2.2\,\text{V}$) and AC NBTI experiments. The AC NBTI stresses were run by using an external pulse generator at frequency of 7.4 Mhz and 50% duty factor with the pulse amplitude selected at the same DC V_{gs} conditions. Figure 4.55 indicates that a similar time slope (0.17–0.20) is observed between the AC and DC NBTI stresses that suggests similar NBTI physics applies in both cases.

2. *AC NBTI has no frequency dependence.* It is observed that AC NBTI seems independent of frequency in the entire frequency range of 1 Hz–2 GHz. The V_{th} shifts due to AC NBTI stress in a single pMOSFET are also shown in Figure 4.56 (a). The result of the corresponding DC NBTI stress obtained on identical devices is also shown for comparison. An AC to DC NBTI ratio of roughly around 0.55 is measured in these measurements. Experimental evidence suggests [85] that this AC/DC ratio is not a universal parameter. It is observed, in fact, to be less for thicker oxides (6.5 nm) (Fig. 4.56b) and have a different value (plateau) at low and high frequency. A simulation seems to explain this behavior based on N_{it} generation–recovery under the reaction–diffusion picture.

3. *AC NBTI duty factor dependence.* ΔV_{th} resulting from an AC NBTI stress ($V_g = -2.0\,\text{V}$) at 10 kHz with varying duty cycles is shown in Figure 4.57. The V_{th} shift after 1000 sec stress as function of duty cycle shows a

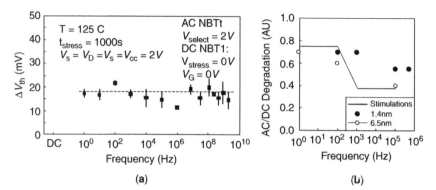

Figure 4.56. (a) The V_{th} shifts due to AC NBTI stress in a single pMOSFET are seen independent of frequency in the entire frequency range of 1 Hz–2 GHz. (b) AC/DC degradation as function of frequency. The existence of two plateaus at low and high frequency is observed for different T_{ox}. Reprinted with permission from Ref. 85, © 2005 IEEE.

monotonic dependence with an apparent plateau at around 50% duty cycle. The dependence is monotonic, with an apparent plateau around duty cycle 50%. The data can be well described by the following dependence:

$$\Delta V_{th} = \left[\left(\frac{DC - 50\%}{50\%} \right)^3 + 1 \right] \times \Delta V_{th}@50\%. \qquad (4.44)$$

Any physical description of the NBTI degradation needs to take these three observations into account.

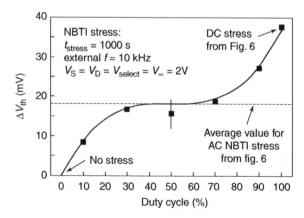

Figure 4.57. ΔV_{th} resulting from AC NBTI at 10 kHz with varying duty. Reprinted with permission from Ref. 94, © 2006 IEEE.

4.8 N_{it} GENERATION BY REACTION–DIFFUSION (R–D) PROCESSES

The reaction–diffusion (R–D) picture is the generally accepted description of interface states generation that contributes to NBTI damage. In this physical picture, it is assumed that the interface states generation rate at the SiO_2/Si interface results from the contribution of two physical processes at a given temperature:

1. A balance between the breaking of the Si–H bond to produce interface states by a gate oxide (F_{ox}) field-dependent electrochemical reaction and an interface state repassivation.
2. A diffusion-limited process where the released hydrogen diffuses in a stable hydrogenated form (H*) away from the interface, leaving behind an interface state (Si*).

In the case of a pMOSFET devices, only donor-type interface states contribute to the NBTI-induced V_{th} shift. This process is schematically described in Figure 4.13, which suggests that interface states generation is not the only possible mechanism responsible for the positive-charge-induced V_{th} shift. As it has being experimentally observed, other mechanisms, such as hole trapping, contribute to the NBTI damage.

The R–D process can be generally described [69] as an example, by Equations 4.45 through 4.47.

$$\{(Si-H) + Y\} + A^+ \leftrightarrow B^+ + Si^* + H \tag{4.45}$$

(Electrochemical reaction at the SiO_2/Si interface)

$$\alpha \times H \leftrightarrow H^* \tag{4.46}$$

(Formation of a stable hydrogenated diffusing species H*)

$$H^* \rightarrow X_{Bulk}^+ \tag{4.47}$$

(Positive fix charge formation by H* trapping in bulk trapping centers)

Equation 4.45 provides a general description of an electrochemical reaction at the SiO_2/Si where, under the presence of A^+ species an interface state [Si*] is produced by hydrogen dissociation of the Si–H bond (part of an electrically inactive interface defect ($\{Si-H\} + Y\}$)) with the resulting reaction by-products H and possibly a positive fixed charge (B^+). H may diffuse in a stable hydrogenated species H* (Eq. 4.46) in the gate oxide or participate in the Si* repassivation process (Si* + H → Si–H). The diffusion of the H* species leaves the unpassivated Si–H bond (Si*) and leads to the possible formation of positively charged oxide defects (X_{Bulk}^+) in the bulk of the gate oxide (Eq. 4.47). The strong NBTI dependence on both FEOL and BEOL CMOS processing makes the details of the R–D process very controversial. There is no general agreement on the nature of the interface defect nor to the reaction by-product H*. It has also not been established whether the formation of positively charged oxide defects originates from the same process. Several authors have shown experimental evidence that different Y and H* species are involved in the electrochemical reaction described by Equation 4.45. Positively charged defects can possibly be generated in the bulk of the gate oxide. At this point the following two mechanisms have been considered to contribute:

(a) Reaction by-product B^+

(b) Direct interaction of the diffusing species H* with bulk trapping sites to produce (Eq. 4.47) the by-product of positively charged oxide defects (X^+_{Bulk}). Both positive charge defect centers contribute to the V_{th} shift through (ΔN^+_f).

Table 4.1 gives some suggested microscopic descriptions related to the R–D process. Most of the proposed microscopic pictures describing the R–D process suggest the following:

1. The A^+ species participating in the electrochemical reaction is made up by holes in the inverted channel in thermal equilibrium with the lattice. As discussed in Section 4.7.2, there is plenty of experimental evidence suggesting that the holes contributing to the NBTI damage are cold holes (to distinguish them from hot holes generated during carrier heating phenomena as such anode hole injection or channel hot carriers).

2. The stable diffusing species H* is probably a hydrogenated complex ($H^* = H^+$, H_0, H_2 etc). Hydrogen can diffuse in its atomic state H_0 ($\alpha = 1$), as molecular hydrogen H_2 ($\alpha = 2$) or as a proton (H^+) being part of the OH, H_3O^+ or HO^- groups.

As suggested by the last reaction listed in Table 4.1, a hole-assisted R–D process may not be the only way by which the NBTI damage can evolve. In this example [12], based on first principles calculations, it is assumed that interface states generation is due to Si–H bond breaking in the presence of protons (H^+) at the SiO_2/Si interface. The H from the Si–H bond reacts with the H^+ to form H_2, which diffuses away leaving behind an interface state. It is assumed that during the

TABLE 4.1. Suggested Microscopic Descriptions of the R–D Process

(Si–H) + Y	B^+	H*	A^+	Reference
(Si–H) + Si–O–Si	Si^+	Si–OH	h^+	[3]
(Si–H) + H_2	–	H_3O^+	h^+	[7] [8] [9]
Si–H	–	H^+	h^+	[9]
Si–H	$(Si^*)^+$	$\frac{1}{2}H_2$	h^+	[10] [11]
Si–H	$(Si^*)^+$	H^0	h^+	[10]
Si–H	$(Si^*)^+$	H_2	H^+	[10] [12]

NBTI stress, H is originated from P-H bonds thermal dissociation in the n-well and the H on the way to the SiO_2/Si interface picks up a hole in the inverted channel to become H^+.

4.8.1 Modeling of N_{it} Generation by Reaction–Diffusion Processes

As described in the previous section, the R–D picture is used to explain the N_{it} contribution to the NBTI damage. The general solution of the R–D model in the NBTI stress phase is very difficult to find for a given MOSFET geometry and diffusing species (H*). In particular, as suggested by Table 4.1, their dependence on the stress bias configuration (V_g, T at stress) strongly relates to the reactants involved in the R–D reaction. In order to describe the general characteristics of the evolution of the N_{it} generation during a constant NBTI stress we consider the commonly discussed case of N_{it} generation controlled by a hole-assisted electrochemical reaction and diffusion of a stable species in the gate oxide and in the poly.

In the rest of this section, we discuss the mathematical description of this N_{it} generation process.

1. N_{it} generation is controlled by a cold-hole-assisted process. To simplify the analysis, we assume that the hole assisted Si–H breaking is described by the electrochemical reaction

$$Si–H + h \leftrightarrow Si^* + H, \qquad (4.48)$$

with consequent diffusion of the released hydrogenated species H in the stable diffusing species H* in the dielectric. The H and H* species are in chemical equilibrium ($\alpha \times H \leftrightarrow H^*$), obeying the law of mass action at the Si/SiO_2 interface ($x = 0$):

$$\text{Const} = [N_H(0)]^\alpha / N_{H^*}(0). \qquad (4.49)$$

Conflicting information is available in the literature about the nature of the H* species. First, principle calculations suggest that H^+ is the most stable H-charge state and reacts directly with the Si–H bond [12], while H_0 (atomic H) is found unstable in both Si and SiO_2 [101]. H_2 diffuses interstitially and its dissociation is very unlikely [$H_2 \leftrightarrow 2H_0$ (1.74 eV), $H_2 \leftrightarrow H^+ - H^-$ (1.34 eV)].

We consider the general case of diffusion in both the gate oxide with the physical oxide thickness T_{ox} and the polygate considered to have infinite thickness. The rate equation that regulates the N_{it} generation then becomes

$$\frac{\partial N_{it}}{\partial t} = K_F(N_0 - N_{it}) - K_R N_{it}[N_{H^*}]^{1/\alpha} (\text{Reaction@}\, x = 0), \qquad (4.50)$$

where $x = 0$ denotes the Si/SiO_2 interface and $x > 0$ is (in the gate oxide) towards the gate, N_{it} is the number of interface traps generated at any given time during the stress, N_0 is the initial surface density of unbroken Si–H bonds, and $N_{H*}(0)$ is the H* concentration at the interface. The reaction kinetics is controlled by the temperature and oxide field dependence of the forward dissociation (K_F), as well as by the repassivation rate (K_R) constants. The value of α gives the order of the reaction. If neutral atomic H_0 is assumed to be diffusing species then $\alpha = 1$. For molecular neutral hydrogen diffusion H_2 $\alpha = 2$. The total number of interface traps (N_{it}) generated is equal to the number of H* released and diffusing in the gate oxide. This assumption implies that total interface states generated at the time, t, is such that

$$N_{it}(t) = \int_0^{+\infty} N_H^*(x, t)dx. \tag{4.51}$$

2. The breaking of the Si–H dipole is a thermally assisted process in the presence of holes in the inverted channel under the gate oxide field F_{ox} at the SiO_2/Si interface. The microscopic details of the N_{it} generation and repassivation at the SiO_2/Si interface are represented by the constants K_F and K_R, respectively, described in Figure 4.58. In this case, the forward

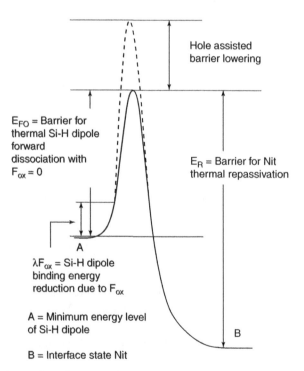

Figure 4.58. Schematics of energy barriers involved in the thermal breaking and repassivation of the Si–H bond under the electric field F_{ox}.

dissociation constant (K_F) should be dependent on the holes concentration (p_h) in the inverted channel channel and the probability (P_T) of a hole tunneling through the Si/SiO$_2$ interface and being captured by the Si–H bond (σ = capture cross section). K_F is also dependent on the probability (B(F_{ox})) of breaking the Si–H dipole whose binding energy is reduced by the hole capture and the presence of the gate oxide field F_{ox}. Assuming a thermal dipole dissociation model we get

$$B(F_{ox}) \approx \exp(-E_{F0} - \lambda F_{ox})/kT, \tag{4.52}$$

where E_{F0} is a barrier for forward thermal dissociation under barrier lowering at a given F_{ox} and λF_{ox} is the reduction of the dipole binding energy due to electric field F_{ox}. The probability P_T is also expected to be exponentially dependent on F_{ox} so that

$$P_T \approx \exp(\gamma_T F_{ox}). \tag{4.53}$$

Under these conditions,

$$K_F \approx p_h \times \sigma \times \exp\left(\gamma_T F_{ox} - \frac{E_{F0} - \lambda F_{ox}}{kT}\right) \tag{4.54}$$

The repassivation coefficient K_R should be dependent on the thermal repassivation barrier (E_R) through

$$K_R \approx \exp\left(-\frac{E_R}{kT}\right) \tag{4.55}$$

Notice that both the forward and repassivation barrier are reduced by the hole-assisted barrier lowering due to the hole image charge.

3. H* diffusion in the gate oxide with physical oxide thickness T_{ox}. The equilibrium of the forward and backward reaction is controlled by the H* density at the interface $N_{H*}(0)$. Thus, the transport mechanism of the H* species away from the Si/SiO$_2$ interface characterizes the degradation mechanism. The diffusive transport of H* in the gate oxide is described by the Fick's law:

$$D_{H*}^{ox} \frac{\partial}{\partial x}\left[\left(\frac{\partial N_{H*}}{\partial x}\right) - \frac{qF_{ox}}{k_T} N_{H*}\right] = \frac{\partial N_{H*}}{\partial t} \quad (0 < x < T_{ox}) \tag{4.56}$$

The presence of the oxide field (F_{ox}) term in this equation takes into account the case that H* is assumed to be positively charged (H$^+$). D_{H*}^{ox} is the H* diffusion coefficient in the gate oxide, respectively.

A controversial argument is the nature of the thermal transport of the H* in SiO$_2$. Typically it has been assumed in describing the N_{it} generation that the H* diffusion in gate oxide is associated with an Arrhenius thermally activated transport so that

$$D_{H*}^{ox} = D_{H*,0}^{ox} \times \exp\left(-E_a\left(D_{H*}^{ox}\right)/kT\right), \tag{4.57}$$

where $E_a\left(D_{H*}^{ox}\right)$ is the corresponding activation energy. It has been long established that the H* diffusion in a disorder dielectric such as SiO_2 is not Arrhenius and should be treated in a dispersive way [92]. This concept has only recently been applied in describing the NBTI-assisted N_{it} generation [11, 40]. As we will see in Section 4.8.3, the dispersive diffusion coefficient of the H* species is given by

$$D_{H*}^{ox} = D_{H*,0}^{ox} \times (vt)^{-p}, \tag{4.58}$$

where v is the jumping attempt frequency and p the dispersion parameter. For completeness we will consider both cases. However, no clear experimental evidence has been provided to support either one of these particle transport models exclusively in SiO_2.

4. H* diffusion in the poly gate oxide. The diffusion of the H* species in the poly gate is regulated by the following two equations:

$$D_{H*}^{ox} \frac{\partial N_{H*}}{\partial x} = D_{H*}^{Poly} \frac{\partial N_{H*}}{\partial x} \quad (x = T_{ox}) \tag{4.59}$$

$$D_{H*}^{Poly} \frac{\partial^2 N_{H*}}{\partial x^2} = \frac{\partial N_{H*}}{\partial t} \quad (x > T_{ox}) \tag{4.60}$$

Equation 4.59 and Equation 4.60 describe the diffusion of the H* species in the poly, assuming constant flux at the poly/SiO_2 interface ($x = T_{ox}$). D_{H*}^{Poly} is the H* diffusion coefficient in the poly. We will assume an Arrhenius thermally activated transport so that

$$D_{H*}^{Poly} = D_{H*,0}^{Poly} \times \exp(-E_a\left(D_{H*}^{Poly}\right)/kT), \tag{4.61}$$

where $E_a\left(D_{H*}^{Poly}\right)$ is the corresponding activation energy.

5. Relative contribution of H* diffusion in SiO_2 and poly to N_{it} generation. Table 4.2 [85] can help us to estimate the relative contribution of the H* diffusion in the gate oxide and in the poly. This can be done by estimating the stress time needed by the H* species, while diffusing in the gate oxide, to reach the poly/SiO_2 interface. Assuming a diffusion front (see next paragraph), this stress time can be estimated from the relation $\left(4 \times D_{H*}^{ox} \times t\right)^{1/2} = T_{ox}$ where T_{ox} is the oxide physical thickness around 1–2 nm (typical of today's advanced CMOS technologies). Considering, for example, that the diffusivity of molecular hydrogen H_2 in SiO_2 is around 1.32×10^2 nm^2/sec at 105°C (second reference in the table), it is easy to estimate that the H_2 profile will reach the SiO_2/poly interface ($T_{ox} = 2$ nm) within 7.5 ms. This time is much shorter than the stress time typically used during DC-accelerated DC-NBTI wafer-level stressing (1000–10000 sec). This means that during

TABLE 4.2. Diffusivities of Different Hydrogen Species in Several Materials

Diffusing Species	Diffusion Medium	Distance Diffused x (x (nm) – $\sqrt{4dt}$ t = s; T = 106°C	Reference
H_2	SiO_2	4.6 E+02	Ogawa et al. *Phys. Rev. B,* 51, 4216, (1995)
H_2	SiO_2	2.3 E+01	B. Fishbein et al., *J. Electochem. Soc.,* 134, 574 (1997)
H	SiO_2	9.4 E+04	B. Tuttle, *Phys. Rev. B,* 81, 4417, (200)
H	SiO_2	1.5 E+04	D. L. Griscom, *J. Appl. Phys.,* 58, 2524 (1988)
H_2	Poly-Si	5.3 E–01	M. H. Nichol et al., *Phys. Rev. B,* 52, 7791 (1955)
H_2	Poly-Si	5.4 E–01	B. L. Soppori et al. *Appl. Phys. Lett.,* 61, 2560 (1992)
H^+/HO	C-Si	3.5 E–01	Y. L. Huang et al. *J. Appl. Phys.* , 88, 7060 (2004)
H_2	C-Si	3.7 E+01	B. L. Soppori et al. *Appl. Phys. Lett.,* 61, 2560 (1992)
H/H_2	SiN	9.9 E–04	S. S. He et al. 4th Int. Conf. Sol. St. and IC Tech., 289, (1988)
H/H_2	SiN	7.8 E+14	A. Bik et al. *Appl. Phys. Lett.,* 60, 2883 (1990)

DC-typical stress conditions, most of the H_2 diffusion should be taking place in the poly gate.

In addition, we can conclude from the same table that the diffusivities of different neutral H* species (H_0, H_2) are such that diffusion in oxide > diffusion in the poly > diffusion in Si substrate ≫ diffusion in nitride. These considerations determine the realistic assumptions to be used in describing the diffusion of a given H* species and regulating the R–D process. In particular, the H* diffusion profile as controlled by the H* constant flux at the SiO_2/Si interface (Eq. 4.59).

4.8.2 R–D Kinetics Controlled By an Arrhenius Diffusion Process

Figure 4.59 gives a graphical representation of N_{it} generation controlled by H* Arrhenius diffusion in a homogeneous dielectric. In this case, the H* diffusion is assumed to proceed by H* hopping in similar potential wells uniformly distributed in the oxide. Under these conditions, the H* diffusion is represented by an Arrhenius thermally activated transport.

Si-H breaking

Si* ⟷ H ⟷ H*

Si-H Passivation

H* diffusion

Figure 4.59. Graphical representation of Arrhenius H* diffusion in the gate oxide.

In order to calculate the N_{it} generation, we assume that the diffusion of H* in both the gate oxide and the poly is associated to an Arrhenius-like-activated transport, so that Equation 4.57 and Equation 4.61 are satisfied. We describe the case of dispersive transport of SiO_2 in the next paragraph.

As assumed, we consider a simplified version of the R–D picture which consists of a two sequential steps process, i.e., an early reaction-controlled regime followed by an H* diffusion-driven regime. Following Alam [67] (see Fig. 4.60), the actual H* diffusion can be approximated by a triangular [(1), (2), (3), (4)] profile while within the gate oxide and a trapezoidal [(5)] profile when diffusing in the poly. This approximation is adequate enough to give the main features of the kinetics of the N_{it} generation. In this case, the distance traveled by the H* diffusion front can be approximated by the following equations:

i. While in the gate oxide

$$x = \left(D_{H*}^{ox} \times t\right)^{1/2} \quad \text{for a neutral } H^* \text{ species} \tag{4.62}$$

$$x = \mu_{H*}^{ox} \times F_{ox}t \quad \text{for a charged } H^* \text{ species} \tag{4.63}$$

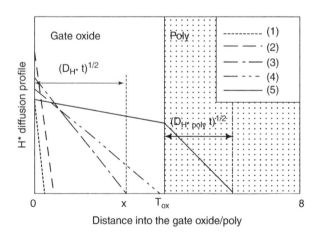

Figure 4.60. H* diffusion profile under both the reaction limited (1, 2), diffusion in the gate oxide (3, 4) and in the poly (5) diffusion regimes. In the case (4), the NBTI stress time is such that the H* profile reaches the gate oxide/poly interface $((D_{H*}t)^{1/2} = T_{ox})$. Case (5) describes the H* profile diffusing in the poly gate at distance $(D_{H*,poly}t)^{1/2}$. Reprinted with permission from Ref. 67, © 2003 IEEE.

ii. While in the poly

$$x = \left(D_{\text{H*}}^{\text{Poly}} \times t \right)^{1/2} \quad \text{for a neutral/charge H* species} \quad (4.64)$$

4.8.2.1 Reaction-Limited Regimes. We will describe two reaction regimes by assuming that the N_{it} generation is less than the initially passivated Si–H bond ($N_{\text{it}} \ll N_0$). In this case, the N_{it} generation rate can be approximated to $K_F N_0$. During the initial reaction-controlled regime the evolution of the NBTI damage is regulated by Equation 4.50. At the very early phase [profile (1)], both N_{it} and N_H are small compared to N_0 (typically $\approx 10^{12} – 10^{13}\,\text{cm}^{-2}$ after forming gas anneal) and hydrogen diffusion, as well repassivation are still negligible. Under these conditions it is easy to get from Equation 4.50

$$N_{\text{it}} \approx K_F N_0 t. \quad (4.65)$$

As more N_{it} is generated at the Si/SiO$_2$ interface, the repassivation rate in Equation 4.50 starts to contribute [profile (2)], but still not enough H* diffusion takes place ($N_{\text{H*}}(0) = N_{\text{it}}$). Eventually, both the forward and reverse rate will equalize $\left(\frac{\partial N_{\text{it}}}{\partial t} \approx 0 \right)$. Under these conditions we get

$$N_{\text{it}} \approx \sqrt{\frac{K_F}{K_R}} N_0. \quad (4.66)$$

This reaction-driven phase is typically short (μs to ms), and therefore is not observed in standard NBTI measurements.

4.8.2.2 H* Diffusion in SiO₂. The NBTI phase that is mostly observed in typical DC NBTI stresses is controlled by the H* diffusion. In this case we have the following processes:

1. The net trap generation is controlled by the H* diffusion and is negligible in comparison to the large fluxes on the right-hand side of Equation 4.50 $\left(\frac{\partial N_{\text{it}}}{\partial t} \approx 0 \right)$ so that

$$N_{\text{it}} \approx \frac{K_F}{K_R} \times \frac{N_0}{(N_{\text{H*}}(0))^{1/\alpha}}. \quad (4.67)$$

Considering the case that $x \ll T_{\text{ox}}$ [profile (3)], the integral in Equation 4.51 is shown by

$$N_{\text{it}}(t) \propto \frac{N_H^*(0) \times \sqrt{D_{\text{H*}}^{\text{ox}} t}}{2} \quad (4.68)$$

and

$$N_{it}(t) \propto \frac{N_H^*(0) \times \mu_{H^*}^{ox} \times F_{ox} \times t}{2} \qquad (4.69)$$

The N_{it} generation can be easily estimated by using Equation 4.67 together with Equations 4.68 and 4.69 respectively, for neutral and charged H* species. The following equations give the N_{it} time evolution expected for different diffusing species H* assuming that during the stress the H* diffusion profile has not reached the poly/SiO$_2$ interface $((D_{H^*}t)^{1/2} \ll T_{ox})$.

$$N_{it} \propto \left[\frac{K_F N_0}{2K_R} \right]^{\frac{\alpha}{1+\alpha}} \times (D_{H^*}^{ox})^{\frac{1}{2(1+\alpha)}} \quad \text{(for neutral H* species)} \qquad (4.70)$$

and

$$N_{it} \propto \left[\frac{K_F N_0}{2K_R} \right]^{\frac{\alpha}{1+\alpha}} \times (\mu_{H^*} F_{ox} t)^{\frac{1}{(1+\alpha)}} \quad \text{(for charged H* species)}, \qquad (4.71)$$

where $\alpha = 1, 2$ for atomic and molecular hydrogen, respectively.

4.8.2.3 H* Diffusion in SiO$_2$ and Poly.

The case [profile (5)] that the NBTI stress is long enough that the H* species reaches the poly/SiO$_2$ interface $((D_{H^*}^{OH}t)^{1/2} \gg T_{ox})$ quickly and mainly diffuses in the poly is probably the most realistic one in advanced CMOS technologies ($T_{ox} \approx 1$–2 nm). As an example of H* diffusion in the poly, we describe the case of a neutral H* diffusion. Using the trapezoidal approximation for the H* diffusion profile, the integral in Equation 4.51 can be approximated by

$$N_{it} = \frac{[N_{H^*}(T_{ox}) + N_{H^*}(0)]T_{ox}}{2} + \frac{1}{2} N_{H^*}(T_{ox}) \times \left(D_{H^*}^{Poly} t \right)^{1/2}, \qquad (4.72)$$

where $N_{H^*}(T_{ox})$ is the concentration at the SiO$_2$/Poly silicon interface.

Taking into account the H* flux continuity at this interface (Eq. 4.59), we have

$$D_{H^*}^{ox} = \left[\frac{N_{H^*}(0) - N_{H^*}(T_{ox})}{T_{ox}} \right] = D_{H^*}^{Poly} \frac{N_{H^*}(T_{ox})}{\left(D_{H^*}^{Poly} \times t \right)^{1/2}}. \qquad (4.73)$$

With stress time, the fraction H* atoms stored in the gate oxide become negligible in respect to the H* atoms diffusing in the poly. Under the condition $\left(D_{H^*}^{ox} t \right)^{1/2} \gg T_{ox}$, taking into account the above two equations, we can approximate N_{it} generation as

$$N_{it} \approx \frac{N_{H^*}(0)}{2} \times \left(D_{H^*}^{Poly} t \right)^{1/2}. \qquad (4.74)$$

From the above equation and Equation 4.67 it is easy to prove

$$N_{it} \propto \left[\frac{K_F N_0}{2K_R}\right]^{\frac{\alpha}{1+\alpha}} \times \left(D_{H^*}^{Poly} t\right)^{\frac{1}{2(1+\alpha)}}, \tag{4.75}$$

where α is equal to 1 for H_0 and 2 for H_2 diffusion. It is important to notice that in the case the NBTI stress time is long enough that $((D_{H^*}^{Poly} t)^{1/2} \gg T_{ox})$, most of the H^* diffusion takes place in the poly where no electric field is applied (equipotential surface) and the same N_{it} evolution is experienced independently of being the H^* a neutral or charged species. In this case, the N_{it} generation is evolving as if the H^* is neutral independently of its charge status.

In summary, assuming $N_{it} \ll N_0$ and mostly H^* diffusion control process in the poly, we can conclude that the N_{it} generation is regulated by a relation similar to Equation 4.75. Using this equation, three important consequences are expected:

1. After an initial linear time evolution ($N_{it} \approx t$), the N_{it} generation is expected to follow a power-law time dependence ($N_{it} \approx t^n$) with the time exponent n as

$$n = \frac{1}{2(1+\alpha)}. \tag{4.76}$$

   The above result suggests $n = 1/4$ for H_0 and $n = 1/6$ for H_2 diffusion. In addition no temperature dependence is assigned to the n exponent.

2. The N_{it} generation temperature dependence is related to the temperature dependence of K_F/K_R ratio and the H^* diffusion coefficient in the poly ($D_{H^*,poly}$). Assuming Equation 4.54 and Equation 4.55 hold, these parameters follow an Errhanius temperature dependence with the measuredactivation energy:

$$E_a = \frac{\alpha(E_a(K_F) + E_a(K_R))}{1+\alpha} + \frac{E_a\left(D_{H^*}^{Poly}\right)}{2(1+\alpha)}, \tag{4.77}$$

   with $E_a(K_F)$ and $E_a(K_R)$ being the activation energy for the generation and repassivation process respectively, while $E_a(D_{H^*,poly})$ is the activation energy associated with the H^* diffusion in the poly. For the quasi-equilibrium condition in Equation 4.77 to hold at different temperature, it is reasonable to assume $E_a(K_F) - E_a(K_R) \approx 0$ so that experimentally measured temperature activation energy (E_{meas}) can be approximated to

$$E_{meas} = \frac{E_a\left(D_{H^*}^{Poly}\right)}{2(1+\alpha)}. \tag{4.78}$$

3. The N_{it} generation dependence on the gate oxide field is determined by the K_F/K_R ratio as

$$N_{it} \approx \left[\frac{K_F}{K_R}\right]^{\frac{\alpha}{1-\alpha}}. \tag{4.79}$$

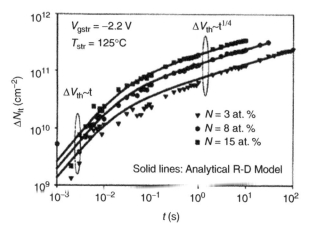

Figure 4.61. The measured and simulated N_{it} generation as function of stress time for nitrided oxides with nitrogen concentrations 3, 8, and 15 at.%. The NBTI stress is $V_g = -2.2\,V$, $T = 125°C$. Reprinted with permission from Ref. 68.

Assuming Equation 4.54 and Equation 4.55 hold, the F_{ox} dependence is given by

$$N_{it} \propto F_{ox} \times \exp^{(\gamma T + \lambda / kT)} F_{ox}. \qquad (4.80)$$

Being the gate oxide field dependence related to the reaction phase through the K_F/K_R ratio, this dependence strongly relates to the expected electrochemical reaction assumed to contribute to the Si–H bond breaking. Even if the N_{it} generation during the NBTI stress is explained by the R–D picture, the assumptions used above need to be taken very carefully, since they may not be applicable for a given CMOS process.

Analytical solutions of the R–D equations are given by Yang et al. [68]. Figure 4.61 shows his comparison between a simulated and measured N_{it} shift (ΔN_{it}) for nitrided oxides with three different nitrogen concentrations. As expected, the initial time dependence with $n \approx 1$ is confirmed for all the nitrogen concentrations. This observation is the direct experimental evidence that the initial phase of the NBTI degradation is controlled by the reaction regime. For sufficiently long stress times ($t > 1$ sec), a power law behavior with an exponent $n \approx 1/4$ is observed, suggesting that in this case the NBTI degradation is controlled by the diffusion regime with the diffusing species being atomic hydrogen (H_0). Notice that the NBTI is enhanced with increasing nitrogen concentration. Considering that, generally, the NBTI enhancement is controlled by the $K_F N_0/K_R$ ratio, it is possible that the incorporation of nitrogen into the gate oxide can increase the initial concentration of passivated interface states (N_0) and/or increase the forward reaction rate (K_F) as well as decrease the passivation rate K_R. Both effects have been observed.

4.8.3 R–D Kinetics Controlled By a Dispersive Diffusion in SiO$_2$

One of the key conclusions of the N_{it} generation by an R–D picture, under the assumption H* diffusion with Arrhenius thermal transport, is the prediction that the time power law exponent (n) is independent of temperature, time, or T_{ox}. However, recent reports in the literature have questioned this conclusion suggesting that the experimentally measured n has a wide range of values that depend on temperature, time, and the oxide thickness. One reason for this large range of n values is due to NBTI recovery that takes place during the time delay between the time at which the stress is stopped for taking a measurement and the time the beginning of the measurement takes place (see Section 4.7.7). Another possible reason for this observed n exponent variation relates to the assumption of some form of dispersive behavior of the NBTI damage. Two possible physical explanations have been given to explain this behavior. It is assumed that, while SiO$_2$ is a disordered material, the possible dispersive nature of the NBTI damage relates to the existence of a distribution of Si–H bonds energies [4] or the random distribution of H* migration barriers [82, 90, 102].

As an example of this dispersive behavior, we attempt to describe the second approach. The situation is illustrated in Figure 4.62. The random distribution in energy of H* potential wells modifies the properties of hydrogen transport to and from the Si/SiO$_2$ interface, which will, in turns, influence the properties of the entire NBTI process.

This disorder translates into an allowed energy distribution of the potential wells that the H* species can jump into. We assume an exponential H* distribution of density of states $g(E)$,

$$g(E) = \frac{N_t}{E_0} \exp(-E/E_0), \qquad (4.81)$$

with the total density of localized states N_t and the typical potential well E_0.

To describe this phenomenon we consider the case of dispersive diffusion of a neutral H* (H$_0$, H$_2$) species in thick SiO$_2$ oxide (such that $N_{H*} = 0$ for $x = \infty$). In agreement with [93], we assume that the dispersive diffusion of neutral H* in an

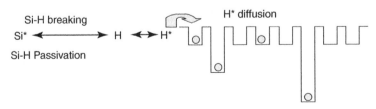

Figure 4.62. Graphical representation of dispersive H* diffusion in the gate oxide.

amorphous oxide can be described by the following equation:

$$N_{H*} = D^{ox}_{H*} \tau(t) \frac{\partial^2 N_{H*}}{\partial x^2}. \tag{4.82}$$

This equation replaces Flick's law in the case of the case of H* diffusion in an homogenous dielectric. In this case τ is given by

$$\tau(t) = \frac{N_c}{v} \left[\int_{k_T \ln(vt)}^{\infty} g(E) dE \right]^{-1}, \tag{4.83}$$

where N_c is the density of shallow H* hopping sites and v is the attempt to jump frequency.

Following as arguments similar to Kaczer [90], it can be shown that

$$N_{it} = \left(\frac{K_F}{K_R} \right)^{1/(1+\alpha)} \times \left(\frac{D^{ox}_{H*}}{v} \right)^{1/2(1+\alpha)} \times \left(\frac{N_c}{N_t} \right)^{1/2(1+\alpha)} \times (vt)^{p/2(1+\alpha)}, \tag{4.84}$$

with N_t as the total density of localized states and p as the dispersion parameter which is linearly proportional to T by the relation

$$p = kT/E_0. \tag{4.85}$$

From this relation it is interesting to notice the following:

1. The effective dispersive diffusion regime is possible only if the characteristic potential well E_0 is larger that kT such that $p < 1$.
2. The time evolution of the N_{it} generation follows a power law time dependence with the n exponent smaller that what estimated under Arrhenius H* diffusion.

$$n = \frac{p}{2(1 + \alpha)} = \frac{kT}{2(1 + \alpha)E_0} \tag{4.86}$$

3. The linear temperature-dependent n exponent implies that the temperature dependence of NBTI cannot follow an Arrhenius dependence.

Figure 4.63 gives experimental evidence of the exponent n increasing with the stress temperature during an NBTI DC stress of a FinFet [91]. The V_{th} shift is measured by using the single point, fast switching measurement described in Section 4.7.7 and the effect of hole trapping is minimized. Assuming H_0 diffusion $n = kT/4E_0$. In this case, $E_0 = 47$ meV with p varying between ~ 0.6 at 25°C and ~ 0.9 at 200°C.

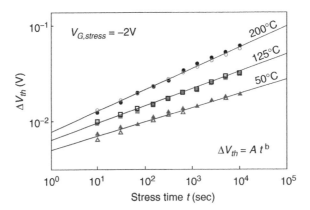

Figure 4.63. V_{th} shifts (ΔV_{th}) as function of stress time at three different stress temperatures. All data can be described with a power law time dependence. It is apparent that the time exponent n is increasing with stress temperature. Reprinted with permission from Ref. 90, © 2004 IEEE.

4.8.4 N_{it} Repassivation Phase

Once the stress is removed, the Si–H breaking that is controlled during the NBTI stress by the forward dissociation coefficient (K_F) is no longer active. Therefore H* species can diffuse back and be the supply of H species needed for the Si* + H → Si–H to take place. We consider for simplicity the N_{it} repassivation of neutral diffusing species H* after a NBTI stress phase with diffusion limited in the gate oxide. Similarly to the N_{it} generation phase the interface states repassivation can be approximated by two processes: a fast reaction-limited process followed by a slow diffusion-limited process. During the fast reaction-limited repassivation phase, the Si* anneal is taking place with depletion of H* species in the neighborhood of the Si/SiO$_2$ interface. Being $K_F = 0$, this process is controlled by the rate equation

$$\frac{\partial N_{it}}{\partial t} = -K_R N_{it}[N_{H^*}]^{1/\alpha} \quad \text{for} \quad t > t_0, \tag{4.87}$$

with N_{H^*} being the H* concentration at the interface. Assuming that during this initial phase the H* at the interface is almost constant and being equal to $N_{H^*}(t_0)$, the solution of this equation is

$$N_{it}(t) \approx N_{it}(t_0) \times \exp(-K_R \times [N_{H^*}(t_0)]^{1/\alpha} \times t). \tag{4.88}$$

This relation implies a very rapid (exponential) reduction of the interface states during this reaction phase. Obviously, the peak concentration of H* species at the interface ($N_{H^*}(t_0)$) does not remain constant and the above relation only can be valid for a very short transient (ms). Following this phase the reaction is controlled by a diffusion phase where the passivation of the interface states near the interface is controlled by the back diffusion of the existing H* species in the gate oxide. Because this reaction is so short, we approximate the beginning of the diffusion phase at around t_0. Following Alam, the diffusion controlled repassivation phase can be treated analytically as follows: We assume that at the end of the NBTI stress phase at the stress time t_0, the number of generated interface states is $N_{it}(t_0)$ and the density of the H* at the Si/SiO$_2$ interface is $N_{H*}(t_0)$ so that $N_{it}(t_0) = (1/2) \times N_{H*}(t_0) \times (D_{H*}^{ox} t_0)^{1/2}$. The right term of this equation approximates the total area associated to the H* diffusion profile at the t_0. We assume this total area is not changed that much in the course of the repassivation phase, so that, calling $N_{H*}^*(t + t_0)$ the H* concentration at the Si/SiO$_2$ interface, we can approximate $N_{H*}^*(t + t_0)/N_{H*}(t_0) \approx ((t + t_0)/t_0)^{1/2}$. During the repassivation phase, N_{it}^* traps are annealed at the time $t + t_0$ with a corresponding decrease of H* concentration $\left(N_{H*}^* \right)$ at the Si/SiO$_2$ interface, so that $N_{it}^* = (1/2) \times N_{H*}^* \times (\xi D_{H*}^{ox} t)^{1/2}$ (ξ is equal to 1/2 for one-sided diffusion). Under these assumptions we have the equation

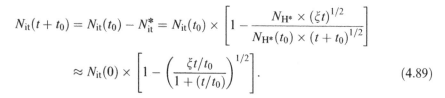

$$N_{it}(t + t_0) = N_{it}(t_0) - N_{it}^* = N_{it}(t_0) \times \left[1 - \frac{N_{H*} \times (\xi t)^{1/2}}{N_{H*}(t_0) \times (t + t_0)^{1/2}} \right]$$

$$\approx N_{it}(0) \times \left[1 - \left(\frac{\xi t / t_0}{1 + (t/t_0)} \right)^{1/2} \right]. \tag{4.89}$$

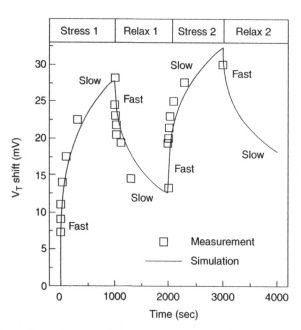

Figure 4.64. Numerical solution of the R–D model for a two cycles of interrupted stress (stress 1000 sec, relaxation 1000 sec). The simulation, carried our by Alam et al., suggests a good agreement with reported experimental data once H* = H$_0$ as well as the dominant role of the diffusion controlled regime during both the N_{it} generation and repassivation phases. Reprinted with permission from Ref. 67, © 2003 IEEE.

Figure 4.64 shows a numerical solution of the R–D model reported by Alam [11] for two cycles of interrupted stress (stress 1000 sec, relaxation 1000 sec) assuming that the H* species is H_0. The simulation (continuous line) will agree with the experimental data reported by Chen et al. [43] using Equation 4.70 and under these assumptions we have the equation ($\alpha = 1$), respectively, for the N_{it} generation and repassivation phase. The parameters used are $K_F = 10^{-2}\,sec$, $K_R = 10^{-18}\,cm^3/s$, $N_{it}(t_0) = 1.24 \times 10^{14}/cm^2$, and $T_{ox} = 1.3\,nm$.

4.8.5 Dynamic NBTI

In this section we explain, by the use of the R–D picture, some of the experimental data shown in Section 4.7.8 regarding the NBTI frequency and AC/DC ratio dependence in terms of N_{it} effective generation in AC conditions. Figure 4.65 supports the experimental observations of frequency independent NBTI. Following the arguments by Alam [67] the explanation is given by the symmetric behavior expected, based on the R–D picture illustrated in the previous section, between the diffusion-driven N_{it} generation and repassivation. The simulations described in this figure apply to the case of H* being the molecular neutral hydrogen H_2. An R–D simulation of H_2 profile as function of distance into the gate oxide is shown, assuming the case of a pulse with 50% duty factor for a given number of cycles (N_c) at two different frequencies, f_1 and f_2 ($f_1 = 2f_2$). As it can be seen from the two leftmost curves a larger diffusion is observed with higher frequency for a given cycle. On the other end, if the stress time is constant ($T_{stress} = N_c/f$), the total H_2 area of the diffusion profile, based on Equation 4.51, is equal to the total N_{it} and essentially frequency independent.

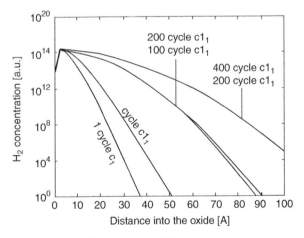

Figure 4.65. The H_2 profiles [N_{H2} (x, $t = N_c/f$)] at the end of a certain number of cycles (N_c) for two different frequencies, f_1 and f_2. Similar profiles are shown for the same stress time ($T_{stress} = N_c/f$). Reprinted with permission from Ref. 67, © 2003 IEEE.

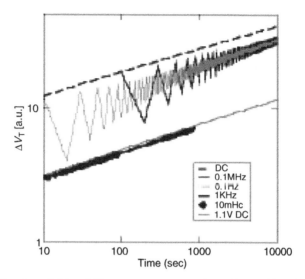

Figure 4.66. Simulated V_{th} shifts versus stress time for different frequencies showing that higher frequencies can be models of lower voltage stress (V_{eff}). Reprinted with permission from Ref. 85, © 2005 IEEE.

At a low frequency, the H_2 diffusion front moves further from the interface with a corresponding increase in N_{it}; with longer recovery phases (off state) H_2, is allowed to diffuse back to repassivate a larger fraction of the interface states with a residual fraction of N_{it} left for each cycle. As higher frequency, both N_{it} generation due to H_2 diffusion and the recovery are limited. In this case, the residual N_{it} generation per cycle is lower. The symmetric behavior of the dynamic NBTI effects requires that, at a given frequency and stress time, the total N_{it} generated and the corresponding AC/DC ratios (defined as T_{AC}/T_{DC}) are frequency independent. An interesting observation is given by [85] that suggests a high frequency operation can be modeled as a lower effective DC voltage stress $V_{eff} = V_{DC} - \log(T_{DC}/T_{AC})$ (see Fig. 4.66). This finding can be used to estimate the end-of-life AC operation by calculating the corresponding DC NBTI for the same stress time at V_{eff}.

4.9 HOLE TRAPPING MODELING

The basic processes responsible for hole trapping are described in Figure 4.67. Hole traps are positively charged when occupied by a hole (missing an electron) and neutral when not occupied. From this point of view, they electrically behave like donor-type interface states. In thermal equilibrium, hole traps with energy below the Fermi level E_F are neutral, while those ones above E_F are activated and contribute to the V_{th} decrease. Figure 4.67 represents the case of strong inversion. In this figure it is assumed that the bandgap is above the Fermi level ($E_v \approx E_F$).

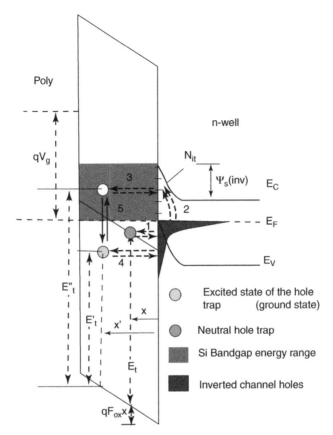

Figure 4.67. Graphical description of three possible mechanisms by which hole trapping can take place at the Si/SiO$_2$ interface. Path 1 $\equiv$ Elastic hole tunneling from the valence band. Path (2 + 3) $\equiv$ Two step interface states assisted process. Path (4 + 5) $\equiv$ Elastic hole tunneling into a hole trap exited state and lattice relaxation/emission multiphonon-assisted process.

When a voltage V_g is applied in inversion, the energy level of a given oxide trap at a distance x from the Si/SiO$_2$ interface will be given by

$$E = E_t + qF_{ox}X, \qquad (4.90)$$

with E_t being the energy level of the trap (relative to the SiO$_2$ valence band) and F_{ox} the applied oxide electric field (V_g/T_{ox}). Thermal equilibrium is reached by two types of charge exchange mechanisms. Oxide traps at a distance x from the Si/SiO$_2$ interface and with trap level ground state ($E_t + qF_{ox}x$) below the Si valence band get neutralized by elastic hole tunneling at the same energy level (path 1). Hole traps at a distance x' from the Si/SiO$_2$ interface states and ground energy level $E_t'' + qF_{ox}x'$ above the Si valence band get activated by the assistance of other mechanisms to supply or lose the required exchange in energy. Here we are describing two possible paths. Paths 2 and 3 are a two-steps process. The inverted channel holes interact thermally with the interface states N_{it} (2), which, in turn, exchange charge with oxide traps placed at the same energy level ($E_t'' + qF_{ox}x'$) by direct tunneling (3). An inverted channel hole can tunnel elastically to an excited energy level ($E_t' + qF_{ox}x'$) of an oxide trap with ground state energy level ($E_t'' + qF_{ox}x'$) and place itself in the Si bandgap at distance x' from the Si/SiO$_2$ interface(4), with subsequent relaxation to the ground state by multiphonon-assisted processes (5) (see path 4 and 5).

Taking all these processes into account, the total time constant τ associated with the hole trapping is given by

$$\frac{1}{\tau} = \frac{1}{\tau_1} + \frac{1}{\tau_{23}} + \frac{1}{\tau_{45}}, \tag{4.91}$$

where τ_1, τ_{23}, τ_{45} are the time constants associated respectively with path 1; 2 and 3; and 4 and 5.

We assume for simplicity that τ is dominated by the elastic tunneling of inverted channel holes to oxide traps with energy $E_t \approx E_v$ through a rectangular

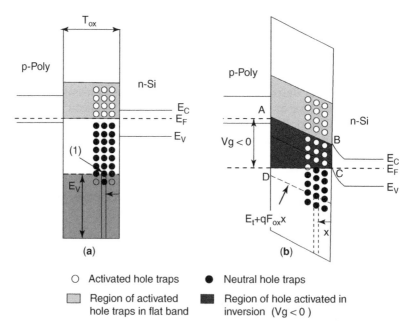

Figure 4.68. Energy band diagrams showing uniform space (*x*) and energy (*E*$_t$) distribution of hole traps at flatband (a) and pMOSFET channel inversion ($V_g < 0$) (b) condition.

barrier ($F_{ox} \approx 0$) so that

$$\tau = \tau_0(V_g) \times \exp(\lambda(E_V, x)) \tag{4.92}$$

with

$$\lambda(E_V, x) = 2 \left(\frac{2m_h^* E_V}{h^2} \right)^{1/2} x \equiv 2\beta x, \tag{4.93}$$

where $m_h^* \approx 10 \, m_e^*$ is the hole effective mass and x is the oxide trap distance from the Si/SiO$_2$ interface. The prefactor τ_0 is weakly dependent on energy and x and can be assumed constant at a given V_g, with a value that is dependent on which path is dominant during the activation phase. An estimate of the parameter β can be given in the case of holes tunneling from the valence band so that $E_t \approx 4.5 \, \text{eV}$.

In the case described in Figure 4.68, the process of hole traps activation involves hole traps that at flatband (a) have energy E_t below E_F (neutral) and with the application of the voltage V_g gain energy $E_t + qF_{ox}x$ that raises them above E_F [area ABCD in (b)]. Hole traps that are above E_F in flatband or with energy $E_t + qF_{ox}x$ below E_F at the applied V_g will not change their charging state.

In the rest of this section, we will describe how hole traps contribute to the V_{th} shift during an NBTI stress. In particular we will describe the activation of preexisting traps during the NBTI stress ($V_g = V_{g,stress}$) and the corresponding modulation of the same traps at test ($V_g = V_{g,test}$). Figure 4.69 illustrates the energy band diagram in this case. In a given test and stress configuration in thermal equilibrium, the hole traps above the associated Fermi level E_F are activated, while those below it are empty. During stress, the Fermi level E_F at the interface gets closer or eventually into the valence band (strong inversion regime) and oxide traps energy levels are increased by the higher $V_{g,stress}$ (larger oxide field F_{ox} in the gate oxide). When the stress is removed, trap energies return to their original levels and trapped holes tunnel back out to the interface.

The V_{th} shift due to the hole trapping/detrapping contribution during a stress phase with $V_g = V_{g,stress}$ for the stress time t_S followed by a recovery phase with $V_g = V_{g,test}$ for the time t_R can be estimated is

$$\Delta V_{th}(t_S, t_R) = \frac{qT_{ox}}{C_{ox}} \int_0^{T_{ox}} \int_{E_F}^{E_G'} D_{ot}(E_t, x) \times \Delta f(E_t, x)$$

$$\times \left(1 - \frac{x}{T_{ox}} \right) \times \left(1 - e^{-t_S/\tau_S} \right) \times \left(e^{-t_R/\tau_R} \right) dE_t dx \tag{4.94}$$

where E_G' is the energy gap of the SiO$_2$ ($\approx 9 \, \text{eV}$), $D_{ot}(E_t, x)$ is the oxide trap density (cm^{-3}eV^{-1}), and

$$\Delta f(E_t, x) = f\left(E_t + \frac{q}{T_{ox}} V_{g,test} x, E_F \right) - f\left(E_t + \frac{q}{T_{ox}} V_{g,stress} x, E_F \right) \tag{4.95}$$

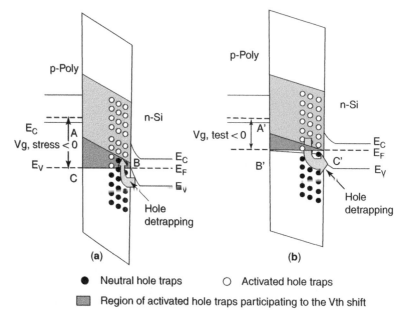

Figure 4.69. Energy band diagrams showing neutral and activated hole traps distribution at two different bias conditions: (a) during the NBTI stress ($V_g = V_{g,stress}$), (b) at test after stress ($V_g = V_{g,test}$). In condition of strong inversion at both V_g conditions. The triangle ABC and A'B'C' represent the oxide traps that contribute to the V_{th} shift between a stress and test phase.

is the difference between the two Fermi functions $f(E, x)$ during stress and test. The parameters τ_S and τ_R are the time constants associated respectively to the stress and recovery phase.

We estimate the integral of Equation 4.93 by making the following assumptions:

1. Heavy inversion at both stressing and recovery conditions ($E_F \approx E_v$).
2. Uniform hole trap distribution $D_{ot}(E_t, x) = D_{ot}$ for $x \leq d \leq T_{ox}$.
3. Step approximation of the Fermi distribution: $F(E, E_F) = 1$ for $E > E_F$, 0 for $E < E_F$.
4. The time constant τ_S and τ_R are determined by direct tunneling processes involving inverted holes in valence band. In this case τ_S and τ_R take the form given by Equation 4.91 with the same β since $E_F = E_v$ at stress and during the recovery, while having different prefactors τ_{0S} and τ_{0R} due to the different V_g conditions.

Assuming an elastic tunneling process and uniform energy and space distribution of hole traps sites, (D_{ot}) the threshold voltage shift due to the hole traps can be approximated to

$$\Delta V_{\text{th}}(t_s, t_r) \cong \frac{q}{C_{\text{ox}}} \text{AD}_{\text{ot}} K_B T \text{e}^{\text{BEox}} \ln\left(1 + \frac{\tau_{0e^ts}}{\tau_{0c^tr}}\right). \tag{4.96}$$

Notice that a clear signature of the direct-tunneling-assisted hole trapping is the log time behavior of the time evolution. This particular feature has been identified by an experimental evidence of hole trapping activation during an NBTI stress.

4.10 NBTI DEPENDENCE ON CMOS PROCESSES

One of the main reasons why there is no general consensus at this time on the nature of the NBTI damage is due to its strong dependence on the various CMOS processes. Many CMOS process steps and chemical species have been shown to have a significant impact on NBTI (hydrogen species, nitrogen, fluorine, thermal budgets, plasma, and BEOL process, etc.). Because of the complexity of this topic, we will give our best description of recent observations.

4.10.1 Hydrogen Species

Hydrogen (H) is the most common impurity found in gate oxides and is incorporated during both FEOL and BEOL CMOS processes. Forming gas anneal is one BEOL process that is an example of H having the beneficial effect to passivate residual dangling Si bonds at the SiO_2/Si interface. This H passivation improves the electronic properties of MOSFET transistors by the reduction of interface states in the Si bandgap, with consequent increase in channel mobility.

Hydrogen plays two main roles during the evolution of the NBTI physical damage. First, interface states are generated by depassivation of the SiH bonds at the SiO_2/Si interface during the NBTI stress. Enhanced trapping of H species in nitrogen Si–N bonds rather than in Si–O bonds is considered to be responsible for the increased NBTI sensitivity in nitrided oxides. Because of the lower activation energy, diffusing H is more likely to be trapped in Si–N than in the Si–O trapping centers; this leads to the observed, enhanced positive-charge formation. The evolution of the NBTI damage strongly depends on the stable form in which hydrogen diffuses in the gate oxide after the SiH dissociation. In particular, hydrogen can exist in its atomic state (H_0), as molecular (H_2), as positively charged hydrogen or protons (H^+) as part of the hydroxyl group OH, as hydronium H_3O^+, or as hydroxide ions OH^+. Recent literature suggests that H_0 is unstable in both silicon and SiO_2 [66]. Its existence as a neutral or ionized species can contribute to the electric field dependence of NBTI. Conflicting information is found in the literature on this topic.

The replacement of Si–H with the deuterium isotope Si–D at the SiO_2/Si interface has been expected to reduce the NBTI sensitivity (similar to what is observed in nMOSFET channel hot carriers [45–47]). The deuterium effect in nMOSFET CHC has been explained in terms of a faster relaxation rate of vibrational energy of the Si–D bond through coupling with the bulk TO phonons in respect to the Si–H bond. Conflicting results are reported on the effect of Si–D passivation to the NBTI degradation. The effect Si–D contributions have on the NBTI damage has been observed to follow the same evolution as for the SiH bond. Kimikuza et al. [48] observed a reduction of the NBTI sensitivity on pMOSFETs exposed to deuterium annealing.

Figure 4.70 shows that pMOSFETs with dry gate oxides, if exposed to D_2 post metal anneal (PMA), have the same time slope ($n \sim 0.25$) as devices undergoing H_2 PMA. The similar time evolution ($n \approx 0.25$) indicates that the measured V_{th} shift is controlled by N_{it} degradation. In this case, the R–D kinetics is controlled by either H_0 or D_0 diffusion in the gate oxide. The NBTI improvement in the hardware exposed to the D_2 PMA is due to the twice-larger mass of deuterium compared to hydrogen. Opposite observations were reported by Hook et al. [49]. A better insight on this topic is given by Mitami et al. [50] by comparing p^+ poly versus n^+ polygate pMOSFET exposed to the deuterated process. As given in Figure 4.71, they reported the increase in charge pumping current (ΔI_{CP}) of both hydrogenated and deuterated p^+ poly and n^+ poly gate pMOSFETs as function of stress time for NBTI stresses at similar F_{ox}. The deuterium incorporation has an effect on the n^+ poly/pMOSFETs, while no substantial difference is observed in the case of the p^+ poly/pMOSFETs. The different NBTI behavior between the n^+ poly and p^+ poly pMOSFETs is explained as follows. In the case of p^+ poly NBTI, the deuterium effect reflects the reduced deuterium diffusion coefficient, which is 0.5 smaller than for hydrogen. For n^+ poly/pMOSFETs the

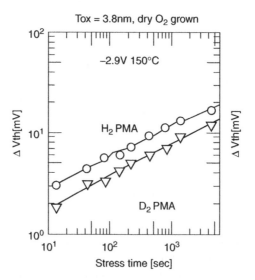

Figure 4.70. pMOSFETs exposed to H_2 and D_2 post metal anneal follow the same time evolution. Reprinted with permission from Ref. 48, © 2000 IEEE.

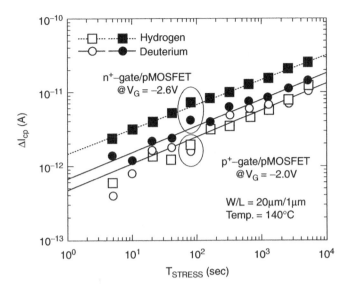

Figure 4.71. ΔI_{CP} increase as function of stress time. Filled symbols correspond to the data for n^+ poly/pMOSFET and open symbols correspond to those for the p^+ poly/pMOSFETs. Circles are hydrogen and squares for deuterium annealed transistors. Reprinted with permission from Ref. 50, © 2006 IEEE.

NBTI damage is controlled by energetic electrons injected from the conduction band of the n^+ gate and produce Si–H and Si–D breaking similarly to what observed for nMOSFET in CHC conditions. These observations lead to the conclusions that the physics of the Si–H bond breaking is definitely different when driven by the presence of holes or electrons.

4.10.2 Nitrogen

Nitrided oxides are used in dual workfunction technologies for two main reasons: to prevent boron penetration in surface channel pMOSFETs and to reduce the leakage current of the nMOSFET gate oxide for a given effective thickness. Figure 4.72a shows the NBTI lifetime dependence on V_g for pMOSFET devices with thermal (dry O_2 grown) and nitrided (NO gas anneal followed by gate oxidation) oxides. The high NO concentration contributes substantially to the reduction of the NBTI lifetime [61]. The high nitridation process also decreases the temperature activation energy as seen in Figure 4.72b. Liu et al. [62] compared the NBTI degradation at similar electric fields for different nitridation processes: thermal oxide grown on nitrogen implanted silicon substrate (NISS), N_2O nitrided oxide, rapid thermal nitric oxide (RT–NO), and remote plasma nitrided oxide (RPN) (see also Fig. 4.73). This work indicates a strong NBTI dependence on the nitridation process used to grow the gate oxide. In particular the RPN seems to exhibit the least NBTI shift.

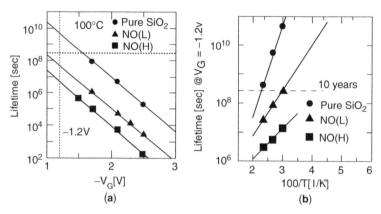

Figure 4.72. (a) Comparison of gate voltage (V_g) dependence on lifetime for dry and NO annealed gate oxides. (b) Dependence lifetime on activation energy (E_a). E_a decreases with increase in nitrogen concentration. Reprinted with permission from Ref. 61, © 2000 IEEE.

The role of nitrogen to the NBTI enhancement is still not clear. We will focus on few key experimental observations that have helped to clarify some of its related physics. The reader is referred to Section 4.6 for more details on the nature of the NBTI damage.

Figure 4.74 suggests that the time evolution of NBTI-induced V_{th} shift is enhanced by increasing nitrogen concentration at the SiO_2/Si interface (N_{int}) [63].

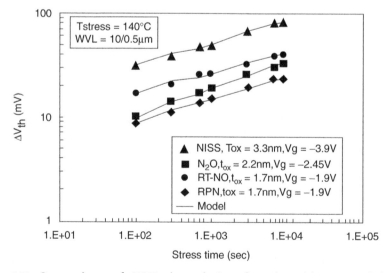

Figure 4.73. Comparison of NBTI degradation for ultra-thin gate dielectrics (1.7–3.3 nm) fabricated with different nitridation processes and stressed at the same gate oxide field. Reprinted with permission from Ref. 62, © 2001 IEEE.

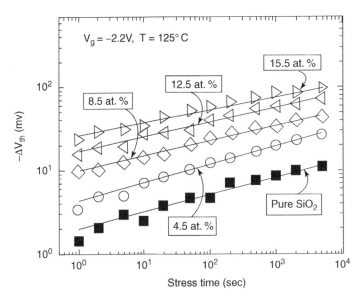

Figure 4.74. Typical threshold voltage shift (ΔV_{th}) as a function of NBTI stress time for various interfacial nitrogen concentrations (N_{int}). For comparison, a control sample (i.e., pure SiO$_2$) is included. Reprinted with permission from Ref. 63, © 2003 IEEE.

In addition, ΔV_{th} follows a power law relation, such as $\Delta V_{th} = At^n$, with the n factor independent of the N_{int} concentration and having a value of about 0.20 in this work. This figure also indicates that both thermal oxides and nitrided gate oxides experience equivalent power-law time evolution.

Following Tan [64], the ΔV_{th} dependence on the N_{int} links the prefactor A value to the interface states (ΔD_{it}) and positive fixed charge (ΔN_f) increase. Both ΔD_{it} (Fig. 4.75a) and ΔN_f (Fig. 4.75b) were found to have a similar time evolution as ΔV_{th}. The linear proportionality between (Fig. 4.75c) ΔN_f and ΔD_{it} (Fig. 4.75d) as a function of N_{int} concentration and the similar n value ($n \approx 0.25$) suggest that both types of physical damage are related to the same root cause. This finding is consistent with the conclusion that the NBTI damage in both thermal and nitrided oxides is governed by a diffusion-controlled electrochemical reaction.

More insight on the mechanism that drives the nitrogen-enhanced NBTI degradation is provided by observation that the activation energy (E_a) of both ΔN_f and ΔD_{it} has the same linear dependence ΔV_{th} on the N_{int} concentration. As can be seen in Figure 4.76, E_a has the largest values ($\approx 0.2\,eV$) for the thermal oxides ($N_{int} = 0$) and decreases linearly with increasing N_{int} [65]. In [63] evidence has also been given that two NBTI regimes are expected with increasing N_{int} concentration. It has been found that the dependence on N_{int} has two linear regions with different slopes separated by $N_{int} \approx 8$ at%. This phenomenon has been explained in terms of the N neighboring effect, and suggests that the N–Si–O$_3$ structure is controlling the NBTI damage for $N_{int} < 8$ at%, while the N$_2$–Si–O$_2$ becomes dominant for $N_{int} > 8$ at%.

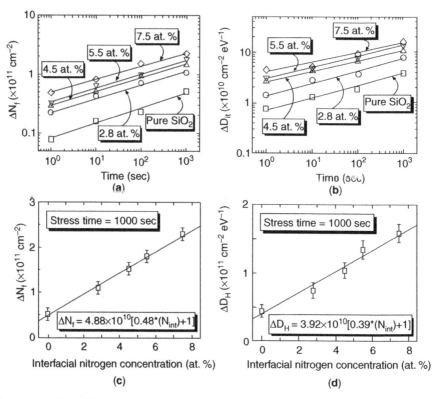

Figure 4.75. Interface states generation (a) and positive charge creation (b) as function of stress time for various nitrogen concentrations. Same N_{it} and N_p dependence is found as function of N_{int} concentration [(c) and (d)]. The NBTI stress was carried out at $T = 125°C$ and $V_g = -2.2$ V. Reprinted with permission from Ref. 64. *Applied Physics Letters*, 82 p. 12. Copyright (2003) American Institutes of Physics.

As seen in Figure 4.77, hole trapping is believed to be the dominant mechanism of positively charged defects in nitrided oxides [41]. In this case, the normalized N_{it} shift ($\Delta N_{it}*e/C_{ox}$) is compared to the V_{th} shift for both thermal oxides (pure oxide in the figure) and nitrided oxides with increasing N content. The lower observed slope with increasing N concentration indicates that measured V_{th} shift can not be completely explained by N_{it} generation (as in the case of thermal oxides), but has a strong contribution from hole trapping activation/ generation at the applied NBTI stress conditions.

4.10.3 Fluorine

Fluorine is known to have some beneficial effects on MOSFET devices. The presence of fluorine in the gate oxide yields an increase in the effective oxide

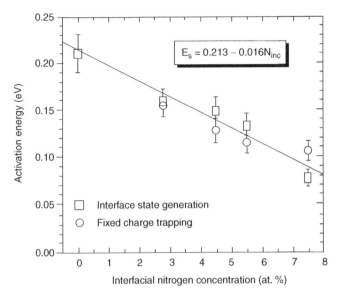

Figure 4.76. Activation energies of both interface states and positive charge formation as function of nitrogen concentration at the Si/SiO_2 interface. Reprinted with permission from Ref. 65. *Applied Physics Letters*, 82, p. 269. Copyright (2003) American Institute of Physics.

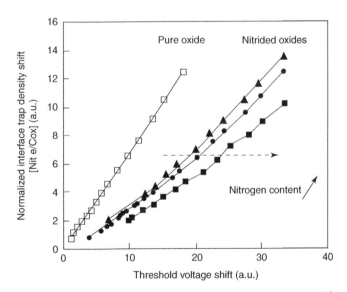

Figure 4.77. Correlation between the V_{th} shift and the normalized N_{it} shift ($\Delta N_{it}*e/C_{ox}$) for a pure oxide (open squares) and nitrided oxide devices with different nitrogen concentration. Reprinted with permission from Ref. 41, © 2004 IEEE.

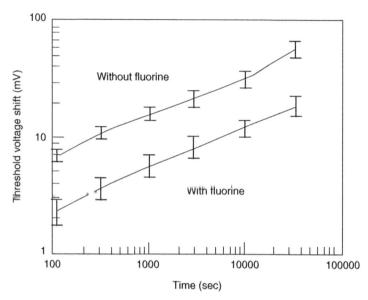

Figure 4.78. NBTI shift as function of time for wafers with and without fluorine implantation in the polygate. After Hook et al. [16]. Reprinted with permission from Ref. 71, © 2001 IEEE.

thickness and an improvement in the nMOSFET hot carrier behavior as well as its TDDB reliability. These observations are consistent with the assumption that Si–F replaces Si–H in the SiO_2/Si interface, reducing the concentration of Si–H sites available for interface state generation. The initial reduction of the Si–H bonds should explain the observed reduced NBTI sensitivity in pMOSFETs exposed to high doses of F implantation in the gate polysilicon as shown in Figure 4.78 from [71]. The conflicting estimate of the NBTI activation energy with and without fluorine found in the literature (Fig. 4.79) suggests that it is not known if the H diffusion kinetics is the same with increased F concentration. The beneficial effects introduced by the F presence in the gate oxide need to be balanced with such detrimental effects as enhanced boron penetration and higher junction leakage.

4.10.4   Boron

Boron is used in dual workfunction technologies as a dopant species for the polysilicon gate as well as for the source and drain shallow junctions. Boron can reach the device channel region by diffusing through the gate oxide from the p^+ poly or from the source and drain regions during the high temperature activation anneals. Yamamoto et al. [31] compared the NBTI sensitivity of surface channel pMOSFETs ($T_{ox} = 65$ A) exposed to BF_2 self-aligned implantation with a p^+ doped polysilicon gate and formed the source and drain shallow junctions with (by

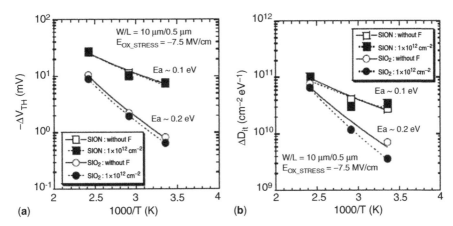

Figure 4.79. Temperature-dependence of both (a) ΔV_{th} and (b) ΔD_{it}. Reprinted with permission from Ref. 26, © 2002 IEEE. The oxide thicknesses (t_{ox}) are 2.3 nm. NBT stress was applied under F_{ox} −7.5 MV/cm for 5120 sec. It is found that the activation energies of both ΔV_{th} and ΔD_{it} for SiON are about 0.05 eV; no effect of fluorine incorporation is observed. On the other hand, in the case of SiO_2 film, the activation energies of both ΔV_{th} and ΔD_{it} are estimated as about 0.1 eV. Furthermore, it should be noted that both ΔV_{th} and ΔD_{it} are improved by fluorine incorporation, maintaining almost the same activation energies.

furnace anneal) and without (by RTA) boron penetration. Figure 4.80 shows that the NBTI time evolution is the same with and without B penetration ($\approx t^{0.25}$) suggesting the in both cases, the NBTI damage follows the same diffusion kinetics. In addition, B penetration will increase the V_{th} shift sensitivity to the NBTI damage, but, at the same time, will reduce the N_{it} generation. These experimental results suggest that B penetration from BF_2 implantation enhances positive oxide defect formation ΔN_p, while suppressing the interface state generation. The lower value of ΔN_{it} can be explained by the formation of Si–F bonds with high binding energy near the SiO_2/Si interface introduced by the BF_2 diffusion. On the other end, the larger ΔN_p expected with B penetration is attributable to enhanced, positive oxide defect formation due to the BF_2 implant in the source and drain regions. The enhancement of the ΔN_p increase in the gate–source/drain overlap region is expected to explain the observed enhanced channel length dependence of the NBTI degradation because in smaller channel length devices, the relative importance of the gate–source/drain overlap area on the channel area is increased. However, this NBTI dependence on channel length has not been typically observed.

The observation that fluorine from the BF_2 species reduces the ΔN_{it} contribution to the NBTI damage somewhat explains why pMOSFET devices exposed to a BF_2 self-aligned implant have lower NBTI lifetimes than if exposed to the B implant.

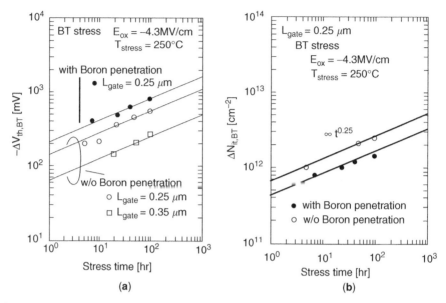

Figure 4.80. Time dependence of NBTI induced V_{th} shift (a) and N_{it} generation (b) with and without boron penetration. Reprinted with permission from Ref. 31, © 1999 IEEE.

4.10.5 NBTI Sensitivity to BEOL Charging

BEOL plasma charging damage has been shown to impact both gate oxide integrity as well as channel hot carrier lifetimes. In particular, it is expected that transistor's lifetime associated to a given wearout is strongly dependent on allowed antenna ratios, gate oxide thickness, as well as the applied back-end processing in the technology. Similar trends have been experimentally observed in the case of NBTI. Here we report, as an example, the findings by Krishnan et al. [73] that give the key features of the NBTI sensitivity to plasma damage. The NBTI enhancement due to plasma charging can be explained as follows: Plasma damage of the gate oxide during the BEOL processing produces an increase in interface states (Si–H unpassivated bonds). The usually adopted BEOL H_2 forming gas anneals passivate these Si*. This means that more passivated Si–H bonds are available at the SiO_2/Si interface as result of the plasma damage. As expected, the increased availability of Si–H bonds at the beginning of the stress does not modify the nature of the NBTI physical damage. Possible NBTI enhancement can also be explained in terms of increase of available neutral traps with increased charging. As described before, the neutral traps will be activated during the NBTI stress, enhancing the hole trapping contribution to the V_{th} shift. These features are experimentally observed by comparing the V_{th} shift as function of stress time for devices with diode protected (control) and exposed to a 120 kμm (antenna) (see Fig. 4.81).

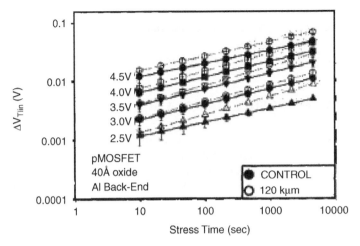

Figure 4.81. NBTI dependence on antenna. Reprinted with permission from Ref. 73, © 2001 IEEE.

A larger degradation in the 120 kμm antenna structure is measured in comparison to the diode protected transistor when both devices were stressed under similar NBTI stress conditions.

Figure 4.82 indicates that NBTI observed enhancement measured in the transistor with 120 kμm antenna does not result in an increase in the factor A (see figure), suggesting the product of the activation energy and the prefactor C are the same in both cases. However, from the previous figure, it can be observed that the time slope (n) of the antenna devices is always larger that the slope of the control transistors. The higher slope values can be associated with a reaction-rather than diffusion-limited Si–H breaking processes or an increase of hole trapping activation due to a possible increase of neutral traps with plasma charging, coupled with a

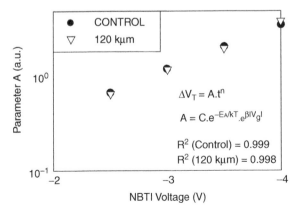

Figure 4.82. Charging does not modify the physical NBTI dependence. Reprinted with permission from Ref. 73, © 2001 IEEE.

slow measurement. This last observation will result in a fictitious value of n, which is strongly dependent on the stress/test methodology rather the differences in N_{it} generation. As seen from this example, a careful NBTI characterization is needed to estimate the effect of BEOL charging on the NBTI physical damage and modeling.

4.11   NBTI DEPENDENCE ON AREA SCALING

It is well known that as MOSFET dimensions are reduced with CMOS scaling, microscopic variations in the number and spatial distribution of dopants in the channel region result in an increased variability of the device electrical parameters [74–79]. These atomic-level, intrinsic fluctuations can not be eliminated by tight control of the manufacturing process and are most pronounced in transistors with minimum geometry. As discussed in the previous sections, the NBTI damage is due to both the contribution of hole trapping activation, fixed charge, and interface state generation. The NBTI-induced N_{it} and N_F^+ (fixed positive charge) generation as well as hole trapping both result in intrinsic charge fluctuations similar to the case of random dopants [80]. The intrinsic fluctuations of the NBTI damage results in a corresponding variation of key device parameters such as the V_{th} shift, affecting the V_{th} distribution after NBTI aging. In the previous sections, we discussed and modeled the effect of the NBTI damage on the mean shift of key device parameters.

It is, however, important to model the distribution of these shifts after stress and estimate distribution parameters such as the variance or sigma since this variation may affect NBTI-induced circuit stability. The case of analog circuits using matched pairs, such as differential pairs or current mirrors, is of particular interest. As a result of the intrinsic NBTI-induced charge fluctuations, the variances of the V_{th} and β (mobility) mismatches are expected to scale inversely with the total device active gate area A_G (given by-product $L_{poly} \times W_{poly}$). These circuits are expected to be insensitive to matched V_{th} and β shifts of the pair, but they are much more sensitive to increases in V_{th} and β mismatches between the pMOSFETs part of the matched pair. Normal NBTI stress can be used to predict the mean V_{th} shift for the matched pair and any mismatch induced by differential stress conditions. But even for identical use conditions and completely identical devices, NBTI will cause mismatch shifts due to random variations in the number of charges formed and their spatial distribution.

A theory that describes the overall effect of the intrinsic fluctuations of the different type of NBTI induced charges is not available. In this section, we attempt to quantify these dependencies in the case of N_{it} generation only for mismatched circuits. As an example, we limit the investigation of these effects to V_{th} fluctuations. We reference [80] for details on β-induced fluctuations.

We label, first, by γ a given device parameter mismatch between the device A and B part of a pair such that $\Delta V_{th}(t) \equiv V_{th,A}(t) - V_{th,B}(t)$ is the corresponding V_{th} at a given stress time. γ, as usual, relates to the shift in time during the stress for a given device. Following [80], we will derive a general quantitative equation to model the dependence of the V_{th} variance of the mismatched pairs [$\text{Var}(\Delta\gamma V_{th})$] and the mean ΔV_{th} [$\text{Mean}(\Delta V_{th})$], assuming that the main contribution to the mismatch is given by interface states generation. In this case, the statistics of the NBTI-induced V_{th} shift distribution are mainly contributed by the generation of the donor-type interface states according to

$$\Delta V_{th} = \frac{q T_{ox,EFF}}{\varepsilon_{ox} \times A_G} \Delta N_{it}^D, \tag{4.97}$$

where $T_{ox,EFF}$ is the electrically measured value of T_{ox}.

We assume that the interface states generation is a random process (similar to what has been done with TDDB dielectric breakdown). Under these conditions, for a given area A_G the random N_{it} generation at a given stress time follows a Poisson distribution such that

$$\text{Var}(\Delta N_{it}^D) = \text{Mean}(\Delta N_{it}^D). \tag{4.98}$$

Therefore

$$\text{Var}(\Delta V_{th}) = \frac{q T_{ox,EFF}}{\varepsilon_{ox} \times A_G} \text{Var}(\Delta N_{it}^D) \tag{4.99}$$

and

$$\text{Var}(\Delta\gamma V_{th}) = \frac{2q T_{ox,EFF} \text{Mean}(\Delta V_{th})}{A_G}. \tag{4.100}$$

The factor 2 in the above equation comes from the observation that we are tracking the NBTI shift of two different pMOSFETs device in the matched pairs. In addition to the contribution of ΔN_{it}^D charge fluctuations to $\text{Var}(\Delta\gamma V_{th})$, we need to consider the contribution due to the random spatial distribution of the interface states. Simulations [81] suggest that this contribution can be quantified by including an empirical constant K_1 such that

$$\text{Var}(\Delta\Delta V_T) = \frac{2K_1 q T_{ox,EFF} \text{Mean}(\Delta V_T)}{A_G}. \tag{4.101}$$

Experimental evidence of the ratio of the $\text{Var}(\Delta\gamma V_{th})$ to be dependent on $\text{Mean}(\Delta V_{th})$ according to Equation 4.100 is given in Figure 4.83. In this figure from [80], matched pairs using three different pMOSFET devices with different T_{ox} thickness and V_{dd} were used. The measured value of K_1 is 2.7, consistent with what predicted by simulation (≈ 2).

Since $\text{Var} = \sigma^2$, the sigma of the V_{th} mismatch relates to the mean V_{th} shift through the equation

$$\sigma(\Delta\gamma V_{th}) = \sqrt{\frac{2K_1 q T_{ox,Eff}}{A_G} \times \text{Mean}(\Delta V_{th})}. \tag{4.102}$$

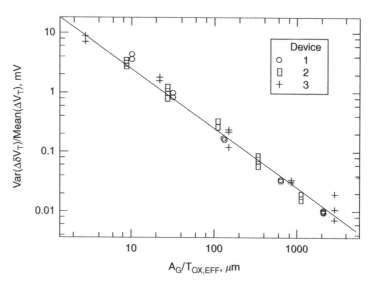

Figure 4.83. Ratio of the variance of the V_T (V_{th}) mismatch shift to the mean V_T shift versus $A_G/T_{ox,EFF}$. The dashed line is the model prediction for $K_1 = 2.7$. Reprinted with permission from Ref. 80, © 2002 IEEE.

This equation implies that during the NBTI aging, smaller active-area devices are expected to experience a larger V_{th} mismatch for the same mean V_{th} shift. This enhanced sensitivity is particularly important in matched pairs circuits using small area devices such as SRAMs. As an example of the critical implications of NBTI aging to SRAM circuits, we report the estimated dependence of end-of-life failure count and its relation to a given SRAM design, the CMOS technology used, as well as the screening methodology adopted. Figure 4.84, taken from [112], describes the dependence of the I_{CRIT} (measure of the SRAM cell static noise marging) final distribution (after NBTI aging) in terms of an initial (T0) distribution screened with an $I_{CRIT,CUT}$ point and a corresponding amount P_{CUT} of the screened out parts.

In this case, it is found that the NBTI-induced mean V_{th} shift ($\mu(\Delta V_{th})$) effect in failure count increase is dominated by a Gaussian distribution with a mean (μ) and sigma (σ) given by

$$\mu \approx I_{CRIT,CUT} - \alpha \times \mu(\Delta V_{th}(t))$$

$$\sigma \approx \sqrt{\alpha^2 \gamma \times \mu(\Delta V_{th}(t))}$$

(4.103)

where the α and γ parameter relate, respectively, to the SRAM design and internal variability of NBTI. This observation suggests that in order to define a pMOSFET V_{th} target more conveniently related to cell stability of a given SRAM design, we can not limit it to the mean V_{th} shift; we need an estimation of the allowed variance (sigma) and initial distribution.

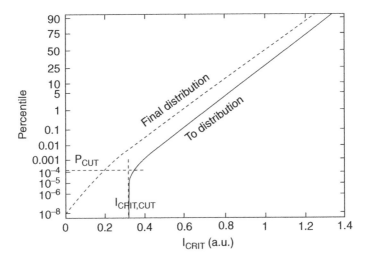

Figure 4.84. Dependence of final I_{CRIT} failure count (% tile) on initially screened failure count for a given V_{tsat} NBTI shift induced in one of the pMOSFETs SRAM devices. Reprinted with permission from Ref. 112, © 2006 IEEE.

Equation 4.101 describes the relation between the NBTI-induced shift of the sigma and mean of the matched pair V_{th} distribution. In the case of a single pMOSFET device, it is easy to prove that the sigma and mean of the ΔV_{th} distribution (assumed Gaussian) during NBTI aging is given by

$$\sigma(\Delta V_{th}) = \sqrt{\frac{K_1 q T_{ox,Eff}}{A_G}} \times \text{Mean}(\Delta V_{th}). \qquad (4.104)$$

The increase in sigma with pMOSFET NBTI aging needs to be taken into account when estimating the worst-case-allowed shift (typically 3σ) for a given circuit application of the device and end-of-life expectation.

4.12 OVERVIEW OF KEY NBTI FEATURES

We would like to finish this chapter by giving an overview of the key features contributing to the NBTI damage and the corresponding kinetics (see Table 4.3). This overview does not pretend to be comprehensive, since information on NBTI is still in the initial phase of research. However, a comprehensive kinetic description of the evolution of the NBTI damage must take all these experimental observations into account. None of the kinetic models so far presented consistently explains all of them. It has been generally accepted that during a technology qualification two key factors need to be taken into account:

1. The stress and test methodology adopted to characterize NBTI at use condition. In particular, it is critical to confirm that the damage activated

TABLE 4.3. Key Dependencies of the NBTI Damage in the Case of Cold Hole Assisted Damage

Key Features	ΔN_{it}	ΔN_{ht}	ΔN_{f}
Kinetics	• Si–H bond breaking by reaction–diffusion driven process • Reversible (Si* repassivation) • (V_g/Temp controlled depending on diffusing species)	• Hole traps activation by hole tunneling ($V_{gtest} \rightarrow V_{gstress}$) • Dependent on V_g • Reversible (hole detrapping) ($V_{gstress} \rightarrow V_{gtest}$)	• By-product of N_{it} kinetics/deep level hole trapping/H release model • Not clear if reversible
Time dependence	• Time power law dependence $$\Delta N_{it} \cong t^n$$ $$n = 1/2(1+\alpha)$$ • Diffusion in homogeneous oxide $$n = kT/2(1+\alpha)E_0$$ with $\alpha = 1$ (H$_0$), 2 (H^2) diffusion • Diffusion in dispersive oxide	• Log of time dependence $$\Delta N_{ht} \cong \ln(1 + (\tau_{0e}t_s/\tau_{0c}t_1))$$ where t_s, t_t = stress, test time, τ_{0e}, τ_{0c} = hole emission, capture time constant	• Expected to be similar to N_{it} $$\Delta N_f = \alpha N_{it}$$
Temperature dependence	• Arrhenius-like $$\Delta N_{it} \cong \exp(-E_a/kT)$$ $E_a = 0.1$–0.3 eV	$E_a = 0.03$ eV (Very small if direct tunneling)	Not clear

F_{ox} dependence	F_{ox} dependence not clear- $\Delta N_{it} \cong \exp(\gamma F_{ox})$ $\Delta N_{it} \cong F_{ox} \times \exp(\gamma F_{ox})$ $\Delta N_{it} \cong (F_{ox})^{\alpha}$	Exponential $\Delta N_{ht} \cong \exp(\times F_{ox})$	Not clear
Frequency/ Duty cycle dependence nitrogen	• No frequency dependence • Duty cycle dependence	Regulated by applied wave form	Not clear
	Yes	Yes	Yes
	• Si–H binding energy reduced with larger dose.	• Hole traps precursors density increases with N dose	
Fluorine	Yes	Not clear	Not clear
Effect on μ	Yes	No	No
Effect to flatband	Yes	Yes	Yes

at accelerated NBTI conditions is representative of what is expected at use conditions (cold holes regime) and to fine-tune the NBTI testing to monitor the damage activated at use conditions.

2. Careful understanding on the NBTI process sensitivities of the technology. In particular, for a given technology it is necessary to look at the intrinsic NBTI FEOL sensitivities by use of gate diode protection schemes and/or appropriate device design with allowed BEOL antenna rules. In addition, a comparison with a fully integrated process is needed to estimate the NBTI sensitivities with BEOL charging as well different BEOL anneals.

The lack of careful monitoring of these effects may lead to the inaccurate understanding of the NBTI damage with consequent modeling of the two components of the NBTI damage (interface states and positively charged oxide defects) and wrong end-of-life NBTI estimation.

REFERENCES

1. B. E. Deal, M. Sklar, A. S. Grove, E. H. Snow. J. Electrochem. Soc., 114: 1967, p 266.
2. A. Goetzberger, A. D. Lopez, R. J. Strain. J. Electrochem. Soc., 120: 1973, p 90.
3. K. O. Jeppson, C. M. Svensson. Negative bias stress of MOS devices at high electric fields and degradation of MNOS devices. J. Appl. Phys., 48: 1977, pp 2004–2014.
4. V. Huard, M. Denais, F. Perrier, N. Revil, A. Bravaix, E. Vincent. A thorough investigation of MOSFETs NBTI degradation. Microelectronics Reliability, 45: 2005, pp 83–98.
5. S. Mahapatra, P. B. Kumar, M. A. Alam. Investigation and modeling of interface and bulk trap generation during negative bias temperature instability of p-MOSFETs. IEEE Trans. on Electron Devices, 51(9): Sep 2004.
6. V. Huard, F. Monsieus, G. Ribes, S. Bruyere. Evidence for hydrogen-related defects during NBTI stress in p-MOSFETs. 41st Ann. Int. Reliability Physics Sym., Dallas, Texas, 2003.
7. C. E. Blat, E. H. Nicollian, E. H. Poindexter. Mechanisms of negative-bias temperature instability. J. Appl. Phys., 69: 1991, pp 1712–1720.
8. C. Gerardi, E. H. Poindexter, P. J. Caplan, M. Harmatz, W. R. Buchwald. Generation of Pb centers by high electric fields: Thermochemical effects. J. Electrochem. Soc., 136: 1989, 2609–2614.
9. C Gerardi, E. H. Poindexter, M. Harmatz, W. L. Warren, E. H. Nicollian, A. H. Edwards. Depassivation of damp-oxide Pb centers by thermal and electric field stress. J. Electrochem. Soc., 138: 1991, pp 3765–3770.
10. S. Chakravarthi, A. T. Krishnan, V. Reddy, C. F. Machala, S. Krishnan. A comprehensive framework for predictive modeling of negative bias temperature instability. 2004 Int. Reliability Physics Sym. Proc.: 2004, pp 273–282.
11. M. A. Alam, S. Mahapatra. A comprehensive model of PMOS NBTI degradation, Microelectron. Reliab, 45: 2005, pp 71–81.

12. L. Tsetseris, X. J. Zhou, D. M. Fleetwood, R. D. Schrimph, S. T. Pantelides. Physical mechanisms of negative-bias temperature instability. Appl. Phys. Lett., 86: 2005, p 142103.

13. A. Stesmans, V. V. Afanasev. Nature of the Pb1 defect in (100) Si/SiO2 as revealed by electron spin resonance 29Si hyperfine structure/ Microelectron. Eng., 63: 1999, pp 5776–5793.

14. A Stesmans, V. V. Afanase. Electrical activity of interfacial paramagnetic defects in thermal (100)Si/SiO$_2$. Phys. Rev., 57: 1998, pp 10030–10034.

15. P. Lenahan. IEEE Rel. Phys. Tutorial Notes, Adv. Rel. Topics. IEEE IRPS: 2002, Section 223.

16. A. Stesmans, V. V. Afanase. Electrical activity of interfacial paramagnetic defects in thermal (100)Si/SiO$_2$. Phys. Rev., 57: 1998, pp 10030–10034.

17. C. Schlunder et al. ESREF 1999.

18. V. Stathis, G. LaRosa, A. Chou. Broad energy distribution of NBTI-induced interface states in p-MOSFETs with ultra-thin nitrided oxide. 42nd Ann. 2004 IEEE Int. Reliability Physics Sym. Proc.

19. V. Huard, F. Monsieur, G. Ribes, S. Bruyere. Evidence for hydrogen-related defects during NBTI stress in p-MOSFETs, IRPS 2003.

20. S. Tsuijkawa, K. Watanabe, R. Tsuchiya, K. Ohnishi, J. Yugami. Experimental evidence for generation of bulk traps by negative bias temperature stress and their impact on integrity of direct-tunneling gate dielectrics. 2003 Sym. on VLSI Technology.

21. S. Tsujkawa, J. Yugami. Positive charge generation due to species of hydrogen during NBTI phenomenon in pMOSFETs with ultra-thin SiON gate dielectrics. Microelectronics Reliability, 45: 2005, pp 65–69.

22. M. Ershov, S. Saxena, S. Minane, P. Clifton, M. Redford, R. Lindley, H. Karbasi, S. Graves, S. Winters. Degradation dynamics, recovery, and characterization of negative bias temperature instability. Microelectronics Reliability, 45: 2005, pp 99–105.

23. H. Reisinger, O. Blank, W. Heinrigs, A. Mühlhoff, W. Gustin, C. Schlünder. Analysis of NBTI degradation- and recovery-behavior based on ultra fast VT measurements. 2006; IEEE Int. Reliability Physics Sym. Proc.

24. V. Huard, M. Denais, C. R. Parthasarathy. NBTI degradation from physical mechanisms to modeling. Microelectronics Reliability, 46: 2006, pp 1–23.

25. C. R. Parthasarathy, M. Denais, V. Huard, G. Ribes, E. Vincent, A. Bravaix. Characterization and modeling NBTI for design-in reliability. 2005 IRW Final Report.

26. Y. Mitani, M. Nagamine, H. Satake, A. Toriumi. NBTI mechanism in ultra-thin gate dielectrics nitrogen originated mechanism in SiON. IEDM 2002.

27. M. Makabe, T. Kubota, T. Kitano. Bias-temperature degradation of pMOSFETs: Mechanism and suppression. IRPS, 20: 2000, pp 205–209.

28. Y. Chen et al. IRW 2000.

29. S. Mahapatra, M. A. Alam. A predictive reliability model for PMOS bias temperature degradation. IEDM 2002.

30. Y. Taur, T. H. Nin. *Fundamental of Modern VLSI Devices*. Cambridge University Press, Cambridge: 1998.

31. T. Yamamoto, K. Uwasawa, T. Mogami. Bias temperature instability in scaled p$^+$ polysilicon gate pMOSFETs. IEEE Trans. on Electron Devices, 46(5): May 1999.

32. C. Schlunder, R. Brederlow, B. Ankele, A. Lil, K. Goser, R. Thewes. On the degradation of P-MOSFETs in anlalog and RF Circuits under inhomogeneous negative bias temperature stress. IRPS 2003, Dallas, Texas.
33. M. S. Liang et al. Inversion level capacitance and mobility of very thin gate oxide MOSFETs. IEEE Trans. Electron Devices, ED-, 33: Mar 1986, p 409.
34. M. Ershov, R. Lindley, S. Saxena, A. Shibkov, S. Minehane, J. Babcock, S. Winters, H. Karbasi, T. Yamashita, P. Clifkon, M. Redford. Transient effects and characterization methodology of negative bias temperature instability in PMOS transistors. 41st Ann. Int. Reliability Physics Sym., Dallas, Texas, 2003.
35. G. Groeseneken et al. TED, 31: 1984, p 42.
36. F. P. Heiman, G. Warfield. IEEE Trans. Elec. Dev., 12: 1965, p 167.
37. S. Tsujikawa, T. Mine, K. Watanabe, Y. Shimamoto, R. Tsuchiya, K. Ohnishi, T. Onai, J. Yugami, S. Kimura. Negative bias temperature instability of pMOSFETs with ultra-thin SiON gate dielectrics. IRPS 2003.
38. D. S. Ang, S. Wang, C. H. Ling. Evidence of two distinct degradation mechanisms from temperature dependence on negative bias stressing of ultrathin gate p-MOSFET, IEEE Electr. Dev. Lett., 26(12): Dec 2005.
39. H. Aono, E. Murakami, K. Okuyama, A. Nishida, M. Minami, Vooji, and K. Kubota. Modeling of NBTI degradation and its impact on electric field dependence of the lifetime. 42nd Ann. Int. Reliability Physics Sym., Phoenix, Arizona, 2004.
40. S. Zafar, B. H. Lee, J. Stathis, A. Callegari, T. Ning. A model for negative bias temperature instability (NBTI) in oxide and high-K pFETS. VLSI 2004.
41. V. Huard, M. Denais. Hole trapping effect on methodology for DC and AC negative bias temperature instability measurements in PMOS, IRPS 2004, pp 40–45.
42. G. Chen, M. F. Li, C. H. Ang, J. Z. Zheng, D. L. Kwong. Dynamic NBTI of pMOS transistors and its impact on MOSFET scaling. IEEE Electron Device Lett., 23(7): Jul 2002, pp 734–736.
43. G. Chen et al. Dynamic NBTI of PMOS transistors. Proc. IEEE Reliab. Phys. Symp 2003, pp 196–202.
44. T. Yang, C. Shen, M. F. Li, C. H. Ang, C. X. Zhu, Y.-C. Yeo, G. Samudra, D.-L. Kwong. Interface trap passivation effect in NBTI Measurement for p-MOSFET qith SiON gate dielectric. IEEE Electron Device Lett., 26(10): Oct 2005.
45. J. W. Lyding, K. Hess, I. C. Kizilyalli. Appl. Phys. Lett. 1996, p 2526.
46. K. Hess, I. C. Kizilyalli, J. W. Lyding. IEEE Trans. Electron Devices 1998, p 406.
47. T. G. Ference, J. S. Burnham, W. F. Clark, T. B. Hook, S. W. Mittl, K. M. Watson, L. K. Han. IEEE Trans. Electron Devices: 1999, p 747.
48. N. Kimizuka et al. VLSI 2000.
49. T. B. Hook, R. Bolam, W. Clark, J. Burnham, N. Rovedo, L. Schutz, Microelectro Reliab.: 2005, p 47.
50. Y. Mitani, H. Satake. Re-examination of deuterium effect on negative bias temperature instability in ultra-thin gate oxides. ICICDT 2006.
51. S. Tsujikawa, J. Yugami. Evidence for bulk trap generation during NBTI Phenomenon in pMOSFETs with ultrathin SiON gate dielectrics. IEEE Trans. on Electron Devices, 53(1): Jan 2006.
52. D. S. Ang, S. Wang. Recovery of the NBTI-stressed ultrathin gate p-MOSFET: The role of deep-level hole traps. IEEE Electron Device Lett., 27(11): Nov 2006.
53. N. Kimizuka et al. VLSI 1999.
54. D. Esseni et al. IEEE Trans. Electron Devices 2002, pp 49–254.

55. J. P. Campbell, P. M. Lenahan, A. T. Krishnan, S. Krishnan. Observations of NBTI-induced atomic-scale defects. IEEE Trans. Device and Materials Reliability, 6(2): June 2006.

56. J. P. Campbell, P. M. Lenahan. Density of state of Pb1 Si/SiO_2 interface trap centers. Appl. Phys. Lett., 80(11): Mar 2002, pp 1945–1947.

57. M. Ershov, S. Saxena, H. Karbasi, S. Winters, S. Minane, J. Babcock, R. Lindley, P. Clifton, M. Redford, A. Shibkov. Dynamic recovery of negative bias temperature instability in p-type metal-oxide-semiconductor field-effect transistors. Appl. Phys. Lett., 83(8): 2003, pp 1647–1649.

58. V. Huard, C. R. Parthasarathy, C. Guerin, M. Denais. Physical modeling of negative bias temperature instabilities for predictive extrapolation. 44th Ann. IRPS, San Jose, California, 2006.

59. A. T. Krishnan, V. Reddy, S. Chakravarthi, J. Rodriguez, S. John, S. Krishnan. NBTI impact on transistor and circuit: Models, mechanisms and scaling effects. IEDM 2003.

60. Y. Mitani, M. Nagamine, H. Satake, A. Toriumi. NBTI mechanism in ultra-thin gate dielectric—nirogen-originated mechanism in SiON. IEDM 2002.

61. N. Kimizuka, K. Yamaguchi, K. Imai, T. Iizuka, C. T. Liu, R. C. Keller, T. Horiuchi. NBTI enhancement by nitrogen incorporation into ultrathin gat oxide for 0.10-μm gat CMOS generation. 2000 Sym. of VLSI Technology.

62. C. Liu et al. Mechanism and process dependence of negative bias temperature instability (NBTI) for pMOSFET devices of ultrathin gate dielectrics. IEDM 2001.

63. S. S. Tan et al. Neighboring effect in nitrogen-enhanced negative bias temperature instability. Int. Conf. on Solid State Devices and Materials, Tokyo, Japan, 2003: pp 70–71.

64. S. S. Tan et al. Nitrogen-enhanced negative bias temperature instability: An insight by experiment and first-principle calculations. APL, 82(12): 24Mar 2003.

65. S. S. Tan et al. Relation between interfacial nitrogen concentration and activation energies of fixed-charge trapping and interface state generation under bias temperature stress condition. APL, 82: 2003, pp 269–271.

66. D. K. Schroder, J. A. Babcock. Negative bias temperature instability: Road to cross in deep submicron silicon semiconductor manufacturing. J. Appl. Phys., 94(1): 20031–18.

67. M. A. Alam. Technical Digest. IEDM 2003, p 345.

68. J. B. Yang, T. P. Chen, S. S. Tan, L. Chan. Analytical reaction-diffusion model and the modeling of nitrogen-enhanced negative bias temperature instability. App. Phys. Lett., 88: 2006, p 172109.

69. C. E. Blat, E. H. Nicollian, E. H. Poindexter. Mechanism of negative-bias temperature instability. J. Appl. Phys., 69(3): Feb 1 1991.

70. V. Huard, F. Monsieur, G. Ribes, S. Bruyere. Evidence for hydrogen-related defects during NBTI stress in p-MOSFETs. IRPS 2003.

71. T. B. Hook et al. The effects of fluorine on parametrics and reliability in a 0.18-m 3.5/6.8 nm dual gate oxide CMOS technology. IEEE Trans. On Electron Devices, 48(7): July 2001.

72. C. Liu et al. IEDM 2001.

73. A. Krishnan et al. IEDM 2001.

74. T. Hagivaga, K. Yamaguchi, S. Asai. Threshold voltage variation in very small MOS transistors due to local dopant fluctuations, 1982 Proc. of the Sym. on VLSI Technology: Digest of Technical Papers: pp 46–47.

75. T. Mizuno, J. Okamura, A. Toriumi. Experimental study of threshold voltage fluctuation due to statistical variation of channel dopant number in MOSFETs. IEEE Trans. on Electron Devices, 41(11): 1994, pp 2216–2220.
76. K. R. Lakshmikumar, R. A. Hadaway, M. A. Copeland. Characterisation and modeling of mismatch in MOS transistors for precision analogue design. IEEE J. Solid State Circuits, SC-, 21: 1986, pp 1057–1066.
77. K. Takeuchi, T. Tatsumi, A. Furukawa. Channel engineering for the reduction of random-dopant-placement-induced threshold voltage fluctuations. IEDM Technical Digest 1997, pp 841–844.
78. A. Asenov. Random dopant induced threshold voltage lowering and fluctuations in sub-0.1 mMOSFETs: A 3D 'atomistic' simulation study. IEEE Trans. on Electron Devices., 45(12): 1998, pp 2505–2513,
79. K. Takeuchi. Channel Size Dependence of Dopant-Induced Threshold Voltage Fluctuations, 1998 Sym. on VLSI Technology Digest, pp. 72–73.
80. Stewart Rauch. The statistics of NBTI-induced VT and K mismatch shifts in pMOSFETs. IEEE Trans. on Device and Materials Reliability, 2(4): 2002, pp 89–93.
81. A. Asenov. Random dopant induced threshold voltage lowering and fluctuations in sub-0.1 μm MOSFETs: A 3D 'atomistic' simulation study. IEEE Trans. on Electron Devices, 45(12): 1998, pp 2505–2513.
82. B. Kaczer, V. Arkhipov, M. Jurczak, G. Groeseneken. Negative bias temperature instability (NBTI) in SiO_2 and SiON gate dielectrics understood through disorder-controlled kinetics. Microelectronic Engineering, 80: 2005, pp 122–125.
83. I. A. Shkrob, A. D. Trifunac. Time-resolved EPR of spin-polarized mobile H atoms in amorphous silica: The involvement of small polarons. Phys. Rev. B, 54: 1996, pp 15073–15078.
84. A. E. Islam, G. Gupta, S. Mahapatra, A. T. Krishnan, K. Ahmed, F. Nouri, A. Oates, M. A. Alam. Gate Leakage vs. NBTI in plasma nitrided oxides: Characterization, physical principles, and optimization. IEDM 2006.
85. A. T. Krishnan, C. Chancellor, S. Chakravarthi, P. E. Nicollian, V. Reddy, A. Varghese, R. B. Khamankar, S. Krishnan. Material dependence of hydrogen diffusion: Implications for NBTI degradation. IEDM 2005.
86. Liu et al. IEDM 2001.
87. S. Tsujikawa, K. Watanabe, R. Tsuchiya, K. Ohnishi, J. Yugami. Experimental evidence for the generation of bulk traps by negative bias temperature stress and their impact on the integrity of direct tunneling gate dilectrics. VLSI 2003.
88. C. Shen, M.-F. Li, C. E. Foo, T. Yang, G. S. Samudra, Y.-C. Yeo. Characterization and physical origin of fast V_{th} transient in NBTI of pMOSFETs with SiON dielectric. Proc. Intl. Electron Devices Meeting, 2006.
89. S. Rangan, N. Mielke, E. C. C. Yeh. Universal recovery behavior of negative bias temperature instability, IEDM 2003.
90. B. Kaczer, V. Arkhipov, R. Degraeve, N. Collaert, G. Groeseneken, M. Goodwin. Disorder-controlled-kinetics model for negative bias temperature instability and its experimental verification. Proc. Intl. Rel. Phys. Symp.: 2005, pp 381–387.
91. M. Denais et al. IEDM 2004.
92. J. C. Dyre. Phys. Rev. Lett., 58: 1987, p 792.
93. V. I. Arkhipov, A. I. Rudenko. Phil. Mag. B, 45: 1982, p 189.

94. R. Fernández, B. Kaczer, A. Nackaerts, S. Demuynck, R. Rodríguez†, M. Nafría, G. Groeseneken. AC NBTI studied in the 1 Hz–2 GHz range on dedicated on-chip CMOS circuits. IEDM 2006.

95. G. La Rosa et al. IRPS tutorial. 2002.

96. T. Nigam, E. B. Harris. Lifetime enhancement under high frequency NBTI measured on ring oscillators. IRPS 2006.

97. V. Reddy, A. T. Krishnan, A. Marshall, J. Rodriguez, S. Natarajan, T. Rost, S. Krishnan. Impact of negative bias temperature instability on digital circuit reliability. IRPS 2002.

98. C. Schlunder, R. Brederlow, B. Ankele, A. Lill, K. Goser, R. Thewes. On the degradation of p-MOSFETs in analog and RG circuits under inhomogeneous negative bias temperature stress. IRPS 2003.

99. Y. Wang, K. P. Cheung, A. Oates, P. Mason. Ballistic phonon enhanced NBTI. IRPS 2007.

100. D. J. DiMaria et al. J. Appl. Phys., 73: 1997, 3337.

101. C. G. Van de Walle et al. IEEE Trans. Electron Devices, 47: Oct 2000.

102. S. Zafar, B. H. Lee, J. Stathis, A. Callegari, T. Ning. VLSI Symp. Proc 2004, p 208.

103. W. Abadeer, W. Ellis. IRPS Prc.: 2003, p 17.

104. R. A. B. Devine, J.-L. Autran, W. L. Warren, K. L. Vanheusdan, J.-C. Rostaing. Interfacial hardness enhancement in deuterium annealed 0.25 µm m channel metal oxide semiconductor transistors. Appl. Phys. Lett., 70(22): 1997, pp 2999–3001.

105. S. Mahapatra, P. B. Kumar, M. A. Alam. A new observation of enhanced bias temperature instability in thin gate oxide p-MOSFET. IEDM Tech. Digest: 2003, pp 337–340.

106. S. Mahapatra, D. Saha, D. Varghese, P. B. Kumar. On the generation and recovery of interface traps in MOSFETs subjected to NBTI, FN, and HCI stress. IEEE Trans. On Electron Devices, 53(7): July 2006.

107. D. Varghese, D. Saha, S. Mahapatra, K. Ahmed, F. Nouri, M. Alam. On the dispersive versus arrhenius temperature activation of NBTI time evolution in plasma nitrided gate oxides: Measurements, theory, and implications. IEDM 2005.

108. M. L. Reed et al. JAP 1988, p 5776.

109. D. J. Dimaria et al. J. Appl. Physics, 86: 1999, p 2100.

110. D. S. Ang, S. Wang. Insight into the suppressed recovery of NBTI-stressed ultrathin oxynitride gate pMOSFET. IEEE Electron Device Lett., 27(9): Sep 2006.

111. D. S. Ang. Observation of suppressed interface state relaxation under positive gate biasing of the ultrathin oxynitride gate p-MOSFET subjected to negative-bias temperature stressing. IEEE Electron Device Lett, 27(5): 2006.

112. G. La Rosa, W. L. Ng, S. Rauch, R. Wong, J. Sudijono. Impact of NBTI induced statistical variation to SRAM cell stability. IRPS 2006.

113. V. Huard et al. Interface traps and oxide charges during NBTI stress in pMOSFETs, 2002 IRW final report.

114. J. P. Campbell et al. Atomic-scale defects involved in the negative-bias temperature instability. IEEE Trans. Device and Materials Reliability, 7(4): Dec 2007.

115. D. K. Schroder. Negative bias temperature instability (NBTI): Physics, materials, process, and circuit issues., Tutorial. IRPS 2005.

116. S. Mahapatra, M. A. Alam, P. Bharath Kumar, T. R. Daleil, D. Saha. Mechanism of negative bias temperature instability in CMOS Devices: Degradation, recovery and impact of nitrogen. IEDM 2004.

5

HOT CARRIERS

Stewart E. Rauch, III

5.1 INTRODUCTION

Carriers (electrons or holes) can gain large kinetic energies from transit through regions of high electric field. Several mechanisms are capable of further increasing carrier energy. When the mean carrier energy is significantly larger than that associated with the lattice in thermal equilibrium ($E_{\mathrm{AVG}} = \frac{3}{2}kT_L$), they are called "hot" because the carriers were historically assumed to be thermally distributed at an effective temperature higher than that of the lattice. Hot carriers can gain enough energy to be injected into the gate oxide or cause interfacial damage, introducing instabilities in the electrical characteristics of a MOSFET device. Hot carrier degradation is a critical reliability concern, particularly when the design of MOSFET transistors allows large electric fields at operating conditions. Hot carriers can be generated in several ways and in different regions of the device; in the gate oxide, in the silicon substrate, or at $\mathrm{Si/SiO_2}$ interface of the channel. In the gate oxide they can be produced by an avalanche process induced by large gate oxide electrical fields or high energy radiation. In the substrate they can be produced by optical illumination or by the substrate electrical fields assisted injection.

This chaper focuses on channel hot carrier (CHC) effects generated in the MOSFET channel regions under channel conduction conditions similar to applications in integrated circuits. In qualifying CMOS technologies two CHC bias stress conditions are typically investigated: the conducting and nonconducting

Reliability Wearout Mechanisms in Advanced CMOS Technologies. By Strong, Wu, Vollertsen, Suñé, LaRosa, Rauch, and Sullivan
Copyright © 2009 the Institute of Electrical and Electronics Engineers, Inc.

CHC. Figure 5.1 gives a general description of the evolution of these two processes in both bias stress conditions for a MOSFET device. Conducting CHC damage is experienced by a MOSFET transistor with the channel in inversion in saturation ($V_{gs} > V_{th}$, $V_{ds} > V_{dsat}$), while the nonconducting CHC condition characterizes the transistor in off state ($V_{gs} = 0$ V) with the drain junction reverse biased ($V_{ds} \geq V_{dd}$). In the conducting CHC case the source of the channel carriers is the drain current in the pinch-off region, while in the nonconducting CHC is the channel off current (I_{off}). In order to produce CHC damage, a channel carrier needs to undergo three sequential processes: energy gain along the MOSFET channel or in the drain region, energetic carrier impact or injection, and production of physical damage or trapping in the interface or gate oxide.

Two mechanisms are responsible for carrier heating: the channel electric field and/or secondary carrier energy gaining or redistribution processes such as electron–electron scattering. The channel carrier energy distribution is described by the energy distribution function (EDF) calculated by an appropriate solution of the Boltzmann transport equation (BTE) in the channel. The majority of the hot carriers are collected by the drain, but a small fraction may be scattered towards the interface with enough energy to get injected into the gate oxide (oxide

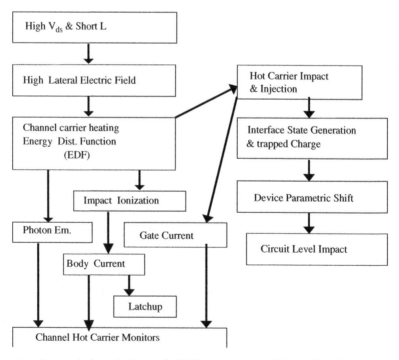

Figure 5.1. General description of CHC processes taking place in an MOSFET device.

bulk trap creation/activation) or to generate interface states. The hot carrier damage will modify the electrical characteristics of the MOSFET device (V_t, I_{on}, etc.). These modifications can impact the functionality of the integrated circuits. If the carrier kinetic energy is larger than the Si bandgap energy E_G (1.12 eV at room temperature), they may generate electron hole pairs by impact ionization (II). In the case of an nMOSFET, the majority of the generated holes makes up the substrate current (I_{sx}), while the generated electrons contribute to the drain current ($I_d = I_s + I_{sx}$). Under these conditions I_{sx} is an indirect measurement of carrier heating and hence nMOSFET susceptibility to hot carriers. At high V_{ds} and high substrate resistance the I_{sx} induced voltage drop between the drain and the substrate can forward bias the source/substrate junction turning on a parasitic bipolar device in parallel with the nMOSFET. The large injection of source carriers into the substrate further enhances II initiating a self-sustaining mechanism called snapback breakdown. Photon emission is a possible effect of the hot carrier generation and is also an indication of the channel carrier heating as well. A similar picture can be given for the pMOSFET device.

The nature of these two CHC bias configurations as well as the methodologies used to establish the CHC sensitivity of a given CMOS technology and their impact to device shift will be discussed in this chapter.

5.2 HOT CARRIERS: PHYSICAL GENERATION AND INJECTION MECHANISMS

The basic physical mechanisms that results in the CHC problem in a MOSFET transistor are illustrated in Figure 5.2 for the case of an NFET device. Five different processes participate in the CHC induced MOSFET degradation:

1. The CHC damage is generated by carriers flowing in the MOSFET inverted channel ($V_{gs} > V_{th}$), which gain significant energy past the channel pinch-off point. Thus, the carrier heating is directly related to the set up of the saturation condition ($V_{ds} > V_{dsat}$). In the linear regime, ($V_{ds} \leq V_{dsat}$), V_{ds} is dropped across the entire channel, and the channel electric field (F) is limited to relatively low values below F_c, the critical field for velocity saturation. As V_{ds} is increased above V_{dsat}, the excess voltage, $V_{ds} - V_{dsat}$, is dropped across a relatively short pinch-off region at the drain end of the device. At this point F is exponentially dependent on position. Under the influence of the high lateral electric fields in short channel MOSFETs, carriers in the transistor pinch-off region reach a nonequilibrium energy distribution.

2. The energetic carriers may lose their energy via impact ionization contributing to the substrate current.

3. Hot carriers can acquire sufficient energy to break H bonds at the interface, generating dangling bonds which have carrier trapping levels in the forbidden gap. These are called interface states.

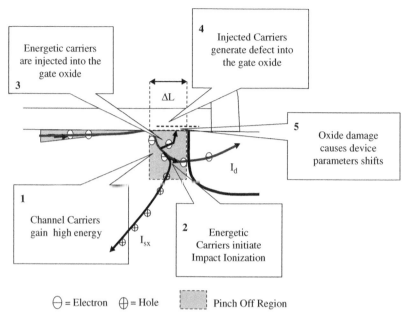

Figure 5.2. Physical mechanisms participating in the CHC induced device degradation in nMOSFET in saturation. The pinch-off region of length L_S is also shown.

4. Hot carriers may also surmount the energy barrier at the Si/SiO_2 interface, and once injected into the gate oxide may be trapped, or generate defects in the bulk oxide.

5. The CHC generated oxide damage induces MOSFET parameter shifts. This instability in the device parameters may produce fails in the long term operation of the circuits using these transistors.

5.2.1   Electric Field in a MOSFET Pinch-Off Region

The electric field is the main driving force of the hot carrier generation in MOSFET devices and modulates the injection of hot carriers in the gate oxide. Depending on the bias conditions in saturation, the lateral electric field in the pinch-off region heats up the channel carriers, while the vertical oxide field can favor or prevent the injection of either hot holes or electrons in the gate oxide by modulating the barrier height (Φ_B) at the Si/SiO_2 interface.

In the pinch-off region of a MOSFET in saturation, the electric field is two-dimensional. A rigorous calculation of the lateral electric field near the drain is a very complex procedure since it requires the simultaneous solution of the current transport equation and the Poisson equation. In addition the estimate of the lateral electric field is further complicated by the fact that with CMOS scaling the different MOSFET design option (LDD, DDD, LATID, etc.) has been implemented to meet a tradeoff between performance and reliability. The estimate of electric field F

distribution in the pinch-off region strongly depends on the different source/drain junction schemes and the well engineering.

Typically a numerical solution of both equations is found by using two dimensional simulators. To obtain analytical expressions for the lateral electric field in the pinch-off region some approximations need to be made to reduce this description to be one-dimensional.

To illustrate the dependence of the lateral electric field on the S/D engineering, we will describe an analytical formulation of the lateral electric field in the pinch-off region respectively of a single drain (SD) and lightly doped drain (LDD) MOSFET structure. This description uses the pseudo two-dimensional approximation (PTDA) which reduces the electric field two-dimensional effects in the pinch-off/drain region into a one-dimensional description. Even if these MOSFET designs are not a complete representation of the ones adopted in aggressive CMOS scaling, we feel that it is important to give the reader a simple insight into the general dependence of the channel electric field dependence on the device design.

5.2.1.1 Electric Field in the Pinch-Off Region of a Single Drain MOSFET.
Figure 5.3 gives a cross section of a SD MOSFET in saturation ($V_{th} < V_{gs} < V_{ds}$, $V_{ds} > V_{dsat}$) under the PTDA approximation [3, 9]. In addition

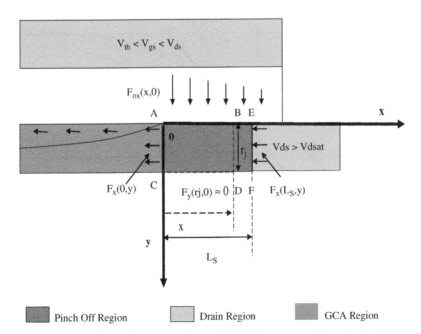

Figure 5.3. Cross section of a conventional MOSFET device with a single S/D implant biased in saturation ($V_{ds} > V_{dsat}$). Velocity saturation is assumed in the pinch-off region (rectangle AECF) with the area $L_s \times r_j$. The Gauss rectangle (ABCD) used in the PTDA is also shown. Reprinted with permission from Ref. 3, © 1977 IEEE.

we will assume that the carriers reach velocity saturation in the channel. This condition is satisfied when the carrier drift velocity (v_d) approaches the average thermal velocity $v_{th} = \sqrt{3k_B T_L/m_{eff}}$, where m_{eff} is the carrier effective mass. Figure 5.4 shows the measured relation between drift velocity (v_d) and the channel electric field (F) for both holes and electrons in bulk silicon [5, 6]. At low electric fields the carriers are almost in thermal equilibrium with the lattice and v_d is linearly proportional to the electric field with the effective mobility being the proportionality constant ($v_d = \mu_{eff} \times F$). At sufficiently high electric fields, the average carrier kinetic energy increases and carriers undergo scattering events by optical–phonon interactions with resulting decrease in mobility with the increasing electric field until v_d reaches the limiting value v_{sat}.

The experimental results represented in Figure 5.4 can be approximated by the following empirical expression for the drift velocity (v_d):

$$v_d = \frac{\mu_{eff} \times F}{[1 + (F/F_c)^\gamma]^{1/\gamma}}. \tag{5.1}$$

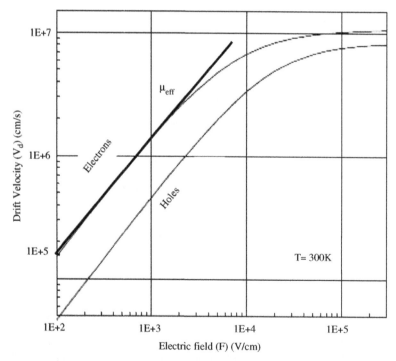

Figure 5.4. Hole and electron drift velocity (v_d) as function of electric field (F) in bulk silicon at $T = 300$ K.

The parameter F_c is the critical field. γ is 2 for electrons and 1 for holes. For $F \gg F_c$, v_d saturates to the value $v_{sat} = \mu_{eff} \times F_c$. Therefore, $F_c = v_{sat}/\mu_{eff}$. Values of v_{sat} such as $v_{sat} \approx 7-8 \times 10^6$ cm/sec and $6-7 \times 10^6$ cm/sec have being reported respectively for electrons and holes in a MOSFET channel at 300 K [7, 8]. Since μ_{eff} in the channel is a function of the vertical field, F_c depends somewhat on V_{gs}. Typical numbers are $2-5 \times 10^4$ V/cm for electrons and $0.8-1.5 \times 10^5$ V/cm for holes in Si inversion layers at 300 K.

The PTDA is based on the following assumptions (see Fig. 5.3):

1. Channel carriers, entering the pinch-off region at the point A, experience velocity saturation ($v = v_{sat}$) at the voltage v_{dsat}. The pinch-off region is approximated by a box bounded by the sides ABEF.
2. The electric field inside the pinch-off region is assumed to have only a lateral component $F_x(x, y)$ ($F_y(x, y) \approx 0$).
3. The source and drain junctions are rectangular in shape with the junction depth r_j.

Two regions are identified in the channel under these conditions: the inverted channel region that can be described by the gradual channel approximation (GCA) region and the pinch-off region, where the carrier velocity is v_{sat}.

To calculate the lateral electric field $F_x(x)$ in the velocity saturation depleted region, we start by applying the Gauss' law to the volume with rectangle ABCD (area $r_j \times x$) and width W (MOSFET width). The bottom of this rectangle is at a certain depth y_j such that the vertical component of the electric field $E_y(x, y_j)$ is negligible. We assume in the following that $y_j \approx r_j$ [10]. The application of Gauss' law in this volume gives

$$
-\int_0^x F_y(x, 0)\, dy + \int_0^{r_j} F_x(0, y)\, dy - \int_0^{r_j} F_x(x, y)\, dy
$$
$$
= -\frac{q}{\varepsilon_0 \varepsilon_{Si}} N_A r_j x + \frac{Q_m}{\varepsilon_0 \varepsilon_{Si}} x,
$$

$$(5.2)$$

where Q_m is the mobile charge density in the drain region and F_{ox} is the gate oxide field at Si/SiO$_2$ interface. Assuming that the channel carriers move in the pinch-off region with velocity saturation, the mobile charge Q_m in the drain region is equal to that a the point A where $V(0) = V_{dsat}$. Therefore, assuming charge neutrality:

$$
Q_m = C_{ox}(V_{gs} - V_{FB} - 2\phi_f - V_{dsat}) + qN_A r_j.
$$

$$(5.3)$$

The vertical component of the electric field at the Si/SiO$_2$ interface is given by

$$
F_y(x, 0) = \frac{\varepsilon_{Si}}{\varepsilon_{ox}} F_{ox}(x, 0),
$$

$$(5.4)$$

where F_{ox} is

$$F_{ox} = \frac{V_{gs} - V_{FB} - 2\phi_f - V(x)}{t_{ox}}. \tag{5.5}$$

Differentiating Equation 5.2 with respect to the lateral dimension x, we get

$$r_j \frac{dF_x(x)}{dx} + \frac{\varepsilon_o}{\varepsilon_{Si}} F_y(x,0) = -\frac{q}{\varepsilon_0 \varepsilon_{Si}} N_A r_j + \frac{Q_m}{\varepsilon_0 \varepsilon_{Si}}. \tag{5.6}$$

Taking into account Equations 5.3, 5.4, and 5.5 from 5.6 we get

$$\frac{dF_x(x)}{dx} = \frac{V(x) - V_{dsat}}{l^2}, \tag{5.7}$$

where

$$l = \sqrt{\frac{\varepsilon_{Si}}{\varepsilon_{ox}} t_{ox} r_j}. \tag{5.8}$$

The solution of Equation 5.8 in the pinch-off region $(0 \le x \le L_S)$ satisfying the boundary condition $V(0) = V_{dsat}$, $F_x(0) = F_c$ is

$$V(x) = V_{dsat} + l \times F_c \times \sinh\left(\frac{x}{l}\right) \tag{5.9}$$

and

$$F_x(x) = -\frac{\partial V}{\partial x} = F_c \times \cosh\left(\frac{x}{l}\right), \tag{5.10}$$

where F_c (as defined earlier) is the critical lateral electric field to reach velocity saturation ($F_c \approx 2$–$5\,V/\mu m$ for electrons in the channel, 5–$15\,V/\mu m$ for holes.) The term l can be thought as a scale factor of the spatial dependence of the electric field. One approximation for V_{dsat} is given by Equation 5.11 [20]:

$$V_{dsat} = F_c(L_{eff} - L_s)\left(\sqrt{1 + \frac{2(V_G - V_T)}{mF_c(L_{eff} - L_s)}} - 1\right), \tag{5.11}$$

where m is the body effect coefficient.

Using the identity $\sinh^2 A + 1 = \cosh^2 A$, $F_x(x)$ can be written as

$$F_x(x) = \sqrt{F_c^2 + \frac{(V(x) - V_{dsat})^2}{l^2}}. \tag{5.12}$$

At the drain end of the channel (E point) we get the following expressions for L_S and F_{max}:

$$L_S = l \times \ln\left[\frac{(V_{ds} - V_{dsat})}{l \times F_c} + \frac{F_{max}}{F_c}\right] \qquad (5.13)$$

$$F_{max} = \left[\left(\frac{V_{ds} - V_{dsat}}{l}\right)^2 + F_c^2\right]^{1/2} \qquad (5.14)$$

Typically, $(V_{ds} - V_{dsat}) \gg F_c \times l$ and F_{max} can be approximated by

$$F_{max} \approx \frac{V_{ds} - V_{dsat}}{l} \qquad (5.15)$$

Figure 5.5 describes the dependence of the calculated extension of the pinched-off region (L_S) and the maximum lateral field (F_{max}) as function of V_{ds} ($2V_{gs} = V_{ds}$) applying the PTDA and assuming device scaling as given in Table 5.1, while keeping the simple SD MOSFET design. This L_S versus F_{max} relation is not expected to be followed by the industry adopted MOSFET structures. With CMOS scaling, in fact, MOSFET design has evolved from the

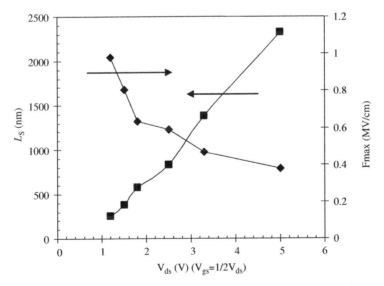

Figure 5.5. Estimated length of the pinched-off region (L_S) and maximum electric field (F_{max}) as function of V_{ds} with $V_{gs} = 1/2\ V_{ds}$ assuming the 2D Poisson approximation and the technology scaling assumptions as listed in the Table 5.1.

TABLE 5.1. Key MOSFET Parameters Used in CMOS Scaling

Year	1980	1992	1995	1998	2001	2004
V_{dd} (V)	5	3.3	2.5	1.8	1.5	1.2
V_{th} (V)	0.8	0.7	0.6	0.5	0.4	0.25
L_{des} (μm)	0.9	0.45	0.25	0.15	0.1	0.07
T_{ox} (nm)	23	12	7	5	3.5	2.5
r_j (nm)	150	100	60	40	25	15

simple SD design. An example of this is given by LDD MOSFET design discussed in the following paragraph. These results, however, give the observed trends in advanced submicron technologies. Under these conditions, MOSFET devices of 1 V technologies are expected to experience an F_{max} larger that 1 MV/cm with L_S approaching values of the order of the carrier mean free path (~ 10 nm for electrons).

5.2.2 Lateral Electric Field in the Pinch-Off Region of an LDD MOSFET

As seen from Figure 5.5 and assuming CMOS scaling by keeping the conventional single source and drain engineering, it is expected that the maximum lateral electric field increases to values that can make the transistor very sensitive to CHC effects. In the industry adopted source and drain (S/D) junction schemes the nMOSFET transistors is often designed with a lightly doped drain (LDD) feature as part of structure. The main objective of the LDD S/D design is to reduce the lateral electric field in the drain region for a given power supply and channel length. This is accomplished by forming the S/D junction using a double implantation scheme. A lightly doped n-region (n-) is first formed by a low energy P or As implantation self-aligned to the gate electrode. Then oxide spacers are formed at the poly silicon gate sidewall. An n^+ As implant is self-implanted to the spacers. The n^+ implant diffuses under the spacers to the edge of the gate. Figure 5.6. gives a schematic cross section of the LDD S/D junction. Characteristic lengths are the length of the lightly doped region L_n and the gate–drain/source overlap region L_g.

An analytic formulation of F_{max} in an LDD nMOSFET can be based on the PTDA approximation as discussed in the previous paragraph for the case of MOSFETs with a single drain/drain junction. In this case both the LDD (n^-) and the drain (n^+) junction implants are assumed to have a constant doping. Figure 5.7 shows the LDD MOSFET cross-section under the PTDA approximation. In this case, F_{max} depends on the relative overlap between the gate edge and the LDD region, which is controlled by both parameters L_n and L_g.

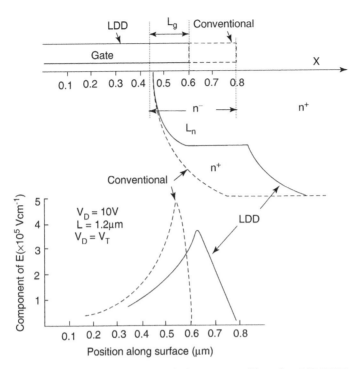

Figure 5.6. Schematic cross section and doping profile of a MOSFET transistor with LDD source and drain junction. Both the lateral length (L_n) of the n^- diffusion and the gate to drain/source overlap (L_g) region are shown. The estimated lateral electric field (F_x) at the Si/SiO$_2$ interface as function of position for both a conventional and LDD nMOSFET design is shown. Reprinted with permission from Ref. 4, © 1980 IEEE.

We discuss the case of full overlap LDD design ($L_g = L_n$) in more detail as an example. Following the same arguments, we get the two differential equations,

$$\frac{dF_x(x)}{dx} = \frac{V(x) - V_{dsat}}{l^2} \tag{5.16}$$

in the pinch-off region ($0 \le x \le L_S$) and

$$\frac{dF_x(x)}{dx} = \frac{V(x) - V_{dsat}}{l^2} - \frac{qN_d}{\varepsilon_{Si}} \tag{5.17}$$

and in the drain region ($L_S \le x \le L_S + L_n$).

Solving these differential equations in these two regions with the boundary conditions at $V(0) = V_{dsat}$ and $F_x(0) = F_{sat}$ and the requirement of potential and

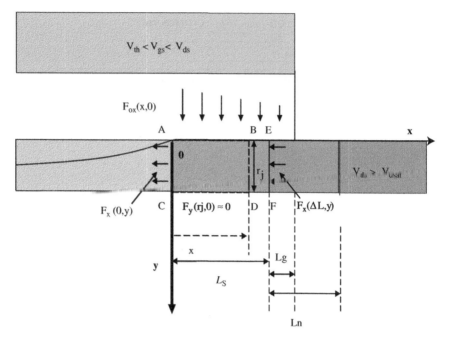

Figure 5.7. Cross section of the drain junction of an LDD nMOSFET, showing the sidewall spacer, the gate overlap length (L_g) and the length of the LDD region (L_n). Reprinted with permission from Ref. 4, © 1980 IEEE.

lateral electric field ($F_x(x)$) continuity at $x = L_S$, we get

$$F_x(x) = -\frac{\partial V}{\partial x} = F_c \times \cosh\left(\frac{x}{l}\right) \tag{5.18}$$

in the pinch-off region and

$$F_x(x) = F_{\text{sat}} \times \cosh\left(\frac{x}{l}\right) - \frac{q\eta N_d \times l}{\varepsilon Si} \sinh\left(\frac{x - L_S}{l}\right) \tag{5.19}$$

in the LDD region.

The decrease on the intensity of the F_x in an LDD nMOSFET is clearly illustrated in the two dimensional simulation results shown in Figure 5.6. The lateral electric field distribution as function of distance along the channel near the drain region is calculated for both a conventional (SD) and an LDD device, both with the same channel length and oxide thickness. In the SD transistor F_x peaks F (F_{max}) at approximately the drain metallurgical junction and drops quickly in the drain region. In the LDD MOSFET design F_x extends across the n- region before dropping to zero in the n$^+$ region. In addition the F_{max} value is reduced.

Generally the LDD devices are able to provide improved CHC reliability respect to the conventional SD MOSFETs. Two features of the LDD design need to be carefully optimized to make sure that the MOSFET will not suffer some key drawbacks compromising the improved CHC resistance. First the n^- dose introduces an extra drain series resistance R_{n-} which in the case of uniform n-concentration is given by

$$R_{n-} = L_{n-}/(q\mu_n N_D r_j) W. \tag{5.20}$$

A sufficiently low n^- dose increases Rn^- degrading the drain current. In addition if F_{max} is located underneath the spacer region, electrons injected during CHC can deplete the sufficiently low n^- dose region producing a parasitic drain series resistance increase (PDSRI) which will further decrease the drain current.

5.2.3 Vertical Electric Field in the Pinched-Off Channel Region

The hot carrier injection into the gate oxide is controlled by the modulation of the vertical gate oxide field (F_{ox}) above the pinch-off region. Under the inverted channel condition F_{ox} is given by

$$F_{ox} = \frac{(V_{gs} - V_{FB} - 2\phi_F - V_{ds})}{T_{ox}}, \tag{5.21}$$

where V_{FB} is the flatband voltage and ϕ_F is the Fermi potential.

From Equation 5.21, the modulation of the vertical gate oxide field (F_{ox}) above the pinch-off region depends on the $V_{gs}-V_{ds}$ difference during the stress and has two important consequences, i.e., favor CHC injection in the gate oxide and modulate the barrier height at the Si/SiO$_2$ interface.

As an example, Figure 5.3 shows the E_{ox} lines in a nMOSFET transistor in saturation with $V_{gs} \approx V_{th}$. In this case the electric field experiences a reversal above the pinch-off region which favors injection of hot holes, while discouraging hot electrons. As the gate voltage is increased beyond the hot hole region, hot electron injection is favored. As the reversal region is weakened with increasing V_g, the probability of hot electrons to be injected in the gate oxide is increased.

This effect is further enhanced by the Si/SiO$_2$ potential barrier lowering due to the image force effect (see Fig. 5.8) The net barrier height is given by

$$\Phi_b = \Phi_{b0} - 2.59 \times 10^{-4} E_{ox}^{1/2} - a_0 E_{ox}^{2/3}, \tag{5.22}$$

where $\Phi_{b0} = 3.2\,eV$ is the Si/SiO$_2$ interface barrier for the electrons, where $a_0 \approx 1.0 \times 10^{-5}$ (cm). The second term of Equation 5.22 accounts for the barrier lowering effect due to the image field, while the third term takes into account the finite probability of tunneling between Si and SiO$_2$.

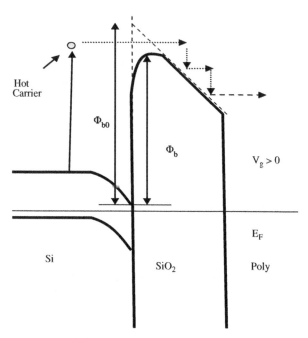

Figure 5.8. Potential barrier lowering due to image force effect.

5.2.4   High Field-Induced Carrier Heating and the Carrier Energy Distribution Function

The phenomenon of carrier heating in a MOSFET in saturation results from the high electric fields appearing in the channel near the drain junction in the pinch-off region. This mechanism can be mathematically described by noticing that energetic carriers traveling in the channel can exchange energy via scattering events (impurity, electron–phonon scattering, etc.) and can undergo energy gaining processes such as carrier acceleration by the channel electric fields. The total effect of these processes is a distribution of the carrier population as function of kinetic energy (E) that is described by the energy distribution function (EDF). EDF is defined by the function $f(E)$ such that $f(E)dE$ gives the number of carriers with energy between E and $E + dE$. Knowledge of the EDF is a necessary condition to study hot carrier effects since hot carrier damage is generated by a subpopulation of the channel carriers with high energies.

5.2.4.1   *EDF in Thermal Equilibrium.*   For small fields, the carriers are in approximate thermal equilibrium with the lattice. In this case carriers both emit and absorb phonons and the net rate of energy exchange is zero. At thermal equilibrium the carrier occupancy follows the Fermi–Dirac distribution [6] with a Fermi level (E_F). To simplify our discussion we focus on electrons in thermal equilibrium. If the electron gas is nondegenerate ($(E_c - E_F) > 3kT/q$), the electron

EDF (EEFD) can be approximated by the Maxwell distribution function (MDF) for $E > E_c$ by

$$f(E) = \frac{2N}{\sqrt{\pi}} \frac{\sqrt{E}}{(k_B T_L)^{3/2}} \times \exp\left(-\frac{E}{k_B T_L}\right), \tag{5.23}$$

where T_L is the lattice temperature (electrons in equilibrium with the lattice), and N is the number of electrons in a unit volume. The probability that an electron can have energy between E and $E + dE$ can be easily defined as $P(E)dE = (F(E)/N)dE$. Figure 5.9 shows the probability $P(E)$ calculated with thermal equilibrium at three different lattice temperatures (30°C, 500°C, and 1000°C).

Two important points can be observed. T_L is associated with the energy "spread" in the distribution. The larger T_L is, the larger is the average electron kinetic energy ($\langle E \rangle = 3/2 k_B T_L$) and the lower is the peak value ($P_{peak} = \sqrt{e/2\pi} \times (N/kT_L)$). At higher T_L values more electrons are available at higher energies. This energy "spread" at high energy exhibits as a "thermal tail" which is controlled by $\exp(-E/kT_L)$ factor. As will be seen, this concept of a thermal tail is extended to the case of nonequilibrium. At thermal equilibrium with T_L in the range of 30°C to 140°C (typical application temperatures), electrons do not gain enough energy from phonon interactions alone to activate hot carrier damage.

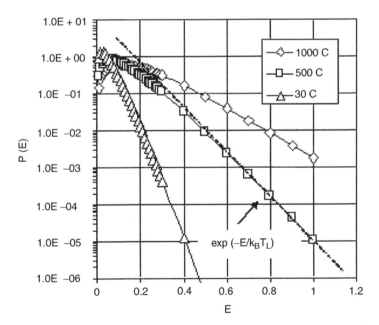

Figure 5.9. Maxwell distribution function at thermal equilibrium ($T_L = 30$°C, 500°C, 1000°C).

5.2.4.2   EDF in Submicron MOSFETs in Saturation. The calculation of the EDF of a MOSFET device biased in saturation conditions is associated with the description of carrier transport in the channel pinch-off region. Carrier heating results from a balance between carrier energy redistribution mechanisms such as scattering events as well as energy gaining processes such as the carrier acceleration by the electric fields in the pinch-off region. The understanding of the mechanisms responsible for carrier heating has evolved with time particularly with the effort of explaining the observed channel hot carrier damage at V_{ds} values lower that the assumed threshold energy for channel carrier injection into the gate oxide ($\Phi_B \approx 3.2\,\text{eV}$). Two regimes have been identified for carrier transport in the channel: the local and quasi-ballistic carrier transport. A parameter that helps to distinguish between those two regimes is the ratio between the carrier mean free path (MFP), λ, and the length of the scale factor l. The carrier MFP is the average distance carriers travel in the channel without undergoing scattering events. In general the MFP is a function of scattering processes channel carriers undergo at a given temperature. Mathematically

$$\frac{1}{\lambda} = \sum \frac{1}{\lambda_i}, \tag{5.24}$$

where λ_i is MFP associated to a given scattering events such as electron–phonon, ionized impurities interactions, etc. In the case that only electron–phonon interactions are active, it has been shown that

$$\lambda = \lambda_0 \tanh(\varepsilon_r / 2kT_L), \tag{5.25}$$

where λ_0 is the low temperature limit of λ, ε_r is the optical phonon energy and T_L the lattice temperature. In this case λ is a decreasing function of lattice temperature. The length of the pinch-off region L_S is the distance along the channel between the point where $V = V_{dsat}$ and V_{ds} (see Figure 5.3). It is a function of the MOSFET design (source/drain engineering, T_{ox}, etc.) as well as bias conditions.

1. *Local Carrier Transport.* This condition takes place if $l \gg \lambda$. In this case carriers reach a quasi-thermal equilibrium while traveling in the channel and carrier transport as well as carrier heating can be described in terms of local value of channel electric field. In this regime a "drift-diffusion" approximation to solve the BTE has been very successful to describe the MOSFET behavior to $L_{eff} \approx 0.25\,\mu\text{m}$. Under these conditions, carriers travel in the pinch-off region in velocity saturation conditions [6]. The EDF is adequately described by a heated MDF with an effective temperature T_{eff} given by the relation

$$T_{eff} = \frac{q\lambda}{k} F_{max}, \tag{5.26}$$

where F_{max} is the peak channel electric field in the channel pinch-off region. The fraction of carriers N_t above a threshold energy E_t is given by

$$N_t = \frac{2}{\sqrt{\pi x}} \exp(-x) + erfc(\sqrt{x}), \qquad (5.27)$$

where $x = E_t/kT_{eff}$. Figure 5.10 shows an estimated fraction of electron population above a given E_t as function of the drain voltages [13] in the case of nMOSFETs stressed at peak I_{sx} condition.

As expected the higher E_t is, the smaller the electron population is with energy larger than E_t. In particular, if $E_t \approx 3.7\,eV$ (assumed electron energy threshold for interface states generation), the fraction of electrons population (part of the thermal tail) available to produce the CHC damage reduces dramatically ($N_t \approx 10^{-8}$ at $V_{ds} = 3\,V$). This carrier heating picture is at the heart of the lucky electron model CHC picture.

2. *Quasi-ballistic carrier transport.* "Ballistic" transport means that carriers are accelerated by the electric field and experience no scattering

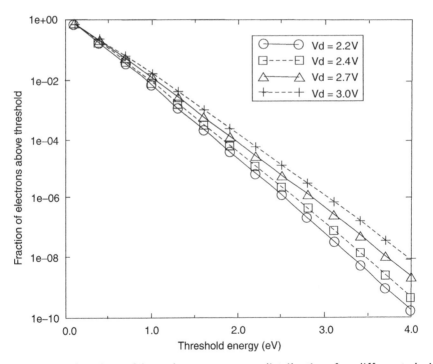

Figure 5.10. Estimation of hot electron energy distribution for different drain voltages. Reprinted with permission from Ref. 13, © 1995 IEEE.

events. Then, all the carriers will have the same energy at a given position, namely, the potential difference between start and end point. The EDF is a delta function at this energy. "Quasi-ballistic" transport is an intermediate case between pure ballistic and local transport. The local electric field is varying too quickly for carriers to reach a steady state EDF, however, total distances (say, L_S) are such that almost all carriers do experience a number of scattering events. This regime is dominant as the scale length of the electric field variation scales down to dimensions of the order of a few carrier MFP (10–20 nm for electrons at room temperature). While traveling in the pinch-off region channel carriers experience large varying electric fields and, because of the reduced number of scattering events, do not have enough time to reach steady state. The aggressive source and drain engineering coupled with the thinning gate oxide allows channel electric fields in access of 1 MV/cm in the pinch-off region, even for low V_{ds}, leading to extremely nonlocal quasi-ballistic transport over all or most of the pinch-off region. The large gradients of the channel electric field produce large carrier energy gradients, which, under appropriate conditions can lead to velocity overshoot.

   Two different approximations of the Boltzmann transport equation (BTE) are successfully used to describe this regime. One is a deterministic approach, called the "hydrodynamic approximation," which is an extension of the drift–diffusion model with the addition of an equation for the energy conservation. This hydrodynamic approximation is used in MOSFET device simulators such as the IBM FIELDAY and TMA MEDICI. An alternative approach is the Monte Carlo (MC) method [14, 15]. This is stochastic in nature and allows the trajectories of an ensemble of channel carriers to be traced individually at microscopic level in a given MOSFET structure subject to the action the channel electric fields and several allowed scattering events. The type of scattering events and the consequent energy exchange these carriers experience are determined by probability functions weighted according to the physical assumptions used. The EDF is defined by making a histogram of the energy and momentum values of each carrier. The obtained EDF can be used to calculate the carrier density, momentum and energy by appropriate integration. The physics can be correctly modeled by including a full-band structure model. Different full-band structure models are found in the literature. Fischetti et al. [14] as well as Bude et al. [16] employ a full-band structure from the pseudo-potential method, whereas Ghetti et al. [27] use an analytical band structure. The probabilities associated to the scattering mechanisms are quantified from experimental data. MC techniques are fairly established and applied to investigate ballistic transport in advance submicron MOSFETs and their CHC sensitivities [28, 29]. Since the outcome of a MC calculation is the EDF, MC techniques have been very useful in estimating the effects of energy gaining processes to the CHC damage particularly in the case of low voltage hot carriers effects ($qV_{ds} < \Phi_b$).

Figure 5.11 gives the typical EEDF [30] calculated at a given point x with voltage V_x along the pinch-off region of a nMOSFET device obtained by using a MC method. The EEDF exhibits the following three regions:

a. This part of the EEDF is populated with electrons in thermal equilibrium with the lattice. These are carriers that have been thermalized while traveling in the inverted channel, but did not yet experience the lateral electric field acceleration in the pinch-off channel region. This part of the EEDF can be treated by a MDF distribution at the lattice temperature T_L. The dashed curve (d) represents the extension of MDF in thermal equilibrium (no electric field applied) with $(-1/kT_L)$ slope.

b. A middle energy "hump" associated with the lateral electric field heated carriers. The electron population in this section of the EEDF can be approximated by a quasi-MDF at the effective temperature $T_{eff}(T_{eff} \gg T_L)$ determined by the accelerating electric field (F) in the pinch-off region as

$$kT_{eff} = q\lambda F. \tag{5.28}$$

This part of the EDF is given by the electrons gaining kinetic energy from the channel electric fields that have not enough time to reach thermal

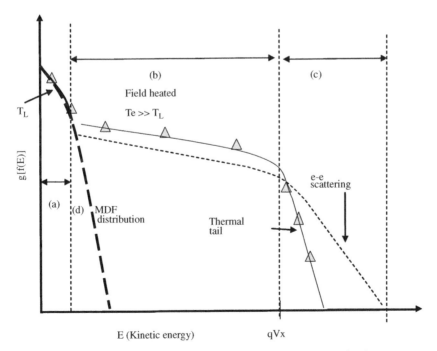

Figure 5.11. Schematic description of EEDF calculated in a submicron MOSFET device.

equilibrium with the lattice. Typically this part of the EEDF function extends to energy value close to qV_x, where V_x is the voltage applied to the carrier at the point x along the channel. The slope of this leg is proportional to $(-1/kT_{eff})$.

c. This energy region is beyond qV_x and is populated by electrons that are quasi-ballistically transported from the pinch-off point into the drain. This high energy region is generated by energy exchange processes such as electron–phonon or e–e interactions. In the case of electron–phonon interactions being dominant, the EEDF in this region can be approximated with a shifted MDF at $T_{eff} = mT_L$. In this case the energy tail has a slope inversely proportional to $mT_L(\approx -1/mkT_L)$ and it is called the thermal tail. A typical value for m is ~1.66 [37]. A different tail shape is obtained if other energy exchange mechanisms such as electron–electron interaction are active.

It is important to note that both parts (b) and (c) of the EEDF are generated from the presence of the pinch-off electric fields heating up the carriers. Without these fields the EEDF will follow thermal equilibrium and be described by the MDF as in (a). Parts (b) and (c) of the EEDF play a critical role in determining the fraction of relevant electron participating to such phenomena as impact ionization ($E > E_G$) to contribute to I_{sx} and current injection in the gate oxide ($E > \Phi_b$) to contribute to I_g. We will refer to the EEDF with a thermal high energy tail as the base EEDF.

Figure 5.12 gives the EEDF along the channel of a 75 nm nMOSFET device at $V_{gs} = V_{ds} = 3$ V bias condition [31]. The EEDF is calculated by using alternative MC algorithms such as MOMC, MOCA, and Damocles (IBM full band Monte Carlo device simulator) and similar results are obtained in all cases. This simulation assumes scattering events due to phonon absorption and emission, impact ionization, ionized impurities as well as surface scattering. It is clearly shown the transition from a MBD function in the inverted channel region of the device (the electric field is very small in the inverted channel) to the EEDF with a thermal tail in the drain region (high lateral electric fields). Notice also two points:

1. Impact ionization and electron injection into the gate oxide may take place in different positions along the channel.
2. The high energy thermal tail is expected to supply electrons with energy E larger than qV_{dd} and contribute to electron injection into the gate oxide.

Considering that Φ_b is around 3.2 V and assuming that the CHC damage is produced by hot carriers injected into the gate oxide then the electrons part of the base EEDF contribute to the CHC degradation only if their energy is larger than Φ_b. This has been assumed to be the case for sufficiently large V_{dd}. This picture, however, is not adequate to explain CHC degradation observed in advanced submicron CMOS technologies with power supplies much lower that 3.2 V.

Several mechanisms are responsible for heating up the channel carrier population and modifying the high energy tail of the EEDF and allowing hot

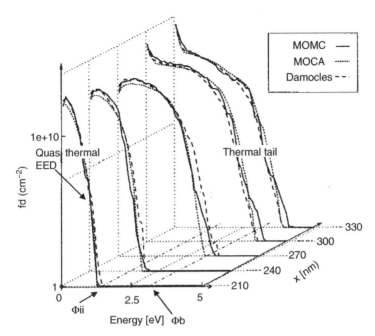

Figure 5.12. Comparison between calculated EEDF using MC procedures. The nMOSFET has 50 nm channel length with $V_{gs} = V_{ds} = 3.0\,V$ (V_s, $V_{sx} = 0\,V$). The transition from the channel to the drain region is at 275 nm. Reprinted with permission from Ref. 31, © 1998 IEEE.

carriers at V_{dd} voltages low enough that CHC can not be explained by injection of carriers. Three energy gaining mechanisms have been reported to play a major role in contributing to an extension of the EEDF beyond the qV_{ds} limit, i.e., electron–phonon scattering electron–electron scattering and secondary impact ionization. We will describe the first two briefly in the next two sections. Secondary impact ionization will be discussed later.

5.2.4.3 Electron–Phonon Scattering.

As described before, electron–phonon scattering (EPS) events are responsible for energy redistribution at high energy ($E > qV_{ds}$). In this energy range can be approximated by shifted MDF at $T = mT_L$ (thermal tail). According to [32] the thermal tail is produced by those electrons with energy around qV_{ds} that loses it by emitting an optical phonon. There is the possibility that this optical phonon can be absorbed by another electron extending this way the EDF beyond qV_{ds}. For electrons beyond qV_{ds} this would mean quasi-equilibrium with the lattice since the phonon absorption and emission are related. Consequently the tail above qV_{ds} is thermal ($T = mT_L$). The energy exchange in EPS is limited to the energy of optical phonons. In silicon the highest optical phonon energy is 63 meV.

5.2.4.4 Electron–Electron Scattering. Electron–electron scattering (EES) is due to energy exchange between electrons as result of Coulomb interaction. Two kinds of EES effects have been identified, namely, short and long range EES. Short range EES (SREES) relates to direct Coulomb interactions between single electrons. Figure 5.12 gives an example of SREES. In this case two electrons travel quasi-ballistically in the channel of the device and enter the drain region with an energy qV_{ds}. In the drain they undergo an EES event where one of the electron may transfer almost all of its kinetic energy to the other electron. Thus the maximum energy is just under twice the maximum energy due to the field heating (Figure 5.13).

The effect of this energy transfer is a broadening of the EDF to values that can be higher than qV_{ds}. The higher qV_{ds} is the larger the broadening of EDF. Contrary to OPS where the energy exchange in controlled by the optical phonon energy, SREED are very effective to broaden the EDF since the maximum exchangeable energy is controlled by qV_{ds} and can be significantly higher than that of an optical phonon. Figure 5.14 shows a comparison between the EEDF without EES (case 1) and with EES (case 2) [33]. In this work using MC simulation, the EEDF of an nMOSFET device ($L_{des} = 0.12\,\mu m$) with $V_{ds} = 2V_{gs} = 1.8$, 3.375 and 4.95 V was investigated. EES seems to have a dominant contribution to the EEDF for energies above 1.8 eV for the 1.8 V stress condition. For the highest stress configuration (4.95 V) EES has a small effect to the EEDF. In particular, at lower V_{ds} (<4.95 V), e–e scattering supplies hot electrons with energy higher than the thermal tail. Notice that in this example assuming that only hot carriers with energy larger that 3.2 eV contribute to the CHC damage, hot carrier injection will not take place at $V_{ds} = 1.8$ V without e–e scattering being effective to extend the high energy tail.

For a given electric field distribution in the channel SREES increases with the electron density n(E) as per Equations 5.29 and 5.30. This is due to the increased frequency of e–e interactions with increasing electron density.

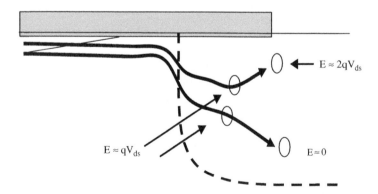

Figure 5.13. Graphic representation of EES scattering event in the drain region of a nMOSFET.

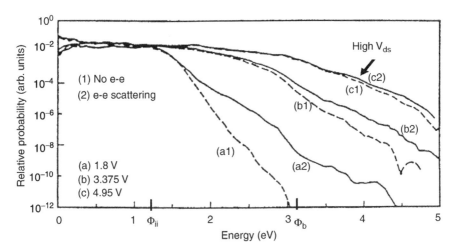

Figure 5.14. EEDF with [case(1)] and without [case(2)] EES for a 0.12 μm nMOSFET with $V_{ds} = 2V_{gs} = 1.8$ V, 3.375 V, and 4.95 V. Reprinted with permission from Ref. 33, © 1996 IEEE.

The probability of this event to take place strongly depends on the electron concentration $n(E)$ in the region where SREES is active. At bias conditions typical of the hot carrier regime, these concentrations are high enough that e–e scattering might occur and produce energetic carriers contributing to the hot carrier damage. Because two electrons are involved in the scattering process, the SREES rate is expected to be given by

$$R_{SREES} \approx C \times n(E1) \times n(E2), \tag{5.29}$$

where E1, E2 are the kinetic energies of the two scattering electrons, respectively.

The e–e scattering event capable of producing the most energetic electrons is the interaction of carriers with energy $\sim q V_{dd}$. In this case,

$$R_{SREES} \approx C \times n(q V_{dd})^2 \approx I_d^2. \tag{5.30}$$

I_d is maximized at the $V_g = V_{ds}$ conditions; this is a contributor to the observation that worst CHC degradation has been observed at this condition in advanced submicron nMOSFET.

The dependence of the R_{SREES} on carrier density may be one of the reasons why worsening of the CHC damage is observed in nMOSFET device with increasing V_{gs}. In this case, the peak of the electron concentration shifts near the Si/SiO$_2$ interface near the drain edge. This effect has two consequences: increase in electron concentration drives and higher EES rate. In addition EES takes place closer to the Si/SiO$_2$ interface making higher the probability of injection into the gate oxide [36].

Long-range electron–electron scattering (LREES) are due to the fact that the Coulomb interaction extends beyond the one Debye length. LREES can be modeled as plasma excitations induced by a high energy electron in an electron gas in thermal equilibrium. An example is the interaction of high energy ballistic electrons interacting with the cold electrons in the drain of an n-channel MOSFET device. The EDF of cold electron gas in the drain is MDF at the temperature T_L (lattice temperature). The high energy electrons entering the drain can gain the energy of the cold holes. Since the average energy of the cold electrons in the drain is $1.5\,kT_L$, this interaction is mostly an energy loss for the hot carriers mechanism. In the rest of this Chapter, LREES events will not be considered as contributors to a high tail region of the EDF.

5.2.5 Impact Ionization Phenomena

The physical process responsible for impact ionization (II) is the generation of electron–hole pairs (e–h pair) due to channel carriers (both electrons or holes) that have kinetic energy larger than a threshold energy (E_G, the silicon bandgap) needed to break the Si valence bond. II can be viewed as a carrier–carrier scattering process. In the case of electrons initiating the II process (primary carriers), the e–h generation proceeds through a screened Coulomb interaction between a valence electron and an energetic conduction band electron.

Figure 5.15 describes the II process in the case that the channel electric field is mainly responsible for carrier heating in quasi thermal equilibrium at the effective temperature T_e ($kT_e = q\lambda F$). Defining λ_{II}, the II mean free path as the average distance a carrier travels between collisions before gaining enough energy to create an e–h pair. If the electric field F is mainly responsible for carrier heating this distance is given by

$$\lambda_{II} = E_{II}/qF, \tag{5.31}$$

where E_{II} is the mean energy lost due to II. Under these conditions an e–h pair is generated if $\lambda_{II} < \lambda$, where λ is the carrier mean free path at the temperature T_e. Each of generated carriers (both electron and holes) may also be accelerated in the electric field and create additional e–h pairs. This process is called avalanche multiplication, which is critical, for example, in the avalanche breakdown of semiconductor junctions, and has a detrimental effect in short channel MOS devices in terms of excess substrate current and decreased reliability.

In this example it assumed that the electric field F is the main source of the carrier heating. This condition may apply in the pinch-off region of a MOSFET transistor in saturation as well in the depleted region of a highly reverse biased junction. Other mechanisms responsible for carrier heating may produce II. As we will describe later, e–e scattering is considered to play a major role in the II in advanced CMOS transistors operating in saturation with $qV_{dd} \leq E_G$.

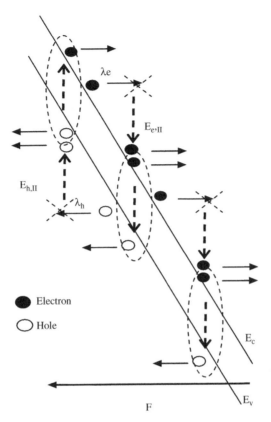

Figure 5.15. Energy bandgap diagram describing e–h pair generation due to applied electric field.

The physical parameter that describes II is the II ionization rate R_{II} defined as the rate of e–h pairs generated by the per unit time. In the general formulation of II, R_{II} can be written as

$$R_{II} \propto \int_{\Gamma} \left(\int_{E_{II}}^{\infty} f(E, \xi) S_{II}(E) dE \right) d\xi. \tag{5.32}$$

This takes into account an appropriate integration in volume Γ where II is effective ($E \geq E_G$).

The function $f(E, \xi)$ is the EDF the heated carriers in a point ξ in Γ and at a given energy E. $S_{II}(E)$ is the energy-dependent II rate, defined such that $S_{II}(E)dE$ gives the II ionization rate per unit time (1/sec) in the energy range E, $E + dE$. Introducing the II coefficient α_{II} (α_e for electrons, α_h for holes) such that $\alpha_{II}d\xi$ is the number of e–h pairs generated in the volume $d\xi$, from Equation 5.32

we get

$$\alpha_{II}(\xi) = \int_{E_{II}}^{\infty} f(E, \xi) S_{II}(E) dE. \tag{5.33}$$

Typically, the energy dependent II rate (averaged over all wave vectors on a constant energy shell) has been expressed analytically in the power law form:

$$S_{II}(E) = P \times (E - E_{II})^{\beta}. \tag{5.34}$$

P and β are used as fitting parameters for both experimental data as well sophisticated Monte Carlo simulations. Figure 5.16 describes the calculated II rate ($S_{II}(E)$) by different authors as reported by Kamakura [18].

Keldysh [19] was first to introduce the power law dependence for $S_{II}(E)$ for electron induced II with $\beta = 2$, and the constant $P = C/E_G$, where $C = 1.19 \times 10^{14}/$ sec. More complete full band structure calculations of the impact ionization rate

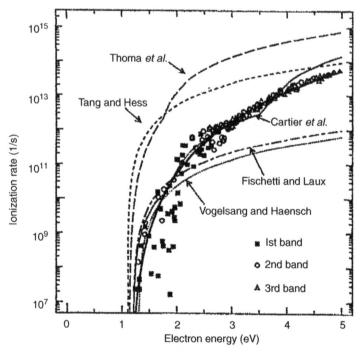

Figure 5.16. Calculated impact ionization rate (S_{II}) as function of carrier energy. The bold solid line represents the best fitted curve to the calculated II rate. This curve is given by Equation 5.34 with $p = 1.0 \times 10^{11}$ and $\beta = 4.6$ [18]. Reprinted by permission from Ref. 18, © 1994 American Institute of Physics.

have been reported for Si. In particular, $\beta \approx 4.6$, $P = 1.0 \times 10^{11}$ represent the Kamakura relation [18], which we will adopt in our discussion.

A parameter used to quantify II is the multiplication factor M defined as

$$M \equiv \frac{I_{\text{out}}}{I_{\text{in}}}, \tag{5.35}$$

where I_{in} and I_{out} are respectively the total primary carriers (carriers initiating II) current entering and exiting the volume Γ where II takes place. In the general case when both types of carriers do have enough kinetic energy to contribute to the II process, M is a function of α_e and α_h. To illustrate this situation we consider the case of II originated by an electrons (primary carriers) current I_e entering a cylindrical volume (Γ) at the circle P with current $I_e(P)$ along z direction (unidirectional flow) as described in Figure 5.17. Both secondary electrons and holes gain enough energy to generate further II events (x), but II is confined within the volume Γ between the two circles P and Q. Flow of the generated carriers is along the z direction as well. This situation is typical of II generation in region with high electric fields as in the depleted region of a strongly reversed biased junction.

Assuming steady state condition, the total current I is constant at any given point ξ inside the cylinder Γ. So that

$$I = I_e(\xi) + I_h(\xi), \tag{5.36}$$

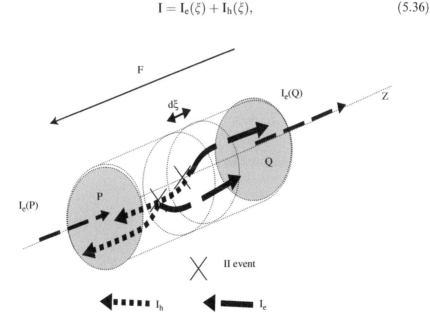

Figure 5.17. Example of impact ionization generated by both electrons and holes in a cylindrical volume (Γ). II ionization is assumed to occur along the z direction. $I_e(P)$ is the primary carrier's current. Both secondary electrons and holes gain enough energy in Γ to produce II.

where $I_e(\xi)$ and $I_h(\xi)$ are the electron and hole current at ξ, respectively. The electron current increases going from P to Q getting the value

$$I_e(Q) = I. \tag{5.37}$$

(No II takes place at the point Q). Similarly, the hole current increases from Q to P ($I_h(Q) = 0$), getting the value $I_h(P) = I - I_e(P)$ at the point P. In this case the electron multiplication factor M_e is such that

$$M_e \equiv \frac{I_e(Q)}{I_e(P)} = \frac{I}{I_e(P)} \tag{5.38}$$

The incremental electron current at the point ξ produced by II events generated per second within the distance $d\xi$ is

$$dI_e = I_e \alpha_e d\xi + I_h \alpha_h d\xi \tag{5.39}$$

where α_e and α_h are respectively the electron and hole II coefficients. Taking into account Equation 5.38, Equation 5.39 becomes

$$\frac{dI_e}{d\xi} = I_e(\alpha_e - \alpha_h) + I\alpha_h \tag{5.40}$$

From the solution of this differential equation under the boundary condition (Equation 5.37) we can find [20] that

$$1 - \frac{1}{M_e} = \int_P^Q \left(\alpha_e \exp\left(-\int_\xi^Q (\alpha_e - \alpha_h) d\xi' \right) \right) d\xi. \tag{5.41}$$

Similarly, if we have II initiated by heated holes the hole multiplication factor M_h is

$$1 - \frac{1}{M_h} = \int_P^Q \left(\alpha_h \exp\left(-\int_P^\xi (\alpha_n - \alpha_h) d\xi' \right) \right) d\xi \tag{5.42}$$

From the definition of electron II multiplication factor it is easy to show that the total number of holes generated within the volume Γ by the $I_e(P)$ entering current is

$$I_h(P) = (M_e - 1)I_e(P) \tag{5.43}$$

Typically, II is not a very efficient process ($I_e(P) \gg I_h(P)$, $I_e(P) \approx I_e(Q)$). Under low level II condition, if II is initiated by electrons, the measured hole current is a direct estimate of II. From Equations 5.42 and 5.43 , we have

$$I_h(P) = I_e(P) \times \int_P^Q \left(\alpha_e \exp\left(-\int_\xi^Q (\alpha_e - \alpha_h)d\xi' \right) \right) d\xi. \tag{5.44}$$

Similarly, if II is initiated by holes, the electron current is

$$I_e(Q) = I_h(Q) \times \int_P^Q \left(\alpha_h \exp\left(-\int_P^\xi (\alpha_h - \alpha_e)d\xi' \right) \right) d\xi \tag{5.45}$$

As we will see, hole generation by II by electron channel carriers is estimated in an nMOSFET in saturation by a direct measurement of the substrate current (I_{sx}). In an equivalent way, the n-well current (I_{nw}) in a pMOSFET device in saturation is a measure of the I_e generated by channel holes driven II in the drain of the device.

5.2.6 Primary Impact Ionization (1II) in a MOSFET in Saturation

Primary impact ionization (1II) is simply the II process activated only by the primary carriers. Figure 5.18 gives a graphical description of this phenomenon for

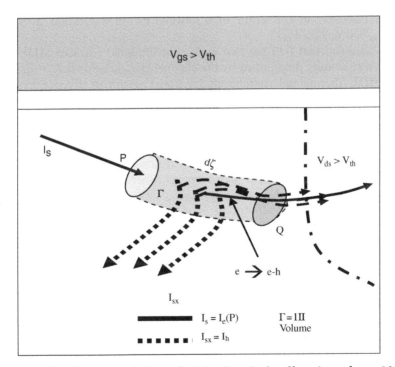

Figure 5.18. Graphical description of 1II in the pinch-off region of an nMOSFET device in saturation.

an nMOSFET device in saturation ($V_{gs} > V_{th}$, $V_{ds} > V_{dsat}$). In this case the primary carriers are electrons entering the II region Γ from the inverted channel. The generated electrons may gain kinetic energy to create additional e–h pairs. Holes are the byproduct of this II process, but are assumed not to contribute into generating further e–h pairs ($\alpha_h \approx 0$).

The substrate current (I_{sx}) is a direct measurement of the hole generated by III ($I_{sx} = I_h$). In the case of low level II it is easy to approximate Equation 5.45 to

$$I_{sx} = I_s \times \int_{\Gamma} \alpha_e d\zeta. \tag{5.46}$$

A similar relation can be established for pMOSFET devices where the n-well current (I_{nw}) relates to the generated electron current. Under this condition, the ratio $R_{II} = I_{sx}/I_d$ is direct measurement of the III rate.

For a given $S_{II}(E)$ functional dependence on carrier energy E, the II coefficient α_e is strongly sensitive to EDF dependence on energy heating mechanisms acting in the volume Γ as well as the carrier transport processes. As such, it depends on the bias conditions as well as device geometry. This observation is critical to understand the different II regimes observed with CMOS scaling. In particular the observed III at V_{dd} less than E_G/q [30].

Figure 5.19 illustrates this point through a schematic representation of the EDF of electrons in the drain of a nMOSFET in saturation at $V_d = V_{dd}$ ($V_s = V_{sx} = 0\,V$, $V_g > V_{th}$).

It is assumed that f(E) has two main contributions: a heated MDF with an effective temperature determined by the channel electric fields ($kT_e = q\lambda F$) and a shifted MDF (thermal tail) dominating for $E \geq qV_{dd}$ with an effective temperature T_L due to electron–phonon interaction. This thermal tail is dominant when e–e scattering is negligible. Two EDF with lattice temperatures T_{L1} and T_{L2} ($T_{L2} > T_{L1}$) and no e–e scattering contributions are shown. An EDF at the lattice temperature T_{L2} with e–e scattering contributing at high carrier energies is also presented. In addition the energy dependent II rate $S_{II}(E)$ is shown schematically.

At this point, we discuss two regimes where II exhibits two opposite behaviors, namely, for $qV_{dd} \gg E_G$ and $qV_{dd} < E_G$. In the case $qV_{dd} > E_G$, the heated MDF mainly contributes to II. Under these conditions it is easy to demonstrate that the II coefficient α_{II} can be approximated to

$$\alpha_{II} \propto \exp(-E_{II}/q\lambda F\,\text{max}). \tag{5.47}$$

The maximum value of α_{II} is modulated by increase of the carrier population in the heated MDF region with lower T_e. This implies that the II rate should be increasing with decreasing lattice temperature. This regime is typical of the lucky electron model (LEM) picture.

Experimental values of the II coefficients have been measured in the case that the lateral electric field is the source of energetic carriers. Under these conditions,

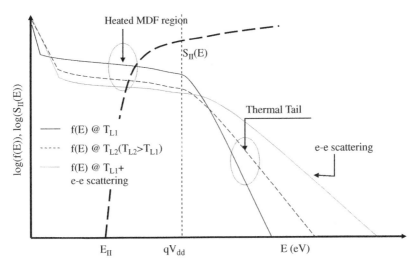

Figure 5.19. Schematic representation of electron $f(E)$ in the drain region of a nMOSFET in saturation at a given drain voltage V_{dd} ($V_s = V_{sx} = 0\,V$, $V_g > V_{th}$). The EDF is assumed to be made up of a MDF heated by the channel high fields and a thermal tail at the lattice temperature T_L dominant for $E \geq qV_{dd}$. Three EDF functions are represented; two $f(E)$ at $T_L = T_{L1}$, $T_{L2}(T_{L1} > T_{L2})$ with negligible e–e scattering and a $f(E)$ at $T_L = T_{L1}$ under the influence of strong e–e scattering (typically $V_g = V_{dd}$). The energy dependent II rate is also shown.

α_{II} is mainly a function of F and has the form:

$$\alpha_{II} = A_{II} \exp\left(-\frac{B_{II}}{F}\right), \qquad (5.48)$$

where A_{II} and B_{II} are II constants and depend on location where II take place. Most reported data for these II constants have being measured in bulk Silicon and a wide range of values have been reported. As a reference, Table 5.2 shows the values of these constants for α_n and α_p given by [22]. Two important observations can be drawn from these experimental data. First the II rate seems to increase rapidly with increase in channel electric field. We have shown from the use of the 2D Poisson approximation to describe F in the pinch-off region of an nMOSFET

TABLE 5.2. Measured Values of the II Coefficient Parameters for Both Electrons and Holes

$F(MV/cm)$	$\alpha_e(cm^{-1})$		$\alpha_h(cm^{-1})$	
	$A_e(MV/cm)$	$B_e(cm^{-1})$	$A_h(cm^{-1})$	$B_h(MV/cm)$
$0.2 < F < 0.24$	2.6E06	1.43	2.0E06	1.97
$0.24 < F < 0.53$	6.2E05	1.08	2.0E06	1.97
$0.53 < F$	5.0E05	0.99	5.0E05	1.32

device with SD and LDD drain engineering that $F_{\max}$ located in a small region of the drain metallurgical junction and that strongly depends on the drain engineering. Thus, to minimize II the S/D designs need to be optimized to reduce $F_{\max}$. In particular doping profile grading schemes like LDD can effectively reduce II. In addition α_e is larger than α_h, particularly at low electric fields. This effect is due to the larger hole effective mass in silicon.

A different physical picture is expected in the case that $qV_{dd} \leq E_G$. Considering that E_G is the Si bandgap energy ($E_G = 1.12\,eV$ @ 300°C) carriers heating by the lateral electric field in the pinch-off region can not by itself explain the observed II at V_{dd} values lower than E_G. Even in the condition of quasi-ballistic transport, the maximum energy gained by channel carriers directly by the pinch-off electric fields is qV_{dd}. At these low drain biases, the II rate is controlled by tails caused by the electron–phonon interactions or e–e scattering [30]. As previously discussed, e–e scattering broadens the EDF to about twice qV_{dd} and may be the dominant energy gaining mechanism at low V_{dd}, particularly at low T_L. Assuming electron–phonon interactions are dominant at these low V_{dd} conditions, the II rate is basically controlled by the Maxwellian tail with an effective temperature $T_e = T_L$. This means that II coefficient α_e can be approximated to

$$\alpha_e \propto \exp(-E_G/k_B T_L), \tag{5.49}$$

and α_e is expected to have a positive dependence on lattice temperature (contrary to what is observed in the LEM regime). The same relation can be applied in the case that e–e scattering is dominant, with T_L to be replaced with T_{e-e} (Maxwell tail associated to e–e scattering). Experimental evidence of the inadequacy of the LEM to explain II effects at low V_{dd} is the reported observation [21] that the substrate and gate currents follow a positive temperature behavior with reduction in V_{dd}.

Depending on the gate bias, the electrons and holes generated by the 1II process can be further heated by the electric field and being injected into the gate oxide contributing to the CHC damage. Hot electrons generated by 1II mainly contribute to the base EEDF since they are heated by the channel lateral field. Hot holes and electrons generated by II are the source of the drain avalanche hot carrier injection mechanism.

The picture just described assumes that the e–h generation from hot holes is negligible in nMOSFET devices in saturation. If this not the case, a new II process called secondary impact ionization (2II) may be active. This phenomenon is also a source of hot carrier damage at bias conditions where direct electric field carrier heating is not expected to produce HC damage. Details are given in the next section.

5.2.7 Channel Hot Carrier Injection Mechanisms

The injection of hot carriers in the gate oxide or in the spacer/drain region is one cause of the CHC damage in MOSFET devices. In a MOSFET in saturation, the majority of the heated carriers simply follow the current flow (I_d) and continue towards the drain. A small number, however, is redirected towards the Si/SiO_2 interface and has enough energy to surmount an effective energy barrier (Φ_B) at the Si/SiO_2 interface and be injected into the gate oxide contributing to the gate current (I_g). The energy barrier (Φ_B) for carrier injection is, to a first order, the Si/SiO_2 conduction band offset (3.2 eV) for electrons and the valence band offset (4.8 eV) for holes, respectively. This energy barrier is modulated by the gate vertical electric field (as previously discussed). Some of the injected carriers may not have energy to cross through the gate oxide; thus, they do not contribute to the measured gate current but may still generate defects in the oxide or at the Si/SiO_2 interface.

Figure 5.20 gives a graphical description of the different processes contributing to the gate current of a MOSFET device in saturation. It is assumed I_g is due to channel carriers (I_s) entering a volume Ω where the carrier kinetic energy is larger that Φ_B/q. Given I_{inj}, the rate of carrier injected in the gate oxide per unit time, the portion of the injected carriers collected in the gate poly is the measured I_g.

In order for the channel carriers to reach the gate and contribute to the gate current the following conditions must be satisfied:

1. The carrier EDF has a portion of the carrier population with energy larger than the potential barrier at the Si/SiO_2 interface. The energy required to surmount the Si/SiO_2 barrier is around 3.2 V for electrons and 4.8 eV for holes the gate current due to hot holes is much smaller than hot electrons.

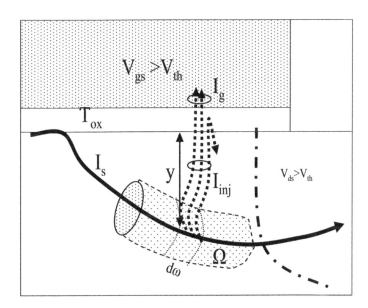

Figure 5.20. Graphical description of the different processes contributing to the gate current in the case of MOSFET device in saturation.

2. Redirection of the momentum normal to the potential barrier by under-going elastic collisions. In the following, λ_r is the mean free path for momentum redirection.

3. Carriers do not experience any inelastic collisions (Probability P_1) when traveling toward the Si/SiO_2 interface.

4. No collision in the gate oxide image potential well (Probability P_2).

Since all of these events are statistically independent of each other, I_g is proportional to the product of probability of each individual process:

$$I_g \approx I_s \int_\Omega I_{inj} P_2 d\omega = I_s \int_\Omega \int_{E_B}^{\infty} f(E, \omega) P_1 P_2 \frac{d\omega}{\lambda_r} dE \qquad (5.50)$$

In this formula, $d\omega/\lambda_r$ is the probability that hot carriers are redirected while traveling a distance $d\omega$ in the volume Ω.

It is important to notice that Equation 5.50 refers to the condition of I_g generated by only one type of channel carrier; we will describe bias conditions where I_g has contribution from both channel hot electrons and holes. In addition, the region in the channel is different from the II region.

The nature of the hot carrier damage strongly depends on the hot carrier injection mechanisms activated at given V_{gs} and V_{ds} bias conditions. We distinguish two types of CHC injection mechanisms depending of the hot carrier generation mechanisms involved.

1. *CHC injection by channel electric field carrier heating.* Figure 5.21 gives a qualitative description of the measured gate current (I_g) of an nMOSFET device as function of V_{gs} for a given V_{ds} in saturation ($V_{ds} > V_{dsat}$). The measured gate current has two components; $I_{g,e}$ and $I_{g,h}$, respectively. $I_{g,e}$ is provided by those electrons injected in the gate oxide ($I_{inj,e}$), which are able to be collected at the poly end. A similar definition is given for Ig,h. Two I_g regimes are typically observed: the drain avalanche hot carrier ($V_{th} < V_{gs} \leq 1/2 V_{ds}$) and channel hot electrons ($V_{gs} \approx V_{ds}$) injection regime respectively.

   Given the observation that the role of the electron and holes in pMOSFET device in saturation is swapped with respect to nMOSFET transistors, a description similar to Figure 5.21 is valid for a pMOSFET, as well with the condition that electron and hole currents are swapped respectively.

   Drain avalanche hot carrier injection. Drain avalanche hot carrier (DAHC) injection denotes the injection into the gate oxide of secondary holes and electrons generated by 1II at the MOSFET drain region. The investigation of the effects of 1II to CHC damage is difficult since both hot electrons and holes may be injected simultaneously in the gate oxide. In the region $V_{th} < V_{gs} \leq 1/2 V_{ds}$ two additional peaks in gate current are observed. The dotted line peak is associated with injection of hot holes ($I_{g,h}$) at $V_g \approx V_{th}$, while the I_g peak (shoulder region) is attributed to hot electrons at

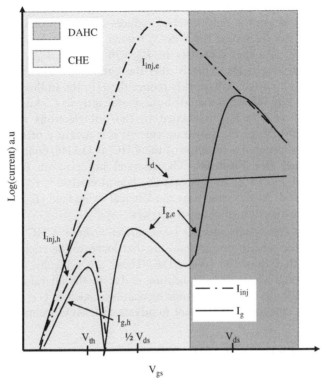

Figure 5.21. Qualitative description of the injection (I_{inj}) and measured gate (I_g) currents in nMOSFET as a function of V_{gs} for a given $V_{ds}(> V_{dsat})$. The injection currents are not measurable directly, but they may contribute to the degradation. The drain (I_d) currents are shown as well. The DAHC and CHE regimes are described.

($I_{g,h}$) $V_g \approx 1/2 V_d$. Both hot carrier species are byproducts of the e–h pairs produced by 1II and made highly energetic by the lateral electric field. In this case, the primary carriers are channel electrons contributing to the source current I_s.

Hole gate current occurs when $V_{th} \leq V_{gs} \ll V_{ds}$. Under these bias conditions, due to the vertical electric field reversal near the drain, the barrier potential for hole injection is reduced because E_{ox} is directed from the drain to the gate. Holes produced by the II are easy to inject, while electrons are repulsed by the vertical electric field. Some "lucky" electrons with sufficient energy may overcome both the field and the barrier and be injected in the gate oxide. The initial rise of $I_{inj,h}$ as the voltage increases is due to an increase in the source current (I_s), which increases 1II. With further increase in V_{gs} $I_{g,h}$ reaches a peak. Further increase in gate voltage reduces the vertical electric field producing a reduction in hole gate current. As the gate voltage is increased passing the hot hole region, CHC injection due to ionized hot electrons is dominant. With the further increase in V_{gs} the vertical electric field reversal reduces and the probability that a hot electron can overcome the Si/SiO_2 barrier increases, leading to a peak at $V_{gs} \approx \frac{1}{2}V_{ds}$. Further increases in V_{gs} will increase V_{dsat} and reduce the lateral electric fields. This is also the condition of maximum substrate current. This effect will quickly reduce the 1II rate. In this regime electrons generated by 1II will not contribute significantly to I_g. Another peak is still observed at $V_g \approx V_{ds}$ associated to the hot electrons regime (see next paragraph). Since the substrate current is a measure of the 1II rate, it is as well considered a monitor of the CHC at DAHC conditions.

Channel Hot Electrons. The channel hot electron (CHE) condition occurs at $V_{gs} \gtrsim V_{ds}$. In this case the lateral electric is reduced since V_{dsat} is increased. At the same time the Vertical oxide field (E_{ox}) gets larger and favor the injection of channel carriers.

2. *CHC injection from multiple steps carrier heating.* CHC injection due to hot carrier generation mechanisms that increase the energy of the carrier further that than the base EDF. These are typically two-step process mechanisms where initially heating is due to the lateral electric field and sequentially to energy exchange mechanisms such as e–e scattering or 2II. This last category is dominant in advanced CMOS technologies.

5.2.8 Lucky Electron Model

The injection of carriers into the gate oxide has been typically described by using the lucky electron model (LEM) [2]. The LEM is based on the idea that the probability of a hot electron to reach the Si/SiO_2 interface and produce damage is based on the following three assumptions:

1. The heated MDF portion of the EDF is the main contributor to carrier heating producing the HC damage and gate current, as well as 1II ($qV_{dd} \gg E_G,\ \phi_B$).
2. A given portion of the heated carriers have enough energy to surmount the Si/SiO_2 potential barrier and elastic collisions redirect a given portion of these heated carriers towards the interface.
3. Carriers reach the interface without suffering any more collisions.
4. CHC damage takes place by a single carrier energy release event.

5.2.8.1 Bulk Current in the LEM Picture.
In this section, we will discuss the case of 1II due to carrier heating from lateral electric fields in the pinch-off region of MOSFET in saturation. This condition represents the case of the electric field heated MDF is broad ($qV_{dd} \gg E_G$) and controls the 1II process. Figure 5.22 gives a schematic representation of the situation at two different lattice temperatures. For simplicity we assume that the tail extending for $E > qV_{dd}$ is thermal (only electron–phonon interactions are active).

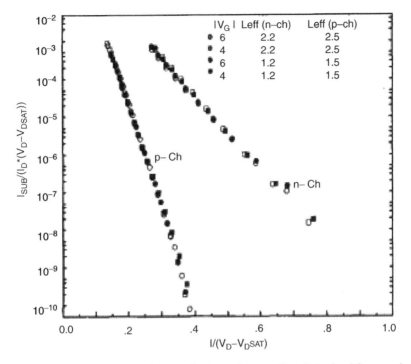

Figure 5.22. Experimental evidence of LEM picture. Reprinted with permission from Ref. 23, © 1990 IEEE.

In this case, the slope of the field heated MDF is $(-1/kT_e)$ where $kT_e = q\lambda F$ (quasi-thermal equilibrium). In addition, the II coefficient is given by the Equation 5.47. A direct measurement of the produced secondary carriers (holes (nMOS-FET)/electrons (pMOSFET)) is the bulk current I_b ($I_b = I_{sx}$(nMOSFET), I_{nw}(pMOSFET)). Here we derive I_b in a SD MOSFET in saturation under the following assumptions:

- 1II occurs in the device pinch-off region.
- The pseudo-2D approximation to solve the Poisson equation is valid to describe the lateral electric field $F_x(x)$ in the pinch-off and under the condition $((V_{ds} - V_{dsat}) \gg 1 \times F_{sat})$.

Under these conditions:

$$I_b = I_s \int_{x=0}^{x=l_{II}} \alpha_{II} dx \tag{5.51}$$

where $x = 0$ is the start of the 1II region and l_{II} (1II length) is the length of the pinch-off region where 1II occurs. α_{II} is the II coefficient for the primary carriers participating to the II event ($\alpha_{II} = \alpha_e$ for nMOSFET or α_h for pMOSFET). This parameter is a function of the incoming carrier energy and in this case a strong function of F. Making a change of integration variable from x to F_x we get

$$I_b = I_s \times A_{II} \int_{F_x=F_{sat}}^{F_x=F_{max}} \exp\left(-\frac{B_{II}}{F_x(x)}\right) \times \frac{dx}{dF_x} \times dF_x. \tag{5.52}$$

F_x is the lateral electric field and a function of the bias conditions at pinch-off. Under the condition $((V - V_{ds})/l \gg F_{sat})$ we get

$$I_b = I_s A_{II} l \times \int_{F_{sat}}^{F_{max}} \frac{1}{\sqrt{F_x^2 - F_{sat}^2}} \exp\left(-\frac{B_{II}}{F_x}\right) dF_x. \tag{5.53}$$

This equation has no closed form solution. Given, however, the α_{II} exponential dependence on F_x we can assume most of the 1II is controlled by F_{max} and Equation 5.53 can be approximated to

$$I_b \approx I_s A_{II} l \frac{1}{\sqrt{(F_{max}^2 - F_{sat}^2)}} \exp\left(-\frac{B_{II}}{F_{max}}\right), \tag{5.54}$$

which under the condition $F_{sat} \ll F_{max}$ reduces to

$$I_b \approx I_s \left(\frac{A_{II}}{B_{II}} \times l \times F_{max}\right) \exp\left(-\frac{B_{II}}{F_{max}}\right) \tag{5.55}$$

Assuming $F_{max} \approx \frac{V_{ds} - V_{dsat}}{l}$, nongraded junctions (no halo implants), R_{II} is given by

$$R_{II} = \frac{I_b}{I_d} \approx \frac{A_{II}}{B_{II}} (V_{ds} - V_{dsat}) \exp\left(-\frac{l \times B_{II}}{V_{ds} - V_{dsat}}\right) \tag{5.56}$$

where

$$V_{dsat} = \frac{E_{sat} \times L_{eff}(V_{gs} - V_{th})}{(V_{gs} - V_{th}) + (E_{sat} \times L_{eff})} \tag{5.57}$$

Figure 5.22 shows the measured $\log(R_{II}/V_{ds} - V_{dsat}))$ versus $1/(V_{ds} - V_{dsat})$ for both nMOSFET and pMOSFET transistors of a single drain CMOS technology [23]. A straight line is obtained for different bias conditions and MOSFET channel length, assuming $F_{sat} = 1.2E05$ (V/cm). According to the LEM picture $B_{II} = E_{II}/q\lambda$. The observation that the constant B_{II} for the pMOSFET device is 2.2 lower than that measured for nMOSFETs is consistent with $\lambda_e \approx 2\lambda_h$. This implies that pMOSFET devices can take 2X the value of $V_{ds} - V_{dsat}$ to generate the same level of I_{nw} current. In addition, Figure 5.22 gives experimental evidence of the validity of LEM picture for these device geometries at a high V_{DD} ($qV_{DD} > E_G$) conditions.

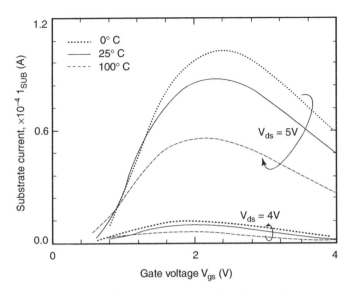

Figure 5.23. Plot of measured I_{sx} versus V_{gs} as function of temperature for an nMOSFET with SD design (12.5/1 device with $V_{ds} = 4.5$). After Arora et al. Reprinted with permission from Ref. 24, © 1991 IEEE.

The typical bell shape of the substrate current as a function of V_{gs} is shown in Figure 5.23 in the case of a nMOSFET device with a SD design [24] at high V_{ds} biases ($V_{ds} = 4, 5$ V). For a given V_{ds}, I_{sx} initially increases with increasing V_{gs} due to an increase in I_d. A further increase in V_{gs} results in a decrease in I_{sx} due to an increase in V_{dsat}, which in turn reduces the lateral maximum electric field (F_{max}). The shape of the I_{sx} curve is controlled by the channel length. At a given V_{ds}, a MOSFET with $L_{des} > V_{ds}/F_{sat}$ with increased V_g changes from saturation to linear mode condition. The I_{sx} peak is then sharp. A short channel device ($L_{des} \ll V_{ds}/F_{sat}$) will experience a broader peak. The peak of I_{sx} current occurs at $V_{gs} \approx 1/2 \ V_{ds}$, which corresponds to the point where the increasing number of electrons and the decreasing field heating balance.

The temperature dependence of the field-heated 1II is mainly controlled by the carrier mean free-path (λ) dependence on temperature. Under these conditions I_{sx} is expected to increase with decreasing temperature at given V_g, given the reduction of phonon scattering at lower temperature. Evidence confirming this trend was given in early works in CHC reliability where the condition $qV_{dd} > E_G$ and the dominance of the field heated part of the EDF to 1II process were both satisfied. Figure 5.23 confirms this dependence as reported by [24]. To understand the observed I_{sx} dependence on temperature, consider two observations illustrated in Figure 5.24.

The II energy threshold (E_G) increases with temperature. The field heated part of EDF has higher occupation at low temperature due to the reduced phonon scattering ($\lambda(T_{L1}) > \lambda(T_{L2})$). At lower temperatures, the channel carrier

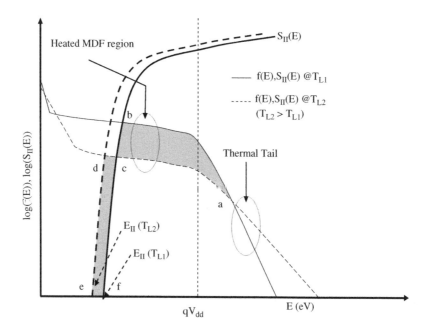

Figure 5.24. Graphical representation of the EDF at two different lattice temperatures ($T_{L2} > T_{L1}$) for a MOSFET device in saturation at V_{dd}.

population in the field heated MDF region is increased since carriers will undergo less scattering events.

The I_{sx} dependence on temperature is controlled the product $S_{II}(E)\,f(E)$ above E_G at that temperature. The gain in I_{sx} due to reduced scattering at lower temperature has the main contribution by the area abc. In addition, the E_G increase at lower temperature (T_{L1}) corresponds to an I_{sx} reduction controlled by area cdef in the figure. The I_{sx} dependence on temperature is controlled by the competing contributions of both areas. If $V_{dd} \gg E_G$, the area abc is larger contribution than the area cdef. In this case, I_{sx} is increasing with decreasing temperature. We will see that if $V_{dd} \approx E_G$, I_{sx} will have a positive dependence on temperature. This phenomenon is the basis of the I_{sx} cross-over effect that we will be discussing.

5.2.8.2 Gate Current in the LEM picture.

The gate current in the LEM picture is given by

$$I_g \propto I_d \frac{T_{ox}}{\lambda_r} \left(\frac{\lambda F_{max}}{E_B} \right)^2 P(F_{ox}) \exp\left(-\frac{\Phi_B}{q\lambda F_{max}} \right) \qquad (5.58)$$

where $\frac{P(E_{ox})}{\lambda_r}$ is the lumped probability that an electron does not suffer an energy-robbing collision per unit length in the Si and gate oxide and $\exp\left(-\frac{\Phi_b}{q\lambda E_{max}} \right)$ is probability that an electron has energy larger than the Si/SiO$_2$ potential barrier Φ_B in a channel electric field E_{max}. The barrier potential Φ_b is lowered by the image force effect. The relation between I_{sx} and I_g is shown here:

$$\frac{I_g}{I_s} = A(V_{gs} - V_{ds}) \times \left(\frac{I_{sx}}{I_s}\right)^{\frac{q\Phi_B}{E_{II}}} \tag{5.59}$$

5.2.9 Bulk Current at Low V_{dd} ($V_{dd} \leq$ or just above E_G/q): The Cross-Over Effect

As discussed in the previous section, I_{sx} in an nMOSFET shows a negative temperature dependence (increases as the temperature decreases) due to the reduction of the mean free path at lower temperature allowing more high energy carriers to produce II. Several authors have reported a positive temperature dependence of the substrate current of nMOSFET devices for sufficiently low V_{ds}. Henning [41] called this change from negative to positive temperature dependence of the I_{sx} with reducing V_{ds} the "cross- over" effect. The V_{ds} value at which I_{sx} is independent from temperature is call the "cross-over" voltage (V_{XOVER}). The value of V_{XOVER} was reported in the literature to vary between 1.75 to 2.6 V. In addition it has been observed that V_{XOVER} increases with the channel length as well as decreases with increasing V_{gs}. An explanation of this phenomenon has lead to a better understanding of the hot carrier physics, particularly on the limitation of the LEM picture to describe the HC regime at low V_{ds}. Two different explanations of this phenomenon have been given so far. An early interpretation relates the cross-over effect to the presence of the thermal tail in the EDF. The thermal tail is the extension of the EDF beyond qV_{ds} and has a slope inversely proportional to the lattice temperature. Since II is caused by hot carriers with energy greater that E_G, the tail is the main contributor to the II process as qV_{ds} approaches E_G. The thermal tail is expected to be suppressed as the temperature is reduced, and hence, I_{sx} has to increase with temperature at low V_{ds}. Fischetti et al. [30] proposed that the I_{sx} cross-over effect is due to the increase of the II threshold energy (E_G) with decreasing temperature. E_G increases by about 40 meV when T is reduced from 300 to 77 K.

5.2.10   Secondary Impact Ionization (2II)

High vertical electric fields in the drain/bulk reverse-biased depleted junction of advanced submicron MOSFETs are responsible for this phenomenon. Figure 5.25 describes the phenomenon of secondary impact ionization (2II) in an nMOSFET transistor. The high lateral electric fields in the pinch-off channel region of the device are responsible for 1II process. Channel electrons produce low energy e–h pairs with a multiplication factor M_1. The produced secondary electrons contribute to the drain current, while the secondary holes diffuse through the drain/substrate junction depleted region W_j. Some of these secondary

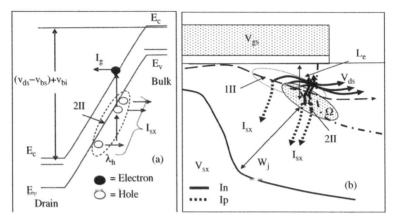

Figure 5.25. Secondary impact ionization in nMOSFET device.

holes may be heated by the vertical fields applied in this depleted region and ionize again deep in the substrate generating e–h pairs with multiplication factor M_2. Holes generated by this secondary II (2II) process contribute to I_{sx}, while some of the generated electrons may flow back along the vertical electric fields underneath the drain and get injected into the gate oxide to produce hot carrier damage. The primary electrons can, in turn, lead to sequential e–h pairs production resulting in the 2II feedback effect. Secondary II takes place as a consequence of two carrier heating processes: carrier heating by 1II, which is controlled by lateral electric fields in the pinch-off channel region. This process is controlled by the $V_{ds}-V_{dsat}$ difference. Carrier heating by 2II is controlled by the vertical electric fields in the depleted region W_j.

Assuming the abrupt junction approximation for the drain/bulk junction of an nMOSFET, the maximum electric field F_{max} is

$$F_{max} = \frac{qN_B W_j}{\varepsilon_s}, \tag{5.60}$$

where

$$W_j = \sqrt{\frac{2\varepsilon_s (V_{bi} + (V_{ds} - V_{bs}))}{qN_B}} \tag{5.61}$$

with

$$V_{bi} = \frac{kT}{q} \ln\left(\frac{N_B N_D}{n_i^2}\right), \tag{5.62}$$

with N_B and N_D respectively the substrate (bulk) and drain doping concentration ($N_D \gg N_B$). V_{bs} is the substrate voltage relative to the source. From the electric field dependence point of view, 2II is controlled by the quantity $(V_{ds}-V_{bs}) + V_{bi}$.

This means that even with no back bias ($V_{bs} = 0$ V), channel carries can get energy larger that V_{dd} by the built-in voltage they experience while traveling in the depleted region W_j. Electrons generated by the 2II process undergo carrier heating because of the vertical electric field they experience in the depleted region W_j. This heating has the following two effects:

1. Electrons generated during the 2II process may gain kinetic energy and interact with the channel carriers while traveling upward towards the drain. This e(1II)–e(2II) scattering process will extend the channel carrier EDF as seen in Figure 5.25.
2. 2II takes place only if 1II occurs in the MOSFET pinch-off region.
3. When arriving to SiO_2/Si interface, 2II electrons may be able to gain enough kinetic energy to be able to surmount the Si/SiO_2 barrier, producing hot carrier damage.

The EMFP (λe) determines the injection efficiency of 2II hot carriers into the gate oxide. λe needs to be larger than L_e (secondary electron travel distance from the 2II e–h pairs generation to the gate oxide interface) for the carriers not to be

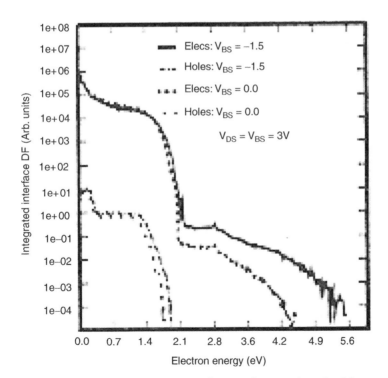

Figure 5.26. Effect of VBS on the EEDF after Bude. Reprinted with permission from Ref. 26, © 1995 IEEE.

thermalized. The dependence of 2II to the V_{sx} bias (<0) in nMOSFET devices is shown in Figure 5.26. In this case, both the secondary electrons and holes EDFs are shown for a nMOSFET ($L_{des} = 0.5\,\mu m$) tested with $V_{gs} = V_{ds} = 3\,V$ and $V_{sx} = 0$ and $-1.5\,V$. Electrons contributing to the EEDF with $E < 2.1\,eV$ are channel carriers heated by the lateral electric field. In this part of the EEDF the electron population does not depend on V_{bs}. 2II contributes with a high energy tail depending on V_{bs}. The condition $V_{bs} = -1.5\,V$ produces more energetic carrier respect to $V_{bs} = 0\,V$. Notice, however, that 2II is still observable at $V_{bs} = 0$, where carrier heating is up to $4.2\,eV$. The different roles of 1II and 2II in heating of electrons and holes are clearly evident from the different dependencies of I_g and I_{sx} on V_{bs}. Since the gate current is contributed by hot electrons with energy larger that $3.2\,eV$, the secondary electrons produced during the 2II determine an increase in I_g and this dependence should be enhanced with increasing V_{bs}. At the same time, the substrate current is made up of channel holes and should not depend on V_{bs}. These observations are confirmed in Figure 5.27 for both I_g (a) and I_{sx} (b). The I_g increase induced by 2II is responsible for increased CHC damage observed as a function of V_{bs} [42].

An estimate of the multiplication factor M_2 is easily obtained by adopting the lucky electron picture. The source of 2II phenomena is the hole carrier generated from the 1II process. This current is the main contributor to the measured I_{sx} current. The multiplication factor M_{II} is given by

$$M_{II} = \int_\Omega \alpha_p^{2II} dV, \tag{5.63}$$

where Ω is the volume in the substrate where holes have enough energy to produce 2II. The impact ionization coefficient α_p^{II} on the junction field F_j is given in this case by

$$\alpha_p^{II} \propto A_p \exp\left(-\frac{\Phi_{II}}{q\lambda_h F_j}\right). \tag{5.64}$$

Since in this volume they experience the maximum drain–substrate electric field F_{jmax} (typically at the drain–substrate metallurgical junctions), M_{II} can be approximated by

$$M_{II} \propto A_p \exp\left(-\frac{\Phi_{II}}{q\lambda_h F_{jmax}}\right). \tag{5.65}$$

It may have a contribution at low V_{ds} ($V_{bulk} = 0\,V$) for very aggressive drain engineering.

The hot electrons generated by 2II modify the EEDF at V_{gs} and V_{ds} bias conditions such the electron contributing to the base EEDF have enough energy to produce 1II. At sufficient energy, the effect of the 2II carrier heating is giving by an extension of the EEDF beyond the thermal tail. Figure 5.27 shows the calculated EEDF using a MC simulation. In this specific example it is clear that even if the electrons in the base EEDF can contribute to the 1II process, they will not have enough energy to surmount the Si/SiO$_2$ barrier. In this case even

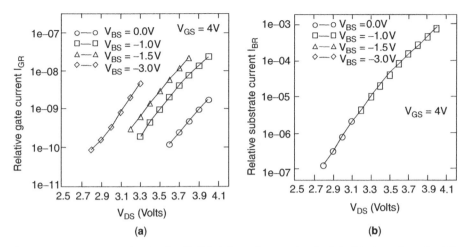

Figure 5.27. Measured I_g ratio (I_g/I_d) (a) and I_{sx} ratio (I_{sx}/I_d) (b). Reprinted with permission from Ref. 42, © 1995 IEEE.

electrons in the thermal tail will not will have energy larger then 3.2 eV. Only the electrons produced by the 2II process contribute to an extension of the high energy tail up to an energy much larger than 3.2 eV.

5.2.11 Limits of the Lucky Electron Model

There are two main LEM assumptions that break down as CMOS devices are scaled down:

1. Energies are limited by electron–phonon scattering; the limit imposed by the applied drain bias is much higher than the energies where impact ionization, or hot carrier damage, occur. In the LEM, these energies are given as $\phi_{ii} \sim 1.3\,\text{eV}$ (for electron initiated impact ionization), and $\phi_{it} \sim 3.7$ eV for interface state generation. This implies that for $V_{ds} < 1.3\,\text{V}$ or 3.7 V, there will no longer be impact ionization or hot carrier damage, respectively. Neither, of course, actually goes to zero.

2. The EDF is dependent only on the local electric field. That is, the carrier energy loss due to phonon scattering is in steady state with the energy gain due to the local electric field. This assumption is only valid if the electric field is slowly varying over the scale of λ, the mean free path. We have seen that the electric field in the pinch-off region grows exponentially with the scale factor l, so this assumption is the same as $l \gg \lambda$. Since the scale length, l (for a given technology) is generally on the order of $L_{min}/10$ ($1/10$ of the minimum channel length), and λ is on the order of 10 nm, technologies with 100 nm or lower channel length will certainly violate this assumption. In actuality, at quarter micron or below, locality begins to

break down, and quasi-ballistic transport becomes important in the high field region.

Therefore, as NMOSFET size and voltage are scaled down, the electron energy distribution becomes less dependent on the maximum electric field and more dependent on the applied bias.

5.2.12 The Energy-Driven Model

Recently, the so-called energy-driven model (EDM) [25] has been proposed to approximately describe the hot carrier behavior at conditions beyond the scaling limits of the LEM. Whereas the LEM describes the behavior at the long channel, high voltage limit, the EDM covers the opposite limit—ballistic transport (at least in the pinch-off region.) In the LEM, the electric field is considered the primary driver of hot carrier phenomena, hence it is a "field-driven model." The basic idea behind the EDM is as follows: Suppose that the hot carrier rates (impact ionization or hot carrier damage) can be written as

$$Rate = \int f(E)S(E)dE, \qquad (5.66)$$

where $f(E)$ is the EDF, and $S(E) = S_{II}$ (effective impact ionization "scattering rate"), or S_{IT} (interface state generation rate.) The integrand of this rate equation will generally peak at one or more points, which are referred to as dominant energies. This occurs when

$$\frac{d\ln f}{dE} = -\frac{d\ln S}{dE}. \qquad (5.67)$$

Mathematically, the dominant energy can be controlled by 'knee' points (points of high curvature) of either $\ln(f)$ or $\ln(S)$. Recently, it is a general consensus that the EEDF within the drain, just past the point of peak electric field in the device, has a significant knee near the maximum energy available from the steep potential drop in the pinch-off region. This is approximately the potential from the drain to the channel pinch-off point (which is referred to by V_{EFF} in this model.) This is shown graphically in Figure 5.28.

If the knee in the EDF determines the dominant energies, then the II and ISG rates are energy-driven, 1) the dominant energies track with bias condition; 2) the hot carrier bias dependencies are due primarily to the energy dependence of the S functions, through the bias dependencies of V_{EFF}. The field dependence of the EEDF is secondary. Figure 5.28 is a conceptual schematic of the energy driven concept for impact ionization. In this figure, the S_{II} used is the modern model of Kamakura et al.: $\propto (E-E_G)^{4.6}$ [18]. Also, electron–electron scattering induces a second, weaker knee at just less than 2 V_{EFF}.

To illustrate the conditions under which the hot carrier behavior is energy driven, let us assume an idealized EEDF that is LEM-like for $E < qV_{EFF}$, but is

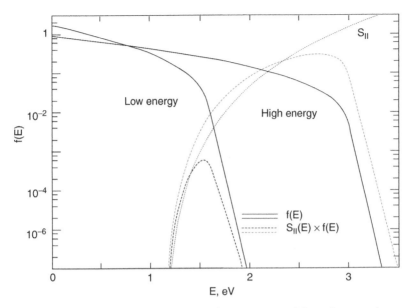

Figure 5.28. A graphical representation of the energy driven hot carrier model applied to II. Reprinted with permission from Ref. 25, © 2005 IEEE.

truncated by a thermal tail for higher energies. That is,

$$
\begin{aligned}
f &\propto \exp(-\chi E/V_{\text{EFF}}), \quad E \le qV_{\text{EFF}} \\
&\propto e^{-\chi} \exp[(qV_{\text{EFF}} - E)/mkT], \quad E > qV_{\text{EFF}}.
\end{aligned}
\tag{5.68}
$$

Under the LEM assumption, $\chi = l/\lambda$, where l = scale length of the high field and λ = mean free path. A scattering rate of the following form is used: $S = A(E-E_{\text{TH}})^p$. In this case, the energy-driven regime can be defined as when the dominant energy = qV_{EFF}. Using Equation 5.68, this can easily be shown to be

$$
E_{\text{TH}} + pmkT \le qV_{\text{EFF}} \le \frac{E_{\text{TH}}}{1 - p/\chi}.
\tag{5.69}
$$

The field-driven regime is when V_{EFF} is above this region. If $\chi \le p$, there is no field driven regime. For V_{EFF} below this region, the dominant energy is in the thermal tail. This might be referred to as the thermally driven regime. To give some approximate numbers as examples, for S = impact ionization rate (S_{II}), $E_{\text{TH}} = E_G = 1.12\,\text{eV}$, $p \sim 4.6$, and using $m = 1.66$, $T = 300\text{K}$, λ (mean free path) = 9 nm,

$$
1.317\,\text{eV} \le qV_{\text{EFF}} \le \frac{1.12\,\text{eV}}{1 - 40\,nm/l}.
\tag{5.70}
$$

The value of l is always less than L_S, the total length of the high lateral field region. Typical values of L_S for 1–2 V devices with halo are 20–70 nm, implying that for these devices the impact ionization will tend to be energy or thermally driven. It will be seen later that the case for ISG turns out to be even more energy driven, because S_{IT} is a steeper function of E (higher p value).

Here we discuss the energy-driven model formulation:

A. *Calculation of V_{EFF}.* V_{EFF} equals effective potential drop from channel to drain:

$$V_{EFF} = V_0 + V_{ds} - V_{dsat} \tag{5.71}$$

where V_0 equals added potential due to halo, and/or source function (total expected to be on the order of several hundred mV) and V_{dsat} equals pinch off voltage.

From [20]

$$V_{dsat} = \frac{2(V_{GS} - V_T)/m}{1 + \sqrt{1 + \frac{2(V_{GS} - V_T)}{mF_c(L - L_s)}}} \tag{5.72}$$

where F_c is the critical field for velocity saturation, $L = L_{POLY}$ or L_{EFF}, L_S is the length of velocity saturated region, and m is the body effect coefficient.

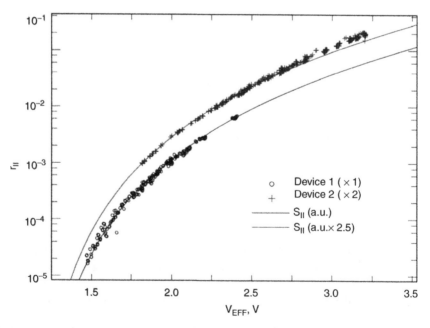

Figure 5.29. Points: Measured impact ionization ratio, r_{II}, for NFETs from two different technologies versus calculated V_{EFF}. Lines: Impact ionization scattering rate, S_{II}, calculated from equation. Reprinted with permission from Ref. 25, © 2005 IEEE.

B. $S_{II}(E)$ function. As given by Y. Kamakura et al. [18],

$$S_{II}(E) = A(E - E_G)^{4.6}. \tag{5.73}$$

C. *Impact Ionization Rate.* Under the energy driven approximation, the II rate is proportional to the S_{II} function at the available energy which is qV_{EFF} for II, if e–e scattering effects do not contribute. Thus,

$$r_{II}(V_{EFF}) \approx BS_{II}(qV_{EFF}), \tag{5.74}$$

where r_{II} is the impact ionization ratio (I_B/I_S) and B is a technology-dependent constant. This is illustrated in Figure 5.29.

5.2.13 Localized Self-Heating Effects

This section describes the origin and impacts of the localized self-heating effect generated by hot carriers [38, 39]. Carriers in the channel before the pinch-off point have relatively low energies close to thermal equilibrium with the silicon lattice, almost all due to the emission and absorption of acoustic phonons. Excess acoustic phonons are very efficient in carrying away the energy of carriers gained

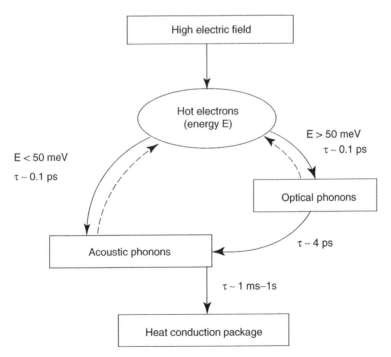

Figure 5.30. Hot carrier driven mechanism of localized self-heating from Rowlette. Reprinted with permission from Ref. 39, © 2005 IEEE.

by traversing the relatively low lateral electric field (less that the critical field) in this region. (This is the dominant mechanism of heat flow in silicon.) Thus, there is very little self-heating (in a bulk device) before the pinch-off point. Beyond the pinch-off point, of course, is where the hot carriers are generated. Carriers of sufficient energy have a new avenue of energy loss—optical phonons. Hot carriers tend to interact with optical phonons (especially the "transverse optical" or TO branch) much more strongly than with acoustic phonons. TO phonons, as opposed to acoustic phonons, are very sluggish and will tend to build-up in and around the region of their creation, until they decay into acoustic phonons that carry the energy away as heat. Figure 5.30 schematically shows this process.

Additionally, there is another mechanism that does not require phonons for the local EDF to acquire a higher effective temperature tail [30]. When hot carriers enter the sea of cold electrons in the neutral drain region, they will transfer energy to the cold electron population via long range Coulombic ("plasmon") interactions. The heated drain electrons will, in turn, raise the tail temperature of the hot electron distribution via plasmon interaction.

5.2.13.1 Impacts. Either mechanism is capable of increasing T_{eff} by at least several hundred degrees per mW/um of localized drain power ($I_d \times V_{eff}$) In a PFET, the effective temperature for NBTI is increased in the drain region, which will induce a local increase in the NBTI damage rate [40]. This mechanism may be somewhat difficult to distinguish from that of a high V_{gs} hole injection one, because they share many features:

1. Since the damage is localized in both cases, there will be a resulting asymmetry in forward versus reverse device I_d–V_g characteristics.
2. A pronounced channel length dependence would be observed in both cases due to the drop in V_{eff} with L (increase in V_{dsat}).
3. Both mechanisms would have a sharply increasing V_{gs} dependence.
4. Typical time slopes for NBTI and for carrier injection mechanisms are coincidentally very similar (0.15–0.3).

Here are the major differences between the two mechanisms:

1. A higher V_{ds} dependence for the hole injection.
2. The local self-heating activated (LSHA) NBTI mechanism would exhibit a thermal activation on the lattice temperature, which would be similar (actually a bit less) than that of isothermal NBTI.

As far as a traditional hot carrier mechanism is concerned, the increased effective temperature of the carrier EDF tends to increase impact ionization and hot carrier damage rates, especially at lower drain biases.

Since self-heating will require a certain build-up time, the onset of these mechanisms may be much slower than that of the hot carrier distribution itself. Although the thermal time constants involved are expected to be much faster than for traditional SOI body self-heating because of the smaller area involved, they may still be much longer than typical CMOS logic switching times (< 100 ps). If this is so, contribution of these mechanisms to actual CMOS circuits may be very small. At present, it is unclear whether localized self-heating is a significant product reliability concern for typical CMOS switching environments.

5.3 HOT CARRIER DAMAGE MECHANISMS

5.3.1 Introduction

A small fraction of the more energetic channel carriers that impact the Si/SiO_2 interface or that are injected into the SiO_2 are responsible for the CHC physical damage resulting in shifts in the device characteristics (V_{th}, I_d, etc.). It is important to notice that shifts in key MOSFET parameters only indirectly correlate with the nature of the CHC damage at the Si/SiO_2 interface. The localization of CHC damage further complicates the relation between the device parameter shifts and the physical damage. Three different types of damage mechanisms have been observed during the CHC stresses of both nMOSFET and pMOSFET devices: interface states generation (N_{it}), electron trapping, and hole trapping. The dominance of each of these mechanisms is strongly related to the carrier injection processes, which, in turn, depend on the bias condition at stress. The injection currents activated at these bias conditions have already been described. It must be remembered that since the Si/SiO_2 barrier heights are ~ 3.2 and ~ 4.5 eV for electrons and holes, respectively, actual injection of carriers into the SiO_2 conduction or valence band becomes rapidly smaller (but does not completely disappear) as the V_{ds} is decreased to below about 2.5–3 V in the case of electron injection and about 3–4 V for holes. Table 5.3 describes the CHC damage processes observed in nMOSFET's under stress.

For recent technology nodes, the dominant damage mechanism is the generation of acceptor-type interface states (ΔN_{it}^A) at the Si/SiO_2 interface. The ΔN_{it}^A generation reduces both the concentration and mobility of electrons in

TABLE 5.3. Injection and Physical Damage Mechanisms Observed in an nMOSFET Device Under CHC Regime as Function of V_{gs} for a Given V_{ds} ($V_{ds} > V_{dsat}$)

V_{gs} Bias Condition	Injection Mechanism	Physical Damage
Low V_{gs} (0 to $\sim V_t$)	Hole injection	Positive charge build-up + Interface state generation
Mid V_{gs} ($V_t +$ to $V_{ds} -$)	Little and compensating injections	Interface state generation
High V_{gs} ($\sim V_{ds}$ and above)	Electron injection; peak I_g conditions	Negative charge build-up + Interface state generation

the inverted channel and produces increases in channel resistance leading to reduction of device parameters such as the drive current and transconductance. The worst case CHC condition for N_{it} generation has been associated historically with the maximum substrate current ($V_{gs} \approx 1/2V_{ds}$) bias condition, which does not correspond to strong hole or electron injection but to the largest number of high energy carriers. In this case, the charge trapping contributions are small and tend to compensate from both HH and HE injection from e–h pairs generated during the II process and injected into the gate oxide. With technology scaling, however, the worst case CHC DC condition has generally moved to $V_{gs} \approx V_{ds}$. There are several reasons for this: 1) When L_{eff} is small, V_{dsat} is small, and both maximum electric field and available energy only drop weakly with V_{gs}. 2) E–E scattering effects increase with increasing V_{gs}.

Su et al. [41] have reported a cross-over point of the worst case CHC at $L_{eff} \approx 0.1\,\mu m$ from V_{gs} at peak I_{sx} to $V_{gs} = V_{ds}$ ($V_{ds} = 2.75\,V$) for a 0.08 μm CMOS technology. This result suggests that at V_{ds} values lower than the energy needed to surmount the Si/SiO$_2$ barrier height new mechanisms in addition to the lateral electric field carrier heating are responsible for hot carriers in the channel (Figure 5.31).

Table 5.4 gives an overview the CHC damage mechanisms observed in pMOSFETs under stress and normal operations. The pMOSFET CHC damage is controlled by the competitive role of hot holes (HH) and hot electrons (HE).

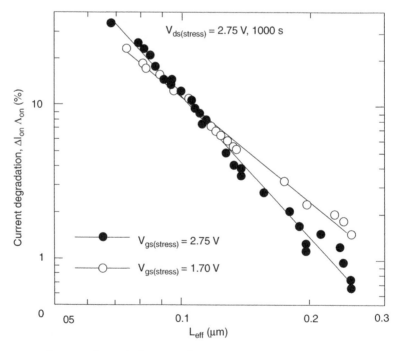

Figure 5.31. Current degradation as a function of L_{eff} Reprinted with permission from Ref. 1, © 1996 IEEE.

TABLE 5.4. Injection and Physical Damage Mechanisms Observed in a pMOSFET Device Under CHC Regime as Function of V_{gs} for a Given V_{ds} ($V_{ds} < V_{dsat}$)

V_{gs} Bias Condition	Injection Mechanism	Physical Damage
Low V_{gs} (0 to $\sim V_t$)	Electron injection; peak I_g conditions	Negative charge build-up + Interface state generation
Mid V_{gs} ($V_t +$ to $V_{ds}-$)	Little and compensating injections	Interface state generation
High V_{gs} ($\sim V_{ds}$ and above)	Hole injection	Positive charge build-up + Interface state generation

The generation of donor-type type interface states (ΔN_{it}^D) is dominant in most of the conductive CHC conditions and is controlled by HH impact. The worst case is observed at $V_{gs} \approx V_{ds}$. This last condition can produce positive charge but at sufficiently high fields. At low gate voltages ($V_{gs} \leq V_{th}$ to slightlty above V_t), HE injection is favored ($I_g \approx I_{inj,e}$).

Under these conditions (peak I_g), electron trapping is localized at the drain side during the stress. The damage is related to the occupation of preexisting traps with negative charge build-up. This negative charge induces the well known channel shortening effect, which masks the influence of the ΔN_{it}^D generated simultaneously in the same region. This phenomenon has been observed in thick gate oxide devices (up to 0.25 μm technologies). The activation of process induced electron traps has less contribution in recent technologies, as electron detrapping dominates in thin oxides and hole trapping becomes the major contributor in nitrided oxides. The dominant role of HH and HE damage in controlling the CHC degradation is further determined by the nature of the gate oxide process. It should be noted that at high V_{gs} stress, especially at elevated temperatures, NBTI becomes a strong degradation mechanism for pMOSFETs and should not be confused with CHC. Later, a hybrid mechanism, the local enhancement of NBTI by hot carrier-driven heating effects, will be discussed.

Figure 5.32 shows the g_m degradation in linear mode for pMOSFET devices processed the same way, but with different gate oxide processes (pure SiO_2 versus N implant before oxidation). Both devices were stressed at different V_{gs} conditions with $V_{ds} = -5.0\,V$. It is clear that the worst case CHC is $V_g = $ Peak I_g and $V_g = V_{ds}$ in pMOSFET devices processed with thermal and SiN oxide, respectively. HH damage is expected to be dominant in the pMOSFET CHC degradation with high V_{ds} conditions.

Figure 5.33 describes the time evolution of V_{th} shift during a peak I_{nw} stress at different V_{ds} for different T_{ox} (2.5 to 5.0 nm). In pMOSFET's with $T_{ox} = 3.4\,nm$, a positive V_{th} shift is measured during the stress with $V_{ds} = -5.0\,V$, while the V_{th} shift is negative at $V_{ds} = -4.0\,V$. This change in V_{th} shift time behavior suggests that CHC damage is controlled by HH injection at $V_{ds} = -5\,V$ and HE injection at $V_{ds} = -4.0\,V$. It is expected that HH will be dominant contributor to the CHC damage at thinner T_{ox}. Turnaround effects in device parameters have been

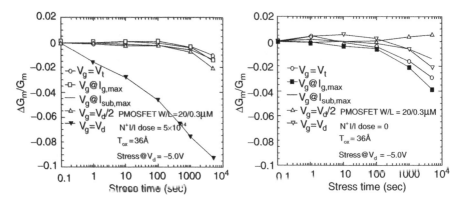

Figure 5.32. Comparison of pMOSFET CH phenomena between similar devices with different gate oxide processes.

observed during hot carrier stressing that can be explained by a dynamic balance between HH and HE damage, as seen in Figure 5.34.

5.3.2   Interface States Generation

Interface states (N_{it}) are trivalent Si atoms with an unsaturated (unpaired) valence electron at the Si/SiO_2 interface. They are usually denoted by $Si_3 \equiv Si_i^*$. The symbol

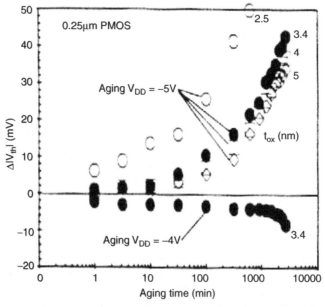

Figure 5.33. Time evolution of the V_{th} shift for various V_{ds} voltages.

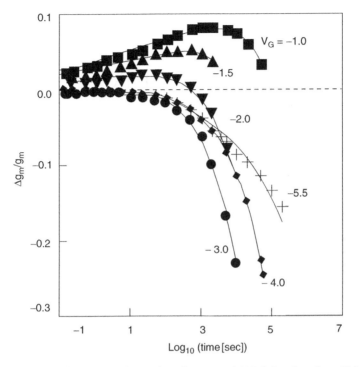

Figure 5.34. Effect of competitive role of HH and HE injection in pMOSFET as function of of V_{gs} Reprinted with permission from Ref. 43, © 1995 IEEE.

$\equiv$ indicates three complete valence bonds to other Si atoms in the crystalline lattice, while $*$ indicates the fourth, unpaired electron in the dangling bond. This electron can pair with the electron from an H ion to make $Si_3\equiv Si - H$ and complete the valence bond. It is well accepted that Nit are generated by channel hot carrier induced breaking the $Si_3Si - H$ bond through the reaction:

$$Si_3\equiv Si - H + (hot - carrier) \leftrightarrow Si_3\equiv Si^* + H^* \qquad (5.75)$$

This reaction involves two processes: the breaking of the $Si_3\equiv Si-H$ bond and the diffusion of the H^* species out of the Si/SiO_2 interface region leaving the $Si_3\equiv Si^*$ unpassivated. At steady state conditions the N_{it} generation rate results from a balance between the rate of $Si_3\equiv Si^*$ formation (reaction controlled process) and $Si_3\equiv Si^*$ repassivation with H^* (diffusion controlled process). Several H^* species have been suggested in the literature to participate to this reaction (interstitial H (Hi) at the gate, dissociation of H_2 into two H by hot carrier, etc.).

Interface states are electrically active defects and have an energy distribution through the silicon bandgap. Their distribution in the bandgap is such that they are acceptor-type in the upper half and donor-type in the bottom half part of the Si bandgap. In a donor-type state, N_{it} are positively charged (N_D^+) when empty

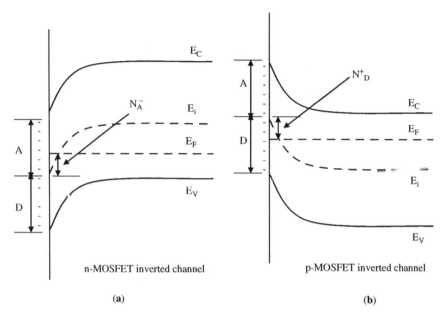

Figure 5.35. Electrical activation of interface states in nMOSFET (a) and pMOSFET (b) devices operated in inversion.

($E_{it} > E_F$), and neutral when filled ($E_{it} < E_F$). Similarly, acceptor-type N_{it} are neutral when empty ($E_{it} > E_F$), and negatively charged (N_A^-) when filled ($E_{it} < E_F$). Figure 5.35 shows the net charge contribution to a pMOSFET and nMOSFET device of acceptor- and donor-type N_{it} with the channel in inversion. During a CHC stress, both acceptor-and donor-type interface states may be generated, but the charge impact to the device characteristics is dependent on the position of the energy level associated to the interface states (E_{it}) relative to the Fermi level (E_F) at test conditions. In nMOSFETs inversion, only acceptor-type $N_{it}(N_A^-)$ contribute with negative charge state (a). Donor-type N_{it}, even if generated, will only be neutral. The opposite situation occurs to pMOSFETs. In inversion, only donor-type Nit (N_D^+) contribute with positive charge (b), while the acceptor N_{it} can only be neutral.

Evidence of N_{it} generation during a CHC stress is given by charge pumping (CP) measurements. We refer to [11] for details of this very powerful measuring technique. Calling I_{cp} the charge pumping current measured at a given stress time it is easy to prove that

$$N_{it} \propto I_{cp} \tag{5.76}$$

The N_{it} generation is proportional to the I_{cp} change. In Figure 5.36, the percent change in CP current as a function of stress time is plotted for several V_{ds} biases at stress.

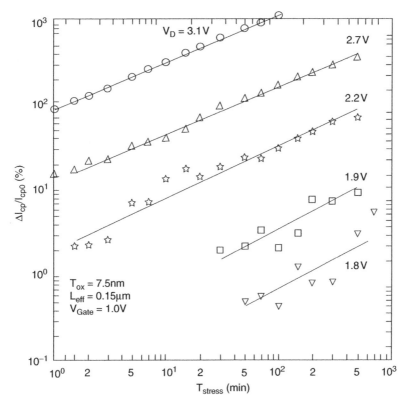

Figure 5.36. Charge pumping current shift ($\Delta I_{cp}/I_{cp}$) as function of stress time for different drain voltages with $V_{gs} = 1.0\,V$, $L_{eff} = 0.15\,\mu m$ and $T_{ox} = 75\,A$. After J. Chung et al. Reprinted with permission from Ref. 12, © 1998 IEEE.

Notice that the I_{cp} degradation follows a power law time evolution with a slope around 0.5 for a V_{ds} range between 3.1 to 1.8 V. The mechanism causing the interface states generation is very controversial in the literature. One of the reasons is that, as seen in the previous paragraph, different carrier injection/impact regimes may produce N_{it} generation. Two pictures have been proposed:

1. The lucky electron model (LEM). In order to generate interface states, channel carriers need to be injected into the gate oxide and have a kinetic energy larger than the threshold energy value (E_{th}) such that E_{th} is the sum of the Si/SiO_2 barrier height and the energy to break the $Si_3 \equiv Si\text{-}H$ bond. This is the origin of the much used phrase hot carrier injection (HCI) to encompass the hot carrier mechanism. The LEM itself suggests a value for E_{th} around 3.7 eV. Taking into account the Si/SiO_2 barrier height (3.2 eV), the energy needed to break the $Si_3 \equiv Si\text{-}H$ bond should be around 0.5 eV. This picture is the basis of the lucky electron model. Strong support is

given by experimental evidence favoring the lucky electron model at higher stress biases ($V_{ds} > 4\,V$.) In particular, this picture explains why the worst case V_{gs} for nMOSFET CHC degradation is near peak I_{sx}. Since this model requires additional energy gaining mechanisms above simple field heating, the "pure" model is not applicable for $V_{ds} < 3.7\,V$. An attempt to include e–e scattering effects into the LEM, called the electron effective temperature model, was made by Rauch, et al. [34]. In any event, it is very unlikely that even with e–e scattering, significant hot carrier effects would persist for $V_{ds} < \sim 2\,V$ under these assumptions. Thus, for modern technologies (at or beyond the 180 nm node), it would seem that the LEM should be only of historical interest; however, use of this model in the industry has persisted for lack of widespread acceptance of newer alternatives, such as the energy-driven model discussed in the following text.

2. The reaction does not necessarily require charge carrier injection into the gate oxide. The channel hot carrier can directly stimulate the reaction since their wave function can overlap the $Si_3 \equiv Si–H$ bond. In this case, a possible N_{it} generation mechanism may involve multiple vibrational excitations of the Si-H bond at the Si/SiO_2 interface. Experimental support is given by the giant isotope effect [17] and by the recently proposed EDM. In this picture, the N_{it} generation is produced directly by channel electrons. Two mechanisms have been proposed to cause the breaking of the $Si_3–H$ bond. A single electron with very high kinetic energy can cause activation, probably with some assist from a single vibrational excitation. This is the scenario in the energy-driven model. Some authors propose that multiple vibrational excitations due to a series of low energy electron–H collisions can accumulate sufficient energy to produce the damage.

After the hydrogen is released (desorbed) from the interfacial state, it diffuses away, or it can recombine (repassivate) with the interface state. The dynamic steady state between hydrogen desorption, repassivation, and diffusion leads to a square root of time behavior.

Interface states affect the MOSFET device characteristics in different ways. Since they act as generation recombination centers they contribute to the junction leakage as well as low frequency ($1/f$) noise. As scattering centers, they reduce channel mobility and in turn the drive current. Because of their charge state, they contribute to the change of the flatband voltage, therefore shifting the threshold voltage of the device.

5.3.3　The Giant Isotope Effect

In regard to interface state generation, it is important to introduce the so-called giant isotope effect [17]. This is, interface states passivated by deuterium (D) show a much reduced generation rate than those containing hydrogen under the same hot carrier stress conditions. Thus, the Si–D bond appears more resistant to breakage than the Si–H bond due to energetic carrier impact, even though the total energy of these bonds would be expected to be almost equal. The difference is in energy levels of the vibrational states, which are bound states in which the H or D atom will vibrate, stretching the bond to silicon. The energy levels of these states depend on the atomic mass of the H or D. Since D has twice the mass, the vibrational energy levels will be on the order of $1/\sqrt{2}$ times than those of H. It was found that the vibrational states of Si–D are very strongly coupled to similar phonon modes of Si–Si in the silicon lattice and Si–O in the SiO_2 [35], whereas those of Si–H are not; this makes them much more stable. The dominant explanation in the current literature is called "multiple vibrational excitation," and basically entails the hydrogen bond's "climbing the ladder" of vibrational states in small jumps caused by electronic impacts until the bond breaks. The Si–D bond, on the other hand, will tend to decay before it can jump to a higher level.

5.4 HC IMPACT TO MOSFET CHARACTERISTICS

5.4.1 I_d–V_{gs} Shifts

As discussed in the previous section, the CHC results from localized channel carrier heating in the region near the drain which causes damage to the gate oxide or gate/drain spacer by charge trapping/creation (ΔN_t) in the bulk oxide and/or interface states generation (ΔN_{it}) at the Si/SiO_2 interface. The HC-induced physical damage impacts of channel parameters such as channel mobility (μ_0) and flatband voltage (V_{FB}), which in turn affect key, electrically measurable MOSFET parameters such as linear (I_{dlin}) and saturation (I_{dsat}) drain current, threshold voltage (V_{th}), linear maximum transconductance (g_m), the subthreshold slope (ss), drain series resistance (R_s) and the drain junction leakage.

We illustrate the relation between the CHC-induced physical damage and device parameters shifts with an example of CHC sensitivity of nMOSFET of a 0.25 µm CMOS technology with source and drain design made up of a single shallow As drain (8.0E14, 10 Kev) and halo BF2 (2.4E13, 55 Kev, 30°). Both the As and BF_2 are implanted before the gate oxide spacer. In particular, we will focus on the I_{dlin} shift (V_{ds} = 50 mV) and its relation to the CHC damage during a peak I_{sx} stress. The observed I_d–V_{gs} characteristics in linear conditions as a function of stress time are given in Figure 5.37.

In this example, it is clear that the CHC damage decreases the subthreshold slope (ss), which in turn increases the one point V_{th} ($V_{th,1p}$) (a). At the same time, I_{on1} ($I_{dlin}@V_{gs} = V_{dd}$) decreases, while the extrapolated V_{th} ($V_{th,ext}$) does experience a small shift during the stress (see case(b)).

It is found that %I_{dlin} follows a power law dependence on stress time with $n \approx 0.5$ (Figure 5.38). This behavior is typically observed in nMOSFET devices and has been associated with the generation of interface states that impact channel mobility. In particular, the %I_d parallel time evolution between testing at $V_g \approx V_{th}$ and V_{dd} is a further indication that in this case the HC damage is controlled by ΔN_{it} generation.

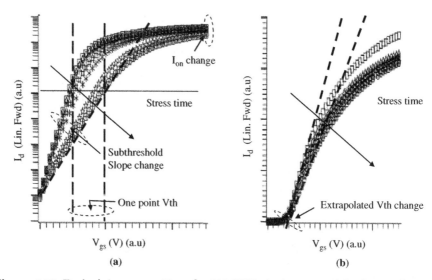

Figure 5.37. Typical I_d versus V_{gs} of nMOSFET device tested in linear forward configuration ($V_{ds} < V_{dsat}$) during a peak I_{sx} CHC stress condition. The CHC damage decreases the subthreshold slope as well as I_{on} (a) and increases the extrapolated V_{th} (b).

To explain the dependence of shifts of these device parameters on the CHC induced-physical damage, we will describe the I_{dlin} versus V_{gs}. Figure 5.39 illustrates the typical I_{dlin} versus V_{gs} characteristics. I_{dlin} is plotted in both a logarithmic and linear scale. We distinguish two separate regimes; the diffusion regime ($V_{gs} < V_{th,ext}$) which describes channel conduction in the subthreshold region and the drift regime ($V_{gs} > V_{th,ext}$) which controls the I_{dlin} in inversion.

We will describe the relation between device parameters in these two regimes assuming the gradual channel approximation; this applies to the case of a long channel device. The basic conclusions apply also for a short channel device.

5.4.1.1 The Diffusion Regime ($V_{gs} < V_{th,ext}$). This is the regime of weak inversion under which subthreshold conduction takes place. The drain current is dominated by charge carrier diffusion through the channel. Following [20] we have

$$I_{dlin} \approx \mu_{eff} C_{ox} \frac{W}{L} \times \left(\frac{kT}{q} \right)^2 e^{q(V_g - V_{th,ext})/mkT} \left(1 - e^{-qV_{ds}/kT} \right). \qquad (5.77)$$

W and L are the device channel width and length, respectively, and C_{ox} is the gate oxide capacitance. μ_{eff} is the effective channel mobility and $V_{th,ext}$ is the extrapolated threshold voltage at the stress time t. The body effect coeffient m is

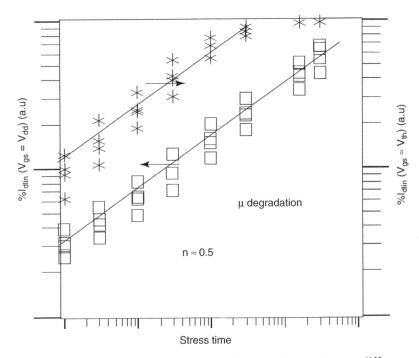

Figure 5.38. %I_{dlin} evolution as function of stress time at two different test conditions ($V_{gs} \approx V_{th}$ and $V_{gs} = V_{dd}$).

given by

$$m = 1 + \frac{C_{dm}}{C_{ox}} \tag{5.78}$$

with C_{dm} being the bulk depletion capacitance per unit area at $\psi_s = 2\psi_B$. Since C_{dm} is contributed by the depletion (C_{Depl}) and interface states capacitance (C_{it}) in parallel we have

$$\frac{1}{C_{dm}} = \frac{1}{C_{Depl}} + \frac{1}{C_{it}} \tag{5.79}$$

In the case of an nMOSFET interface states capacitance C_{it} is defined by

$$C_{it}(\psi_s) \equiv q \times \frac{dN_{it}^A(\psi_s)}{d\psi_s}, \tag{5.80}$$

where $N_{it}^A(\psi_s)$ is the acceptor-type N_{it} at the Si/SiO$_2$ interface. The increase of interface states during a CHC stress will increase $C_{it}(\psi_s)$ with consequent decrease

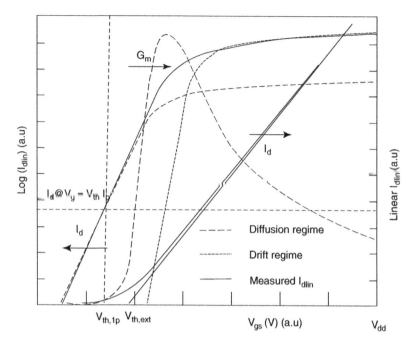

Figure 5.39. Typical I_{dlin}–V_{gs}. The same current is plotted on both logarithmic and linear scale. Both the diffusion and drift regime are shown. The $V_{th,1p}$ and $V_{th,ext}$ V_{gs} points are also indicated.

in C_{dm}. From Equation 5.77, the subthreshold swing (ss) defined as $ss = \left(\frac{d \log_{10}(I_{dlin})}{dV_{gs}}\right)^{-1}$ is given by

$$ss = 2.3\,\frac{mkT}{q} = 2.3\,\frac{kT}{q}\left(1+\frac{C_{dm}}{C_{ox}}\right) \tag{5.81}$$

since the subthreshold swing is dominated by the bulk depletion capacitance (C_{dm}). With the increase in donor-type interface states $N_{it}^{A}(\psi_s)$ during a CHC stress on nMOSFET device, C_{it} increases with consequence decrease in ss.

From Equations 5.80 and 5.81 it is easy to show that the $V_{th,1p}$ ($\Delta V_{th,1p}$) and $V_{th,ext}$ ($\Delta V_{th,ext}$) shifts are related to each other as follows:

$$\Delta V_{th,1p} = \Delta V_{th,ext} \times \%ss \tag{5.82}$$

A large $\%ss$ shift due to an increase in N_{it} during a CHC damage results into a large ($\Delta V_{th,1p}$) for a small ($\Delta V_{th,ext}$), as observed in Figure 5.39. A similar behavior is expected in a pMOSFET device degraded by interface states during a CHC stress, but in this case, the $\%ss$ is due to the increase in the donor-type interface states (ΔN_{it}^{D}). In the case where no N_{it} are generated, $C_{it} = 0$, and according to Equation 5.79, $C_{dm} = C_{Depl}$. Under these conditions, it is easy to demonstrate that

$$\Delta V_{th,1p} = \Delta V_{th,ext} \qquad (5.83)$$

5.4.1.2 Drift Current Regime ($V_{gs} > V_{th,ext}$). This is the regime of strong inversion; the drain current is dominated by charge carrier drift through the channel. In this case [20], we have

$$I_{dlin}(t) \cong \frac{W C_{ox}}{L} \times \mu_{eff}(t) \times (V_{gs} - V_{th,ext}(t)) V_{ds}^{*}(t), \qquad (5.84)$$

associated to the voltage V_{ds}^{*} which takes into account the drain to source voltage drop due to a possible CHC-induced increase in drain series resistance (R_d) at a given stress time t. V_{ds}^{*} is defined as follows:

$$V_{ds}^{*}(t) \equiv V_{ds} - I_{dlin}(t) \times R_d(t) \qquad (5.85)$$

In general we assume that the initial source $R_s(0)$ and drain $R_d(0)$ series resistances are neglegible and, due to localized damage in the drain region, only R_d increases during the stress. Combining Equations 5.84 and 5.85 we get

$$Id_{lin}(t) = \frac{C_{ox}(V_{gs} - V_{th,ext}) \times W \times \mu_{eff} \times V_{ds}}{L + C_{ox}(V_{gs} - V_{th,ext}) \times W \times \mu \times R_d}. \qquad (5.86)$$

Assuming negligible increase in drain series resistance, this equation reduces to

$$I_{dlin}(t) = \frac{C_{ox}(V_{gs} - V_{th,ext})) \times W \times \mu_{eff} \times V_{ds}}{L} \qquad (5.87)$$

Calling $I_{dlin}(0)$ the initial (before stress) linear drain current, from Equation 5.87 it is easy to demonstrate that the fractional I_{dlin} shift (δI_{dlin}) at a given stress time is such that

$$|\delta I_{dlin}| = \left| \frac{\Delta I_{dlin}}{I_{dlin}(0)} \right| \approx \left| \frac{\Delta \mu_{eff}}{\mu_{eff}(0)} \right| + \left| \frac{\Delta V_{th,ext}}{V_{gs} - V_{th,ext}(0)} \right|, \qquad (5.88)$$

where $\Delta V_{th,ext} = V_{th,ext}(t) - V_{th,ext}(0)$. This equation suggests two observations:

1. The I_{dlin} shift is controlled by both the mobility degradation and the extrapolated V_{th} shift dependence on the physical damage (ΔN_{it}, ΔN_t) generated during the CHC stress.
2. In the case that $\Delta V_{th,ext}$ contributes to the CHC damage, the effect to the $\% I_{dlin}$ shift at given V_{gs} is controlled by the overdrive difference ($V_{gs} - V_{th,ext}(0)$).

The relation between the physical damage (ΔN_{it} and ΔN_t) and the channel mobility as well as the $V_{th,ext}$ in linear bias conditions is described in the rest of this section.

5.4.1.3 Channel Mobility Degradation. Under linear conditions ($V_{ds} < V_{dsat}$), a common expression for the the effective channel mobility in strong inversion ($V_{gs} > V_{th,ext}$) is given by this relation:

$$\mu_{eff} \cong \frac{\mu_0}{1 + (\theta(V_{gs} - V_{th,ext}))}, \tag{5.89}$$

where μ_0 is the channel mobility factor and θ the channel mobility attenuation factor. Also, the degradation of channel mobility factor relates to the interface states generation ΔN_{it} through

$$\frac{\Delta \mu_0(t)}{\mu_0(0)} \approx \frac{1}{1 + \Delta N_{it}(t)}, \tag{5.90}$$

5.4.1.4 V_{th} Shift Degradation. The extrapolated threshold voltage ($V_{th,ext}$) in MOSFET device is given [20] by

$$V_{th,ext} = V_{FB} + 2\Phi_B + \frac{\sqrt{4\varepsilon_s q N \Phi_B}}{C_{ox}}. \tag{5.91}$$

where $\Phi_B = (k_B T/q) \times \ln(N/n_i)$ and C_{ox} is the oxide capacitance per unit area. V_{FB} is the flatband voltage which is given as

$$V_{FB} = \Phi_{MS} - \frac{Q_f + Q_{it}}{C_{ox}}, \tag{5.92}$$

with Φ_{MS} being the poly to Si workfunction difference, Q_f the fixed charge density, and $Q_{it}(\varphi_s)$ the interface states induced charge density at the surface potential φ_s. Only the change of Q_f and Q_{it} during the stress contribute to the $V_{th,ext}$ shift as follows:

$$\Delta V_{th,ext}(t) = q\left(\frac{\Delta N_t(t) + \alpha \Delta N_{it}(t)}{C_{ox}}\right), \tag{5.93}$$

where α is equal to $+$ or $-$ for acceptor- or donor-type N_{it}.

The contribution of both the μ_{eff} and $V_{th,ext}$ shift into the device degradation is clearly observed in the measured transconductance (g_m) degradation as function of V_{gs} at different stress times. Figure 5.40 shows the observed g_m shift in the case of the nMOSFET under discussion. The g_m peak value is μ_0. Calling V_{peak} the V_{gs} value corresponding to $g_{m(max)}$, it is easy to show that $\Delta g_{m(max)} = \Delta \mu_0$ and ΔV_{peak} is proportional to $\Delta V_{th,1p}$. Both $\Delta g_{m(max)}$ and ΔV_{peak} shifts are controlled by the N_{it} increase in this case.

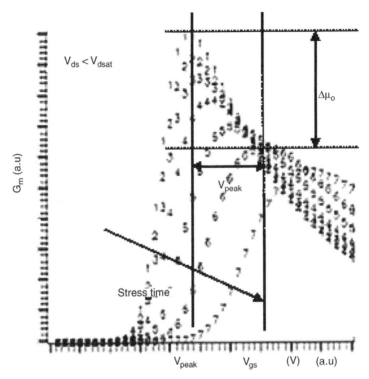

Figure 5.40. G_m-V_{gs} at different stress times measured for the nMOSFET device under investigation.

5.4.2 Localization of Channel Hot Carrier Damage in the Drain Region

One of the key aspects of the CHC damage is that it is mostly localized in the drain region of the device. This has important implications for the device parameters sensitivities to the bias configuration during testing. Physically, this is explained by the relation between the spreading of the HC damage induced during the stress and its extension to the inverted channel during test. In particular this effect is shown by testing in saturation conditions ($V_{ds} > V_{dsat}$). Figure 5.41 describes this situation for the I_d shift of a nMOSFET device tested in saturation.

It is typically observed (Fig. 5.41(c)) that the I_d shift in reverse saturation conditions (source and drain at test are swapped with respect to the stress configuration) is more sensitive to the CHC damage than if measured in forward saturation conditions (source and drain are the same at both stress and test). This difference is explained considering that the I_d shift is proportional to the change in mobility, which in turn is modulated by a portion ($\Delta N'_{it}$) of the total ΔN_{it} generated during the stress that is activated by the inverted channel during test. In the saturation forward-bias configuration $\Delta N'_{it} < \Delta N_{it}$ (Figure 5.41a), while testing at

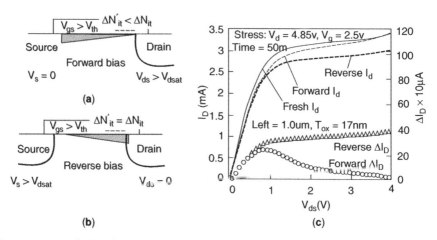

Figure 5.41. Relation between testing and stressing bias configurations. Assuming that the drain is stress at a given V_{ds} we have (a) forward bias characterization if testing is done using the same drain as during stress; (b) reverse bias characterization if at test the source and drain are swapped respect to the bias configurations at stress; (c) comparison of the I_d degradation at saturation condition both in forward and reverse bias configuration for an nMOSFET device.

reverse saturation conditions $\Delta N_{it}' \approx \Delta N_{it}$ (Figure 5.41b). The extent of the damage region in the channel is a function of the MOSFET geometry, the duration of the stress and biases, and the spatial distribution of the oxide and interface states.

5.4.3 CHC-induced Increase in Parasitic Drain Series Resistance

Several drain junction designs have been proposed for scaling down device dimensions; they are valid schemes to reduce CHC degradation and still meet the circuit performance requirements while keeping the nMOSFET short channel effects under control. The typical approach to reduce the CHC sensitivity is to use an optimized drain graded junction doping profile to reduce the maximum lateral channel electric field (F_{max}) in the channel pinch-off region. One of the most popular approaches is the LDD design. It has been observed, however, that the increased hot carrier immunity in LDD structures comes with the introduction of an additional degradation mechanism. This mechanism is associated with electron trapping and/or interface states generation in the spacer oxide/drain interface during operation, which results in a parasitic drain series resistance increase (PDSRI).

The physical mechanism responsible for the PDSRI activation is the drain junction depletion at oxide spacer/drain interface due to the oxide damage. Figure 5.42 illustrates qualitatively the formation of the PDSRI effect in an LDD NMOFET device subject to a conductive channel hot carrier stress.

In this case, the region under the spacer can be subject to electron trapping or interface states generation by channel hot carriers produced by impact ionization

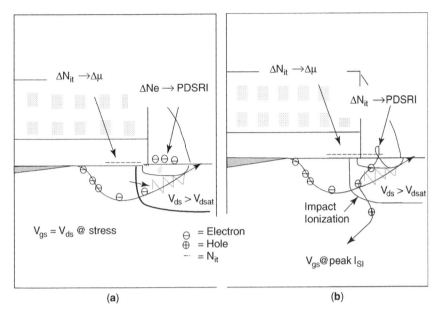

Figure 5.42. PDSRI effect in an LDD NMOFET.

and injected locally in the oxide spacer. The negative effective charge formed at the oxide spacer/drain interface produces a drain depletion, which restricts the electron current flux lines to funnel into a smaller drain region increasing the drain series resistance. The increase in drain series resistance results into a reduction in drain current.

Typically, a two-stage degradation mechanism is attributed the LDD NMOSFET degradation due to channel hot carrier. The channel hot carrier-induced device degradation is due to a combination of an increase in drain series resistance underneath the LDD spacer region and a simultaneous or sequential reduction of carrier mobility in the channel inverted region.

To describe the CHC induced PDSRI damage we investigate the CHC sensitivity of nMOSFET device with source/drain with a Double Diffused Drain (DDD) design made up of of a n^- Ph (5.0E13, 15 Kev) region out-diffused outside the periphery of an n^+ As (1.0E14, 15 Kev) shallow drain. The Ph profile produces a graded junction that reduces the lateral electric field and the related hot carrier generation. Both the As and P implants are done before the oxide spacer.

A good nMOSFET parameter to monitor the activation of the PDSRI damage is the ratio $I_d(0)/I_d(t)$ in linear mode. The $I_d(0)/I_d(t)$ at different V_{gs} biases during the stress suggests the following two-stage physical picture. At very early stages of the degradation, this ratio is controlled by the R_d which increases to a maximum value R_{dmax}. During this phase, the maximum $I_d(0)/I_d(t)$ shift is measured at $V_{gs} = V_{ds}$. The $I_d(0)/I_d(t)$ peak observed at $V_{gs} \approx V_{th}$ during the following stress is an indication that the damage is controlled by mobility degradation.

Mathematically, these observations can be explained by the following considerations. Assuming that in the initial stages the R_d increase is controlling the HC damage, we get

$$\frac{I_d(0)}{I_d(t)} = 1 + \frac{I_d(0) \times R_d}{V_{ds}} \tag{5.94}$$

This means the $I_d(0)/I_d(t)$ follows $I_d(0)$ with a scaling factor of $R_d(t)/V_{ds}$ during the stress. To reach the value we use the following equation:

$$\frac{I_d(0)}{I_d(t)} = 1 + \frac{I_d(0) \times R_d \max}{V_{d0}}. \tag{5.95}$$

In the second stage controlled by mobility degradation, it is easy to demonstrate:

$$\frac{I_d(0)}{I_d(t)} = \frac{(V_{gs} - V_{th}(0)) \times \mu(0)}{(V_{gs} - V_{th}(t)) \times \mu(t)} + \frac{I_d(0) \times R_d \max}{V_{ds}}. \tag{5.96}$$

This means that under these conditions, the $I_d(0)/I_d(t)$ ratio at $V_g = V_{dd}$ condition is directly proportional to the R_d increase during the stress. At later times, the channel mobility degradation becomes dominant, particularly at low V_g biases. In the DDD NMOSFETs, the hot carrier damage follows a two-stage process. An initial PDSRI activation is followed by channel mobility degradation. The increase of the drain series resistance saturates very quickly, while the channel mobility degradation follows a power law time dependence with $n \approx 0.4$–0.5. Initially, the PDSRI activation is the dominant degradation mechanism, while the channel mobility degradation becomes the dominant mechanism at later times.

5.5 HOT CARRIER SHIFT MODELS

5.5.1 Note About "Device Lifetime"

When we refer to "hot carrier lifetime" from a device point of view, this must be viewed strictly as a mathematical concept. This must not be confused with an actual product lifetime. There are several reasons for this:

1. Hot carrier (and also NBTI) induced device shifts that are gradual in time (in fact, sublinear in time). Therefore, any definition of "device failure" is arbitrary. The corresponding "device lifetime" tends to be highly dependent on the defined failure point.
2. A failure definition based on percentage shift in I_{on}, for example, ignores the fact that there is generally a fairly wide distribution of I_{on} for the devices before stress. A "failed" device after stress may have more remaining current drive than another device before stress.
3. Product performance shifts during use will occur due to all of the gradual device shift mechanisms, NFET hot carrier, PFET hot carrier, and NBTI.

Because their contributions add, it follows that an individual mechanism "lifetime" has little predictive power as far as actual product lifetime is concerned.

4. The contribution of device shift to product shift depends on the product sensitivity.
5. Determination of product lifetime is a statistical problem dependent on process-induced variations, power supply, use temperature variations, test margin, etc., in addition to the hot carrier and NBTI shifts.

5.5.2 Lucky Electron Model–Peak I_{sx} CCHC

In this section, the nMOSFET hot carrier shift extrapolation technique based on the lucky electron model [2] is given. This is the most widely used method in the industry, even though, as we have seen, it has little physical justification for $V_{ds} < 4\,\mathrm{V}$ in its "pure" form, and, even for proposed extensions (which would require some modifications to the equations), below $V_{ds} = 2\,\mathrm{V}$. This model is based on the following assumptions: Electron energy gain is solely due to the drain lateral electric field. The main energy relaxation process is phonon scattering. The EEDF is determined by a dynamic equilibrium between field driven energy gain, and phonon scattering. As discussed previously, under the LEM assumptions, the EEDF is given by

$$f(E) = C \times e^{-E/q\lambda F_{\max}};\qquad(5.97)$$

then,

$$\frac{I_{sx}}{I_s} = C_1 e^{-\phi_{II}/q\lambda F_{\max}}\qquad(5.98)$$

$$I_g = C_2 I_d e^{-\phi_b/q\lambda F_{\max}},\qquad(5.99)$$

where ϕ_{II} is the impact ionization threshold energy ($\sim 1.3\,\mathrm{eV}$), ϕ_b is energy for an electron to surmount the electron Si/SiO_2 barrier height ($3.2\,\mathrm{eV}$), and λ is mean free path of the channel carriers in velocity saturation region (mainly controlled by scattering with optical phonons). Note: A value of $1.3\,\mathrm{eV}$ for ϕ_{II} is needed for consistency with experimental results.

Thus, if ϕ_{IT} is the interface state generation threshold energy,

$$\Delta N_{it} = C_4 \left[t\,\frac{I_d}{W}\,e^{-\phi_{IT}/q\lambda E_m} \right]^n = \left[t \times \frac{I_d}{I_W} \times \left(\frac{I_{sub}}{I_d} \right)^m \right]^n\qquad(5.100)$$

and the time, τ, to reach a given ΔN_{it} is

$$\frac{\tau I_d}{W} \propto \left[\frac{I_{sub}}{I_d} \right]^{-\phi_{IT}/\phi_{II}}\qquad(5.101)$$

$$\Delta(\text{Device Parameter}) \propto L_{\text{eff}}^{\alpha} \times \Delta N_{\text{it}}(t)$$

$$\approx L_{\text{eff}}^{\alpha} \left[t \times \frac{I_d}{W} \times \left(\frac{I_{\text{sub}}}{I_d} \right)^m \right]^n \exp(-\Delta H/kT) \tag{5.102}$$

where Δ(Device Parameter) is the change in the device parameter measured at use conditions, I_d is the drain current of fresh device at stress condition, $I_{\text{sub}} = I_{\text{sx}}$ is the substrate current of fresh device at stress condition, T is stress temperature, L_{eff} is the measured L_{eff} from the technology L_{eff} extraction method, t is stress time, and m is a constant given by the ratio $\phi_{\text{IT}}/\phi_{\text{II}}$. From experimental results, $m \sim 2.9$. (Hence, $\phi_{\text{IT}} \sim 3.7\,\text{eV}$) The value of the time slope, n, has the typical range for interface state generation, 0.4–0.6, and usually ~ 0.45–0.5.

The common way of using this model is as follows:

1. Stress wide, minimum-design L devices at typically 3–5 V_{ds} conditions, with V_{gs} set at the peak I_{sx} points.
2. Determine τ (time of $\Delta I_{\text{on}}/I_{\text{on}}$ to a certain shift, typically 10%) for each stress by stressing to the lifetime point, or extrapolating/interpolating the time behavior.
3. Plot τ times I_d (at stress condition) versus the substrate current ratio I_{sx}/I_d at stress on a log–log scale.
4. Fit this with a straight line (the slope should be close to the "accepted" LEM value of 2.9)
5. Extrapolate to use condition (having measured I_d and I_{sx} at the defined use condition) to determine the "hot carrier lifetime" for the technology.

A simplified version also often used is based on the approximation, $F_{\text{max}} \approx \frac{V_{\text{DS}}}{l}$, and further approximating,

$$\tau \approx C \times e^{\phi_{\text{IT}}/q\lambda F_{\text{max}}} \approx C \times e^{\beta/V_{\text{DS}}} \tag{5.103}$$

The theoretical value of $\beta \sim (l/\lambda)(\phi_{\text{IT}}/q)$, however, in practice, β is determined experimentally. In this case, extrapolation is simply to $1/V_{\text{dd}}$. No I_{sx} measurement is needed.

5.5.3   The Electron-Effective Temperature Model

An intermediate model called the electron effective temperature model (EETM) was proposed by Rauch et al., in 1999 [34] and 2001 [36] to extend the LEM into the 2–4 V V_{ds} regime. While not considering quasi-ballistic transport explicitly, it included the EEDF tail contributed by e–e scattering. Under these assumptions, the basic LEM equation,

$$\tau \propto I_d^{-1} r_{\text{II}}^{-m} \tag{5.104}$$

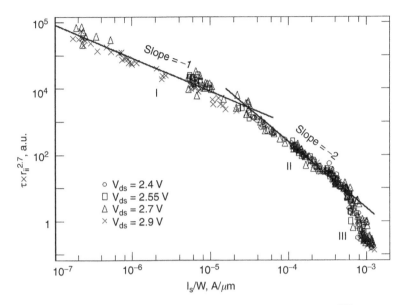

Figure 5.43. Measured NMOSFETs HC lifetime (τ) normalized by $r_{ii}^{2.7}$ versus source current (I_s/W). τ is the stress time needed to reach a given shift in ($1/g_{mlin,max}$). Note the three distinct regimes I, II, and III described in the text. Reprinted with permission from Ref. 36, © 2001 IEEE.

becomes

$$\tau \propto f(I_d) r_{II}^{-m} \tag{5.105}$$

$f(I_d) = I_d^{-1}$ for direct field heating at low I_d (the LEM condition), but becomes I_d^{-2} for the e–e scattering tail contribution (mid-V_{gs} regime), and decreases even faster as V_{gs} approaches V_{ds}. The value of m is somewhat modified from the LEM value by the relative temperatures (slopes) of different parts of the EEDF. These three V_{gs} regimes (low, medium, and high) for nMOSFET hot carrier degradation can be seen in Figure 5.43. Here $f(I_d)$, the I_d dependence of hot carrier lifetime is shown by plotting $\tau\, r_{II}^m$ versus I_d.

Regime III is attributed to the fact that as V_{gs} approaches V_{ds}, electrons become restricted to the interfacial inversion layer, even in the pinch-off region. This is supposed to increase e–e scattering and decrease the distance that an electron must travel between the e–e scattering event and the interface. Since this model shares many of the limitations of the LEM with scaling, its important contributions are the experimental demonstration of e–e scattering effects, and the three V_{gs} regimes for NMOSFET hot carrier.

5.5.4 Energy-Driven Model: nMOSFET CCHC

Under the energy-driven approximation, the ISG rate is proportional to the S_{IT} function at the available energy. The available energy is $\sim q V_{EFF}$ for ISG in the

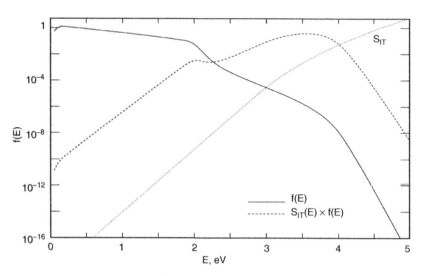

Figure 5.44. A representation of ISG damage rates incorporating an e–e scattering tail to the EEDF. The first peak of $S_{IT}(E)f(E)$ is the linear regime due to electrons in the base distribution, and the second peak is the quadratic regime due to the e–e scattering induced tail. Reprinted with permission from Ref. 25, © 2005 IEEE.

linear (low V_{gs}) regime and is $\sim qm_{EE}V_{EFF}$ for ISG in the quadratic (mid and high V_{gs}) regimes. The parameter m_{EE} is the ratio of the knee energy of the e–e scattering-induced tail to the base. Simulations suggest a nearly constant value (nearly independent of V_{EFF}) of ~ 1.8. The contributions due to the two knees are shown in Figure 5.44.

Thus,

$$1/\tau_{IT}(V_{EFF}) \approx CI_S S_{IT}(qV_{EFF}) \qquad (5.106)$$

for the first knee at qV_{EFF}, and

$$1/\tau_{IT}(V_{EFF}) \approx DI_S^2 S_{IT}(qm_{EE}V_{EFF}) \qquad (5.107)$$

for the second knee due to e–e scattering, where τ_{IT} = ISG hot carrier lifetime, I_S = source current per width, m_{EE} = e–e scattering energy multiplication factor, and C and D are technology-dependent constants.

Figure 5.45 shows the hot carrier data for the two regimes (and two technologies) combined into one graph.

This curve represents an approximation to the $S_{IT}(E)$ function fit to the following empirical form:

$$\begin{aligned}
S_{IT} &\propto \exp(aE), & E \le \phi_{IT} + \frac{p}{a} \\
&\propto (E - \phi_{IT})^p, & E > \phi_{IT} + \frac{p}{a}
\end{aligned} \qquad (5.108)$$

with the parameter values: $\phi_{IT} = 1.6\,\text{eV}$, $p = 14$, $a = 11\,\text{eV}^{-1}$.

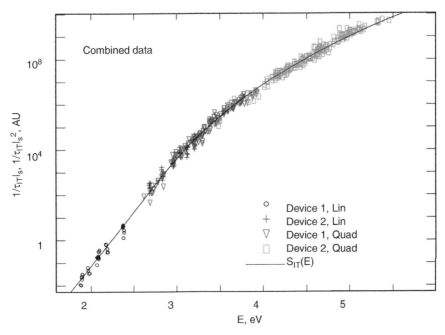

Figure 5.45. Experimental ISG data versus energy. Reprinted with permission from Ref. 25, © 2005 IEEE.

This model predicts a nearly universal description of hot carrier behavior for nMOSFET devices in terms of the variable V_{EFF}, as shown in Figure 5.45.

5.5.5 Modeling of PFET Conducting Channel Hot Carrier

The situation for PFETs is considerably more complex than for NFETs for the following reasons: The dominant damage mechanism for NFETs has been interface state generation with a time slope ~ 0.5. This was true when the LEM was introduced (early 1980s); it is increasingly the case as scaling has progressed since then. However, for PFETs in the 1980s time frame, electron trapping (ET) at low V_g was the dominant mechanism, and time slopes were low (~ 0.2) or even logarithmic time behavior was observed. This is because the barrier height for electron injection is so much lower than for holes and at mid-V_g, where ISG would peak (as in NFETs), the maximum lateral electric field (and the available energy, V_{EFF}) is lower for PFETs due to the 2.5–3x higher F_c (critical field). This makes V_{dsat} higher for PFETs. With scaling, the ET is no longer the dominant mechanism, but observable electron injection persists even at relatively low V_{ds} (down to $<2V$ for modern devices.) This distorts the observed time behavior due to ISG. Since ET will increase I_d, it subtracts from the decrease of I_d due to ISG damage, and measured time slopes can be very high (>1!) Meanwhile, with the advent of NBTI and scaling, the local self-heating-enhanced (LSHA) NBTI mechanism has emerged as the dominant hot carrier mechanism in many cases under DC stress conditions. This latter mechanism will peak at high V_g conditions, measured time slopes will be typical of NBTI (0.15–0.3), and high ambient temperature will be the worst case. However, as stated in Section 5.2.13, the impact of this to an actual CMOS switching circuit is uncertain due to the unknown delay time needed to cause the self-heating. As of the time this is written, there are no published physical models.

All three mechanisms have been modeled with the lucky electron model kinetics [43]. However, for modern devices this makes little physical sense for the same reasons as for NFETs. Additionally, the LSHA NBTI mechanism incorporates by definition a thermally driven damage process, not an energy driven one, and as such, has nothing to do with either the LEM or the ED models. The ET effects are relatively small, and can generally be safely neglected in comparison to the other effects.

1. The "true" CHC mechanism should be modeled with the ED model (the same as for NFET, except with different parameters.) Even though PFET I_{ON} shifts due to CHC have been observed to be significantly less (about one order of magnitude or so) than for NFETs, this may prove to be the most important PFET hot carrier mechanism in actual CMOS switching applications.

2. The LSHA NBTI mechanism should be handled by a modified NBTI model, as described in the next chapter. Even if it is unimportant to fast switching conditions (the self-heating time constants are unclear at the present time), DC stress results can still be fit by a sum of (2) and (3).

REFERENCES

1. L. Su, S. Subbanna, E. Crabbe, P. Agnello, E. Nowak et al. A high-performance 0.08 μm CMOS. VLSI Sym. Digest: 1996, p 12.

2. C. Hu, S. C. Tam, F.-C. Hsu, P.-K. Ko, T.-Y. Chan, K. W. Terrill. Hot-electron-induced MOSFET degradation–model, monitor and improvement. IEEE Trans. Electron Devices, 32(2): Feb 1985, pp 375–385.

3. Y. A. Elmansy, A. R. Boothroyd. A simple two-dimensional model for IGFET operation in the saturation region. IEEE Trans. Electron Devices, 24(2): Mar 1977, pp 254–261.

4. S. Ogura, P. J. Tsang, W. W. Walker, D. L. Critchlow, J. F. Shepard. Design and characteristics of lightly doped drain-source (LDD) insulated gate field-effect transistor. IEEE J. Solid State Circuits, SC-15(4): Aug 1980.

5. D. M. Caughey, R. E. Thomas. Carrier mobilities in silicon empiricallly related to doping and fields. Proc. IEEE, 55: 1967 p 2192.

6. S. M. Sze. *Physics of Semiconductor Devices*, 2nd ed. John Wiley and Sons, New York: 1981.

7. W. Coen, R. S. Muller. Velocity of surface carriers in inversion layers on silicon. Solid-State Electron., 23: 1980, p 35.

8. Y. Taur, C. H. Hsu, B. Wu, R. Kiehl, B. Davari, G. Shahidi. Saturation transconductance of deep-submicron-channel MOSFETs. Solid State Electron. 36: 1993, p 1085.

9. P. K. Ko, R. S. Muller, C. Hu. A unified model for hot-electron currents in MOSFET. IEDM Tech. Digest: Dec 1981, pp 600–603.

10. K. Mayaram, J. Lee, C. Hu. A model for the electric field in lightly doped drain structures. IEEE Trans. Electron Devices, 34(7): July 1987, pp 1509–1518.

11. J. S. Brugler, P. Jespers. Charge pumping in MOS devices. IEEE Trans. Electron Devices, 16(3): 1969, pp 297–302.

12. J. Chung, M.-C. Jeng, J. Moon, P.-K. Ko, C. Hu. Low-voltage hot electron currents and degradation in deep-submicrometer MOSFET's. IEEE Trans. Electron Devices, 37: 1990, p 1651.

13. S. Aur. Low voltage hot carrier effects and stress methodology. Int. Sym. on VLSI Technology, Systems and Applications: 1995, pp 227–280.

14. K. Hess. *Monte Carlo Device Simulation: Full Band and Beyond*. Kluwer Academic Publishers, Amsterdam, 1991.

15. A. Abramo et al. A comparison of numerical solutions of the Boltzmann transport equation for high energy electron transport in silicon. IEEE Trans. Electronic Devices, 41(9): Sep 1994, pp 1564–1646.

16. M. V. Fischetti, S. E. Laux. Monte-Carlo analysis of electron transport in small semiconductor devices including band-structure and space-charge effects. Phys. Rev. B, 38(14): Nov 1988, pp 9721–9745, 14–15.

17. K. Hess, I. Kizilyalli, J. Lyding. Giant isotope effect in hot electron degradation of metal oxide silicon devices. IEEE Trans. Electron Devices, 45(2): 1998.

18. Y. Kamakura et al. J. Appl. Phys. 7575(7): 1994, pp 3339–3506.

19. L. V. Keldysh. Soc. Phys. JETP, 10(3): 1960, pp 509–518.

20. Y. Taur, T. Ning. *Fundamentals of Modern VLSI Devices*. Cambridge University Press, Cambridge, MA, 1998.

21. D. Esseni, L. Selmi, E. Sangiorgi, R. Bez, B. Ricco. Temperature dependence of gate and substrate currents in the CHE crossover regime. IEEE Electron Devices Lett., 16: 1995, pp 506–508.

22. W. H. Grant. Electron and hole impact ionization rates in epitaxial silicon at high electric fields. Solid-State Electron., 16: 1973, pp 1189–1203.

23. T-C. Ong, P-K. Ko, C. Hu. Hot carrier modeling and device degradation in surface-channel p-MOSFETs. IEEE Trans. Electron Devices, 37(7): July 1990.

24. N. D. Arora, M. S. Sharma. MOSFET substrate current model for circuit simulation. IEEE Trans. Electron Devices, 38(6): 1991.

25. S. E. Rauch III, G. La Rosa. The energy driven paradigm of NMOSFET hot carrier effects. IEEE Trans. Device and Mat. Reliabil., 5(4): 2005, pp 701–705.

26. J. D. Bude, M. Mastrapasqua. Impact Ionization and distribution functions in sub-micron nMOSFET technologies. IEEE Electron Device Lett., 16(10): 1995, pp 439–441.

27. A. Ghetti, L. Selmi, R. Bez. Low voltage hot electrons and soft programming lifetime prediction in non-volatile memory cells. IEEE Trans. Electron Devices, 46(4): Apr 1999, pp 696–702.

28. N. Sano, M. Tomizawa, A. Yishii. Temperature dependence of the hot carrier effects in short channel Si-MOSFETs. IEEE Trans. Electron Devices, 42(12): Dec 1995, pp 2211–2216.

29. B. Fischer, A. Ghetti, L. Selmi, R. Bez, E. Sangiorgi. Bias and temperature dependence of homogeneous hot-electron injection from silicon into silicon dioxide at low voltage. IEEE Trans. Electron Devices, 44(2): Feb 1997.

30. M. V. Fischetti, S. E. Laux. Monte-Carlo study of sub-band gap impact ionization in small silicon field-effect transistors. IEDM Technical Digest: 1995, pp 30–308.

31. J. Jakumeit, T. Sontowski, U. Ravaioli. Iterative local Monte Carlo technique for simulation of Si-MOSFETs. Extended Abstracts of 1998 Sixth International Workshop Computational Electronics, Oct 19–21, 1998, pp 92–95.

32. A. Lacaita. Why the effective temperature of the hot electrons tail approaches the lattice temperature. Appl. Phys. Lett., 59(13): Sep 23, 1991, pp 1623–1625.

33. J. J. Ellis-Monaghan, R. B. Hulfachor, K. W. Kim, M. A. Litteljohn. Ensemble Monte Carlo study of Interface-State generation in low-voltage scaled Silicon MOS devices. IEEE Trans. Electron Devices, 43(7): July 1996.

34. S. E. Rauch III, G. La Rosa, F. J. Guarin. Role of E-E Scattering in the enhancement of channel hot carrier degradation of deep nMOSFETs at high V_{gs} conditions. IEEE Trans. Device and Mat. Reliabil., 1(2): June 2001.

35. M. Budde, G. Lüpke, E. Chen, X. Zhang, N. H. Tolk, L. C. Feldman, E. Tarhan, A. K. Ramdas, M. Stavola. Lifetimes of hydrogen and deuterium related vibrational modes in silicon. Phys. Rev. Lett., 87: 2001, 145501.

36. S. Rauch, G. La Rosa, F. Guarin. Role of E-E Scattering in the Enhancement of Channel Hot Carrier Degradation of Deep Sub-Micron NMOSFETs at high V_{gs} Conditions. IEEE Trans. Device and Mat. Reliabil., 1(2): 2001, pp 113–119.

37. P. A. Childs, D. W. Dyke. Hot carrier quasi-ballistic transport in semiconductor devices. Solid State Devices, 48: 2004, pp 765–772.

38. E. Pop, K. Banerjee, S. Sinha, P. Sverdup, R. Dutton, K. Goodson. Localized heating effects and scaling of sub-0.18 micron CMOS Devices. IEDM Tech. Digest: 2001, pp 984–987.

39. J. Rowlette, E. Pop, S. Sinha, M. Panzer, K. Goodson. Thermal phenomena in deeply scaled MOSFETs. IEDM Tech. Digest: 2005, pp 984–987.

40. Y. Wang, K. P. Cheung, A. Oates, P. Mason. Ballistic phonon enhanced NBTI. Proc. IEEE IRPS 2007: 2007, pp 258–263.

41. A. Henning, N. Chan, J. Watt, J. Plummer. Substrate current at cryogenic temperatures: measurements and a two-dimensional model for CMOS technology. IEEE Trans. Electron Devices, 34: 1987, pp 64–74.

42. J. D. Bude. Gate current by impact ionization feedback in sub-micron MOSFET technologies. Proc. Sym. VLSI Tech.: 1995, p 101.

43. R. Woltjer, G. Paulzen, H. Pomp, H. Lifka, P. Woerlee. Three hot-carrier degradation mechanisms in deep-submicron PMOSFETs. IEEE Trans. Electron Devices, (42): 1995, pp 109–115.

6

STRESS-INDUCED VOIDING

Timothy D. Sullivan

6.1 INTRODUCTION

6.1.1 Overview

One of the more odd failure mechanisms in microelectronics metallizations is stress-induced voiding (or simply, stress voiding). It was first reported in 1984 [1] and was observed in reliability tests on 64 k dynamic random access memory (DRAM) chips with sputtered Al-Si metallization and minimum line widths between 2.5 and 3.5 μm. Chip failures were traced to open metal lines that occurred during electromigration testing, discovered because the failure distributions were inconsistent with electromigration failure projections. Subsequent experimentation showed that the failures could be introduced by extended exposure to elevated temperature alone. A scanning electron micrograph (SEM) of a typical void in an Al line is shown in Figure 6.1.

The oddity in the failure mechanism arises from the apparent fluidity of the metallization. Al, Si, and SiO_2 all appear to have a solid, stable appearance at room temperature; it is unexpected to see evidence that the Al would somehow move at relatively low temperatures. But when wafers that eventually show voiding are baked ($\sim 200°C$) and periodically examined, small voids appear that grow with time, until they sever the line. The metal (Al in particular) seems to "spontaneously vanish" (to the uninitiated), leaving a hole behind. The usual question someone asks the first time they encounter it is "Where did the metal go?" Because the metal lines

Reliability Wearout Mechanisms in Advanced CMOS Technologies. By Strong, Wu, Vollertsen, Suñé, LaRosa, Rauch, and Sullivan

Figure 6.1. SEM photo of a void in an Al line, viewed from the top down.

are encapsulated in a dielectric (usually SiO_2), there does not seem to be a way for the metal to get out of the chip—among the wilder ideas that have been put forward is the proposal that the metal was sublimating through the overlying SiO_2.

In subsequent papers [2, 3], a wide variety of behaviors was reported. The optimal temperature for observing failure from voiding ranged from 165°C to 300°C, and optimal line widths for failure ranged from 1 to 3.5 μm. A variety of different Al alloy compositions was implicated, and voiding was observed in both narrow and wide lines. The role of impurities was unclear, and the ranges of reported behavior lead to considerable confusion as to the actual causes. Industrial investigators conducted most of the studies, and often omitted critical information in the published accounts, either because the information was thought to be proprietary or because it was not considered important.

As metallizations continued to develop, they began to employ a layered or sandwich structure in which the Al layer (primary conductor) was coupled with layers of refractory metals (Ti, Hf) or metal combinations (TiW). The refractory layers are typically much thinner than the Al layer, may reside either under and/or over the Al layer, and are included for such reasons as adhesion, as a barrier to Al penetration at contacts, and as an antireflective aid for photolithography. The occurrence of a void in such a structure is no longer such a great concern because the refractory layer provides an electrical shunt, or redundant conductor, around the void, preventing the occurrence of an electrical open circuit and avoiding chip failure. (This is also the reason layered metallization is sometimes termed redundant metallization).

Nonetheless, stress voiding is still a reliability concern for layered metallizations. The steady reduction in feature sizes needed to enable greater densification leads to greater current densities in the metal interconnects and causes reliability testing to focus more on electromigration. As early as 1985, Yue and colleagues [4] demonstrated that the presence of voids in narrow Al interconnects produced both a shorter median time-to-failure (t_{50}) and an increased distribution width (σ), as compared to similar lines containing no voids. A larger σ is usually taken as an indication of increased early failure. Although the work of Yue et al. was done on nonredundant metallization, the trends in failure statistics are assumed to apply, to first the order, to redundant metallizations as well. The failure of a line built with a redundant layer occurs by the formation of a void in the Al layer and then the growth of the void to such a size that it produces an unacceptable resistance shift. The main difference is that the void size needed to reach the specified resistance shift will be much larger than that needed to sever the Al, and the time-to-failure will be correspondingly longer.

An additional concern for redundant metallizations is the interaction of defects. The presence of voids in a redundant metal line may produce only a resistance shift that is within acceptable ranges for circuit operation; however, the occurrence of a void under or over an already-defective via (interlevel connection) could combine to produce a reliability failure before the intended end-of-life. In the case of damascene technologies, which usually have no upper redundant layer, a void under a via will cause an open circuit failure.

6.1.2 Conceptual Basis

In the initial publication identifying stress voiding [1], the cause of metal opens was hypothesized to be due to high mechanical stress in the Al–Si metallization, which combined with grain boundary diffusion to create intergranular fracture and metal line opens. The authors of the initial report suggested that the physical process could be a temperature-assisted mechanical deformation process such as creep. This turns out to be fundamentally correct, if the creep occurs through diffusion.

In order to pursue this idea, the concepts of stress and strain need to be introduced. The relative deformation of an object due to an applied force or load is called *strain*. The applied load, in force per unit area, is called *stress*. These definitions are analogous to the deflection of a spring, x, under an applied force, F, introduced as Hooke's law in elementary physics. The force and deflection are related according to the equation below.

$$F = kx. \tag{6.1}$$

Here, k, the proportionality constant, is called the spring constant. The stronger the spring, the larger k, and the more force needed to move the end of the spring by an amount x. The analogous situation for a long rectangular bar of metal (Fig. 6.2), described in terms of stress, σ, and strain, ε, is given by the

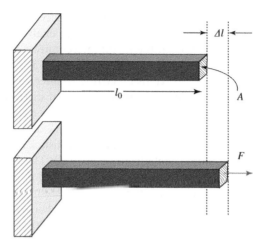

Figure 6.2. Long, rectangular bar with length l_o and area A, shown in its unstressed, or relaxed, state and stretched by a force F.

following equation:

$$\sigma = \varepsilon E,$$
$$\varepsilon = \frac{\Delta l}{l_o}. \tag{6.2}$$

Here, the proportionality constant, E, is the modulus of elasticity of the material, also known as Young's modulus. Strain is defined further in terms of the original length of the bar, l_o, and the deflection, Δl. Stress is called tensile when the bar is being stretched, and compressive when the bar is being compressed (imagine that).

In the present discussion, strain can also be caused by a temperature change, ΔT, through the coefficient of thermal expansion (CTE), α, of a material, according to the following relationship:

$$\varepsilon = \alpha \Delta T. \tag{6.3}$$

The larger the value of α, the greater the thermally induced strain in the material.

The fundamental quantity needed to understand stress-induced voiding in microelectronics interconnects is the coefficient of thermal expansion (CTE) of a material. An every day example of thermal expansion and contraction at work is the incidence of pops and creaks that are audible when the indoor heating system turns on or off. For hot water heating, the introduction of hot water into the circuit causes the pipes to expand and move slightly within the partitions of the house and within the holders and surrounding heat fins that carry the heat into the air. The same thing happens to the air ducts for hot air heating. The exhaust system of a car generates similar noises while cooling immediately after being driven.

Thermal expansion and contraction alone do not generally cause problems. But if the object that is expanding and contracting is constrained, the tendency to expand or contract will push or pull on the constraint and may cause the expanding part to bow under compression or to pull loose under tension. Vinyl house siding has a much larger CTE than wood, which means it expands more when heated. For this reason, the nails holding vinyl siding to a house are not driven into contact with the side of the house. Instead, a gap is left between the nail heads and the house. Also, the holes for the nails are long slots that allow the siding panels to move as they expand and contract with temperatures that can range from -50 to $+120°F$. Without the ability to move, the siding would buckle in the summer and would crack in the winter. The problem arises, then, because of the *difference* in the thermal expansion between two materials that are attached to each other.

In the case of narrow Al lines surrounded by and bonded to the oxide passivation, the CTE for the Al can be as much as 50 times greater than that of the oxide. Furthermore, the oxide is deposited onto the Al at nearly 400°C, when the Al is in an expanded state. Then the Al/oxide/Si combination is cooled to room temperature. The oxide, which expands and contracts very little, retains about the same dimensions it had at 400°C. The Al wants to shrink about 1% along each dimension (length, width, and height), but it is prevented from doing so by the surrounding oxide to which it is strongly bonded. The Si wafer, which has an intermediate CTE of about 2.9 ppm/K, actually governs the final stress state of the lines in the axial direction because it is usually several hundred times thicker than the combined film thickness forming the active circuitry. This means that the oxide is in a state of biaxial compression, and the tensile stress in the Al lines is somewhat reduced in the axial direction. The stress in the vertical and lateral line dimensions is dominated by the mismatch between the Al and the oxide.

To put this magnitude of strain in perspective, a mechanical load that produces a strain of about 0.2% commonly defines the tensile yield strength of a metal. A long, thin wire is usually used for such tests, such that no constraints are applied to the surfaces of the wire during the test. The lack of surface constraints means that the cross-section of the wire is free to shrink during elongation and allows the strain to occur at a lower load than would be the case if the surface were constrained. An Al interconnect surrounded by SiO_2, in comparison, is very strongly constrained by the SiO_2, because it is strongly bonded to the SiO_2 at the surfaces and the SiO_2 is as stiff as the Al. The magnitude of the strain in each direction of the Al can be up to five times greater than the tensile yield limit of the Al. The Al is held in a state of tensile stress (i.e., it is stretched out) in all three dimensions, and therefore occupies a volume greater than it would have if it were allowed to contract freely. It is this difference in volume between the stressed state and the unstressed state that ultimately appears as voids in the metal. The metal does not "go" anywhere. It simply contracts. These relationships are shown in Figure 6.3.

However, the contraction does not occur in quite the same manner as it would if the Al surfaces were unconstrained. In the case of an unconstrained line,

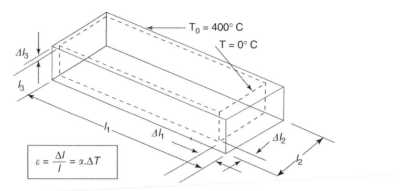

Figure 6.3. Conceptual effects of thermal strain in a rectangular metal bar.

expansion and contraction occurs continuously at an atomic level along each volume element of the line. The atoms in metals are all bonded together in an ordered array called a lattice, with each atomic site located at a three-dimensional point (x, y, z). All of the atoms move a little further apart at higher temperatures and a little closer together at lower temperatures, such that the whole three-dimensional network grows a little or shrinks a little. During this movement, each atom essentially stays in the same location with respect to its nearest neighbors. However, if the surfaces of the line are firmly bonded to oxide, this small atomic movement is mostly prevented.

In order for a void to be formed and grow, atoms from the location where a void is nucleating must be moved from that location and be placed in another location. We call this process mass transport. If the metal line were formed of one perfect lattice in which every atomic site were occupied, and if all the atoms on the surfaces of the line were bonded to Si and O atoms in the oxide, atomic movement of Al atoms within the Al would not be possible. But the Al is filled with imperfections such as vacancies, dislocations, grain boundaries, and the interface between the Al and the oxide.

Vacancies are empty atomic sites that are almost always present in crystalline materials at temperatures significantly above absolute zero. These vacancies allow one of the atoms from the neighboring sites to move one lattice position, allowing the vacancy, in turn, to move one lattice position in the opposite direction. Vacancies move much faster than the atoms do because there are far fewer vacancies than atoms (the fraction of vacant sites depends on temperature and is less than 10^{-5} at room temperature for Al). This is easily envisioned by thinking of a single vacancy traveling along a row of 100 atoms: The vacancy moves 100 positions, while each atom moves only one lattice position in the opposite direction. Usually, vacancy movement is random, such that the net mass transport is zero. However, in the presence of a stress gradient, vacancies will move toward a region of lower tensile stress, allowing the atoms to move toward the region of higher tensile stress. This mechanism of mass transport is called diffusion and is similar to the spreading of dye in a liquid or of odors in air. Because atoms must jump to vacant atomic sites in Al, this diffusion is called substitutional diffusion. Naturally, the diffusivity increases as the number of vacancies increases (with temperature).

Dislocations are linear mistakes in the lattice order. The type of dislocation of importance here can be thought of as extra partial planes of atoms (edge dislocations). The line defining the dislocation can be thought of as a linear array of vacancies, and atoms diffuse along dislocations much faster than by individual vacancies (pipe diffusion).

Grain boundaries are the interfaces where two grains are joined to each other, defined in part by the different orientations of the crystals on either sides of the boundary and in part by angles between the plane of the boundary and each of the two crystal lattices. If the misorientation between the two grains is large, the boundary can have considerable disorder, and the density of bonds at the boundary is considerably reduced compared to that within the grain. Grain boundaries can be thought of as planes of vacancies interrupted by arrays of bonded atoms. There is a great range in the concentration of vacancies and in the order of the bonded regions within the grain boundary because of the wide range of orientation differences between grains and the range of angles between the boundary and either of the two lattices. Clearly, atomic diffusion along grain boundaries is likely to be much faster than through either individual vacancy motion or along dislocations, but it will vary considerably from one grain boundary to another.

Finally, the interface between the Al and the oxide can be thought of as a kind of grain boundary where the bonded regions are less ordered, and the bonds are between Al and O or Al and Si atoms. Diffusion along the interfaces is probably similar to that in grain boundaries, but possibly less rapid due to the random nature of the bond locations.

Figure 6.4 represents a top-down view of a metal line that shows multiple grains, grain boundaries, and dislocations within the grains. If the line were made up of only grains that completely spanned the line, such as that just above the word "interface," the structure is termed bamboo structure. Ideally, the grains

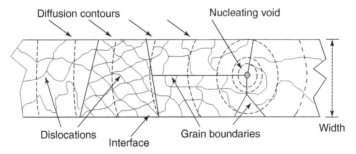

Figure 6.4. Representation of a top down view of a metal line, showing different diffusion paths: grain boundaries, interfaces with confining materials, and dislocations. A nucleating void is also shown with associated diffusion contours.

would be longer than they are wide to better resemble the segments of a bamboo stalk, but the terminology is applied even if the grains are short and fat.

Given the four paths of mass transport (vacancy, dislocation, grain boundary, and interface), atoms can be moved from one location to another provided there is a benefit and provided that there is a place to put the new atoms at the new location. The benefit, in this case, is a reduction in the overall energy of the system (in the strain energy in the metal and in the surrounding oxide [6]). The locations that can accommodate an influx of atoms include grain boundaries, dislocations, and interfaces.

Imagine that a void nucleates at the intersection of grain boundaries or between a boundary and a corner of a line. As soon as the nucleation occurs, strain in the lattice in the immediate vicinity is relaxed, while strain farther from the void remains high. The first path of mass transport would most likely be along the boundary on which the void had nucleated. (Void nucleation is more likely to occur in boundaries with greater disorder because the bonding is weaker; atomic diffusivity will be greater along these boundaries for the same reason). Atoms from the surface of the microscopic (on the order of 40 A in diameter) [7] void will flow towards the center of the line and would be added in layers to one or the other of the two grains on either side of the boundary. With the addition of atoms at the boundary, the spacing of the atoms along the line axis can be relaxed, and the longitudinal strain in the vicinity of that boundary is relaxed.

Similarly, atoms diffusing along the interfaces between the line and the oxide can be added in layers to the outside of the grain, allowing strains in the lateral and vertical direction to be relaxed. For lines wide enough to contain more than one grain across the line width, grain boundary diffusion parallel to the line axis becomes possible, and additional relaxation across the width is possible. Some atomic diffusion along dislocations is also likely to occur in order to grow those atomic planes oriented perpendicular to any given axis of stress. The speed of an atom along a dislocation is not very different from that of an atom along a grain boundary, but the number of atoms that can diffuse along the dislocation is much smaller than the number that can diffuse along a grain boundary or an interface. Thus we would expect the greatest quantity of mass transport to occur initially along grain boundaries and interfaces, and then along dislocations. Some mass transport will occur by lattice vacancies, but only at higher temperatures will the contribution from lattice diffusion become comparable to that due to grain boundary diffusion. The contribution due to dislocation core diffusion will depend on the dislocation density; that varies considerably from grain to grain and depends (among other things) upon the grain orientation relative to the substrate plane.

6.2 THEORY AND MODEL

Although stress-induced void growth has been modeled by several investigators [3, 6, 8–10], attention was originally focused upon growth of an isolated void in order to obtain the most conservative estimate for failure. This is an important objective when evaluating the reliability of a single component metallization and is the primary perspective for evaluating Cu damascene metallizations. However, it neglects some of the statistical aspects of voiding that explain many of the initial observations.

At least one model for electromigration failure [11] indicates that coalescence of voids is an important mechanism for failure. And according to Ho [12], voids can migrate alone in a metal line under the influence of the electron wind, with smaller voids traveling faster than larger ones. The smaller voids join the larger voids when they catch up, until eventually a void large enough to cause failure is developed. Knowledge of the initial void distribution clearly would be helpful for predicting electromigration failure by this mechanism. Voids can also form in short interconnects leading to longer lines. The presence of a void in a short interconnect can cause substantial local resistance shifts, which produce changes in chip timing and (in some cases) failure. Under these circumstances, the information of interest is the distribution in void sizes and the distribution in frequency along the line.

The following approach to modeling stress-induced voiding [13] includes many of the ideas proposed by the researching community and provides both a sound mathematical basis and a satisfying intuitive explanation behind the behavior. Referring back to Figure 6.2, a block of Al is shown with length, width, and height designated by l_1, l_2, and l_3, respectively. Supposing that these are the dimensions of the block at 425°C and that the block is unconstrained, reduction in temperature to 25°C will cause the block to shrink by an amount $\Delta l = \alpha_{Al}\Delta T l$ in each direction, where α_{Al} is the thermal expansion coefficient of the Al. If the Al block is constrained (as by adhesion to a rigid container) during the temperature change so that it cannot shrink, it will experience a strain $\varepsilon = \Delta l/l = \alpha_{Al}\Delta T$ equal and opposite to the linear shrinkage it would have experienced in the unconstrained state. The key idea here is that it is the confinement, or constraint, of the metal to an essentially fixed volume that provides the circumstances that cause voiding.

For Al, $\alpha_{Al} = 25 \times 10^{-6}/°C$, such that for $\Delta T = 400°C$, $\varepsilon \sim 1\%$. The volumetric strain would then be given by

$$\frac{\Delta V}{V} = \frac{3\alpha\Delta T \cdot lwh}{lwh} = 3\varepsilon, \tag{6.4}$$

the volume fraction necessary to totally relieve the strain. This calculation assumes that the container is absolutely rigid, such that no relaxation whatsoever occurs in the Al. This will be referred to in this chapter as the rigid box approximation or model. It is not completely correct, but it allows the theory to be developed in a straightforward way and illustrates the important points about void growth. Deviations from the rigid box model will be discussed later.

Figure 6.5 shows a segment of Al line with width, w, and thickness, h, and having a refractory underlayer such as Ti or TiW. Two voids are shown in the line

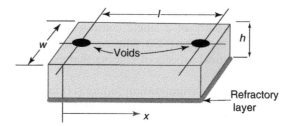

Figure 6.5. Coordinate system and dimensions for Equations 6.5–6.8.

segment, separated by a distance, l. The coordinate system is chosen such that x lies along the line axis, y is in the line width direction, and z is in the vertical thickness direction. One then applies the one-dimensional diffusion equation,

$$\frac{\partial \eta}{\partial x} = D \frac{\partial^2 \eta}{\partial x^2} \tag{6.5}$$

with the boundary conditions

$$\eta(x,0) = N_0 \quad \text{for } 0 < x < l$$
$$\eta(0,t) = \eta(l,t) = 0,$$

where

$$N_0 = \frac{\Delta V}{V} \cong 3\varepsilon_0 = 3\alpha\Delta T. \tag{6.6}$$

N_0 is the fraction of metal required to totally relieve stress in the metal line. The solution to Equation 6.5, found by separation of variables [14, 15], is

$$\eta(x,t) = \sum_{m=1}^{\infty} C_m \sin\left(\frac{m\pi x}{l}\right) e^{-\left(\frac{m\pi}{l}\right)^2 Dt}.$$

The particular solution applicable to the present case is found by representing $\eta(x, 0)$ on $(0, l)$ as an odd function on $(-l, l)$, and then evaluating the Fourier coefficients, C_m. The result is

$$\eta(x,t) = \frac{4N_0}{\pi} \sum_{m \, odd}^{\infty} \frac{1}{m} \sin\left(\frac{m\pi x}{l}\right) e^{-\left(\frac{m\pi}{l}\right)^2 Dt}. \tag{6.7}$$

This is an infinite series of sine functions on x modified by an exponential decay in t. Figure 6.6 is a two-dimensional graph of the first term exponential for a constant diffusivity ($\pi D = 97\,\mu m^2/hr$) as a function of time, t, and of spacing between voids, l. As one would expect intuitively, this term decays rapidly to zero

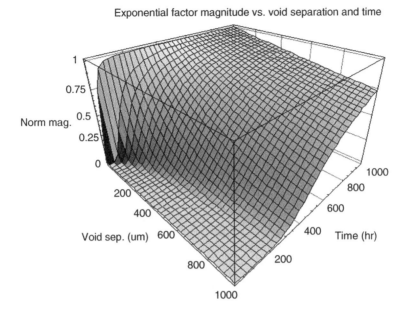

Figure 6.6. 2D graph of the behavior of the exponential factor in Equation 6.7.

for small void spacing, but remains relatively constant for large void spacing. At the diffusivity given above, all stress is relieved after 1000 hr at 200°C for void spacing less than 100 μm, but has decreased only by about 5% over the same time period for voids separated by 1000 μm or more.

For small separations between voids, the magnitude saturates rapidly with time. For large void separation, the magnitude grows much more slowly, and may not reach saturation at all during the test duration. For this graph, $m = 1$, and $D \sim 31\,\mu\text{m}^2/\text{hr}$.

In order to calculate void growth, the metal flux away from the void must be found by differentiating Equation 6.7 with respect to x and multiplying by the diffusivity (Fick's first law) [16].

$$J(0,t)|_{T=\text{cons}\tan t} = D\frac{\partial \eta}{\partial x}\Big|_{x=0}$$

$$= \frac{4N_0 D}{\pi}\sum_{m\,odd}^{\infty}\frac{1}{m}e^{-\left(\frac{m\pi}{l}\right)^2 Dt}\left(\frac{m\pi}{l}\right)^2\cos\left(\frac{m\pi x}{l}\right)\Big|_{x=0}$$

$$= \frac{4N_0 D}{l}\sum_{m\,odd}^{\infty}e^{-\left(\frac{m\pi}{l}\right)^2 Dt},$$

where J is the metal flux. The contribution to the void volume due to diffusion of atoms to the right $(+x)$ is then the integral over time of the flux through the line cross section,

$$v_+(t) = \int_0^t J(t) A dt, \quad A = wh$$

$$= \frac{4N_0 Dwh}{l} \sum_{m \, odd}^{\infty} \int_0^t e^{-\left(\frac{m\pi}{l}\right)^2 Dt'} dt'.$$

After performing the integral, the expression becomes

$$v_+(t) = \frac{4N_0 whl}{\pi^2} \sum_{m \, odd}^{\infty} \left[\frac{1 - e^{-\left(\frac{m\pi}{l}\right)^2 Dt}}{m^2} \right] \tag{6.8}$$

for the time-dependent volume contribution to the right.

Three variables, D, t, and l, occur simultaneously in the exponential and can be combined in a conventional way to simplify and elucidate the mathematics. Letting $z = Dt/l^2 = t/\tau$, where $\tau = l^2/D$ is the relaxation time characteristic of void formation at a given temperature, we can now write $(m\pi/l)^2 Dt$ as $m^2\pi^2 z$. Furthermore, noting that $N_0 whl$ is just the maximum volume required to relieve stress between the two voids and half of that volume will be contributed to each void, we can write

$$v_{f+} = \frac{N_0 whl}{2}.$$

That is, v_{f+} is the maximum contribution to the void volume, which can be made due to atomic diffusion to the right. Equation 6.8 can now be rewritten

$$v_+(t) = v_{f+} \left(\frac{8}{\pi^2}\right) \sum_{m \, odd}^{\infty} \frac{1 - e^{-m^2\pi^2 z}}{m^2}$$

$$= v_{f+} \left\{ \frac{\left[\sum_{m \, odd}^{\infty} \frac{1 - e^{-m^2\pi^2 z}}{m^2} \right]}{\sum_{m \, odd}^{\infty} \frac{1}{m^2}} \right\} \tag{6.9}$$

$$= v_{f+} S(z),$$

where

$$\left(\frac{8}{\pi^2}\right) = \sum_{m \, odd}^{\infty} \frac{1}{m^2}.$$

$S(z)$ will be referred to as the diffusion growth function in the following text. As can be clearly seen from the form of Equation 6.9, this function defines the growth of the void due to atomic diffusion to the right, and contains all the information of interest related to growth. $S(z)$ is shown in Figure 6.7, which is a

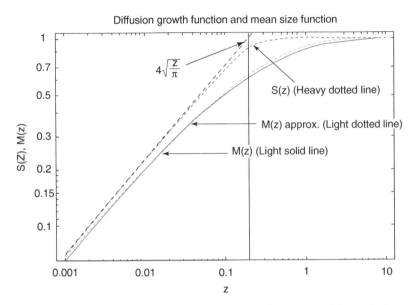

Figure 6.7. Graph showing the diffusion growth function, $S(z)$, and the mean growth function $M(z)$ versus z. Also shown are the straight line approximation to $S(z)$ for $z < \pi/16$, and the resultant approximation to $M(z)$.

log–log plot of $S(z)$ versus z. For small z [short times, large separations between voids, and low diffusivity (low temperature)], the diffusion growth function is proportional to $4\sqrt{z/\pi}$. For large z (long times, small separations between voids, and high diffusivity (high temperature)), $S(z)$ goes to 1. For $z = \pi/16$, the straight line for $S(z) = 4\sqrt{z/\pi}$ terminates at $S(z) = 1$. This point will be of interest shortly.

Since z is a global variable, the horizontal axis must be converted to either units of time or distance to envision how $S(z)$ behaves for a specific case. Although we have focused on diffusion to the right, it is obvious that a contribution to void growth must also come from the left, such that

$$v(t) = v_+(t) + v_-(t). \tag{6.10}$$

Equation 6.10 is an expression for void growth as a function of time when the void has specifically located neighbors. Though valuable for understanding the physics of void growth, this expression is not very useful for predicting failure of metal interconnects. For that purpose, a distribution in void sizes is required, and based upon the development above, a distribution in void separations is needed. If the voids are assumed to be randomly separated, the probability density function for void separations is an exponential distribution:

$$P(l) = \frac{1}{l_m} e^{-\frac{l}{l_m}}, \tag{6.11}$$

where l_m is the mean void separation. The mean void size is then obtained by convolution of Equations 6.10 and 6.11:

$$
\begin{aligned}
\overline{v(t)} &= \int_0^\infty P(l)V(l,t)dl \\
&= \int_0^\infty \frac{1}{l_m}e^{-\frac{l}{l_m}}(N_0whl)S(z)dl \\
&= (N_0whl_m)\int_0^\infty S(z_m/y^2)ye^{-y}dy.
\end{aligned}
\tag{6.12}
$$

Letting $\mu_f = N_0whl_m$ and letting $M(z_m)$ represent the integral on y, the mean void size can be written as

$$
\overline{v}(t) = \mu_f \cdot M(z_m).
\tag{6.13}
$$

This quantity is the mean final void volume that can be attained for a mean void separation of l_m. $M(z_m)$ will be referred to as the mean size function, and is also graphed in Figure 6.7.

The following simplification can be introduced in order to more easily evaluate the integral in Equation 6.12:

$$
S(z) = \left\{ \begin{array}{cc} 4\sqrt{\frac{z}{\pi}} & \text{for} \quad z \le \frac{\pi}{16} \\ 1 & \text{otherwise} \end{array} \right\}.
\tag{6.14}
$$

Then

$$
M(z) \cong 1 - e^{-4\sqrt{\frac{z}{\pi}}},
\tag{6.15}
$$

a very good approximation to $M(z)$, as seen from the dashed curve in Figure 6.7. Combining Equations 6.9 and 6.14, accounting for diffusion to both sides, and substituting $3\varepsilon_0$ for N_0, we can write

$$
v(t) = 12\frac{\varepsilon_0 wh}{\sqrt{\pi}}\sqrt{Dt},
\tag{6.16}
$$

which describes void growth before the stress reservoir becomes saturated.

6.2.1 Relating Time-to-Failure to Void Size

If a void with a defined volume v_f causes failure, the time-to-failure can be obtained from Equation 6.16, and is given by

$$
t_f = v_f^2 \frac{\pi l}{16DN_0^2w^2h^2}.
\tag{6.17}
$$

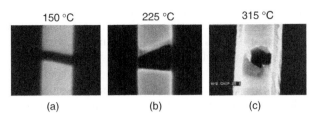

Figure 6.8. Examples of different void shapes in AlCu (0.5%) Si(2%) observed in failed samples baked at (a) 150°C, (b) 225°C, and (c) 315°C.

But, in order to obtain a failure time, the volume of the void at failure must be known. For some metallizations, in which a variety of void shapes are found, this creates substantial uncertainty in the prediction of failure.

Figure 6.8 shows SEM photographs of voids in AlCu(0.5%)Si(2%), observed in failed chips after baking at three different temperatures, 150°C, 225°C, and 315°C for about 1000 hr. The void shapes are quite different in all three cases. At 150°C, the void is narrow with parallel sides, and is commonly referred to as a slit void. At 225°C, the sides of the void are still flat, but not parallel to each other, giving the void a wedge shape. At 315°C, the void walls are no longer flat, and the void has become spherically shaped.

The volume of the void at the time of failure varies considerably from case to case. For example, if the sides of a slit shaped void are absolutely parallel, the absence a single atomic plane of atoms would be enough to cause failure. In actual

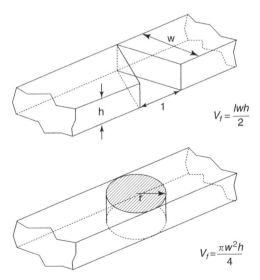

Figure 6.9. Comparison of void volume for wedge-shaped and cylindrical voids that completely span the line.

practice, a separation of several hundred angstroms is probably necessary to accommodate a variety of imperfections in the material, but the volume to failure would still be very small. This is illustrated in Figure 6.9, which shows drawings of a wedge-shaped void and a cylindrically shaped void in lines of identical cross section.

If the length l of the wide end of the wedge-shaped void is taken to be a tenth of the line width, then the volume of the cylinder is about 15 times larger than that of the wedge. Equation 6.17 shows that the time to failure is proportional to the square of the void volume at failure, so this difference would result in a failure time 225 times longer for the cylinder than for the wedge. Since voids with very irregular shapes are commonly observed, even larger differences in void volumes and failure times can be observed. When void shape varies with temperature, as shown in Figure 6.8, determination of activation energies from failure times at different temperatures becomes more difficult.

Okabayashi [17] proposed using the wedge shape to model the thermal behavior of voiding, as this shape was often observed in samples with AlSi metallization annealed over a wide range of temperatures. The volume of the wedge-shaped void shown in Figure 6.10 is given by

$$v(t) = h[w(t)]^2 \tan \Psi. \tag{6.18}$$

By equating Equations 6.16 and 6.18, and by realizing that failure occurs for the wedge void when $w(t) = w$, an explicit expression for the time-to-failure is obtained.

$$t_{fail} = \underbrace{\left[\frac{w\tan\Psi}{12\alpha}\right]^2 \frac{\pi}{D_0} \frac{1}{\Delta T^2}}_{C} e^{\frac{\Delta h}{kT}}. \tag{6.19}$$

The first two factors in Equation 6.19 are constants, and embody the geometric parameters of the void and materials properties. The last two factors embody the temperature dependence of the failure time; they arise from the thermal expansion and the diffusivity and can be thought of as the stress component and the diffusive components, respectively.

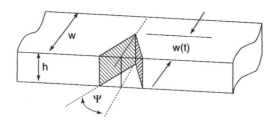

Figure 6.10. Drawing of a wedge-shaped void showing the parameters for calculation of the time-dependent volume.

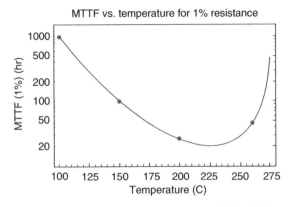

MTTF vs. temperature for 1% resistance

Figure 6.11. Fit of Equation 6.19 to published data.

Any void shape could have been used to obtain an expression for the failure time as a function of temperature, and would only cause a change in the value of the first factor in Equation 6.19. In fact, if the void shape at a given temperature is known in advance, and if the shape changes with temperature as shown in Figure 6.8, the ratio of these factors will give the offset in failure times due to the changes in void shape.

The temperature at which the parts will fail the most quickly is given by finding the temperature at which the derivative of Equation 6.19 with respect to temperature is equal to zero.

$$\frac{dt_{fail}}{dT} = \frac{Ce^{\frac{\Delta h}{kT}}}{\Delta T^2} \left[-\frac{\Delta h}{k} \cdot \frac{1}{T^2} + \frac{2}{(T_0 - T)} \right] = 0.$$

$$T(t_{\min}) = \frac{\Delta h}{4k} \left[-1 \pm \sqrt{1 + \frac{8k}{\Delta h} T_0} \right].$$

(6.20)

Figure 6.11 shows the fit of Equation 6.19 to published data [18] for AlSi metallization, where the failure time at 1% failure (rather than at 50% failure) is used. The fit yields an effective activation energy of 0.8 eV, and a T_0 of 280°C, and conforms well to the data. Note here that the number of data points is small (only four). This is a typical problem with data for stress voiding experiments. The reason is that the stress is expensive and slow, and most industrial laboratories are not equipped to handle large samples at multiple temperatures for months at a time.

6.2.2 Stress Contribution

The development thus far has shown how voids in Al lines can grow with time and temperature, and how this growth can be related to line failure. The driving force presented above is thermal strain. Why then, is the phenomenon referred to as stress voiding, instead of, say, strain voiding? Since one-dimensional stress and strain are linearly related to each other by the modulus of elasticity (Young's modulus), and by elasticity tensors in the more complex three-dimensional cases, one could reasonably argue that they are the same thing, and the nomenclature has arisen either by choice of the first investigators or by convention.

But several other aspects of the phenomenon give the stress aspect emphasis. The first is the mechanical properties of the materials. Deformation in metals depends on the specific mechanical properties of the metal being tested. For example, if the interconnect lines were made of W rather than Al, considerably more stress would be required to cause voiding under the same applied temperatures. However, because of its much greater elastic modulus, the W would be more likely to generate stress cracks in the surrounding oxide rather than oxide constraints causing stress voiding in the W. This is because (among other factors) the yield strength of W is considerably greater than that of Al. One could argue that if the same amount of strain had been applied to the W, a similar result would occur, but W also has a much higher melting point than Al, and the self diffusivity of W at 300°C is extremely small compared to that of Al.

In fact, the deformation behavior observed in the Al lines is a form of deformation known as creep. Such behavior has been known for several decades and is commonly measured on loaded wire samples. A weight is suspended from a wire that passes through a tubular furnace. The position of the load is recorded before the furnace is turned on, and then monitored periodically as the furnace is held at the desired temperature. The mass is observed to sink slowly with time, and if allowed, will eventually break the wire and fall to the floor. The rate at which the mass sinks is the strain rate of the wire, and it increases with increasing mass and with increasing temperature. The strain in the wire occurs by a variety of mechanisms including dislocation movement, grain boundary sliding, and diffusional flow, depending on the applied load (stress) and the temperature. Graphs of this behavior, which show strain rate as a function of applied shear stress and temperature, are called deformation mechanism maps and have been extensively catalogued [19].

The resistance to deformation by creep depends on a number of materials properties, including the yield stress, cohesive strength (which is usually related to the melting point), crystal structure and grain size of the material. The yield stress depends on the cohesive strength, the crystal structure and the shear stress (through dislocation glide). The most obvious way to think about this is in terms of how much weight a given cross section of material can support at room temperature without breaking. Studies of the breaking process have shown that granular materials often begin to fail by forming small cracks or cavities at grain boundaries, and then the cracks grow as the remainder of the grain boundaries are torn apart. Weight per unit area is stress, and cavitation is very similar to forming voids. Thus it is intuitive and also conventional to think of deformation and cavitation in terms of stress.

The unusual aspects of stress voiding in microelectronic interconnect wiring are the way in which the stress is applied (hydrostatic tensile stress) and the magnitude of the stress (several times the yield stress).

Stress can also affect long-range mass transport through diffusion. In an unstressed metal, the vacancy concentration is given by

$$N_v = N_0 e^{-\frac{\Delta h_f}{kT}}, \tag{6.21}$$

where N_0 is the atomic density and h_f is the energy of vacancy formation. In a metal under an applied stress σ, Nabarro [20] showed that the vacancy concentration is increased (tensile stress) or decreased (compressive stress) by an exponential factor to become

$$N_v(\sigma) = N_v e^{\frac{\sigma \Omega}{kT}}, \tag{6.22}$$

where Ω is the atomic volume. Since the diffusivity depends on the vacancy concentration,

$$D = CN_v e^{\frac{\sigma \Omega}{kT}} e^{-\frac{\Delta h}{kT}} = D_0 e^{-\frac{\Delta h - \sigma \Omega}{kT}}, \tag{6.23}$$

where C is a constant. The implication of Equation 6.23 is that the applied stress can alter the intrinsic Al diffusivity through a change in the activation energy for self-diffusion. And in fact, the activation energies reported in the literature for stress voiding are sometimes less than the values expected for grain boundary diffusion in Al. This also implies that the diffusivity may change with time as the diffusion fronts from neighboring voids meet and reduce the maximum stress in the metal.

6.2.3 Accelerated Testing for Stress Voiding

As for all wearout mechanisms, we want to cause the parts to fail in a short period of time under accelerated conditions so we can predict the lifetime of the larger population of parts under intended usage conditions. A necessary condition for projecting from the accelerated conditions to use conditions is to have a physical relationship that describes how the lifetime changes as the conditions are changed. A commonly used relationship is the acceleration factor (AF).

An acceleration factor for stress voiding failure can be calculated, just as for electromigration, by the ratio of the median time to failure at use conditions to that obtained under stress conditions. Using Equation 6.19, the constant factor drops out, leaving a stress contribution, in the form of the ΔT factor, and a diffusion contribution in the exponential. The subscripts s and u refer to stress and use conditions, respectively. Note here, that the activation energy is the effective activation energy for stress voiding, which may be modified by the stress as described in Equation 6.25 above.

$$AF = \frac{t_{50u}}{t_{50s}} = \left(\frac{\Delta T_s}{\Delta T_u}\right)^2 e^{\frac{\Delta h_c}{k}\left(\frac{1}{T_u}-\frac{1}{T_s}\right)}. \tag{6.24}$$

A graph of the AF versus stress temperature, relative to a use temperature of 100°C, is shown in Figure 6.12, for $\Delta h = 0.8\,\text{eV}$ and $T_o = 280°\text{C}$. Rather than increasing continuously as the stress temperature increases, the stress voiding AF has a maximum at a temperature below T_o. This peak is the result of two competing influences, one due to stress, and the other due to diffusion. The stress is proportional to the strain in the Al, (see Section 6.1.2), and is given by

$$\sigma = \frac{E}{(1-2v)}\alpha\Delta T, \tag{6.25}$$

where v is Poisson's ratio and $E/(1-2v)$ is the bulk modulus of the metal. As can be seen, the stress is zero when $\Delta T = (T_o - T) = 0$, or when the temperature equals the stress-free temperature, and it increases as the temperature decreases. The stress factor is represented in Figure 6.12 by the long-dashed curve. In contrast, the diffusivity increases steadily with increasing temperature, and is represented by the short-dashed curve in Figure 6.12. The maximum in the AF is located where the two dashed curves intersect. (These dashed curves in Fig. 6.12 have been scaled for convenient display in the figure, and in reality have greatly different magnitudes.) For the activation energy and zero stress temperature values used here, the peak in

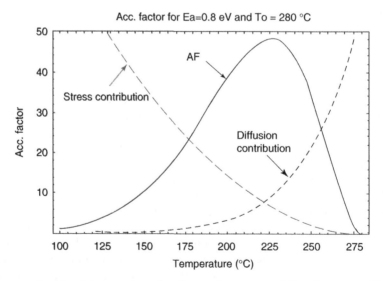

Figure 6.12. Graph of the acceleration factor for stress voiding for a metallization with $\Delta h = 0.8\,\text{eV}$ and a zero stress temperature $T_o = 280°\text{C}$.

the curve in Figure 6.12 occurs at about 225°C, where its maximum value is near 50. The AF has a value near zero at 100°C (arbitrarily chosen as use temperature), since then there is little thermal acceleration, and also at 280°C, since then there is no stress.

The acceleration factor (as explained in Chapter 7 on electromigration, as well as in Chapter 1) is the increase in lifetime expected in going from a high temperature environment, where voiding is accelerated, to a use environment, as would be encountered in an operating computer. In Figure 6.12, for example, the lifetime at 100°C should be about 50 times longer than at 225°C. This acceleration is quite small in comparison to that usually seen in electromigration, where the factors can be on the order of 2000 or more, and it means that a metallization that displays stress voiding at stress temperatures in the range shown in Figure 6.12 will probably fail before the intended end of life of the product. For example, suppose that the median time-to-failure occurs at 500 hr under stress (accelerated) conditions. With an AF = 50, this would project to a median lifetime of 25,000 hr at best (about 3 years), which would be unacceptable for product intended to last 40,000 hr.

Accelerated testing for stress voiding is relatively straight forward. Wafers from the population of interest are separated into several groups, each group to be baked at a different temperature. The resistances of appropriate test structures are measured before baking. The total duration for baking is conventionally 1000 hr, but with resistance measurements recorded at shorter intervals, e.g., at 48, 100, 250, 500 and 750 hr. The wafers are cooled to room temperature before each resistance measurement and then returned to the baking temperature. Resistances of the test structures that are obtained after each exposure to baking are compared to the initial resistances, and any failures are recorded. Failure is defined as a resistance increase greater than a predefined amount, such as 20%, and a median time-to-failure is obtained from the distribution of failures. Failure times for stress voiding are usually lognormally distributed, as for electromigration.

However, in contrast to electromigration, where the entire test sample can be forced to fail, stress voiding structures frequently exhibit only partial failure of the stress sample. Often only a part of a wafer will show failure. Among the factors that can contribute to this behavior are variations in line width, variations in overlying oxide thickness, and processing conditions that are not uniform over the wafer. In order to compensate for this behavior, a lower criterion for failure may be selected, such as 5% or 1% shift in resistance in order to increase the population of failing parts (more data). But sometimes the non-uniformity in behavior is so extreme that no adjustment in the failure criterion will help, and the failing population must be considered as a subset of the total population.

From a practical viewpoint, these difficulties are important for modeling and projection. But there are also practical limits to the amount of accelerated testing (stressing) that can be performed. Large sample sizes require considerable oven space and tester time, as well as engineering time to analyze the data. Tests that last 1000 hr require an overall duration of about six to eight weeks, including test time and data analysis. As mentioned earlier, because stress voiding cannot be accelerated very much, if voiding is observed to any significant extent during the 1000 hr test duration, reliability targets may not be achievable, and the metallization process will have to be improved.

The most difficult situation occurs when limited failure is observed late in the 1000-hour interval, and does not appear uniformly on all wafers. If reducing the failure criterion cannot provide sufficient data for purposes of projection, the test duration can be extended to collect more data. The amount of data that can be collected by doubling the stress duration is limited because of the logarithmic time scale, but the additional time sometimes can produce sufficient failure data to characterize the failure distribution and to enable failure rate projection to use conditions. Because of these difficulties, fabricators often choose to improve the process until no measurable stress voiding behavior is evident in the metallization. This does not mean that the metallization is void free. Rather, it means that no void, or void population, grows large enough to cause a measurable resistance shift by the end of life.

6.2.4 Alloying and Impurity Effects

One implication of Figure 6.12 is that the effects of void size must be kept in mind when comparing data from different temperatures. The differences in void shapes are partially a result of surface diffusion. Surfaces have an energy associated with them that is proportional to the surface area [21], and the metal will behave so as to reduce the surface area and hence the surface energy whenever possible. However, at lower temperatures, surface diffusion is apparently comparable to grain boundary and interfacial diffusion, so atoms at the surface of the void move along the surface just quickly enough to feed the demand for atoms at the grain boundaries and interfaces. At higher temperatures, more atoms move along the void surfaces than are needed to satisfy the demand at the boundaries and interfaces, allowing the surface to reconstruct itself.

It is important to mention here that these statements may not apply to all cases. In fact, the presence of alloying elements often plays a significant role in the kinetic behavior of voiding. For example, Si diffuses very rapidly in Al and alters the general voiding behavior of the alloy. Cu forms Al_2Cu precipitates at the grain boundaries in Al and can substantially impede grain boundary diffusion. As little as 0.1% (wt) of Cu in Al has been shown to be effective in slowing stress voiding [22].

However, the behavior of Cu in Al is complex. The solid solubility of Cu in Al varies substantially over the temperature range where voiding can occur (25 to 350°C). Fabricators using reactive ion etching to define the interconnect lines often find that the etch process cannot effectively remove the Cu precipitates from between the lines, and the precipitates can lead to line-to-line shorting and chip failure. So the Cu content of Al films is often limited to 0.5% (wt) Cu. Five-tenths weight percent Cu is fully soluble in Al at around 317°C. Cu precipitates form preferentially at grain boundaries and grain boundary triple points in Al, and thus inhibit grain boundary diffusion. As the temperature increases, and the Cu dissolves into the Al, the number, density and size of such precipitates will decrease, leaving these sites progressively more denuded of Cu, and making them more vulnerable to void nucleation.

Because narrow lines often have a bamboo-like grain structure (long grains that occupy the full width of the line with grain boundaries perpendicular to the axis of the line) interspersed with polycrystalline segments, there are fewer sites available for precipitates to reside. But as width is increased, the frequency and extent of the polycrystalline segments increases and so does the number of precipitate sites. Thus one can observe void nucleation and growth in lines at $1.5 \times$ or $2 \times$ the width of the narrowest line width, while the narrowest lines remain void free.

Two examples of somewhat unexpected results are shown in Figure 6.13. Figure 6.13 (a) shows the results of baking AlCu(0.5%) metallization for 7000 hr at 225°C. No fails ($>20\%$ resistance increase) were observed over the entire stress duration, while fails in a similar structure for AlCu(0.5%)Si(2%) appeared in less than 100 hr at the same temperature. Examination of Figure 6.13 (a) shows that there are clearly voids present in the lines (darker notches along the edges of the lines), so voids did nucleate. However, when the contrast of the image in the SEM was increased, it revealed that regions of bright contrast appeared adjacent to each void, and X-ray analysis of the bright regions showed that they were Cu precipitates. The reason for the lack of failure seems to be that a Cu precipitate occupied each grain boundary where a void had nucleated. The Cu precipitates were not very evident in the as-deposited films or in the finished chips, so the

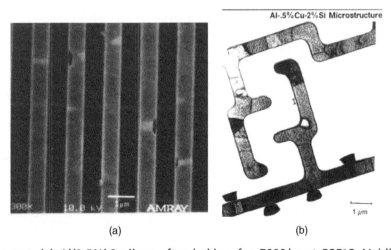

(a)   (b)

Figure 6.13. (a) Al(0.5%)Cu lines after baking for 7000 hr at 225°C. Voiding is evident; adjacent to each void is a Cu precipitate (light contrast). (b) Al(0.5%)-Cu(2%)Si lines after 100 hr at 225°C; voids appear at corners of extensions off parent line, but rarely appear in parent lines. Reprinted with permission from Ref. 32, © 1998 IEEE.

arrangement of precipitates visible in Figure 6.13 (a) is a result of the extended anneal at 225°C. Therefore, in this instance, with this composition and these fabrication conditions, Cu precipitates have migrated to and ripened in the same grain boundaries in which voids preferentially nucleated. Furthermore, the precipitates were not transported away by the same mechanisms that transport the Al and blocked further void growth.

This behavior is consistent. Void nucleation would be expected to occur preferentially at boundaries with the greatest disorder, i.e., those with the least ordered bonding between adjacent grains. At the same time, one would expect Cu precipitate nucleation to occur in the same locations because the additional disorder introduced by the interfaces with the precipitates would be relatively less than if the adjacent grains were more closely aligned. But it is probably fortuitous that enough Cu was available to block every void. A combination of events all had to work out together. The initial grain size distribution and precipitate distribution had to be in the proper range; the diffusivity of the Cu versus that of the Al had to be balanced; the amount of Cu available for precipitation had to be sufficient to block all the voiding sites; the number and distribution of vulnerable grain boundaries had to have been in the right range; and the stress in the AlCu had to be in the proper range. In fact, for wider line widths and other temperatures, these structures produced failures.

This result might never have been observed if the film deposition parameters (sputter rate, temperature, power, vacuum) had been different, or if the substrate were treated differently, or if post-deposition processing such as anneals or passivation thickness and composition had been different.

Figure 6.13 (b) shows a transmission electron microscope (TEM) photograph of the results in AlCu (0.5%) Si (2%) in a memory cell contact chain after 1500 hr at 225°C. Failure in this structure was much less frequent than in one without the appendages used to contact the cells. The difference was that voids preferentially nucleated at the corners of the contacting shape where one or more grain boundaries were present. While shear stresses in the Al may have been higher at these corners, the presence of multiple boundaries provides a more effective local diffusion path for initial void growth. The voids in these appendages acted as sources for atomic diffusion and reduced the incidence of void formation in the straight portion of the line. Voids in these appendages would not influence the resistance of the rest of the chain, and so were not detectable during the stress.

Finally, the precipitation of Cu in AlCu metallizations also produces a subtle effect in the resistance behavior of the stress voiding (SV) structures. Cu atoms in solution in the Al grains raise the resistivity of the grains in proportion to the Cu concentration. As the Cu comes out of solution and adds to the volume of Al_2Cu precipitates, the resistivity of the Al grains decreases slightly, usually on the order of 1–2%. This effect is usually evident in the first 100 hours of stress, and then ceases to be noticeable. If it is not taken into account, then it will delay detection of failure and can add significant error to the voiding analysis, especially when a low failure criterion on the order of 1% is used.

6.2.5 Relating Resistance Changes to Void Growth

As mentioned in the introduction, SV can lead to catastrophic failure in single component metallizations for which the primary conductor is the only component of the interconnect. The threat is much less severe for layered metallizations in which refractory layers above and/or below the primary conductor are part of the interconnect. In these metallizations, a void that causes a measurable resistance shift is usually about the size of a void that would cause catastrophic failure in a single-component metallization. Since the void needs to grow even larger to cause a 10–20% resistance change, the failure time will be considerably longer than that for the single-component case. Furthermore, the rate of resistance increase with time will be strongly influenced by the thickness and resistivity of the refractory layers. SV and EM robustness can therefore be controlled by the design of the metallization stack. (But the process engineer needs to consider manufacturability and cost in the process as well, so that the metal stack design will not be determined by reliability alone.) In addition, many layered metallizations can tolerate void sizes five times or more that needed for a 20% resistance shift. So the specific resistance criterion chosen as failure is somewhat flexible and will depend on factors such as chip performance or the nature of the sensitive structures. In any case, it is possible to design a metallization that will be completely robust to stress voiding failure, but process, design, and cost factors generally push the process to a compromise that ensures sufficient reliability, process window, and low cost.

Just as is the case with electromigration, a side benefit to using layered metallization is that after the initial void is detected, void growth can be monitored electrically. Figure 6.14 shows a drawing of a layered interconnect line of width, w, and length, L_o, and that has a void of length, L_v, at one end. The top layer is the primary conductor, such as Al, and the bottom layer is the refractory layer, such as $TiAl_3$ or TiW, which usually has a resistivity substantially ($\sim 10 \times$) greater than that of the primary conductor. The resistance R of the entire line, with the void, can be calculated as follows, (with the quantities illustrated in Fig. 6.14).

$$R = \underbrace{\frac{\rho_R L_v}{A_R}}_{R} + \underbrace{\frac{\rho_A (L_o - L_v)}{A_A}}_{Al} = \frac{L_v}{w}\left[\frac{\rho_R}{t_R} - \frac{\rho_A}{t_A}\right] + \underbrace{\frac{\rho_A L_o}{t_A w}}_{\substack{Unvoided \\ Line}} . \qquad (6.26)$$

Here, the subscripts R, A, and v refer to redundancy, aluminum, and void, respectively, and L_o is the length of the line. L, w, t, ρ and A refer to the length, width, thickness, resistivity, and the cross-sectional area (when viewed along the line axis) of either the redundancy or the aluminum. The first two terms on the left represent the resistance of the redundancy under the void and the resistance of the Al in the remainder of the line. The parallel resistance of the redundant layer under the Al has been ignored. By rearranging, the two terms on the right are

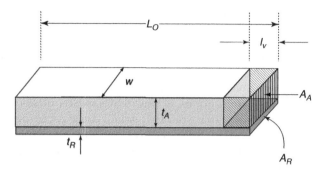

Figure 6.14. Drawing of a line of width, w, length, l_o, and thickness, $t_A + t_R$, with a void at one end having length L_v. The area of the Al and of the redundancy are A_A and A_R, respectively.

obtained. The second term represents the resistance of the unvoided line, leaving the first term to represent the resistance increase due to the presence of the void.

The fractional resistance change (often represented as a percentage change) can then be calculated by taking the ratio of the first and second terms on the right.

$$\frac{\Delta R}{R_o} = \frac{R - R_o}{R_o} = \frac{L_v}{L_o}\left[\frac{\rho_R t_A}{\rho_A t_R} - 1\right] \simeq \frac{L_V}{L_o}\frac{\rho_R t_A}{\rho_A t_R}. \qquad (6.27)$$

The first term inside the brackets is much larger than 1, allowing the 1 to be dropped from the final result. For example, if the redundant layer is Ti, which has a resistivity between 40 and 70 $\mu\Omega$-cm, and a thickness of 500 A, and if the Al is 0.5 μm thick, this term would be equal to about 140, justifying our neglecting of the contribution of the parallel resistance of the redundant layer. The first fraction is just the ratio of the void length to the length of the wire, and it is multiplied by the ratio of sheet resistances, where sheet resistance is given by $Rs = \rho/t$.

The length of the void, L_v, is given by the void volume divided by the cross-sectional area,

$$L_v(t) = \frac{v(t)}{wt_A} = \frac{12}{\sqrt{\pi}}\varepsilon_o\sqrt{Dt}, \qquad (6.28)$$

where Equation 6.16 was used for $v(t)$. Substituting Equation 6.28 into Equation 6.27, we obtain

$$\frac{\Delta R}{R_o} = \frac{\rho_R t_A}{\rho_A t_R}\frac{12\varepsilon_o}{L_o\sqrt{\pi}}\sqrt{Dt}, \qquad (6.29)$$

which expression gives the fractional resistance change as a function of time in a line containing a single void.

As Equation 6.29 shows, the resistance should increase as the square root of time. Figure 6.15 shows a plot of resistance versus time for an 800 μm-long, 0.40 μm-wide Al(0.5%)Cu(2%)Si line during 3000 hr of baking at 225°C. Examination of the line by scanning electron microscopy showed that only a single void was present. The solid line through the data points is a fit of the data by square root of time, which is quite good and validates Equation 6.29.

6.2.6 Factors that Complicate SV Data

Although the agreement between Equation 6.29 and the data in Figure 6.15 is reasonably good, this is not often the case. A wide variety of factors can influence resistance data for stress voiding, including composition of the metallization, reaction with refractory layers, variations in void shape, variation in metallization microstructure and mechanical effects of the overlying oxide layer. Extensive physical analysis is often required to sort these effects out, but some degree of understanding can be provided by general numerical observations.

For example, for an $Al/TiAl_3$ line 0.5 μm in width, a minimum void length needed to generate a 20% resistance shift might be 1.5 μm. The minimum length of line needed to produce such a void at 200°C would be given by

$$\Delta l = 3\varepsilon_o l = 3\alpha\Delta Tl,$$

$$\Rightarrow l = \frac{\Delta l}{3\alpha\Delta T} = \frac{1.5\ \mu m}{3(25e^{-6}/°C)(280 - 200°C)} \approx 250\ \mu m, \tag{6.30}$$

assuming complete relaxation. So in a line four times as long, it might be possible to have four voids in the line at once, and to observe four times the resistance shift.

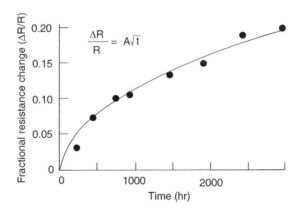

Figure 6.15. Fractional resistance change versus square root of time for a 0.4 μm-wide line, 800 μm-long with a single void, illustrating the behavior predicted by Equation 6.29.

Behavior of this type was observed in 800 μm Al(0.5%)Cu(2%)Si/Ti lines with widths of 0.36, 0.40, 0.48 and 0.72 μm, after baking at 225°C for 3000 hr [23]. Maximum fractional resistance shifts were observed to increase with line width from about 0.25 to about 0.6. Examination of these lines with a scanning electron microscope (SEM) after the overlying oxide was removed revealed that there were more voids per unit length in the wider lines than in the narrow lines.

Figure 6.16 shows resistance versus time curves for four individual lines, one from each width. In increasing order of width, these four lines were found to contain 1, 1, 3, and 4 voids (by SEM imaging), which corresponds to the different magnitudes in resistance shift. In fact, the slopes of the lines tangent to the individual curves are very close to 1:1:3:4. To obtain the correct resistance shift by Equation 6.28, the right-hand side of the equation would have to be multiplied by the number of voids. In Figure 6.16, the magnitude of the resistance shift is indicative of the number of voids per unit length of the line. By the estimate in Equation 6.30, about four voids should be able to grow to appreciable length in an 800 μm-long line, and the maximum amount of resistance shift would have been

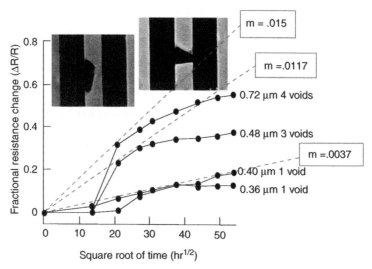

Figure 6.16. Fractional resistance change versus the square root of time for four 800 μm-long AlCuSi lines with widths of 0.36, 0.40, 0.48, and 0.72 μm. The first two lines had one void each, and the last two had three and four voids, respectively. Resistance shift is proportional to the number of voids. Slopes of tangent curves (dashed lines) for the last two lines are very close to three and four times that of the first two. Only the 0.4 μm-line shows a constant resistance increase with the square root of time. The others show delayed initial shifts due to void shape, where the Al conductor is not severed until the void reaches substantial size.

reached in the 0.72 μm-wide lines. In fact, the magnitude of the resistance shifts began to decrease for even wider lines.

But although the magnitude of the resistance shifts can be explained by the number of voids, other discrepancies in the resistance traces cannot. Three out of the four curves show no shift for the first 150 hours, a feature frequently observed in SV data. Examination of the void geometry for each case shows that void shape can have a strong influence on the onset of resistance shifts by allowing a void to achieve a substantial size before completely severing the Al conductor. The SEM photographs inset into Figure 6.16 illustrate two such cases. The areas of the two voids are similar, yet the one on the left would not be electrically detectable, and the one on the right would not appear electrically until the right hand edge finally separated. This is the case both for wedge-shaped voids, like the one on the right, and irregularly shaped voids like the one on the left. Only slit-like voids and wedge-shaped voids with very small angles can produce the type of resistance trace shown in Figure 6.15.

Another feature of the resistance curves in Figure 6.16 is the flattening of the curve at longer times. This resembles the shape of the S(z) and M(z) curves shown in Figure 6.7. Therefore, one explanation for this behavior is that the strain reservoir is being depleted and void growth is saturating. This explanation is consistent with the linear behavior of the 0.4 μm-wide line, since there is only one void present in the 800 μm line, and saturation has not yet begun to occur. It is also consistent with the shallower slope of the 0.72 mm line, because the resistance shift in that line due to all four voids and should follow the M(z) curve, which flattens out more gradually than the S(z) curve.

The curve for the 0.36 μm wide line flattens very quickly even though it has only one void and should not be showing saturation. SEM observation with enhanced contrast revealed the presence of Cu precipitates throughout the entire line and on either side of the void, which would block Al diffusion to regions of the line beyond the precipitates. The strain reservoir for this particular void was thus constrained to the region between the two precipitates, and that limited the growth of void located between them. Although Cu precipitates were present in the wider lines, they appeared more frequently and were not as large, and thus did not block diffusion along the line.

An additional factor, only visible with transmission electron microscopy, is the Al grain size distribution. Grains for the 0.36 μm lines tend to be long and have grain boundaries perpendicular (bamboo structure). The Cu precipitates are mostly constrained to occupy some of these boundaries in the narrow line. But for wider lines, some of the bamboo grains are replaced by collections of smaller grains containing grain boundary triple points, which are favored nucleation sites for Cu precipitates. The formation of additional Cu precipitates at these triple points cause a decrease in average precipitate size so that fewer are able to completely block Al grain boundary diffusion. At the same time, triple points not occupied by Cu precipitates are likely sites for void nucleation and lead to more voids. For even wider lines, sufficient triple points area available for the density of voids to be great enough that none can grow large enough to completely sever the line.

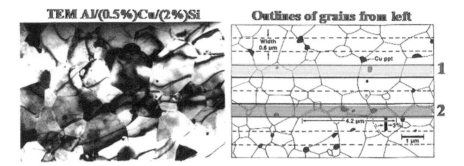

Figure 6.17. Left: TEM photo of thin film AlCuSi grain distribution, including Cu precipitates. Right: array of 0.6 μm-lines randomly imposed on the grain structure to illustrate the degree of variation that can be expected in line segments cut from the film. Reprinted with permission from Ref. 13.

Variation in microstructure can also be the cause of why some structures fail and others do not. Figure 6.17 shows a transmission electron micrograph of an Al(0.5%)Cu(2%)Si film (left) with grain boundaries clearly visible. Smaller dark particles are evident as well, and are the Cu precipitates, which reside most frequently at grain boundary triple points. The right side of Figure 6.17 shows a tracing of the grain boundaries and precipitates. An array of 0.6 μm lines has been randomly imposed on the grain structure to illustrate the degree of microstructural variation that is often observed in narrow lines. Two bands in this single photograph demonstrate extremes in microstrucure. The band labeled 1 contains mostly grains with boundaries perpendicular to the line axis, and three triple points on the left end. The band labeled 2 contains 10 or more triple points, and has an almost unbroken grain boundary path along the length of the segment. On the scale of these lines, the length of the picture is about 15 μm, so that the total line length represented here is about 87 μm. Imagine that a single line is made by placing the 10 segments end to end. Some of the "line" thus formed will have bamboo segments, and some will have segments with two or more grains across the width of the line. As the "width" of the segments is decreased, more of the "line" will be composed of bamboo grains and less will be composed of segments with multiple grains across the width. In practice, the lines are created by etching away unwanted parts of the film, and then the remaining pattern is annealed. During the anneal, many of the boundaries and partial grains are consumed by growth of neighboring grains, making the narrowest lines even more bamboo.

6.3 ROLE OF THE OVERLYING DIELECTRIC

The presence of the overlying dielectric layer (often called passivation) is essential to the generation of stress induced voids. In the model presented earlier in this chapter, the oxide was assumed to be a rigid container that retained the same dimensions regardless of the temperature or the material contained within. From that point of view, the role of the oxide is to define the volume of the enclosed metallization at an elevated temperature, and then to maintain that volume as the temperature is decreased to lower temperatures. However, no material is perfectly rigid, and this approach was used in order to simplify the modeling. All dielectrics give a little under applied stress, and the amount of movement is determined by a number of factors including the applied force, the geometry of the system, and the mechanical and thermal properties of the dielectric.

During the first couple of years after stress voiding was initially observed, some investigators indicated that voiding was correlated with the stress in the overlying passivation [5]. This created considerable confusion in the industry and caused process engineers to closely monitor the stress in the passivation layers in an attempt to avoid stress voiding. Later work [24] demonstrated that there was no necessary correlation between passivation stress and Al stress voiding. Finite element modeling performed at about the same time [25] showed that stress in the Al depended on materials properties (elastic modulus, thermal expansion coefficient), geometry (line width, Al, and passivation thickness) and processing parameters (passivation deposition temperature). The modeling generally treats the Al as a homogeneous material, ignoring microstructure and composition, and metallizations and oxide processing parameters (such as metallization deposition temperature, annealing, etc.).

Part of the reason for the confusion over the relevance of the dielectric stress was that the dielectric stress is measured relative to the Si substrate, and depends on the temperature at which the dielectric is deposited. So if a dielectric layer were deposited at a higher temperature, not only would its compressive stress be greater at room temperature, but also the zero stress temperature of the metal encapsulated by that dielectric would be greater. In addition, if two dielectric films with different elastic moduli are deposited at the same temperature, they will produce different compressive stresses at room temperature; the dielectric with the greater elastic modulus will have the greater stress. Room temperature dielectric stress can therefore be an indication that stress voiding should be checked, but is not a predictor of SV. In addition, dielectric properties can depend on the deposition temperature of the dielectric because the deposition temperature can affect the film composition and density. While computer modeling is really necessary to obtain detailed relative stress information for a specific geometric configuration, some insight can be gained by the following calculation, which illustrates how the passivation thickness and elastic modulus can affect the stress in the enclosed Al.

Referring to Figure 6.18, consider an Al line on a substrate with the sidewalls encased in glass. The top of the line is covered by a rigid plate that adheres to the Al and the glass. Suppose that the system was assembled at an elevated temperature T and then cooled to room temperature. The Al wants to contract to a smaller thickness, but is constrained by the pillars of oxide at the sidewalls and the rigid top plate. A balance of forces then exists in which the downward force, F_A, exerted by the Al trying to contract, equals the combined force, F_P,

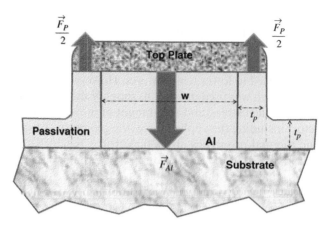

Figure 6.18. Cross-section of an Al line showing some stress variables for the Al line with oxide sidewalls and a rigid top plate.

exerted by the sidewalls, which are resisting compression. The total strain, which is the thermal strain, is

$$\varepsilon_0 = \alpha_e \Delta T = \varepsilon_A + \varepsilon_P \tag{6.31}$$

where ε_A is the strain in the Al and ε_P is the strain in the passivation.

Since force is given by the stress multiplied by the area, we can write $F = \sigma A = \varepsilon E A$, where E is Young's modulus, and then express the force balance as

$$\varepsilon_P E_P 2 t_P = \varepsilon_A E_A w.$$

Solving this expression for the strain in the passivation in terms of the strain in the Al, we obtain

$$\varepsilon_P = \varepsilon_A \frac{E_A w}{2 E_P t_P},$$

and by using Equation 6.31 for ε_P, we finally have

$$\varepsilon_A = \frac{\varepsilon_0}{\left[1 + \frac{E_A w}{2 E_P t_P}\right]}, \tag{6.32}$$

which gives the strain in the Al in terms of a reduction in the thermal strain. The reduction, in the denominator, depends on the ratio of the width of the Al to the thickness of the oxide, and upon the ratio of the bulk modulus of the Al to that of the oxide. Thus, for a very stiff or a very thick oxide, the reduction goes to zero, and the strain in the Al is the full thermal strain. On the other hand, the wider the

TABLE 6.1. % of Structures Failing After 3500 hr of Baking

Temperature	Passivation Thickness (μm)		
	0.25	0.50	3.00
150	2	1	82
225	0	1	92
285	86	36	46
315	97	98	100

Al line in comparison to the oxide thickness, the more the Al compresses the oxide, and the less strain the Al experiences.

Table 6.1 shows the percentage of structures failing after baking for 3500 hr at various temperatures. The structures were made from 0.6 μm Al(0.5%)Cu(2%)Si lines of different lengths, passivated with 0.25, 0.5, and 3.0 μm of oxide. Only the thickness of the oxide passivation was varied; its composition and deposition parameters were the same. As seen in the table, for temperatures of 150°C and 225°C, and passivation thicknesses of 0.25 and 0.5 μm, a very small percentage of the structures failed, while the majority failed with 3 μm passivation. Therefore, for the lower temperatures, the increase in the rigidity of the constraint due to the thickness of the oxide introduced considerably more voiding. At 285°C and 315°C, a greater percentage of the structures failed at the higher the temperature. The lack of regular progression of the percentage failure with increasing oxide thickness at 285°C is attributed to wafer-to-wafer variation; otherwise, these results show that voiding increases with increasing temperature and with increasing passivation thickness.

6.4 SUMMARY OF VOIDING IN AL METALLIZATIONS

So far in this chapter, we have presented an analytical model that describes stress-induced void growth, and explains the temperature and stress dependence. The model is somewhat simplistic in that it ignores influences due to the elasticity of the surrounding dielectric and factors introduced by the microstructure of the Al, but it provides the fundamental relationship for more accurate modeling.

We have used this model to describe an analytical expression for resistance increases as a function of time that can be used to fit selected data obtained from layered metallization employing refractory layers as secondary or redundant conductors. This model shows that resistance increases in narrow lines should follow a square root of time dependence as long as the voids are not close to saturation. Deviations from this behavior can be introduced by and influenced by the Al microstructure and composition, which can alter void shape, void nucleation, atomic diffusivity, and the size of the available stress reservoir. Void shape can alter the interconnect lifetime and change the failure time distribution.

Finally, the passivation properties and thickness can influence the degree of voiding.

Al was the conductor of choice for chip interconnects for more than three decades. But the RC delay limitations in circuit speed have driven chip providers to Cu because of its lower resistivity. In the next sections, voiding behavior in Cu damascene metallization will be discussed. Void growth in Cu obeys the same equations that apply to Al. But Cu has different materials properties and a different microstructure from Al, and these differences have significant implications on how voiding manifests itself in Cu interconnects, and therefore on the lifetime of chips using Cu metallization.

6.5   STRESS VOIDING IN CU INTERCONNECTS

As stated above, Cu obeys the same void growth equations as Al. It is has the same crystal structure as Al (face-centered cubic) and self-diffusion is substitutional; that is, it depends on the presence of vacant sites in the array of atoms in order to proceed. However, a number of other Cu properties (Table 6.2) make the manifestation of voiding in Cu interconnects much different than in Al. The elastic modulus of Cu is substantially greater than that of Al, the atomic volume of Cu is less than that of Al, the activation energy for atomic diffusion for Cu is greater than that of Al, and the coefficient of thermal expansion (CTE) for Cu is less than that of Al.

The smaller CTE means that the thermal strain in the Cu for any ΔT will be smaller than it would be for Al (see Fig. 6.3). The greater elastic modulus means that the Cu sees a reduction in strain compared to Al when encapsulated in the same material with the same geometry, as can be seen by substituting E_{Cu} for E_A into Equation 6.32. The smaller atomic volume means that a smaller void will result from the same atomic flux as would be the case for Al. And the greater activation energy for diffusion means that the number of atomic jumps per unit time in Cu will be less than for Al at the same temperature. Taken together, these factors alone should make Cu more resistant to stress voiding than Al.

TABLE 6.2. Comparison of Selected Cu and Al Materials Properties

Property	Cu	Al	Reliability Impact
T_{melt} (°C)	1083	660	SV, EM, Mech. Props, ESD, Hi-temp EM
Elastic Modulus (GPa)	120	70	SV
CTE (1/K)	17 ppm	25 ppm	SV
Resistivity ($\mu\Omega$-cm)	2	3	EM, EOL, ESD, Hi-temp EM
Hardness (Brinnell)	35–415	19–160	CMP
Atomic Volume (cm^3)	1.18×10^{-23}	1.66×10^{-23}	SV, EM

However, the advantage of the Cu properties is not obvious if one uses the rigid box model. In that case, the stress in the metal is given by

$$\sigma = E\alpha\Delta T,$$

where $\varepsilon = \alpha\Delta T$ is the thermal strain due to the mismatch in the CTE between the metal and the oxide. Calculating the ratio of the stress in Cu to the stress in Al that arises due to the same change in temperature, the result is

$$\frac{\sigma_{Cu}}{\sigma_{Al}} = \frac{E_{Cu}\alpha_{Cu}}{E_{Al}\alpha_{Al}} \cong 1.17X!$$

Apparently the reduction in the CTE from Al to Cu is more than balanced by the increase in elastic modulus. Using a temperature change of 400°C, the equivalent values for stress in the Cu and Al are about 700 MPa and 816 MPa, respectively. However, the rigid box approach is misleading because the elasticity of the oxide allows both metals to relax, and this alters the stress in each substantially.

Figure 6.19 illustrates how the metal and surrounding oxide interact with respect to stress. Consider the situation at the top of Figure 6.19, where a composite bar is fastened between two rigid walls. Half of the bar is made of oxide and the other half is made of metal, either Cu or Al. In this configuration, there is a force balance between the metal and the oxide. Regardless of the temperature, the forces will balance, which means that the dimensions of the bars perpendicular

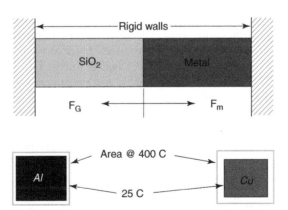

Figure 6.19. Top: bar composed of half SiO_2 and half metal attached between two rigid walls. As temperature is either increased or decreased, the materials will expand or shrink, exerting equal and opposite forces on the walls and on each other at the interface. The resulting strain will depend on the stiffness of each material. Bottom: Since Cu is stiffer than Al, the deflection in the oxide will be greater when the metal is Cu than it will be when it is Al.

to the walls will adjust themselves to balance the force. This can be expressed as

$$F_G = F_M,$$

$$E_G \varepsilon_G A = E_M \varepsilon_M A, \qquad\qquad (6.33)$$

$$\varepsilon_M = \varepsilon_G \left(\frac{E_G}{E_M} \right).$$

So the strain in the metal is equal to the strain in the glass multiplied by the ratio of the elastic modulus of the glass to the elastic modulus of the metal. Using values of 70 GPa for oxide and Al, and 120 GPa for Cu, the strain in the Al equals the strain in the glass, while the strain in the Cu is only 0.58 that of the glass. Using these values of strain to calculate the stress, the stress in the Al is then found to be about 350 MPa, while that in the Cu is only about 300 MPa (compared to values of 700 MPa for Al and 816 MPa for Cu obtained above). Thus, while for the rigid box approximation the Cu sees more stress, when passivation elastic relaxation is taken into account, the Al sees more stress.

In addition to these fundamental materials properties, a number of chemical and metallurgical properties that depend on specifics of the way the metal is integrated into the circuit chip have important influences on the manifestation of stress voiding in Cu. The methods used to define the wiring pattern are very different between Al and Cu. Al patterns are defined by subtractive reactive ion etching (RIE), that removes regions of the metal film and leaves the desired pattern behind. The entire wafer is then heated to around 400°C for deposition of the passivating dielectric. Cu lines and vias are defined by a dual damascene process in which holes, and grooves or trenches, are etched in an oxide layer. An adhesion layer (liner) is then deposited by physical or chemical vapor deposition (PVD or CVD) over the entire surface, followed by a PVD seed layer. Finally, an electroplated Cu film is deposited over the seed layer at room temperature, to a thickness beyond that needed to fill the etched features. The deposited films are then annealed, and a chemical/mechanical polishing process is used to remove all of the excess Cu and liner films, leaving the embedded Cu pattern.

The volume for the Al wires is defined at the high deposition temperature of the overlying dielectric, while the volume of the Cu wires is defined at room temperature when the bulk of the Cu is deposited in the trenches. This difference provides another means to reduce the stress in the Cu. Figure 6.20 shows diagrams of a cross section of both a RIE Al wire and a damascene Cu wire during integration into the chip. In the case of the Al, the base of the wire adheres to the underlying oxide at the Al deposition temperature and is constrained by it during heating and cooling. During deposition of the overlying oxide, the temperature is substantially elevated, causing the wire to expand at the top and sides prior to the onset of oxide deposition. The lateral expansion for temperatures above the metal deposition temperature will be partially constrained by the underlying oxide. This vertical and lateral expansion is frozen in by the layer of oxide deposited on top of the metal layer.

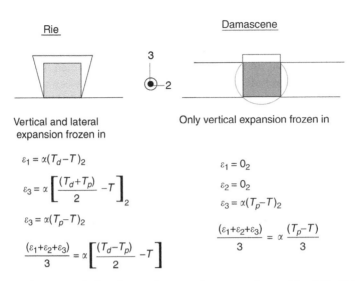

Figure 6.20. Diagram of the thermal expansion experienced by RIE Al (left) and damascene Cu (right) metallization during integration on a chip.

In the case of the damascene Cu, the wire also expands vertically and laterally. However, the sidewalls and bottom of the damascene line adhere to the surrounding oxide and are constrained by it. Again, deposition of the overlying insulator is performed at an elevated temperature, and the wire expands both laterally and vertically. Lateral expansion is partially constrained at the bottom by the oxide to which it is bonded, and is accommodated laterally by lateral elastic (recoverable) compression in the oxide. The vertical expansion is partially constrained by the oxide bonded to the sidewalls. No constraint exists at the top surface, and the vertical expansion at the bottom surface is accommodated by vertical elastic compression of the underlying oxide. Whatever expansion occurs upward at the surface of the Cu is captured, and frozen in, by the overlying insulator. When the temperature is returned to room temperature, all of the lateral and vertical expansion accommodated by elastic compressive stresses in the oxide is returned to zero. Axial compressive stresses arise in both cases (due to thermal expansion) constrained by the substrate and contribute partially to the lateral and vertical expansion through Poisson's ratio. But most of this stress is recovered after the metal is cooled. So, most of the expansion in the Cu is recovered elastically, while most of the expansion in the Al is captured by the oxide.

A quantitative estimate can be made of the relative magnitude of the strain captured in each case. In the following discussion, the directions 1, 2, and 3, represent a right-handed coordinate system with the x-direction 1 pointing out of the page, as shown in Figure 6.20. Strains for Al and Cu interconnects are shown below each case in the figure when the line is returned to room temperature after passivation deposition. For this estimate, contributions to the strain in the 2- and

3-directions from the compressive stress in the 1-direction are ignored. Thus for Cu, all the strain in the axial direction is elastic and is fully recovered when the line is returned to room temperature, as is the case for lateral strain. However, at worst case, the strain in the vertical direction is assumed to be completely captured, and its magnitude depends on the difference between room temperature and the passivation deposition temperature, T_p. For the Al case, the strain can depend on both the passivation deposition temperature and the Al deposition temperature, T_d. The strain in the axial direction depends on the difference between room temperature and the metal deposition temperature, the lateral strain depends on the average of the metal and oxide deposition temperatures, and the vertical strain depends only on the passivation deposition temperature. If $T_d = 150$, $T_p = 400$ and $T = 25°C$, then the average strain in the Al is about twice that in the Cu, as can be seen by inserting these values into the bottom expressions in Figure 6.20.

6.5.1 Microstructure of Cu

Cu is annealed after plating into the trenches, and before deposition of the capping dielectric layer, and considerable change occurs in the Cu microstructure during the anneal. The as-deposited grain size of the plated Cu film is generally very small, on the order of a couple hundred angstroms. The high surface energy of these small grains drives recrystallization and grain growth in the Cu film. In fact, the energy is so high that the grain growth takes place at room temperature over the course of several hours or days, and manifests as a transient resistance decrease [26]. Since the resistance of various wiring structures is measured as part of the fabricator quality monitoring process, a changing resistivity adds an unacceptable variability to the sources already present. Annealing the film imposes a controlled grain growth and stabilizes the resistivity.

However, the grain growth that takes place during the annealing process also causes an increase in stress in the Cu in the trenches. Because of the mismatch of the atomic lattices on either side of a grain boundary, atomic bonding and arrangement is more disordered, leaving a larger percentage of atomic sites vacant. This is extra volume in the metal. When grain growth occurs during annealing, most of these grain boundaries vanish, and the extra volume is eliminated. If the Cu were unconstrained, it would have a somewhat smaller volume than it had before. But since the Cu adheres to the liner surrounding the trenches, and since the volume of the trenches remains unchanged, the atomic bonds are stretched to accommodate the volume and a tensile stress arises in the Cu. This stress then becomes the driving force for stress voiding in Cu.

One can estimate the strain introduced into the Cu by grain growth from a simple model. An estimate of the maximum percentage volume of voiding may be obtained from grain size limits. Suppose the initial, as-deposited grain size is d and the grains have a cubic shape. Grain boundaries on three of the six sides of the cube can be allocated to each grain. Supposing the grain boundary width is δ, then

TABLE 6.3. Stress Values for Unpassivated and Passivated Lines Fabricated by Three Different Methods

Fabrication Method	X-Stress (MPa)	Y-Stress (MPa)	Z-Stress (MPa)	Hydrostatic (MPa)	Max Shear (MPa)
Unpassivated					
RIE Al	338	83	19	147	160
Damascene Al	111	95	117	108	10
Damascene Cu	48	38	98	60	30
Passivated					
RIE Al	706	488	415	537	146
Damascene Al	484	442	308	411	88
Damascene Cu	442	223	192	286	125

Source: Ref. 27. Reprinted with permission.

the ratio of grain boundary volume to grain volume is

$$\frac{\Delta V}{V} = \frac{3\gamma d^2}{d^3} = \frac{3\delta}{d}. \tag{6.34}$$

For $d = 100\,\text{Å}$ (10 nm) and $\delta = 0.20\,\text{Å}$ (0.02 nm), the maximum fractional volume contained in the grain boundaries would be $3(0.2)/100 = 0.6\%$. This quantity can be compared directly with that for Al, given in Equation 6.4. The one-dimensional strain here is then given by $\varepsilon = \delta/d$. Additional volume allowances can be made for removal of other Cu lattice defects (such as dislocations), but these contributions generally would be expected to be less than that of the grain boundary volume.

Each of the above factors, relaxation in the oxide, integration differences and grain growth must be considered in conjunction with the others in order to determine accurately the stress in the Cu by modeling. Fortunately, the stress can also be measured by X-ray diffraction. Table 6.3 provides a comparison of measured stresses [27] for RIE Al, damascene Al, and damascene Cu lines, for lines with and without passivation. The values given represent all deposition and annealing effects. As expected, stresses are uniformly higher in passivated lines than in unpassivated lines, regardless of the fabrication technique. For purposes of understanding the implications for stress voiding, the hydrostatic stress is the key quantity, and here the trend is clear. The RIE Al has the highest hydrostatic stress, damascene Al has the next highest, and damascene Cu has the lowest, validating the points in the discussion above.

6.5.2 Role of Dielectrics in Cu Voiding

Assuming no reaction occurs between Cu and the liner material, the only other contribution to Cu strain would be upward elastic expansion of the Cu at the surface of the line during the overlying dielectric deposition. Then the volume of the Cu would be defined by the expanded volume of the Cu at the temperature of dielectric deposition, in much the same way as it occurs in Al metallization. But this only applies to the vertical dimension. Because the Cu is plated into the trenches at room temperature, when the Cu expands as it is heated to the dielectric deposition temperature, it compresses the dielectric around it in an elastic manner. The top of the line is constrained by the overlying dielectric, so that when it cools, some of the strain remains, but the sides and the bottom can contract to their original dimensions simply be releasing the elastic compressive stress in the oxide. The principle role of the dielectric surrounding the Cu wires on the sides and the bottom is to define the volume of the wire prior to the anneal.

Adhesion of the dielectric layer to the metallization is another important consideration for stress voiding. Al forms its own passivating oxide, Al_2O_3, very rapidly in air. The oxide is only a few tens of angstroms thick and forms an excellent dielectric. If Al comes into contact with SiO_2, without the benefit of the intervening Al_2O_3 layer, Al will bond readily with the oxygen and Si atoms at the surface of the oxide. This property tends to inhibit extrusion of Al, even when driven by electromigration, with the result that a relatively large crack is required to drive an extrusion out of an Al line. The case is otherwise with Cu. Cu oxide is not self-passivating to the extent that Al_2O_3 is for Al, and Cu does not bond strongly to Si and oxygen atoms. So if, during the process of forming the Cu lines, Cu is left in contact with oxides, that location becomes a potential nucleation point for voiding.

The Cu does bond very well to Ta, and most liners are Ta-based. But the top of the Cu line is capped by some type of dielectric that is designed to adhere as well as possible to the Cu surface. This interface has the weakest bonding, and therefore, as is the case for electromigration, becomes the dominant avenue for mass transport. Void nucleation at this interface is thus very likely, and voids that grow to a size to affect circuit performance will most often be associated with this interface.

6.5.3 Microstructural Effects

The microstructure of Cu and Al films is quite different. Figure 6.21 shows TEM photographs of Al(0.5%)Cu(2%)Si and Cu films. Grains in the Al film have a relatively uniform appearance, with a average size on the order of 1 μm, with little evidence of dislocations or other defects other than Cu precipitates (small black spots). Grain boundaries are easily identified, and few or no twins are visible. In contrast, the grains in the Cu film vary widely, appearing to range from a fraction of a micron to more than four microns in diameter. Delineation of grain boundaries is often obscured by the complex and dense twin and dislocation networks. A pair of lines is drawn on each micrograph to represent a 0.5 μm-wide line, so that the variability of the microstructure compared to a line width can be better seen. The variation for the AlCuSi film is shown in greater detail in Figure 6.17, where even within the one photograph, segments of line have nearly

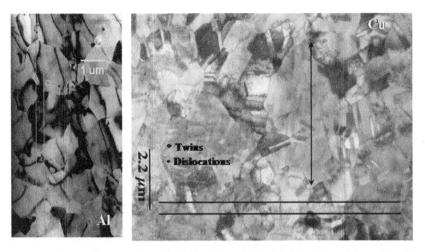

Figure 6.21. TEM micrographs of Al (left) and Cu (right) films. Magnification is approximately the same. Grain size in the Al film is relatively uniform and exhibits few defects or twins. Grains are easily discernable. Grain size is highly variable in the Cu film, and determination of individual grains is made very difficult by the presence of dense twin structures and dislocation networks. Variation in grain size is substantial; the largest grain in the photo is shown by the vertical line.

complete bamboo structure in one extreme, and almost continuous grain boundary paths at the other extreme. In Figure 6.21, many of the Cu grains are large enough to block any long range grain boundary diffusion, and only short lengths of line would be likely to have a continuous grain boundary path. And since many of the boundaries visible seem to be twin boundaries, which introduce only a rotation in the lattice instead of the bonding disorder found in general grain boundaries, they would not be likely to enhance mass transport very much.

Although the AlCuSi film is a dilute alloy, the impurities do not cause the Al grains to be very defective, mostly because both elements precipitate out, leaving the Al lattice relatively undisturbed. But while Cu is unalloyed, it is not pure. Organic additives to the plating bath can become trapped in the as-plated Cu and cause irregularities in the crystalline lattice. Even after annealing and the corresponding recrystallization and grain growth, these impurities are present in the Cu and the grain boundaries with a concentration that varies with location in the film. This fact has some implications for stress voiding in Cu lines.

Two of these impurities in the plating bath are called "brighteners" and "inhibitors" [28]. The purpose of the brightener is to limit grain growth in the film during deposition. Plating bath compositions for microelectronics are derived from commercial plating baths for more macroscopic applications, such as depositing patterns of wiring on printed circuit boards. The quality of the plating included its brightness, partially because tarnishing meant oxidation of the surface of the film, which oxidation caused solder bonding problems. However, if the film surface was very rough, it would appear dark instead of shiny, and could be confused with oxidation. The roughness was caused by uneven grain growth, which left a very uneven surface that reflected little light. To eliminate this problem, materials called brighteners were added to the plating bath to limit the grain growth. Inhibitors, in contrast, regulate the plating rate in order to reduce non-uniformities in film thickness.

Controlled grain growth is needed in microelectronics metallization as well, which means some form of brightener is retained. In fact, the additives play a very important role in producing uniformly filled vias. When plating begins on a surface containing vias as well as trenches, the concentration of the additives is essentially uniformly distributed in the bath. But as plating advances, the inhibitor concentration in the trenches and via holes becomes depleted because the bath circulation is insufficient to replenish it as it is consumed during plating. This means that the plating rate in the trenches, and especially in the vias, is faster than at the surface, and results in much faster filling of the trenches and vias [29]. Because of this feature, the formation of excess Cu at the top corners of the vias and trenches is minimized and no seam forms in either the vias or the trenches, as has been observed for features filled with PVD processes.

Although the vias and trenches are filled very uniformly, they still retain the brightener and thus the small grain sizes. And when the plated wafers are annealed, the small grains in the trenches are the last to grow. If incomplete grain growth occurs in the trenches and vias of the narrowest lines, and the lines are encapsulated, subsequent grain growth can be induced by subsequent dielectric depositions and metal anneals for subsequent metal levels. This grain growth will lead to an increase in the tensile hydrostatic stress in the Cu film, and if it becomes great enough, it will lead to the occurrence of stress voiding.

6.5.4   Structural Influences

Whereas the effect of stress-induced void formation in layered Al metallizations is limited to resistance increases, voids in both single and dual damascene Cu metallizations, as they have been practiced to date, can cause catastrophic open circuit failure. This is because no metal capping layer exists after the CMP polishing step, and the via from the succeeding metallization layer can land on a simple Cu surface. This situation is most likely to occur when a single via from the metal layer above contacts a line of width somewhat greater than the minimum design width. In the case of minimum width lines in the layer below, the via width from the metal level above is at least as wide as the line beneath it, and the liner from the via usually contacts and forms an electrical connection to the liner of the line below. This prevents failure by a catastrophic open circuit, but may not prevent failure by resistance shift. Sometimes the contact between the metal liners is tenuous enough that it is very resistive, and as soon as the void severs the Cu/via connection the resistance shift rise is much greater than 20%.

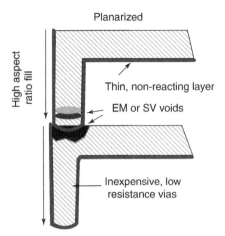

Figure 6.22. Vulnerability of lines and vias made with the damascene process. Lack of contact between liners can lead to open circuit failure when a void is located immediately under a via.

Another mode of failure that can occur in dual damascene Cu metallization that does not occur in multi-level Al metallizations is the formation of a void in the via. There are a number of possible ways this can occur, but only one description will be given here. If the etch process for the via produces a barrel-shaped via, for which the center of the via is somewhat wider than the top and bottom, the Cu seed layer deposition can become too thin, or even be missing, due to shadowing in the central portion of the via. Where the seed layer is inadequate, insufficient current is delivered to that location during plating and the plated layer is not continuous there. When the plating process is complete, it can leave small voided regions around the perimeter of the via. These voids act to reduce, or even eliminate, the stress locally in the via, and consequently become a sink for vacancies created in the via and in the attached line during grain growth. The void(s), which have arbitrary shapes initially, can re-form through surface diffusion into a single plate-like void in the middle of the via, as shown in Figure 6.22, and will cause a resistance increase. Because the liner is likely to be thinner on the via sidewalls, even a relatively small plate-like void through the Cu can cause a large resistance shift.

Via voids do not have to completely sever the Cu in the vias in order to become a reliability issue. If such a void has grown large enough to block a substantial part of the via area, and if the via is in a high-current path, electromigration can cause the void to grow and open the Cu before the intended end of life. However, such voiding problems tend to be systematic, and structures are purposely placed in the dicing channel of product wafers in order to detect them. For example, the resistance of a chain containing hundreds of vias might be measured, and resistance increases beyond specified limits would initiate process examinations. If enough vias contained voids, even though they occupied only a portion of the via, they would cause enough of a resistance shift to be detectable.

6.5.5   Structures and Models for Cu Voiding

Relatively few papers have appeared in the literature on stress voiding is Cu metallizations. One reason for this may be that much less stress voiding in encountered in practice than is the case for Al metallizations. As has already been discussed above, when Cu metallizations is properly made, the propensity for stress voiding is considerably less in Cu than in Al. When it does appear, it tends to be associated with vias, and there is often a fabrication process adjustment, such as metallization annealing conditions, that can be implemented to eliminate it. Because of this tendency to be associated with vias, long serpentine structures are not very effective for detecting stress voiding in Cu.

When voiding is detected during SV testing of Cu metallizations, the affected fraction of the population of parts is often small, making difficult the acquisition of sufficient data to extract kinetics information. So, in order to detect stress voiding better, special structures are designed that are especially sensitive. Examples of such structures are large plates above and below isolated vias and arrays of narrow lines connected to single vias. Structures involving vias are important because voiding under vias can cause open circuit failure in Cu metallizations. The large plates provide a reservoir of vacancies as well as a large volume of Cu within one diffusion length of the via, thus optimizing the availability of vacancies within range of the via. Often a range of plate sizes is used, since the larger the reservoir size, the larger the quantity of vacancies that can be drawn to the via. As a result, voiding may be detected for plates greater than a certain size but not for smaller plates. If this does occur and if process changes are not effective at eliminating the voiding, products made by the process being tested can be protected from stress voiding by disallowing vulnerable sizes from product designs.

The population of test structures that can be reasonably tested is bound to be only a small part of the huge population of vias present on a single chip in contemporary advanced technologies (let alone in the entire population of chips to be made), even for large arrays of vias in chains. Therefore, any voiding in the test population should probably be taken seriously. When process changes are not adequate to eliminate the voiding, another technique that can be used to reduce the risk of voiding is to employ two or more vias. Similar test structures having the same general layout, but having two or more vias in one case and only a single via in the other case, provide a means to compare the effectiveness of using more than one via. Two or more vias may completely suppress electrical effects of voiding, because once the void forms under one via, vacancies will be attracted to that void, making it larger, rather than acting to form a new void under the second via.

Two models have been advanced in the literature, the exponential model [30] and the Eyring model [31]. The exponential model contains no explicit stress dependence, and has the form

$$t_f = Ae^{-\frac{\Delta h}{kT}}. \tag{6.35}$$

This model was used to fit the voiding behavior recorded for a specific vintage of hardware. One explanation for why this model described the data could be that insufficient grain growth had been promoted during hardware fabrication. High-temperature storage stress led to additional grain growth, which caused the resultant voiding. Since grain growth is thermally activated, void growth kinetics would depend on a combination of grain boundary diffusivity and grain growth, and would depend very little on residual stress. In fact, applying temperatures that would cause the Cu with small grains to experience compressive stress would also act to stimulate grain growth through the minimization of strain energy; grain growth causes a volume reduction, which serves to relieve the compressive stress. So, after the grain growth saturates at a given temperature, an increase in temperature will generate more compressive stress and more grain growth until it is energetically unfeasible for more growth to occur.

Another explanation could be that the effective stress-free temperature for that hardware was substantially higher than any of the temperatures used for stress, but this contradicts the assumption that Cu plated at room temperature is essentially stress free.

The Eyring model, which was proposed originally for Al SV [3], does include stress dependence in a form very similar to that in Equation 6.19, except that the

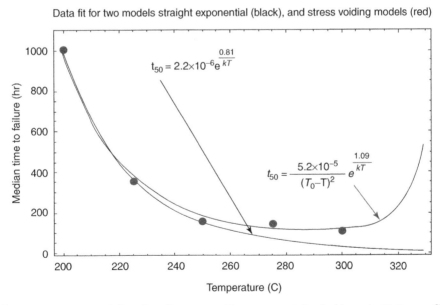

Data fit for two models straight exponential (black), and stress voiding models (red)

$$t_{50} = 2.2 \times 10^{-6} e^{\frac{0.81}{kT}}$$

$$t_{50} = \frac{5.2 \times 10^{-5}}{(T_0 - T)^2} e^{\frac{1.09}{kT}}$$

Figure 6.23. Stress voiding data for narrow line array terminated in a via. Data was fit by exponential model (lower curve) and Al stress voiding model (upper curve). Both different approaches show a similar fit for the lower temperatures, but the stress voiding model fits the high-temperature data somewhat better. Effective stress-free temperature $\sim 340^{\circ}C$ for the effective stress-free temperature.

value for the exponent to the ΔT factor, which is 2 in Equation 6.19 is replaced by the more general value of n. Ogawa, et al.[30] extracted a value of 3.5 for n from their failure data. The value of 2 for the ΔT factor is a direct consequence of assuming a diffusive nature to the mass transport, and provides a link to a physical process. However, leaving the value of n variable allows other stress related process to be included, without specifying their nature.

As a practical matter, both the exponential and Eyring models project to similar failure rates at use temperatures. Figure 6.23 shows the average time to fail for a structure intended to be sensitive to SV effects due to the incomplete grain growth. The data has been fit using both the exponential model and the Eyring model with $n = 2$. Both fits match the data relatively well for the lower temperatures. But the Eyring model with $n = 2$ fits the high temperature points better and gives and activation energy of $\Delta h = 1.09\,\text{eV}$. Projecting the failure rate to use conditions with the two models gives very similar results.

6.6   CONCLUDING REMARKS

Stress voiding is a failure mechanism in which voids are nucleated and grown in the primary electrical conductor on integrated circuit chips as the result of excessive stress created in the metallization. For conductors without refractory redundant layers and for Cu metallizations with no top redundant conductor, these voids can cause a chip to fail catastrophically. The stress is created by the unique processing in integrated circuits, where materials with very different thermal expansion coefficients are joined together at temperatures very different from product use temperatures.

For Al metallizations where abundant data exists, voiding can be modeled effectively assuming a diffusive mass transport mechanism, and most interesting features of voiding behavior can be included. For Cu metallizations, unless metallizations processing issues exist, voiding data is not so abundant. Voiding in Cu metallizations presumably follows the same physical processes as for Al metallizations, but some observed failure data behaves differently from that for Al. This difference is probably due in part to the differences in microstructure of the two metallizations, the differences in deposition processes, and the differences in metallic properties.

Voiding in Al metallizations was observed to occur at random locations throughout the metal, such that long serpentine structures were often suitable to detect a voiding susceptibility. Voiding in Cu tends to occur around vias. Large area pads and arrays of narrow lines connected to vias have been used to detect voiding in Cu. With each new generation of technology, additional sensitivities are being revealed, and new structures are likely to become necessary.

Stress voiding is observed because the magnitude of the stresses generated in microelectronics metallizations is large on the scale of conventional engineering stresses. It is an unusual failure mode because the stress state, tensile hydrostatic, is an unusual state of stress relative to conventional engineering problems.

In addition, with dimensions of the smallest Cu conductors approaching several hundred angstroms, conductors are only a few hundred atoms in width and can have aspect ratios of three or more. Both the reduction on width and the increased aspect ratio are likely to provide additional challenges for dealing with stress voiding.

REFERENCES

1. J. Curry, G. Fitzgibbon, Y. Guan, E. Muollo, G. Nelson, A. Thomas. New Failure Mechanisms in Sputtered Aluminum-Silicon Films. 22nd Ann. Proc. Reliability Phys. IEEE, NY: 1984, pp 6–8.

2. N. Owada, K. Hinode, M. Horiuchi, T. Nishida, K. Nakata, K. Mukai. Stress induced slit-like void formation in a fine-pattern Al-Si interconnect during aging test. Proc. IEEE VLSI Multi-Level Interconnect Conference:, 1985, p 173.

3. J. W. McPherson, C. F. Dunn. A model for stress-induced metal notching and voiding in very large-scale integrated Al-Si (1%) metallization. J. Vac. Sci. Technol. B 5(5): 1987, 1321–3125.

4. J. T. Yue, W. P. Funston, R. V. Taylor. Stress induced voids in aluminum interconnects during IC processing. 23rd Ann. Proc. Reliability Phys. 1985, Orlando. IEEE, New York: 1985, 126–137.

5. R. O. Simmons, R. W. Balluffi. Measurements of equilibrium vacancy concentrations in aluminum. Phys. Rev. B 117(1): 1960, 52–61.

6. T. D. Sullivan. Thermal dependence of voiding in narrow Al microelectronic inter-connects. Appl. Phys. Lett. 55(23): 1989, 2399–2401.

7. C.-Y. Li, R. D. Black, W. R. LaFontaine. Analysis of thermal stress-induced grain boundary cavitation and notching in narrow Al-Si metallization. Appl. Phys. Lett. 53(1): 1988, 31–33.

8. F. G. Yost, D. E. Amos, A. D. Romig. Stress-driven diffusive voiding of aluminum conductor lines. 27th Annu. Proc. Reliabilty Phys. 1989, Phoenix. IEEE, New York: 1989, pp 193–201.

9. W. D. Nix, A. I. Sauter. Modeling void growth and failure in passivated metal lines under stress and electromigration conditions. In: *Stress-Induced Phenomena in Metallization, First Int. Workshop, Ithaca, NY, AIP Conf. Proc. No. 263*, edited by Li, Totta and Ho. AIP, NY: 1991, pp 89–104.

10. S. M. Hu. Stress-driven void growth in aluminum interconnection lines. Appl. Phys. Lett. 59(21): 1991, 2865–2687.

11. C. Y. Li, P. Borgeson, T. D. Sullivan. Stress-migration related electromigration damage mechanism in passivated, narrow interconnects. Appl. Phys. Lett. 59(12): 1991, 1464–1466.

12. P. S. Ho. J. Appl. Phys., 41(64): 1970.

13. S. Rauch, T. D. Sullivan. Modeling stress-induced vod growth in Al-4wt% Cu lines. In: Submicron Metallization: Challenges, Opportunities and Limitations. Int. Soc. for Optical Eng. (SPIE), Bellingham, WA, Vol. 1805: 1992, p. 197.

14. M. Braun. *Differential Equations and Their Applications.* Springer Verlag, New York: 1975, p 453.

15. S. Ghandhi. *The Theory and Practice of Microelectronics*, App. A. John Wiley and Sons, New York: 1986.

16. P. Shewmon, *Diffusion in Solids*, Second Edition, TMS, Warrendale PA: 1998.

17. H. Okabayashi. IEEE Trans. Electron Devices 40(4): 1993, pp 782–788.

18. T. Yamaji, Y. Igarashi, S. Nishikawa. In: 29th Ann. Proc. Reliability Phys. IEEE, New York: 1991, pp 84–90.

19. H. J. Frost, J. F. Ashby. *Deformation-Mechanism Maps*. Pergamon, New York: 1982.

20. F. R N. Nabarro. In: Rept. Conf. Strength Solids. Physical Society, London: 1948, p 75.

21. J. M. Blakely. *Introduction to the Properties of Crystal Surfaces*. Pergamon, New York: 1973.

22. J. Mayumi, T. Umemoto, M. Shishino, H. Nanatsue, S. Ueda, M. Inoue. 1987. The effect of Cu addition to Al-Si interconnects on stress induced open-circuit failures. 25th Ann. Proc. Reliability Phys. 1987, San Diego. IEEE, NY: 1987, pp 15–21.

23. C.-K. Hu, K. Rodbell, X. Lee, T. D. Sullivan, D. P. Bouldin. IBM J. of Res. and Dev.

24. K. Hinode et al., A study on stress-induced migration in aluminum metallization base on direct stress measurements. J. Vac. Sci. Technol. B 8(3): 1990, 495–498.

25. A. Sauter, W. Nix. Finite element calculations of thermal stresses in passivated and unpassivated lines bonded to substrates. Mat. Res. Soc. Symp. Proc. Vol. 188. Materials Research Society: 1990, pp 15–20.

26. J. M. E. Harper. et al., Mechanisms for microstructure evolution in electroplated copper thin films near room temperature, J. Appl. Phys. 86 (5): (1999), pp 2516–2525.

27. P. R. Besser. Mechanical Strains and Stresses in Aluminum and Cu Interconnect Lines for 0.18 μm Logic Technologies in *Stress-Induced Phenomena in Metallization, Fifth International Workshop, Stuttgart, Germany, AIP Conference Proceedings 491*, Ed. O. Kraft, E. Arzt, C. Volkert, P. Ho, H. Okabayashi, AIP, New York: 1999, pp 229–239.

28. P. C. Andricacos, et al., Damascene copper electroplating for chip interconnections, IBM J. of Res. And Development (1998): pp 567–574.

29. M. E. Gross. Considerations for electroplated copper for sub-micron interconnects in advanced integrated circuits, Proc. Of the AFSF, SURF/FIN '99 Annual International Technical Conference, American Electroplaters & Surface Finishers Soc., Inc., 1999, p 1; M.E. Gross, K.M. Takahashi, Transport phenomena that control electroplated copper filling of submicron vias and trenches, J. Electrochem. Soc. 146 (12) (1999): pp 4499–4503.

30. A. H. Fischer, A. E. Zitzelsberger. The quantitative assessment of stress-induced voiding in process qualification, 39th Ann. Proc. Reliability Physics, 2001, Orlando, IEEE, New York: 2001, pp. 334–340.

31. E. T. Ogawa. et al., Stress-Induced voiding under vias connected to wide Cu metal leads, 40th Ann. Proc. Reliability Physics, Dallas, IEEE, New York: 2002, 312–321.

32. T. D. Sullivan. Reliability in copper metallization and low k dielectrics for ULSI interconnects. IEEE IEDM Short Course, Reliablility for Logic and Memory Technologies. A. Oates, Organizer, J. L. Hoyt and D. B. M. Klaassen, Chairpersons, 1998.

7

ELECTROMIGRATION

Timothy D. Sullivan

7.1 INTRODUCTION

The reliability of integrated circuit chip metallization has had a long history. With almost every generation of chips, the metallization has produced new behaviors that needed to be characterized and quantified in order to project the metallization reliability. One might wonder why the testing can not be performed just once to determine the reliability. The answer is complex. The first reason is that advances in integrated circuits (which have been occurring for several decades) depend upon continual reduction in the size of the minimum features used to define circuit components. Physical properties of the materials used to fabricate circuit chips, well defined for macroscopic applications, are altered in the thin films and small dimensions used in integrated circuits, and materials with unlike properties are bonded together and combined in unusual ways such that substantial differences in physical properties occur in distances ranging from a few angstroms to a few microns. Each film is deposited at its own characteristic temperature, and when neighboring films with greatly different thermal expansion coefficients and/or elastic moduli are subjected to large temperature excursions, unusually large mechanical stresses can arise.

These high stresses exacerbate deleterious effects, and the interfaces between different materials become of paramount importance. In order to accommodate these effects, new materials and processes are introduced, and each generation of circuits requires its own fabrication process. We call the particular combination of

Reliability Wearout Mechanisms in Advanced CMOS Technologies. By Strong, Wu, Vollertsen, Suñé, LaRosa, Rauch, and Sullivan

dimensions, materials and fabrication process a *technology*, and for a given fabricator, the technology is often identified by its minimum feature size. Thus, for each technology, a different combination of materials, processes, and dimensions exists, and the devices made with this combination need to be tested to determine their behavior so that we can predict their reliability over the intended lifetime.

7.2 METALLIZATION FAILURE

The wiring on microelectronics circuit chips is used for a number of purposes, including routing signals into and out of the chip or from one part of the chip to another, forming networks of devices on the chip, routing power to various devices on the chip, forming capacitors and inductors, and serving as interfaces to external connections such as solder bumps and wires. On-chip wiring is referred to by many names including metallization, interconnects, lines and wires. Failure in on-chip wiring falls into four categories: open circuit failure, elevated resistance failure, short-circuit failure, and leakage. Open-circuit and short-circuit failures are obvious, because either the line becomes discontinuous and can not transport power or signals, or the signal or power are lost through inadvertent connection to another line. Relatively elementary tests of chip functions are sufficient to detect electrical opens and shorts. Elevated resistance and leakage failures are less obvious because neither a clear electrical discontinuity nor an obvious short is apparent, yet the chip fails to perform some operation successfully. Elevated resistance can cause failure in timing (because the signal is retarded by the increased resistance), failure in memory retention (insufficient current to charge up a capacitor), or erratic device switching. Leakage can cause failure by decreasing the signal supplied to a device to below the threshold needed to trigger the device, or by allowing the stored charge in a memory cell to drop below that needed for detection. Such failure modes are often frequency dependent and require relatively sophisticated chip tests to detect.

Interconnect-induced failures have several different causes including corrosion, mechanical fracture (generally introduced by differences in thermal expansion between adjacent materials during thermal cycling), interfacial degradation (between different metal layers, or between metal and dielectric layers), and void and extrusion formation through metal self-diffusion. All of these can occur during chip life, after the chip has been installed in a machine and has been functioning properly, and so all fall into the category of reliability failures. Corrosion, mechanical fracture, interfacial degradation, and the formation of extrusions or filaments may have a variety of causes that occur for a short duration during manufacturing, but then they do not reappear. These are classified as defects introduced during the manufacturing process. Although they can influence failure mechanisms and may impact reliability, defects are not, in themselves, failure mechanisms. For failure mechanisms, extensive testing is performed during the development of the metallization in order to detect process weaknesses, and process adjustments or revisions are made to eliminate them if they occur.

Except when caused by electromigration-induced extrusions, leakage of charge through the dielectric is more properly a property of the interlevel dielectric (ILD), and will not be addressed here. The formation and growth of voids in metallization can occur by two different failure mechanisms, electromigration (EM) and stress voiding, both of which have reasonable physical models. In this chapter, electromigration will be discussed.

7.3 ELECTROMIGRATION

In most electronics and physics courses, conductors are viewed as conduits for transmission of electrical current forced by a voltage. The conductor itself is thought of as remaining unaffected by the flow of current through it; for most cases most of the time, this is true. However, in the case of wires formed out of the thin-film metallization on microelectronic chips, the current per unit area (current density) is very large, especially when compared to common house wiring, and the dimensions are very small (on the order of $1 \, \mu m^2$ or less for chip wiring versus mm^2 for house wiring).

Electrical failure for most macroscopic applications occurs when applied currents are high enough to cause wires to overheat and burn out or cause adjacent materials to catch fire. The heating, known as Joule heating or resistive heating, is given by the power dissipated in the wire, I^2R, where I is the applied current and R the resistance of the wire. The same phenomenon can also occur in the microscopic environment, but because the wiring is embedded in a hard dielectric, such as SiO_2, and attached to a thermally conductive Si substrate, the power dissipated in the wire is rapidly conducted away and the wiring is kept from burning out until much higher current densities are reached.

For reasons that will become clear in the following pages, Joule heating in the metallization of semiconductor circuit chips is carefully controlled. Failure by wire burn-out is unusual, and only occurs when the chips are subjected to conditions beyond their application specifications. Aspects of Joule heating will be discussed at the end of this chapter. However, even when insignificant Joule heating occurs ($< 1 °C$) electrical current can still cause metal lines to fail.

An important consequence of the higher current density in chip metallization is the phenomenon of electromigration, which is the movement of metal atoms (Al, AlCu or Cu for our purposes) in response to an applied current. When extensive electromigration occurs, holes (voids) or extrusions form in the wires and eventually lead to open-circuit or short-circuit failure of the wire.

Extrusions take the form of filaments or plates of metal that are pushed into the dielectric surrounding the wire under test. Whether extrusions appear in the bulk dielectric or at dielectric interfaces, the dielectric is usually cracked by tensile stress in the dielectric, which grows in response to the buildup of compressive stress in the metal. Once a crack is formed, metal can be forced out of the line into the crack and progressively widen and expand it.

Extrusions are formed as the result of EM but not as a result of stress voiding. Chip failure can occur as a result of extrusions shorting out neighboring circuits when the metal source is a relatively large reservoir containing multiple contacts or vias. Extrusions generally form long after voids, such that for electromigration test structures with a single via at the cathode end of the line, failure will occur by an increase in resistance rather than by shorting. However, although common practice is to design EM test structures such that both failure modes can be monitored, some types of extrusion failures are difficult to detect electrically without practically encasing the test line in metal. Figure 7.1 shows a cross section of an AlCu layered metallization with extrusions that angle upward from the corners of the line. An overlying metal layer, much wider than the failing line, would have been necessary to detect this extrusion electrically. As a practical matter, covering test structures with metal layers increases the difficulty in performing physical failure analysis (PFA, or simply FA), because the failure site is not visible optically.

We use current density, or current per unit area, as the quantity of preference because it is normalized over all line widths and thicknesses. Although current is defined as the motion of positive charge, in metals the negative charge (electrons) actually moves, and the resultant atomic migration is usually in the direction of electron flow. The term "electron wind" is a common term used to denote the force causing the movement of atoms without precisely describing the actual details of the interaction between the electrons and atoms.

Simple ballistic calculations seem to indicate that a single electron does not have enough mass to influence the motion of an atom (an Al atom has a mass about 54,000 times as great as that of a single electron, and a Cu atom is even more). Additional calculations of the number of electrons that would hit an atom, even at the elevated current densities found in electromigration, seem to be insufficient to cause the effect as well. Nonetheless, it occurs.

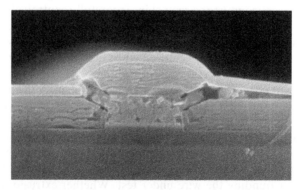

Figure 7.1. EM extrusion in a layered AlCu metal line. The extrusion would not have been detected by a monitor line at the same level as the test line.

TABLE 7.1. Diffusion Parameters for Common Metallization Components[a]

Element	Prefactor		Activation Energy	
	δD_{ob} (cm^3/sec)	D_{ob} (cm^2/sec)	Q_b (kJ/mole)	Δh (eV)
Al	5.0e-14	3.3e-6	84	0.87
Cu	5.0e-15	3.3e-7	104	1.08
β-Ti	5.4e-17	3.6e-9	153	1.59
Ta	5.7e-14	3.8e-6	280	2.90
W	3.3e-13	3.3e-13	285	3.99

[a]Values for δD_{ob} and Q_b taken from *Deformation-Mechanism Maps, The Plasticity and Creep of Metals and Ceramics,* by Harold J Frost, and Michael Ashby, Pergamon Press, New York, 1982.

At a temperature of absolute zero, crystalline materials are in their lowest energy state in which no translational motion of atoms occurs, and for which all positions within the crystal lattice are filled. However, at temperatures above absolute zero, a small fraction of atoms is replaced by vacant sites (vacancies) into which neighboring atoms can jump. When an atom jumps into the vacancy, the location formerly occupied by that atom becomes vacant; both atom and vacancy move. (This is analogous to electrons and holes in electrical conduction through semiconductors.) Over a long enough period of time, a single atom can move a significant distance from its initial location, but the motion is random, so that no net displacement of atomic mass occurs. This method of atomic transport is known as substitutional diffusion, and in a pure metal is called self-diffusion. It is a property of the material, as shown in Table 7.1, and depends on temperature, because both the population of vacancies and the likelihood of an atom making a jump depend on temperature.

The atomic movement caused by electromigration is generally agreed to occur by diffusion, with diffusivity equal to, or at least proportional to, that of the inherent self-diffusivity of the metal. In order to better understand this idea, a short discussion of the atomic arrangement of crystalline materials is needed here.

We often think of crystalline materials as being like precious stones, such as diamonds, rubies, sapphires, etc., and therefore associate the term "crystal" with the property of translucence. However, the term crystalline applies more broadly to any material whose atomic arrangement follows a three-dimensional repeating pattern or matrix, and metals are in this class. But metals generally occur in polycrystalline form. That is, instead of the metal being one single crystal, it is composed of a collection of smaller crystallites (called grains) that are bonded to each other at interfaces called grain boundaries. Each grain has a different orientation from its neighbors, such that the atomic arrangement is discontinuous across the grain boundary (see Fig. 7.2). This discontinuity disorders the bonding of atoms across the boundary and means that the atomic packing at the boundary is less dense than in the interior of the grain. Also, the vacancy concentration is greater at the boundary than in the interior of the grain. Both of these factors act to increase the diffusivity at the grain boundaries (grain boundary diffusion)

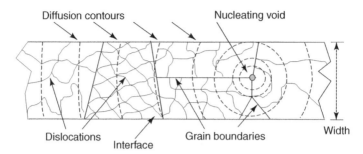

Figure 7.2. Sketch of a cross-section through a metal line showing grains with grain boundaries, dislocations, and interfaces with neighboring materials.

relative to that in the grain interior (lattice diffusion). Since most metallization lines are at least partially polycrystalline in structure, the values for diffusivity in Table 7.1 are all for grain boundary diffusion.

Mass transport (net movement of atoms) can occur by four different paths in metal lines on chips. The paths are lattice diffusion, pipe diffusion, grain boundary diffusion, and interfacial diffusion. Lattice diffusion is diffusion in the interior of the grain, as just discussed. Pipe diffusion refers to diffusion along a dislocation (a linear defect in the lattice). It is much more rapid than lattice diffusion, but the actual mass transport depends on the density of dislocations. Grain boundaries, at least when the angles of misorientation between the grains on either side of the boundary are less than eight degrees or so, are composed of arrays of dislocations. So the mass transport along grain boundaries can be very rapid. Above eight degrees, the dislocations are no longer distinguishable, but the order is even worse, so the diffusivity is greater. Interfacial diffusion is similar to grain boundary diffusion, except that the vacancy concentration depends on the bonding density and order between the metal and the neighboring material (such as SiO_2). In chip metallizations, two common interfaces are 1) between the metal of the primary conductor and the insulator or 2) between the primary conductor and a refractory metal.

In many cases, all four diffusion paths are present and active. Then the fastest diffusion path usually dominates the net mass transport. The migration of metal atoms by itself is generally not a problem. For all temperatures above absolute zero, some degree of translational atomic motion always exists, but since it is random, the bulk metal remains essentially unchanged. However, under the influence of a constant, unidirectional current (DC), a preferred direction to the motion is produced by the electron wind.

Al is an ideal metal to use for semiconductor chip wiring. It is a very good conductor (only Ag, Au, and Cu are better), it is inexpensive and readily available, and has a relatively low melting point, which makes it easy to evaporate and deposit. In addition, it rapidly develops a thin native oxide that protects it from further oxidation and from attack by a variety of otherwise corrosive substances including moisture. And for the initial current densities and wire dimensions used in integrated circuits, Al performed very well.

However, the combination of progressively shrinking dimensions coupled with the combination of materials and process steps used in semiconductor chip fabrication worked together to reveal weaknesses in pure Al conductors. As a result, small amounts of other elements such as Si, Cu, Ti, Hf, and V were added to Al to produce different alloys with properties that would remove the weaknesses of pure Al. Each of these elements had an effect on the electromigration performance of the lines [1], and different alloys were employed by different fabricators. Those containing Si and Cu were most common.

Typically, the Al alloys are evaporated as a single film. For purposes of this discussion, an Al alloy film will be classified as a single component metallization, even though there may be several elements contained in the alloy film.

7.3.1 Single-Component Metallization

When electromigration was initially observed in integrated circuit chip metallization, only one level of metallization was being used for chip wiring, and the wiring width was considerably larger than the film thickness. It was only natural, therefore, for test structures for EM to be made of single level metal, and to consist of long narrow lines attached to much wider segments at either end. The wider segments were connected in turn to bond pads, as shown in the so-called NIST[1] structure in Figure 7.3. The wider line segment is tapered in transition to the narrow line to avoid current crowding effects. Application of a DC current in excess of about $0.5\,\text{mA}/\mu\text{m}^2$ to such structures results in a net mass movement in the direction of electron flow that causes a mass depletion at the negative (cathode) end of a line (producing voids) and a mass buildup at the positive (anode) end (producing hillocks). Formation of these voids and hillocks at opposite ends of the line means that considerable mass transport along the narrow line segment must occur. Voids that form in the narrow segment of the line can grow until they become large enough to make the line discontinuous, creating open-circuit failure.

Characterization of void growth is difficult for single-level, single component metallizations. In electromigration tests, either the voltage or the current is applied, depending on the configuration of the power source, and resistance is

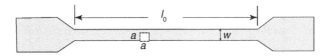

Figure 7.3. Single layer metallization electromigration test structure layout, commonly referred to as a NIST (see footnote) structure, having length l_0 (narrow section) and width w, and a square void of side "a".

---

[1] NIST stands for National Institute of Standards and Technology, formerly the National Bureau of Standards.

measured as a function of time. Very little change in resistance is detectable until a void grows to well over half the line width. The change in resistance as a function of void size is difficult to calculate exactly for a void having an arbitrary shape, but the general behavior can be easily understood with a simple example.

Consider the line of length l_o and width w shown in Figure 7.3, and imagine that a square void (with side a) has grown in the line. The overall resistance of the unvoided line is $R_o = \rho\, l/A$. The voided line will have two segments, one of length (l_o-a) and width w, and the other of length a and width $(w-a)$. The resistance will be the sum of the resistances of the two segments (see Eq. 7.1).

$$R = \frac{\rho(l-a)}{tw} + \frac{\rho a}{t(w-a)} = \frac{\rho}{t}\left[\frac{l}{w} + \frac{a^2}{w(w-a)}\right]$$

$$= R_o + \frac{\rho l}{tw}\frac{a^2}{l(w-a)} = R_o\left(1 + \frac{a^2}{l(w-a)}\right)$$

$$(7.1)$$

The first term in the first line of Equation 7.1 is the resistance of the line without the segment containing the void. The second term in the first line is the additional resistance caused by the void. For values of $a = 0.8\, w$, and $l = 800\,\mu m$, the increase in resistance is about 0.4%. A common criterion for line failure is a 20% increase in resistance. To achieve this would require that the void be about 160/161 of the width of the line (see Fig. 7.4). Since void shapes tend to be irregular and rounded, even more of the width would have to be voided to register a 20% resistance increase. In practice, as the void diameter approaches the line width, the current density in the remaining filament increases tremendously, raising the local temperature and vastly accelerating the migration process. In addition, for long narrow lines, several voids are often present at once, such that more of the line width is still present at failure than would be the case for a single void. For both of these reasons, characterization of the void growth rate in single level, single component lines is difficult.

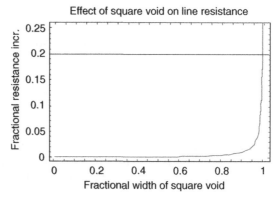

Figure 7.4. Fractional resistance change versus the width of a square void in 800-μm line.

The regular movement of Al atoms in the direction of the electron wind would be harmless in an infinitely long line, or in a line attached to a large metal reservoir at the cathode end of the line, because in any given element of volume, new atoms would be constantly arriving to replace atoms that are carried away. But mass transport by the electron wind is not uniform at all locations along a metal line. This is partially due to the polycrystalline nature of the metal microstructure. The metal is composed of grains joined together at grain boundaries. Grain shape, grain boundary orientation and grain boundary properties vary considerably on a microscopic scale. Atoms travel more rapidly along grain boundaries than within grains, and the grain boundaries themselves allow different degrees of mobility, depending on the orientation of the grains on either side of the boundary.

We define the atomic flux, J, as the number of atoms that pass through a specified cross sectional area in a unit of time (at/cm^2-sec), as shown in Figure 7.5. In the event that more atoms arrive in the volume element than are carried away, metal volume will build up and eventually cause a visible hillock or extrusion to develop. Conversely, if more atoms are carried away than arrive at the location, a void develops and grows until the Al becomes discontinuous. The difference between the atomic flux into the volume element and the atomic flux out is called the flux divergence and is given by

$$divj = \frac{j_{in} - j_{out}}{\Delta x} = \frac{dj_x}{dx}. \tag{7.2}$$

In single layer metallizations, such as AlCu or AlSi layers, grain boundaries and grain boundary triple points can be flux divergence sites. For example, at a grain boundary triple point, such as that shown in Figure 7.6, a single grain boundary enters the triple point, and two grain boundaries exit the triple point. Supposing that the electron wind "blows" to the right, and assuming that the atomic flux has the same magnitude in all three grain boundaries, more atoms will leave the triple point than enter it, and a void will form. If the single boundary were on the right and the two boundaries were on the left, then a hillock would form.

In situ studies of wide EM lines, performed in scanning electron microscopes, show void/hillock pairs separated by only a few microns. The void is always

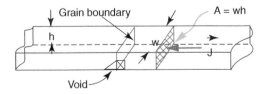

Figure 7.5. Atomic flux, J, is defined as the number of atoms that flow past an area $A = wh$ per unit of time. In the case of electromigration, the atoms generally flow in the direction of electron flow.

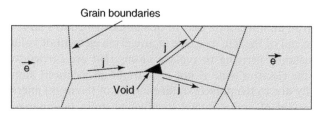

Figure 7.6. An example of a grain boundary triple point that serves as a flux divergence site. If electrons flow to the right, and if the magnitude of the atomic flux along the grain boundaries indicated is the same, more atoms flow out of the triple point than into it, and a void appears.

upstream (on the negative side) of the hillock. In metallizations such as Al, AlSi, or AlCu, such voids move upstream, opposite the electron wind, because atoms traveling along the interior surface of the void are eroded away from the upstream side and accumulate on the downstream side. Smaller voids travel faster than larger voids (since less time is required for atoms to travel from one end of the void to the other), and overtake and combine with larger voids. This process eventually produces a void large enough to kill the line [2].

The atomic flux is generally given as the product of the atomic mobility and the applied force due to the electron wind. The atomic mobility is given by the Einstein relationship, D/kT, and the electron wind force is given by $Z^*eEj\rho$, so that

$$J = n\mu F = \frac{nD}{kT} Z^* eE\rho j, \qquad (7.3)$$

where n is the number of atoms in motion, μ is the atomic mobility, F is the force due to the electron wind, D is the atomic diffusivity, k is Boltzmann's constant, T is the temperature in Kelvin, Z^* is the effective charge of the metal, e is the electronic charge, E is the electric field, j is the current density and ρ is the resistivity of the metal. An excellent treatment with considerable data for several metals is available in Chapter 8 of *Thin Films: Interdiffusion and Reactions*, edited by J. M. Poate et al. [3].

7.3.2 Layered Metallizations

In contemporary Al technologies, which use layered or sandwich metallizations such as Ti/AlCu/Ti, or lined metallizations, such as damascene Cu with refractory liners in the trench, the growing void generally will not cause an immediate open circuit when it severs the primary conductor (Al or Cu), because the refractory layers do not migrate and continue to provide a conductive path. However, the resistance of such refractory layers is considerably greater than that of Al or Cu (e.g., 40 μΩ-cm versus 3 μΩ-cm for Ti and Al, respectively), so a large void will cause a substantial increase in the line resistance, which can cause circuit failure by slowing signals so much that they reach their destinations too late to interact with other signals within the chip.

Anything that increases the translational motion of metal atoms will decrease the time-to-failure. Therefore the failure time decreases as the current density increases, since a higher current density will cause a larger atomic flux. Likewise, we expect failure time to decrease as temperature increases, because the diffusivity increases with temperature. Diffusivity is given by

$$D = D_o e^{\frac{\Delta h}{kT}}, \tag{7.4}$$

where D_o is the exponential prefactor in cm^2/sec, Δh is the activation energy for diffusion in eV, k is Boltzmann's constant, and T is temperature in degrees Kelvin. Therefore, one might expect that

$$t_{\text{fail}} \propto \frac{1}{jD_o} e^{-\frac{\Delta h}{kT}}, \tag{7.5}$$

where j is current density.

An empirical relationship, known as Black's law, is usually used to describe failure by electromigration, and is given by

$$t_{50} = \frac{A}{j^n} e^{\frac{\Delta h}{kT}}, \tag{7.6}$$

where t_{50} is the median time-to-failure (time at which 50% of a tested population has failed), n is the current density exponent (which had the value of 2 in Black's original model), and A is a constant. The origin of this equation can be found in Black's original paper [4]. In practice, the value of n depends both on the metallization and the structure being tested.

Excepting an extra factor of j, Equation 7.6 is very close to Equation 7.5, especially if $A = 1/D_o$. The value of $n = 2$ was empirically determined for the case of single level lines (NIST-type structures) and does not necessarily apply to all cases. For Al and AlSi lines where the line width is larger than the average grain size, the additional factor of j may arise from void motion. Voids move during EM through diffusion along the void surface, as mentioned above. Atoms from the negative (upstream) side of the void flow to the downstream side, causing the void to move in the direction opposite that of the electron (and atomic) flow. Atomic displacement from one side of a void to the other is faster in a small void (since the distance traveled is smaller) so that small voids move faster than large voids. As the smaller voids catch up to a larger void, they often merge with it to form an even larger void, and this process eventually leads to failure [5]. As will be discussed in the following section, in AlCu metallizations, the additional factor of j can be related to the behavior of Cu in the Al. However, in pure Al or pure Cu, the current density exponent is closer to one.

With the addition of refractory layers to the metallization, characterization of void growth becomes considerably less difficult, because the refractory layer provides a secondary conductive path when the Al is severed. Refractory layers were added for reasons other than to improve EM performance (but have been a benefit to electromigration performance nonetheless). For example, metallization contacts to devices (transistors, etc.) in the Si were formerly made directly through the oxide by etching a small hole, or via, in the oxide to permit the deposited Al to contact the diffusions of the devices. One of the first problems encountered with such a contact was the phenomenon of "alloy penetration" or "spiking" [6]. Si is sufficiently soluble in Al that some of the Si in the contact area was absorbed into the Al, leaving behind small pyramidal pits formed in the diffused region of the Si. Al atoms diffused into the pits to occupy the spaces left by the Si. This process initially caused no problems because the depths of the diffused junctions were much deeper than the Al spikes. But as chip features were decreased, so were the diffusion depths, and eventually the pits containing Al extended below the diffusions and caused leakage and shorting. When these parts were delayered, the etch pits were discovered, and the phenomenon was called spiking. In order to prevent the spiking, 1–2% Si was introduced as an alloying element into the deposited Al.

Feature sizes continued to shrink, and the contact areas shrank as well. Eventually, the contact resistance began to increase beyond acceptable levels, and when the contacts were delayered, the cause was found to be the presence of Si precipitates that had grown epitaxially on the exposed Si surface. The precipitates were undoped, and therefore resistive rather than conductive, and effectively reduced the contact area, which correspondingly increased the contact resistance. A solution to this problem was to insert a refractory layer, such as Ti or TiW, between the AlSi and the Si. These layers are highly resistive, but because they are thin, they do not increase the contact resistance very much. Additional advantages of adding a refractory layer are to improve the adhesion between the Al and the underlying oxide and to provide secondary conductive layer (redundancy) to the interconnect.

The presence of this redundant layer also enables the characterization of void growth during electromigration, since it permits electrical conductivity to be maintained once the Al layer has been completely severed. But because multiple voids can form in long, single-level structures (NIST-type structures), and because the voids move, plots of resistance versus time of lines being tested are very noisy and irregular before failure. It is not uncommon for the plot to show sharp increases of resistance and subsequent decreases to near the initial resistance, and for this to happen several times before the resistance finally grows continuously to catastrophic failure. And depending on the thickness of the redundant refractory layer, catastrophic failure may occur at 100% resistance increase or more.

A diagram of a cross-sectional view through a layered single level metal line is shown in Figure 7.7. Using the same approach applied to a single-component metallization (see Eq. 7.1) for evaluating the resistance increase due to the presence of a void, the resistance vs. time behavior of a layered structure can also be calculated. The total resistance is given by the sum of the resistance due to the void and the resistance of the remaining line. Strictly speaking, the resistance

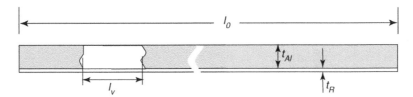

Figure 7.7. Diagram of a cross-sectional view of a layered metal interconnect of length l_o. The interconnect comprises an AlCu layer of thickness t_{Al} deposited on a thin refractory layer of thickness t_R. A void in the AlCu layer of length l_v is shown on the left end of the line.

of the unvoided sandwich metallization is found by adding the conductances of the two layers and inverting, and will reduce the overall resistance of the line. The result is shown in Equation 7.7. The factor in parentheses in the third expression gives the resistance reduction provided by the refractory underlayer. However, because the refractory cladding layer is both thin and very resistive, this factor is small. For an Al thickness of $0.5\,\mu m$, cladding layer thicknesses of $100\,\overset{\circ}{A}$, a resistivity for AlCu of $3\,\mu\Omega$-cm, and that for the cladding layer of 40 (for Ti for example), the value of this factor is about 0.9985. In order to simplify the following discussion, therefore, the resistance of the unvoided portion of the line will be taken to be that of the AlCu.

$$R' = \left(\frac{t_R w}{\rho_R l} + \frac{t_{Al} w}{\rho_{Al} l}\right)^{-1} = \frac{l}{w}\frac{\rho_R \rho_{Al}}{(t_R \rho_{Al} + t_{Al}\rho_R)} = \frac{\rho_{Al}l}{t_{Al}w}\left(\frac{1}{1 + \frac{t_R \rho_{Al}}{t_{Al}\rho_R}}\right). \tag{7.7}$$

Then the additional resistance due to the void is just the resistance of the refractory layer, $R_R = \rho_R l_v / t_R w$, while the resistance of the remaining line is $R_{Al} = \rho_{Al}(l_o - l_v)/t_{Al}w$, so that the change in resistance is given by

$$\frac{\Delta R}{R} = \frac{(R_{Al} + R_v - R_{Al0})}{R_{Al0}} = \frac{l_v}{l_o}\left(\frac{t_{Al}\rho_R}{t_R \rho_{Al}} - 1\right). \tag{7.8}$$

Here, the first term in the parentheses on the right is much larger than 1, so that that the 1 usually can be neglected, and the resistance change is proportional to the length of the void. Note here that the resistance depends on both the properties of the materials (ρ) and the structure geometry (t, l).

$$\Delta R \cong R_o \left(\frac{\rho_R}{\rho_{Al}}\right)\left(\frac{t_{Al}}{t_R l_o}\right)l_v \tag{7.9}$$

Assuming that the void has the same width as the line, the volume of the void is equal to the product of the atomic flux, J, out of the void, the line cross-section, A, the atomic volume, Ω, and the time, t, taken for the void growth.

$$V = l_v w t_{Al} = JA\Omega t = Jwt_{Al}\Omega t, \tag{7.10a}$$

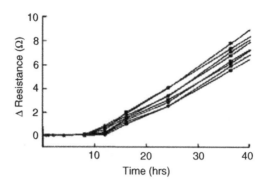

Figure 7.8. Resistance change versus time plot for electromigration in a layered Al line. After an initial interval of no activity, the plot is linear in time for short times. Reprinted with permission from Ref. 26, © 1998 IEEE.

and so

$$l_v = J\Omega t. \tag{7.10b}$$

Substituting the right-hand side of Equation 7.3 for J, we obtain

$$l_v = n\frac{D}{kT}Z^*e\rho\Omega t\, j, \tag{7.11}$$

which shows that the void length is linear both in current density and in time. This resistance versus time behavior is, in fact, observed, as shown in Figure 7.8. The delay between the beginning of the test and the onset of the resistance increase is the time needed for the void to grow large enough to sever the Al layer.

7.3.3   Short-Length Effect

An interesting effect that was first observed by Blech [7] occurs for short lines. Blech studied the rate of void growth in Al lines with a TiN underlayer as a function of line length. He found that, for a given current density, there is a line length below which no electromigration would occur, or alternately for a line of a given length, there is a current threshold below which no electromigration would occur. The length is commonly known as the Blech length, and the product of the Blech length and the current density threshold is a constant,

$$jL_b = \text{const.} \tag{7.12}$$

where j is the current density and L_b is the Blech length. The Blech length is thus inversely proportional to the current density. The ΔR versus t plot in Figure 7.9 illustrates the electrical behavior due to the short length effect. Initially, resistance increases linearly with time. Then the rate of resistance increase slows, and eventually saturates. This behavior corresponds to the nucleation and growth of the void until it saturates.

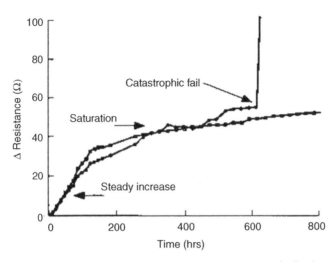

Figure 7.9. Resistance change versus stress time showing both the linear increase region and the saturation region. Reprinted with permission from Ref. 26, © 1998 IEEE.

In Blech's experiment, the Al line was made shorter than the TiN underlayer, such that the applied current would flow initially into the TiN. When it reached the Al/TiN couple, then nearly all of the current would flow through the Al because the resistivity of the Al is much less than that of the TiN. Electromigration was measured by observing the displacement of the cathode end of the Al line with respect to its initial position.

The physical explanation for the short length effect given by Blech is that a compressive stress builds up in the anode end of the conductor that opposes additional atomic movement. Such a stress has actually been measured [8] *in situ* on wide lines by using high intensity X-rays focused by a capillary. In order to envision the creation of the stress, one can imagine that the electron flow acts to push all atoms into adjacent vacancies until all the vacancies initially present in the line are swept toward the cathode end of the line. Additional vacancies are generated by forcing atoms located next to grain boundaries and dislocation cores (of edge dislocations) to jump into the boundaries and dislocation cores. When atoms are inserted into a grain boundary (emitting vacancies from the boundary), the grains become compressed. (When the orientation of the boundary is appropriate and the grain diameter is small, this process can lead to the formation of a hillock or a whisker, which continues to grow out of the film as long as metal atoms are supplied to the boundary.) When atoms are forced into a dislocation core, the half plane associated with the dislocation grows and compresses the surrounding lattice. Eventually, forcing additional atoms into the dislocation becomes energetically unfeasible.

For a constant current density, the saturation level of the resistance shift decreases as the line length is shortened, because the back stress reaches its maximum value more quickly. This, in turn, limits the size of the void that can be grown, and that limits the size of the resistance shift. If the line is short enough, or the current density low enough, the saturation void size can be smaller than that needed to produce an easily measurable resistance shift. Experiments on unpassivated lines show that current density thresholds exist below which no Al migration whatsoever is seen. This behavior is more difficult to envision from the perspective of back stress, since only a thin layer of native oxide in the Al is present to prevent hillock and extrusion growth. Detection of the back stress is considerably more difficult in unpassivated short line segments.

An alternate way of looking at the short length effect is in terms of the vacancy flux and concentration. Vacancies have to reach a minimum necessary concentration at the cathode end of the line before a void will nucleate. But vacancies travel a lot farther and faster than atoms do, with the ratio being roughly equivalent to the inverse of the ratio of the vacancy concentration to the atomic concentration. Since vacancies respond both to electron flow and to a concentration gradient and since the diffusion length of vacancies is much larger than that of atoms, the vacancy concentration gradient will work against the current flow to deliver vacancies back toward the anode end of the line. When the line is short enough, the vacancy backflow will be greater than, and in the opposite direction to, the vacancy flow toward the cathode end of the line. Only when the current is great enough to deliver more vacancies at the cathode end of the line can a void nucleate and grow.

The short-length effect is an important phenomenon for two reasons. First, if an interconnect is designed to be short enough, its electromigration lifetime may be effectively infinite. Second, even for lines longer than the Blech length, lifetime is extended [9] because the ΔR versus t curve may begin to saturate before the failure criterion of 20% is reached. As a practical matter for electromigration testing, test structures must be long enough that the lifetime will not be influenced by the short length effect. This can have important implications since some chip designs have long lines (millimeters in length) connecting widely separated functional blocks. EM lifetime projections for these lines will be optimistic if based on test structures that are too short. Although the short-length effect can be observed in lines without any overlying oxide, the jL product is generally thought to be greater for passivated lines, such that for a given current density, the Blech length will be longer. The presence of the passivation helps suppress the formation of hillocks and extrusions.

7.3.4 Multilevel Metallizations

Further decreases in minimum feature size eventually made it impossible to connect all of the devices together using a single wiring level. For line widths below about 0.7 μm, an additional level of wiring was added on top of the final dielectric layer covering the first wiring level, and connections were made by forming tapered holes, or vias, in the oxide, similar to those used to connect the first level of wiring to the underlying Si. The oxide covering the first level of wiring replicates the topographical features of the wiring below and because the second level wiring is usually laid out orthogonally to the first level wiring, it is deposited on an undulating surface or topography. The steepness of the hills and valleys of this surface depend on the thickness and shape of the underlying metal and also on the thickness of the passivation dielectric.

Two important consequences are caused by the topography. First, because the metal deposition process is directional, where the deposited atoms arrive in a direction predominantly normal to the wafer surface, the actual film thickness and microstructure over the steeper slopes of the topography are different from those in flat regions of the chip. While the same thickness of metal is deposited in the vertical direction on a steep slope, the metal thickness in the direction normal to the slope is proportional to the cosine of the angle between the normal to the slope and the normal to the wafer, and will thus become thinner as the slope becomes steeper. In addition, when the angle is large enough, oxygen contamination in the deposition chamber can cause the metal grain growth to become columnar and brittle. Both of these features (Fig. 7.10) shorten the electromigration lifetime in the lines and within the vias.

The second consequence of the topography is that lithographic images in the resist that define the via openings over the first level of metal are not as sharp because the uneven surface presents a variable focal plane, and via sizes and shapes are more difficult to control. In some etch processes used to form the vias, the metal layer is used as the "etch stop." That is, the etching process easily removes oxide, but is relatively ineffectual at removing metal, so that when the metal layer is exposed, the removal of material stops. If the via is positioned so that it extends beyond the edge of the metal below, then the etch process could remove the oxide below the first wiring level and could expose the underlying Si. Metal subsequently deposited can then short out the circuit or device. Consequently, the tapered vias enabling the second metal level to contact the first metal level have to be "landed," that is, placed over an underlying rectangle of metal (landing pad) that is large enough to ensure that the entire via opening will have metal under it. Vias and minimum line widths are generally at the minimum

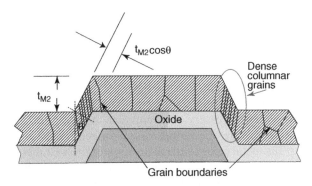

Figure 7.10. Topographical effects on metallization: thinning of metal traversing sidewalls and generation of dense columnar grains at the sidewalls.

lithographic feature size capability for the metal level. This means that the actual via size is likely to be larger than the minimum line width, which in turn means that the landing pad will be wider still than the minimum line width. Photolithographic effects also cause the width of the minimum lines to vary as they follow the undulating contours over the lines of the metal level below, becoming wider in the spaces between lines and narrower as they go over the lines. Thus, in addition to the changes in microstructure on the slopes of the topography, the line width varies.

A result of the change in microstructure to smaller grains is to provide more grain boundary triple points for void and hillock formation. One result of greater oxygen content is to raise locally the resistivity of the metal, which in turn increases the heat dissipation in the short segment; variations in line width and thickness cause variations in current density. All of these factors can contribute to more variation in EM performance and suggest that topography in metallization tends to be detrimental.

The presence of a via between two levels of layered metallization imposes a new limitation on metallization EM robustness. To understand the influence on the EM lifetime of a structure containing two levels of metal and a via between them, we also need to consider an additional effect of refractory underlayers. Typically, refractory metals have much higher melting points than Al alloys or Cu, such that they do not exhibit electromigration at the currents employed in integrated circuits at use conditions. So at a via between metal levels, movement of Al atoms between metal levels is blocked, and the via becomes a flux divergence point. For the case where electrons are flowing from M2 into M1, atoms in M1 will flow away from the via, and will eventually leave a void behind, while atoms flowing toward the via at M2 will eventually form an extrusion that may break the oxide. However, because of the landing requirement, the ends of the lines will also provide metal reservoirs that will delay the occurrence of a measurable resistance shift. In fact, if the reservoir is large enough, electromigration failure may sometimes occur in the narrow portion of the line and sometimes in the wide portion. The resulting failure distribution is then likely to exhibit a bimodal shape with the failures in the line forming the early part of the distribution (because the void volume is smaller), and the failures due to voids under the via forming the later part of the distribution.

Addition of a third level of metal over the second begins to present substantial challenges to step coverage because the topography prior to the deposition of the third metal level includes that due to the two previous metal levels. Cleaner deposition systems are required to prevent Al oxidation on steep slopes, and thicker metal layers are necessary to counteract the thinning due to the steeper slopes. In order to eliminate these concerns, manufacturers began to planarize the oxide layers over the metal prior to the deposition of the next metal layer. This was accomplished by depositing a thicker dielectric layer, then polishing it back with a chemical/mechanical polishing (CMP) process. Initially, this process produced substantial thickness variation over an entire wafer, such that the etch process, used make holes in the oxide for contacting the metal below, would be run long enough to ensure that vias would penetrate through even the thickest part of the oxide. Use of the conventional tapered via required that the landing pads be quite large, and this impacted how densely packed the devices could be.

The solution to this problem was to develop a new via process in which via sidewalls were made very steep (87–89°), and the holes were filled with a conformal layer of tungsten (W), deposited by chemical vapor deposition (CVD). Naturally, the W layer was deposited over the entire wafer, and had to be removed by another CMP process. However, the advantage of this method of making vias was that the size of the landing pads could be greatly reduced, and allowed much greater wiring and device density.

With the introduction of planarization and the use of W vias, the landing requirements could be reduced substantially. The effect on electromigration lifetimes was enormous, causing nearly a 50X reduction [10]. Some lifetime reduction may have been present in two-level metal structures with layered metallization and tapered vias, but the effect was not as noticeable because of the large reservoir provided by the landing pad and the presence of two failure modes (in the landing pad and in the line itself). Additional mechanical flexibility may also be present in tapered via structures, both because of the geometry of the structure and because the elastic moduli of Al and SiO_2 are nearly the same. Such flexibility could reduce the magnitude of the tensile stress in the metal just under the via, perhaps delaying the nucleation of a void. No such flexibility would be present under a W via, since the elastic modulus of W is much greater than either Al or SiO_2. And with the elimination of the reservoirs provided by landing pads, EM-induced voids tended to nucleate immediately under and over the W via. When the via is placed at the end of a line, nucleation and growth of a void under the via is the same as having the void grow from the cathode end of the line toward the anode end (see Fig. 7.11). Little resistance increase is measured while the void is under the via, but as the void begins to clear the via, the resistance rises very rapidly until it reaches the failure criterion. In order to prevent this type of failure, a redundant refractory layer is often added at the top of the metallization layer. Then when the void clears the via, the resistance again grows linearly with time.

Although both the elastic properties of the materials and the flux divergence aspects of a two-level metal/via structure apply equally to electron flow in either the M1/M2 or the M2/M1 directions, the median time to failure is often somewhat

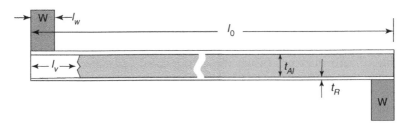

Figure 7.11. Electromigration structure terminated with W vias.

earlier when electrons flow into a line from above rather than from below. There are at least two reasons for this. First, the redundant refractory layers are not usually symmetric in the metallization, and are often thinner on top. Second, the etch process used in making vias in the insulator overlying a metallization level often drills into the metal as well. Thus, the details of the via/metallization interface are different below the via than they are above it, and void nucleation below a via may occur more easily there.

However, there are exceptions. An interesting experiment was conducted by Hu [11], which used a cross structure with M1 and M2 connected by a W via. Electrons flowed in on M2, traveled down through a W via, and out M1. In this configuration, a void should nucleate and grow below the W via. However, sometimes the void was found some distance away from the via in the M1, in the leg with no current flowing. A conclusion that can be drawn from this result is that vacancies overshoot the location of flux divergence. The average distance of the void from the via is greater for greater current. This phenomenon can have implications for electromigration ground rules for lines terminated with W vias; lifetime can be improved if the line extends beyond the via because the void will nucleate behind the via and will require more time to become large enough to pass by the via and generate the 20% resistance increase required for failure.

Alignment of successive layers of metal lines and via connections (often termed overlay) is rarely perfect, such that the position of the via with respect to the end of the line varies in a Gaussian manner. So although the end of a line is aligned with a via by design, the via will sometimes extend beyond the end of the line somewhat, and at other times the line will extend beyond the via by varying amounts due to overlay tolerance. In these cases, the time needed to produce a void large enough to cause a 20% resistance increase will vary in the same manner, such that for the case when the via extends beyond the end of the line, the lifetime will be shorter and will result in a lower allowed use current. This phenomenon may require a special electromigration ground rule for lines designed with no extensions, allowing less current than when extensions are present.

7.3.5 Incubation Time

The ability to track electromigration void growth rate, brought about by testing structures with one or two levels of metallization connected by a W via, enabled the void incubation process to be more closely studied. After the void clears the via, or when the void is located in the line away from the via, resistance increases linearly with time in the initial stages of an electromigration test as shown in Figure 7.8. Ideally, the time at which the void is nucleated should correspond to the time of zero intercept (when the extrapolated resistance versus time curve crosses the time axis).

But this ideal case is almost never observed for a number of reasons. First, the newly incubated void would have to have a flat rectangular shape that would span the cross-sectional area of the line. But such a void would have a very large surface area relative to its volume, and so would have a correspondingly large surface energy. To minimize this surface energy, the nucleating void is much more likely to have a rounder shape with a much smaller surface to volume ratio. Using the same reasoning employed in Figure 7.3, we can approximate this round void by a cube; then the resistance increase caused by the presence of a small cubic void located at the end of a line can be shown to be negligible. The void only begins to cause a sizeable resistance increase when it reaches dimensions on the order of the line width.

First, as shown in Figure 7.11, little resistance shift will be seen until the void length, l_v, becomes larger than l_w, the length of the W stud (via). Strictly, this time should not be counted as incubation time because the void is growing. Careful measurements of the resistance increase and correlation with void lengths obtained from the cross-sectional image of voided lines show that the void growth begins at a time measurably later than the start of the test. It is this time, after the test starts and before the void begins to grow, that is the incubation time.

In addition, the rate of resistance increase (slope of the resistance–time curve) is proportional to the current density, such that the failure time is inversely related to the current density. If void nucleation began as soon as the current was applied, the median time to failure would obey Equation 7.5 instead of Equation 7.6. However, the onset of the resistance increase for AlCu metallizations is delayed from the beginning of stress, and the length of the delay is too long to be attributed to the presence of the via or a line extension. This additional delay is sometimes referred to as the incubation time for the onset of electromigration damage.

The explanation for this phenomenon depends on the behavior of Cu in the line, as described by C.K. Hu [12]. Only a few tenths of a percent (wt) of Cu is soluble in Al at typical chip operating temperatures. For example, for a common sputtering target composition of Al 0.5(wt)% Cu, the Cu is fully dissolved in the Al at about 317°C. Below this temperature, the Cu begins to come out of solution and form Al_2Cu precipitates at grain boundaries and grain triple points. Cu precipitates block the motion of Al atoms along the grain boundaries and prevent Al mass transport that would produce a void at the cathode end of the line. However, Cu precipitates themselves are eroded away by electromigration, and reform farther along the line until there is a denuded zone at the cathode end of the line. When this Cu-free zone is short, the short-length effect prevents Al electromigration damage, but as the length of the Cu-free zone reaches the Blech length, L_b, Al atoms also begin to move and a void is nucleated. Since the Blech length is inversely proportional to the applied current density, a second factor of $1/j$ is introduced to account for the observed dependence of median time to failure on current density. In practice, the current density exponent, n, is not 2 for AlCu metallizations, but assumes values between 1 and 2. The actual value seems to depend at least partially on the Cu concentration.

At this point, we have a relatively complete picture of the electromigration process that occurs in modern integrated circuits using Al metallization. Voids nucleate at flux divergence sites (usually under vias) within the metallization structure and grow in volume linearly with time and linearly with current density. In Al metallizations containing Cu, void nucleation and growth is delayed because

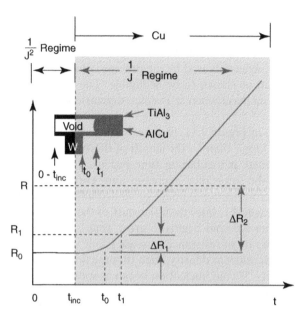

Figure 7.12. Illustrative plot of resistance versus time for a void growing at the end of a line over or under a W via. The first part of the delay in resistance shift is due to nucleation of the void (incubation time) and the second part is due to the time needed for the void to grow past the end of the via.

Cu precipitates block grain boundaries and prevent Al diffusion. When Cu has been swept away from the cathode end of the line by a distance equal to the Blech length, Al electromigration begins. These relationships are shown in Figure 7.12.

7.3.6   Electromigration in Cu Lines

The advent of Cu wiring has brought with it several changes in electromigration behavior. The most notable and expected change was that initially the median lifetime increased over that seen in Al by about 100X [13]. This data was reported at 300°C, and when Cu use was relatively new; later measurements showed improvements over this number. In addition, calculations of the comparative advantage of Cu over Al at use conditions—at temperatures near 100°C—yield a much greater number. This increase is due primarily to Cu materials properties, including a lower diffusivity, larger grain sizes, a longer electron mean free path, and a smaller atomic volume. Whereas Al EM stresses were carried out at around 250°C, Cu EM stresses are carried out at 300°C or more in order to keep the stress duration within workable limits. However, other effects cause a substantial broadening of the failure distribution and offset this advantage to some extent.

First, Cu wiring contains no intentional alloying elements to provide an incubation time as Cu does for Al. Thus the $1/J^2$ regime is absent in Cu electromigration, and void growth begins as soon as current is applied. This is illustrated by the shaded region in Figure 7.12. However, the resistance versus time plots for Cu still show a delay before a measurable resistance increase is observed. This delay is caused mostly by void shape effects [14]. Similar void shape effects are sometimes observed in Al wiring, but because the voids grow much more slowly in Cu metallization and because the liners can be more resistive, void shape effects are more important in Cu metallization. Perhaps the most important aspect of void shape is that the void frequently grows from the top of the line downward, eroding the surface of an entire grain at once.

No significant resistance shift is detected until nearly the entire grain is gone, at which time the current is transferred into the more resistive liner. As the last of the Cu is swept away, the resistance rises very rapidly to a value determined by the length, thickness and resistivity of liner left carrying the current. If the current density in the liner is low enough for the liner to carry it without melting or evaporating, the resistance will subsequently increase in a linear manner. Both the time of this resistance jump and the magnitude of the jump depend on the size of the grain which was eroded away. Liners are often deposited by physical vapor deposition (PVD) into a trench/via structure etched in the oxide. The PVD process is directional, such that the liner is thinner on the sidewalls of the via and the line than in the bottom of the trench and the bottom of the via. Because the via hole has a higher aspect ratio than the trench, the sidewall liner near the bottom of the via will be even thinner than that on the trench sidewall. Thus while the dual damascene process provides less expensive vias with lower resistivity than does the Al/W metallizations system, it is possible for the Cu in the vias to migrate. When this occurs, the current is shunted to the via sidewall liner, and due to the liner thinness there, can burn out the liner and create an open failure.

Second, Cu wiring is made differently from Al wiring. Al wiring is defined by a subtractive etch (metal removed from a blanket film) of a sputtered film. The film thickness is usually controlled to within 10% or less of the intended thickness by crystal thickness monitors, such that the major variation in cross-sectional area is due to photolithographic line width control. The use of refractory layers both over and under the Al metallization virtually eliminates the risk of catastrophic failure due to an electrical open circuit. Cu metallization is formed by electroplating Cu into trenches previously etched into an oxide layer. Prior to electroplating the Cu, a refractory liner is deposited on the wafer to isolate the Cu from the surrounding insulator, and a seed layer of Cu is deposited to provide sufficient conduction for the plating to work. After the final Cu is deposited, the excess Cu and liner are removed from the wafer surface by chemical-mechanical polishing (CMP), leaving only the metallization contained in the trenches. Across-wafer variation in initial Cu thickness and in the removal rate of Cu during CMP cause the effective line thickness to vary more than it does for subtractive etch metallizations and this leads to variations in current density for the same applied current.

Third, the Cu surface is coated with a thin layer of Si_3N_4 or other dense dielectric, both to protect the exposed Cu surface from corrosion, and to act as the lid on the container provided by the refractory layer needed to keep the Cu from diffusing into the surrounding dielectric. Bonding between the Cu and the refractory layer in the trench is usually good, which means that interfacial diffusion between the Cu and the refractory layer is poor. But the bonding between the Cu and the capping layer is not as strong, and mass transport is therefore dominant along that path. There is no refractory layer on the upper surface of the Cu metallization, so that when a via from the metallization above contacts a line somewhat wider than the minimum dimension, it lands only on Cu and does not contact the refractory liner. The significance of this is that a flat void of only a couple of monolayers in thickness and slightly larger than the via diameter, located just under the via can result in an open circuit failure; or a very large void may cause failure. This large range in void sizes also creates a large spread in failure times. The intersection of a via from a metal level above to a wide metal feature below is the weakest part of the metallization for both electromigration and stress voiding.

In the process of forming the trenches, vias are also etched, either before or after the trenches are etched, and both the trenches and the vias are filled at the same time. This procedure is often referred to as "dual damascene." The via holes connecting successive layers of metallization must therefore be bored through two types of insulator: the bulk intralevel dielectric (such as SiO_2, TEOS, FSG, etc.) and the cap layer over the Cu level below. Since the cap layer is usually made of a denser dielectric than the bulk dielectric, the etch process must be altered to handle the denser material at the final stage of penetration. This can result in selective etching of the dense dielectric in such a way that it becomes recessed back under the bulk dielectric above it.

These features can cause different failure modes to appear in the same structure. For example, if the cap dielectric becomes recessed enough at the bottom of the via, liner material may not be continuous across the resulting gap, and bonding between the plated Cu and the liner becomes non-existent. During EM stress, some of the voids can grow across the bottom of the via and generate a very early failure distribution due to the small volume of the void causing failure. When the Cu becomes discontinuous, the test structure shows catastrophic failure. For vias with good liner coverage, failure occurs in a more conventional way with larger voids and longer fail times. Identification of bimodal failure distributions is important in analyzing failure data correctly and will be discussed in greater detail following the section on electromigration testing.

7.3.7 Electromigration Testing

Referring back to Black's law (Eq. 7.6), we find that n and Δh appear to vary with structure, metallization, and microstructure. Therefore, both n and Δh are determined experimentally. This is usually done with a minimum of four separate groups of parts using three different values of current density (typically I_{max}, 0.7 I_{max}, and 0.5 I_{max}) and two temperatures. Typical values for aluminum metallizations are $J = 20$, 14, and 10 mA/μ^2 and $T = 250$ and 200°C. The four cells then would be as shown in Table 7.2.

TABLE 7.2. Typical A1 EM Test Conditions used to
Determine the Values of n and Δh in Black's Equation

T (°C)	J (mA/μ^2)
200	20
250	20
250	14
250	10

From the median times-to-failure of two cells at $20\,\mathrm{mA/\mu^2}$, and at two temperatures, we obtain the activation energy from a plot of log failure time versus $1/T$. This defines the temperature dependence. From the median times-to-failure of the three cells with different current densities tested at 250°C in this case, we obtain the current density dependence from the slope of a plot of log failure time vs. log current density. This is the minimum sample set that can be used. Using two cells with the current density can be very misleading because the current density is varied only by about 2X, and the support of a third data point can be significant. Most people would be uncomfortable with so few data points and would recommend at least one more temperature and one more current, if possible. From a statistical perspective, several more data points (current and temperature conditions) are required to achieve a reasonable degree of confidence in the values of Δh and n. From a practical perspective, each test condition means additional test capacity, stress hardware and analysis time. And for tests with the lower temperatures and currents, test time becomes longer. Together, these constitute a substantial expense to a business. Thus there is always a tradeoff between confidence bounds and the expense of additional data.

7.4 GENERAL APPROACH TO ELECTROMIGRATION RELIABILITY

The overall approach to metallization reliability involves sampling, accelerated testing, data analysis, projection of failure rates to use conditions, and specification of current densities allowed under use conditions in order to meet specified failure rate criteria. Practically, we begin with the specified failure rate in fractional failure at the intended end-of-life (EOL), which is ultimately determined by the customer.

For an example failure calculation and projection, let us assume an electromigration (EM) failure target of one part per million (ppm) = 10^{-6} [2] at the product EOL. Since EM is a wearout mechanism, we want to ensure that no failure by wearout is seen before the EOL. In older technologies, "no wearout" meant less than 1% (10^{-2}) of the target chip cumulative fraction failing (ff). Multiplication of the initial 10^{-6} by this criterion brings the ff to 10^{-8}. In addition,

---

[2] Such a goal was established during a time when "6-sigma" programs were popular. The 6-sigma concept will be discussed in greater detail in the section below on distributions.

each chip has many thousands of structures similar to those we are testing and we must relate the failure per chip to the fraction failing per structure. To do this, we need an estimate of the number of structures that could be present on a chip. For a $1 \, \text{cm}^2$ logic chip with $0.5 \, \mu\text{m}$-wide lines separated by $0.5 \, \mu\text{m}$, we estimate an upper bound for the number of lines by dividing the chip width by the minimum line width to obtain about 10^4 lines. Let our standard test structure be $400 \, \mu\text{m}$ long, such that we can fit 25 such structures in the length of the chip. We then multiply the number of lines per width by 25 to obtain the number of test lines per metal level, and multiply also by the number of levels (say four) to obtain a total of 10^6 structures. However, less than 50% of the chip may be covered by metal lines for any given metal level, not all of those lines will be of minimum width, and not all lines carry the maximum current density. For discussion, assume a 30% metal coverage, 70% of which are narrow lines, and 5% of which carry near the maximum current. The result is that 1% of the 10^6 lines are at risk, which leaves a population of 10^4 lines. If any one of these lines fails, then the chip fails. Thus, we include an additional factor of 10^{-4} in the failure rate to produce a final target of 10^{-12} fraction failing on an individual line basis.

More recently, the trend has been to allocate a greater percentage of the chip failure rate for wear out. For example, if 100 ppm is the actual allowed chip failure rate from all causes, and half of that is allowed for five wearout mechanisms, then 10 ppm would be allowed for electromigration. That may be offset somewhat by the increased metal line density due to further miniaturization of the chip features, but for the duration of this discussion, we will use 10^{-12} as the EOL target.

Given the target failure rate, the next step is to find the failure distribution at the end-of-life. In principle, this can be done directly by testing a large sample of the population to end of life under use conditions, but it would be both costly and of little use. It would be costly because of the time and resources that would be consumed to stress a large population of parts, and results would be largely useless because they would lose their predictive value—that is, the full population of parts would also be experiencing failure on the same time scale.

To overcome these difficulties, a small but representative sample of the larger population of parts is subjected to the same type of wear as the total population, but the wear is intensified in order to produce failure earlier (or at an accelerated rate). The distribution of failure is assumed to be the same under accelerated conditions as under normal use conditions, with the exception that the entire time scale has been collapsed. Thus if the expected EOL is 10 years, and the 99% of the failure times are grouped within three months of the end of the tenth year, then if under accelerated conditions the EOL is 10 days, 99% of the failure times will be grouped within six hours of the end of the tenth day. The magnitude of acceleration is given by an acceleration factor (AF), which is the ratio of time to 50% failure under use conditions to that under accelerated conditions,

$$\text{AF} = \frac{t_{50u}}{t_{50a}}. \tag{7.13}$$

In Equation 7.13, the acceleration factor in days would be $3650/10 = 365$. The AF was calculated using the median time-to-failure (time at which 50% failure has occurred). Again, in principle, any equivalent time could be used, such as the time for 1% failure, or 80% failure, as long as it is the same percentage of the population in both cases. However, use of either a smaller (e.g., 1%) or larger percentage of failure requires a larger sample size to determine the time as accurately as for the median time to failure. For practical reasons, therefore, the median failure time, or $t50$, is most commonly used.

Testing an entire population of parts is impractical for several reasons. Most obviously, if we test to failure, the entire population will be ruined and unable to be used for its intended purpose. Second, the cost of the test is proportional to its size. Third, the entire population may not be present at the beginning of the test, as is the case for semiconductor chips that may be produced for years after the test is performed. The sample chosen for testing should be representative of the entire population, which means that it should be a random sampling of parts that are typical of those produced. Parts from a few (e.g., 3–5) random lots of hardware (produced under the target process) are required for qualification testing (this requirement may still be inadequate from a statistically sound viewpoint). For a large enough sample, the statistical description of the failure behavior then will be a relatively faithful description of the failure behavior of the entire population.

We have the failure target at EOL, and we have a sample of parts from the overall population that we have tested to failure. Next, a theoretical distribution curve is fit to the data, the mean time-to-failure is found from the fit, and the failure time at the target value (10^{-12}) is calculated. This provides a relationship between the t_{50} and the $t(10^{-12})$. The ratio of $t(10^{-12})/t_{50}$ gives the fraction of the median time to failure needed to obtain 10^{-12} failure. When this is known for the distribution of fails under the accelerated test conditions, it also applies (by assumption) to the distribution of failure under use conditions. Suppose that $t(10^{-12})_a/t_{50a} = 1/8$ for the accelerated test, and that the end-of-life (EOL) under use conditions is 10^5 hours (or about 11.4 years). The EOL would correspond to $t(10^{-12})_u$, and then the t_{50u} would be 8×10^5 hr. The actual numbers can only be obtained after we have the failure distribution and the actual relationship of t_{50a} to $t(10^{-12})_a$. These relationships can be seen in Figure 7.13.

So far we have related the failure data for accelerated conditions to the target for failure at the end-of-life. The last step is to specify the allowed use conditions based on the failure data and the EOL target. To do this we need to know the way in which the relevant physical forces are intensified during the accelerated testing, and this is supplied by Black's law (Eq. 7.6) which is the most commonly used physical model. Using the definition of acceleration factor given in Equation 7.13, and inserting Equation 7.6 for both use and accelerated median times-to-failure, we obtain

$$\text{AF} = \frac{t_{50u}}{t_{50a}} = \left(\frac{j_a}{j_u}\right)^n e^{\frac{\Delta h}{k}\left(\frac{1}{T_u} - \frac{1}{T_a}\right)}. \tag{7.14}$$

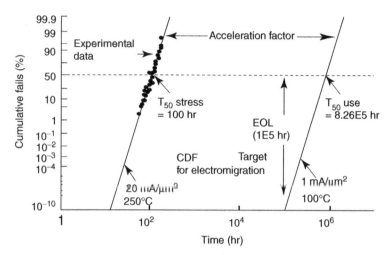

Figure 7.13. Relationships between stress data and extrapolation to use conditions. Accelerated test failure data is on the left and projected failure under use
conditions is on the right. Reprinted with permission from Ref. 26, © 1998 IEEE.

We can now solve Equation 7.14 for the allowed use current j_u. The result is

$$j_u = j_a \left[\frac{t_{50u}}{t_{50a}} \right]^{-\frac{1}{n}} e^{\frac{\Delta h}{nk}\left(\frac{1}{T_u} - \frac{1}{T_a}\right)}. \qquad (7.15)$$

The use temperature, T_u, is defined by the intended use conditions. The
temperature and current density of the accelerated test, T_a and j_a, are known test
conditions. The median time-to-failure under accelerated conditions (t_{50a}) is
obtained from the results of the test. Values of n and Δh are obtained from tests
at different temperature and current conditions, and Boltzmann's constant, k, is
equal to 8.6174e−5 eV/°C. The value for the median time to failure, t_{50u}, is
determined from the failure distribution, and depends on the distribution width
(lognormal sigma) and the target for fraction failing at EOL. This will be
addressed in the next section. The value calculated for j_u is then used to determine
the allowed current densities that are specified in the technology ground rule
manuals.

7.4.1 Projection Methodology Statistical Considerations

All of the analysis is based upon the concept of a statistical distribution of failure
times. Not all parts fail at exactly the same moment in time. If the parts have been
made very carefully and are very nearly identical, then the failure times can be
expected to be closely grouped in time, but a spread, or distribution, will still exist.
If the parts differ very much from each other, the failure times are likely to be
much less tightly grouped, or more widely distributed.

So far, we have been able to circumvent the details of the variation, or distribution, in failure times by using the median time-to-failure (t_{50}). The t_{50} is that failure time for which half the population has smaller failure times and half has larger failure times. The t_{50} is a convenient quantity to use, since it is independent of the way failures are distributed over time. When the data can be fit by a well known distribution, the mean, rather than the median, is extracted from the distribution and can be used for projecting failure rates under use conditions. In addition to the median or mean failure time, we need to know how widely the distribution is spread, and how the fails are distributed in time. The spread in failure times is measured by the standard deviation, or sigma (σ) of the distribution.

Several statistical distributions exist which are suitable for use in describing failure times. The three distributions that are most commonly used for electromigration wear out are the normal distribution, the lognormal distribution and the Weibull distribution. These distributions have all been introduced in Chapter 1, and, for our purposes here, only the normal and lognormal distributions will be discussed. Two representations of these distributions are necessary for the following discussion, the probability density function (PDF) and the cumulative distribution function (CDF). The probability density function describes (oddly enough) how densely the failure times are distributed over the range of failure times. For example, if 100 parts have failed over 20 hours, as shown in the histogram in Figure 7.14, more parts have failed in the eighth hour than in the fifth hour. The cumulative distribution function, or CDF of this distribution (Fig. 7.15), is the integral (cumulative sum) of the PDF and describes (also oddly enough) cumulative failures as a function of time (the sum of all failures accumulated up to a chosen time). In the event that either of these distribution functions is described by a well known function (i.e., the distribution fits the data

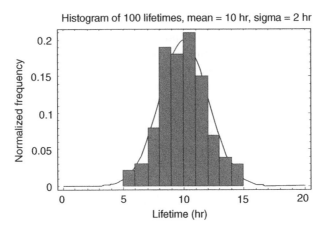

Figure 7.14. PDF of normal distribution of 100 lifetimes; mean = 10 hr, sigma = 2 hr.

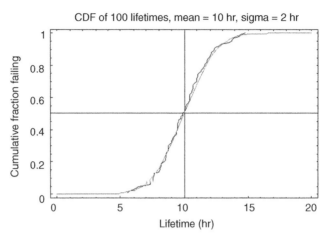

Figure 7.15. CDF of a normal distribution of 100 lifetimes; mean = 10 hr, sigma = 2 hr.

well), we use the well known function to predict failure times in a larger population of parts not yet subjected to wear.

7.4.2 Normal Distribution

One of the most common distributions is the normal distribution, also called the Gaussian distribution or the bell curve (also see Section 1.4.5). The density function for the normal distribution is given by

$$f(t) = \frac{1}{\sigma\sqrt{2\pi}} e^{-\frac{(t-\mu)^2}{2\sigma^2}}, \tag{7.16}$$

where μ is the mean of the time to failure, and σ is the standard deviation of the time to failure. A plot of this function is shown in Figure 7.16, underlying the histogram, with the lifetime (time-to-failure) plotted on the x axis and the fraction failing plotted on the y axis. The curve has a symmetric shape that resembles a bell (hence the name, Bell curve), with the top, or center, of the bell located at μ. The width of the bell at about half the height is 2σ. Both μ and σ have the same units as the values of the distribution. So, for a distribution of failure times in hours, the mean and the standard deviation are also given in hours.

The CDF of a Normal Distribution (shown in Fig. 7.17) is given by

$$F(t') = \int_0^t f(t')dt'. \tag{7.17}$$

The integration is performed from the left side of the curve up to the value t and gives the cumulative failure up to the time t. The CDF must have a value of

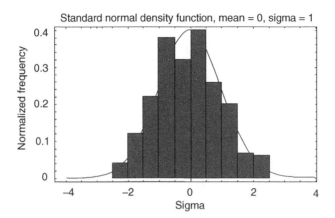

Figure 7.16. Standard normal density function, and a histogram of a randomly chosen sample from the distribution.

one (1) for t = infinity; that is, when summed over all time, all parts must fail, and the fraction failing therefore equals one. The CDF for the normal distribution can not be expressed in closed form, i.e., as an analytic function of t. The integral must be evaluated numerically.

However, because normal distributions are widely used for a variety of populations and for distributions in other variables besides lifetime, tables for the standard normal distribution, in which $\mu = 0$ and $\sigma = 1$ have been calculated for values of the distribution several σ's from the mean. The standard normal

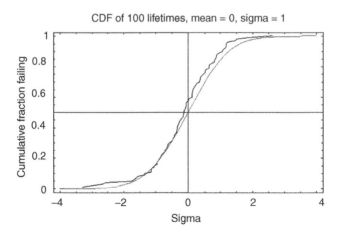

Figure 7.17. Standard normal CDF, corresponding to the density function in Figure 7.16.

density function, shown in Figure 7.16, is given by

$$\phi(z) = \frac{1}{\sqrt{2\pi}} e^{-\frac{z^2}{2}}, \tag{7.18}$$

and the standard normal CDF, shown in Figure 7.17, is given by

$$\Phi(U) = \int_{-\infty}^{U} \phi(z)dz. \tag{7.19}$$

The factor of $1/\sqrt{2\pi}$ must be included in Equations 7.15 and 7.18 in order for the CDF to have a value of 1 as z goes to infinity.

The normal density function is commonly used in statistical line control. Manufacturing processes are assumed to follow the normal distributions for variation about a target taken as the mean of the distribution. Rejection limits are set at each end of the distribution (e.g., at 3σ limits, at about 0.1% and 99.9%). In the "6-sigma" approach, which was popular in the early 1990s, these rejection limits are set at the 6σ points on the distribution. This does not mean that the process rejection criteria have been made very loose. Rather, the production process must be controlled well enough to produce a distribution so tight that the 6σ limits coincide with the boundaries of the process window.

From the tables, one finds that the cumulative fraction failing for 6σ criteria at either end of the distribution would be 0.002 ppm (parts per million). This is a rather severe requirement, but the purpose is to allow "centerline variation" in the process. That is, while the width of the distribution is supposed to remain constant, the mean is allowed to travel as much as 1.5σ to either side of the nominal mean, which would allow the mean to approach to within 4.5σ of the process window boundary on the side towards which the mean moved. A rejection rate of 3.4 ppm corresponds to $z = 4.5\sigma$ in the standard normal distribution., hence the occasionally heard assertion that 6σ means 3.4 ppm failure.

From Figure 7.17, it is evident that the vertical scale must be substantially expanded to make failure rates visible beyond 4σ. This has been done in Figure 7.18 to demonstrate that 3.4 ppm is the value of the standard normal distribution at negative 4.5σ. As a practical matter, care must be taken that the size of the increment of integration ($d\sigma$) is small enough to faithfully reproduce the distribution for large values of σ. (Increments of 1e-4 over the range $(-4.6 < \sigma < -4.4)$ are adequate.)

Any normal distribution whatever may then be expressed in terms of the standard normal distribution as

$$f(y) = \frac{1}{\sigma} \phi \left[\frac{(y - \mu)}{\sigma} \right], \tag{7.20}$$

where

$$z = (y - \mu)/\sigma, \tag{7.21}$$

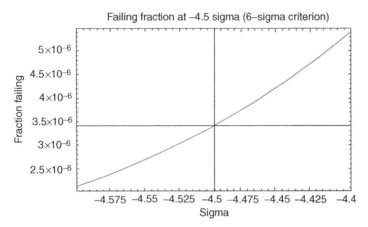

Figure 7.18. Portion of the standard normal CDF showing "6-sigma" criteria.

as in Equation 7.20 above and is the measure of how many σ's away from the mean the value of z is. The factor of $1/\sigma$ arises from the change in the integration increment caused by the change in variables:

$$\frac{d}{dy}\left(\frac{y-\mu}{\sigma}\right) = \frac{d}{dy}\left(\frac{y}{\sigma}\right) = \frac{1}{\sigma} \Rightarrow d\left[\frac{y-\mu}{\sigma}\right] = \frac{1}{\sigma}dy.$$

The tables for the standard normal CDF give the value of the CDF for values of z in units of σ in increments of 0.001σ. Understanding these details is critical to the understanding of the lognormal distribution that is used in analyzing electromigration failure.

7.4.3 Lognormal Distribution

The lognormal failure distribution is conceptually easy to understand if it is simply considered to be a normal distribution of logarithms of the failure times. In other words, we simply replace the variable t in Equation 7.15 by $\tau = \log t$, to give

$$f(\tau) = \frac{1}{\sigma_{\ln}\sqrt{2\pi}}e^{-\frac{(\tau-\mu_{\ln})^2}{2\sigma_{\ln}^2}}, \tag{7.22}$$

where $\mu_{\ln}$ and $\sigma_{\ln}$ are now the mean of the log of life and the standard deviation of the log of life, respectively, rather than the mean of life and the standard deviation of life. Whereas the mean for the normal distribution is given by

$$\mu(t) = \int_0^\infty tf(t)dt, \tag{7.23}$$

The lognormal mean is given by

$$\mu_{\ln}(\tau) = \int_0^\infty \tau f(\tau) d\tau. \tag{7.24}$$

Whereas the standard deviation for the normal distribution is given by

$$\sigma = \left[\int_0^\infty (t - \mu)^2 f(t) dt \right]^{\frac{1}{2}}, \tag{7.25}$$

the lognormal sigma is given by

$$\sigma_{\ln}(\tau) = \left[\int_0^\infty (\tau - \mu_{\ln})^2 f(\tau) d(\tau) \right]^{\frac{1}{2}}. \tag{7.26}$$

The normal distribution for $\tau = \ln t$ is then written

$$f(\ln t) = \frac{1}{\sigma_{\ln} \sqrt{2\pi}} e^{-\frac{(\ln t - \mu_{\ln})^2}{2\sigma_{\ln}^2}}, \tag{7.27}$$

which is identical to Equation 7.15 except for the variable substitution. Unlike the normal distribution, the lognormal density function adopts different shapes depending on the value of σ. For very small values of σ (e.g., 0.1), the lognormal distribution looks very much like the normal distribution. For larger values of σ, the distribution becomes obviously asymmetric with a tail to the right (Fig. 7.19)

The CDF for this distribution is

$$F(\ln t) = \int_0^\infty f(\ln t) d(\ln t). \tag{7.28}$$

However, we usually want to know the cumulative distribution function with respect to lifetime, not the log of life, so we change the increment of integration in the same way we did above in the explanation for Equation 7.14:

$$\frac{d}{dt}(\ln t) = \frac{1}{t}, \quad \text{such that } d(\ln t) = \frac{1}{t} dt. \tag{7.29}$$

Equation 7.28 may now be written

$$F(\ln t) = \int \frac{1}{t} f(\ln t) dt, \tag{7.30}$$

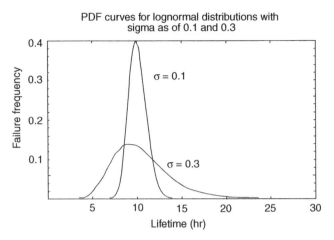

Figure 7.19. Lognormal density function plotted for σ = 0.1 and 0.3. The σ = 0.1 distribution is nearly symmetric like the standard normal distribution, while that for the 0.3 is clearly asymmetric, with a tail to the right.

where

$$f(t) = \frac{1}{t\sigma_{\ln}\sqrt{2\pi}} e^{-\frac{(\ln t - \mu_{\ln})^2}{2\sigma_{\ln}^2}} = \frac{1}{t} f(\ln t)$$

is the lognormal density function with respect to lifetime. This shows mathematically how the lognormal distribution is generated from the normal density function, and should help to make an otherwise confusing formula somewhat less mysterious.

The lognormal CDF is shown in Figure 7.20. We use the lognormal distribution for electromigration fails primarily because it describes the observations well. Exactly why electromigration data is well described by the lognormal distribution is not at present well understood, but having a well defined distribution to describe the data allows us to obtain our target median time to failure at the EOL analytically. We use Equation 7.21 to produce

$$z_{EOL} = \frac{\tau_{0\ln} - \mu_{\ln}}{\sigma_{\ln}} = \frac{\ln \tau_0 - \ln t_{50u}}{\sigma_{\ln}}, \tag{7.31}$$

where $\tau_{0\ln} = \ln \tau_0$ and $\tau_0 = $ EOL is the required or target end of life, t_{50u} is the median time to failure at the end of life, z_{EOL} is the value in standard deviations (σ's) corresponding to the allowed failure at the end of life (z-score), and $\sigma_{\ln}$ is the lognormal standard deviation found from an accelerated test with a small sample of the hardware. Solving Equation 7.31 for t_{50u}, we obtain

$$t_{50u} = \tau_0 e^{-z\sigma_{\ln}}. \tag{7.32}$$

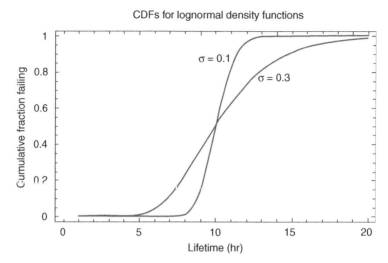

Figure 7.20. Lognormal CDFs corresponding to the lognormal density functions in Figure 7.19.

Substituting Equation 7.32 back into Equation 7.15, we obtain an analytic expression for the allowed use current j_u,

$$j_u = j_a \left[\frac{\tau_0 e^{-2\sigma_{\ln}}}{t_{50a}} \right]^{-\frac{1}{n}} e^{\frac{\Delta h}{nk} \left(\frac{1}{T_u} - \frac{1}{T_a} \right)}. \tag{7.33}$$

Equation 7.33 contains all of the relevant variables needed to calculate the allowed current. The use temperature, the target end of life and the allowed failure at the end of life are all business goals, often determined by the customer. The temperature during test and the applied current are experimental parameters. The median time-to-failure during test, the lognormal sigma, the activation energy, and the current density exponent are all results of the experimental data.

Example Problem 7.1

Assume the following values apply to electromigration structures on a particular type of chip.

$\Delta h = 0.8\,\text{eV}$	$\sigma_{\ln} = 0.2$
$T_a = 250°C$	$T_u = 100°C$
$t_{50a} = 9.46\,\text{hr}$	$\tau_0 = \text{EOL} = 1\text{e}5\,\text{hr}$
$n = 1.7$	$\text{CDF} = 1\text{e}{-}12$
$J_a = 20\,\text{mA}/\mu\text{m}^2$	$k = 8.6174\,\text{e}{-}5\,\text{eV/K}$

What is the use current?

First we calculate t_{50u} from Equation 7.32. To do this, we need a value for z, which we obtain from the tables for the CDF of the standard normal distribution (or calculate with a program which can integrate), and find to be $z = -7.035$. With this value for z, $t_{50u} = 4.084\text{e}5$ hr. As an aside, we note that since the AF is just the ratio of the use median time to failure to that of the accelerated median time to failure, i.e., since t_{50a} here is 9.46 hr, the $AF = 4.32\text{e}4$. The acceleration due to temperature,

$$e^{\frac{\Delta h}{k}\left(\frac{1}{T_u} - \frac{1}{T_a}\right)}, \tag{7.34}$$

is equal to $1.26\text{e}3$, and the allowed current density under use conditions, j_u, is $0.125\,j_a$, which is about $2.5\,\text{mA}/\mu\text{m}^2$.

Example Problem 7.2

As an example of how these relationships might be used, consider the following problem. An etch process has been operating outside of specified limits for some time because post-process measurements were not sensitive to localized variations. As a result, the width of some of the narrowest lines is only a fraction of the design width and ranges from missing lines up to lines of proper width. Some of the parts made in this way have actually been built into machines and will need to be replaced. Customer representatives want to know when to expect the parts to start failing as a function of line width. How can the above relationships be used to answer the question?

In order to answer a failure rate problem, we need to find z at the end of life as a function of the line width. To do this, we can solve for z in Equation 7.33, which provides a functional relationship between z and all the other variables in the current density equation. The result is

$$z = -\frac{1}{\sigma_{\ln}}\left[-n\ln\frac{j_u}{j_a} + \ln\frac{t_{50a}}{(\text{EOL})} + \frac{\Delta h}{k}\left[\frac{1}{T_u} - \frac{1}{T_a}\right]\right]. \tag{7.35}$$

The line width dependence is contained in the current density in the first term. Current density, j, is the current per cross-sectional area of conductor line,

$$j = \frac{I}{A} = \frac{I}{tw}, \tag{7.36}$$

Ordinarily, we expect the line widths to be the same for the parts tested under accelerated conditions as well as those produced for sale. However, in the unusual situation described above, the line widths under use conditions will be smaller than those on the parts used for accelerated testing. Therefore, we can write

$$\frac{j_u}{j_a} = \frac{I_u/t_u w_u}{I_a/t_a w_a} = \frac{I_u}{I_a}\cdot\frac{w_a}{w_u}. \tag{7.37}$$

As just mentioned, $w_a/w_u = 1$, normally, such that Equation 7.36 would show that the ratio of current densities in a given line would equal the ratio of the currents in that line. Thus, to find the line width effect on the failure rate, we need only replace the current

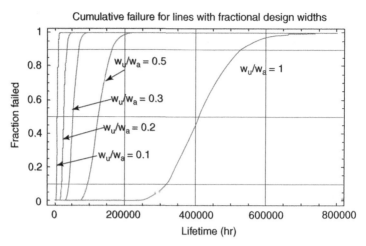

Figure 7.21. Lognormal CDFs of electromigration lifetime for lines with different fractions of the design width.

density ratio in Equation 7.34 by the right-hand side of Equation 7.36, to produce

$$
\begin{aligned}
z &= -\frac{1}{\sigma_{\ln}}\left[-n\ln\frac{I_u}{I_a}\cdot\frac{w_a}{w_u} + \ln\frac{t_{50s}}{(EOL)} + \frac{\Delta h}{k}\left[\frac{1}{T_u} - \frac{1}{T_a}\right]\right] \\
&= -\frac{1}{\sigma_{\ln}}\left[-n\ln\frac{I_u}{I_a} + \ln\frac{t_{50a}}{(EOL)} + \frac{\Delta h}{k}\left[\frac{1}{T_u} - \frac{1}{T_a}\right]\right] + \frac{n}{\sigma_{\ln}}\ln\frac{w_a}{w_n}.
\end{aligned}
\tag{7.38}
$$

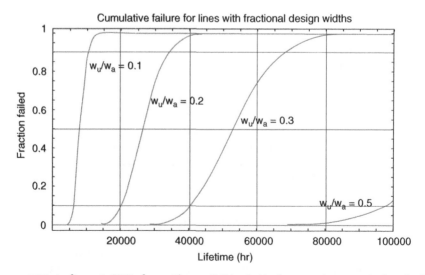

Figure 7.22. Leftmost CDFs from Figure 7.21 plotted on and expanded scale for better visibility.

Note that the first term in the second line of Equation 7.37 is the same as Equation 7.35, except j_u/j_a is replaced by I_u/I_a. Inserting the values listed in the example above into Equation 7.38 produces a simple equation for z

$$z = -7.035 - 8.5 \ln \frac{w_a}{w_n} \qquad (7.39)$$

which can be used to find the failure rate of lines at the end of life.

The results for $w_a/w_u = 0.1, 0.2, 0.3, 0.5$, and 1.0 are shown in Figure 7.21. Median lifetime decreases from 4.23×10^5 hr for full line widths to a little under 10^4 hr for lines 0.1 times as wide as designed. Figure 7.22 is an expanded view of the left part of Figure 7.21, from 0 to 10^5 hr.

7.4.4 Bimodal Distributions

All of the distributions examined previously were nominally monomodal distributions. That is, failure occurred, or was assumed to occur the same way (in the same mode) for all parts. This is not always the case. For example, with Al/W stud metallizations with refractory layers above and below the Al, the location of the void produced by EM can make a significant difference in both the failure time and in the shape of the resultant distribution. Consider a void that completely severs the Al layer and which is located in the line away from the W stud at the cathode end of the line. Although the Al is missing, current is still carried by the refractory layers above and below the void. The resistance increase caused by the void will be equal to the inverse of the sum of the conductance of the two refractory layers. Now consider a void of the same size located over a W via at the extreme end of the line. In that case, the only path the current can travel is through the single refractory layer at the bottom of the line, and the resistance will be equal to the resistance of the exposed portion of the bottom refractory layer. This resistance (per unit length of void) will be greater than that recorded when the void is out away from the via where there are two conductive paths past the void. Since the failure time depends on the fractional resistance increase, the failure time will occur sooner for the second case than for the first.

When the two modes occur in tests of the same structure, the failure distribution assumes a staggered appearance, in which the early and late modes are often clearly defined and offset from each other, and connected by a few fails in the transition region. The number of parts in the population that fail by each mode is usually not evenly divided. In the case of a smaller fraction of parts failing by the early mode, and when the early mode fails in substantially shorter time than the late mode, the early mode is generally not clearly defined, and may appear simply as scattered early fails due to defects. In order to ensure that the early fails are not part of a second failure mode, additional parts must be tested until the early failure mode becomes apparent.

Identification of a second failure mode in such distributions is very important. If the distribution is fit as though it were a single failure mechanism, the σ of the distribution will be much larger, and the allowed use current density, as calculated by Equation 7.33, may be unacceptably low. Instead, the data needs to be fit as a bimodal distribution, and the σ and t_{50} of the early failure mode used to project the allowed use current density.

Fischer [15] has illustrated several ways that bimodal failure distributions can combine, and has shown how such distributions can be analyzed. The fit is a five-parameter fit, and yields σ and t_{50} for each failure mode and f, the fraction of parts in the first failure mode. Since the early failure mode is generally the one of most interest, and since the early failure mode generally constitutes the smallest fraction of the tested population, the fit to the σ and t_{50} usually has a lower confidence than the fit of the later fail mode. The older EM experiments utilized ~ 15 samples to produce a reasonable level of confidence for t_{50}. It seems reasonable, therefore, to require at least 15 samples in the early failure mode to produce reasonable confidence in the projected reliability.

In practice, one often finds that the value of σ for the two modes of a distribution is the same, which reduces the number of fitting parameters to four, and allows a higher degree of confidence in the parameters produced by the fit. Nonetheless, definition of the early failure mode requires a much larger number of parts to be tested than is required for the same level of confidence as for the single mode failure. Suppose that the sample size for an EM test is 24 and that 15% of the parts (~ 3 or 4 parts) belong to the early failure mode. To achieve the same level of confidence in t_{50} for the early mode as for a single mode distribution with 24 parts, the same number of parts is required (24). In order to get those parts, a

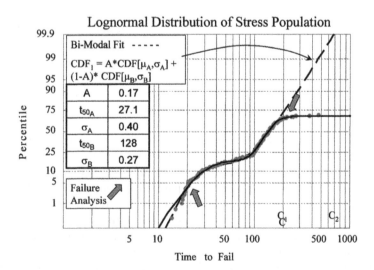

Figure 7.23. Example of a bimodal distribution in EM failure for Cu lines, and a bimodal fit where the late fails were censored out. Early fail mode caused by a void at the via bottom. Reprinted with permission from Ref. 16, © 2002 IEEE.

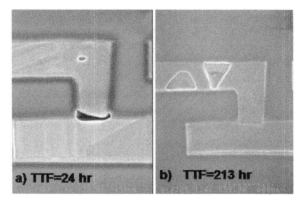

Figure 7.24. (a) EM void at via bottom, corresponding to early fail in bomodal distribution shown in Figure 7.23. (b) Late fail from distribution in Figure 7.23, showing larger void than in (a), located in the line rather than in the via. Reprinted with permission from Ref. 16, © 2002 IEEE.

random sample of the population six to eight times larger is needed, or between 144 and 192 parts.

Bimodal EM failure distribution in Cu metallizations can be produced by void positioning as well [16], as shown in Figure 7.23, but the reason for the differences in failure time can be due to differences in void size. If a void appears at the bottom of a dual-damascened via, it only needs to cover the bottom of the via to cause failure (Fig. 7.24), and this volume is relatively small compared to that of a void in the line. Because the void growth process is much slower in Cu than in Al metallizations, the time-to-failure is much different. In addition, whereas the refractory layers in Al RIE metallizations are of essentially uniform thickness throughout the metallizations, the liner for dual damascene Cu metallizations is thicker on the bottom of the trenches than it is on the sidewalls of the trenches or on the sidewalls of the vias. Thus, when the Cu is removed from a section of a via, it produces a much larger increase in resistance per void length than would be the case if a void with the same volume were produced in the line. Furthermore, small void sizes in Cu lines are less likely because of the way a void grows. Mass transport in dual damascene Cu metallizations occurs mostly along the top interface between the Cu and the overlying dielectric capping layer, so susceptible grains often erode from the top down, and eventually produce a void that has the size of the original grain. Grains in the line are usually larger than grains in the via, so the voids in the line are larger.

7.4.5 Three-Parameter Lognormal Fit to a Single Distribution

The lognormal failure distribution, as applied to electromigration, inherently contains the possibility of contradictory predictions of reliability. Imagine that two lots of hardware produce two different sets of data when the same structure is tested. The distribution from the first lot has a $\sigma = 0.25$ and a $t_{50} = 50$ hr, and the distribution from the second lot has a $\sigma = 0.35$ and a $t_{50} = 75$ hr. Clearly, as the two distributions are extrapolated to much smaller percentages, the curve describing the second distribution will intersect and cross over that for the first distribution, because its slope is shallower. The physical implication of the cross-over is that, because the second distribution is wider (larger σ), if enough parts are tested, some will have failure times that are shorter than those observed for the first distribution, even though the median time to failure is 1.5X greater for the second distribution. Furthermore, in general, the underlying assumption for the lognormal distribution is that as the population of tested parts grows, the earliest failure times become arbitrarily small.

The last assumption is contrary to the physical case. For example, in Cu single-damascene metallizations, where the voids grow from the top of the line downward, no significant resistance shift is seen until the void has almost severed the Cu part of the line. Then, in a very short time, the last of the Cu is removed and the line resistance jumps by a significant amount, often beyond the 20% limit that defines failure. If the Cu line is connected to a W via below, and the void begins growing at the end of the line, then there is a minimum amount of Cu that must be removed before failure can be detected, and this removal requires a minimum amount of time that is greater than zero. Thus the assumption of arbitrarily small failure times breaks down and the lognormal distribution fails to describe early failure in very large populations of parts.

This was first addressed by Filippi [17] with Al metallizations. Filippi showed that the lognormal distribution for a sample size of nearly 500 parts was curved downward towards the left (shorter times and lower percentages away from the straight line of the traditional lognormal fit). Filippi explained this behavior by postulating a minimum failure time, t_0, for failure. The z-score of an individual failure is then given by a modification to Equation 7.21, where τ_0, the end of life, is replaced by $(t_f - t_0)$ to produce the following relation:

$$z = \frac{\ln(t_f - t_o) - \ln(t_{50})}{\sigma_{\ln}}. \tag{7.40}$$

Equation 7.40 fits the data well. The minimum failure time, t_o, or the threshold failure time, is the asymptotic value for failure. As t_f approaches t_o, the quantity $\ln(t_f - t_0)$ becomes large and the lognormal sigma approaches zero (which means the curve approaches verticality). The earliest failure times in Filippi's data were still substantially above the threshold value, so the curvature was only pronounced, but did not begin to show the rapid vertical behavior. Part of the reason for this is that the σ for Filippi's data was already small ($\sigma \sim 0.2$), and the slope of the plot already rather large, which reduces the contrast between the plots of the two fits. However, Li [18] et al. show failure distributions for which the lognormal sigma is greater than 1, where both simulated data and test data approach very close to the threshold value, and clearly illustrate the differences between the traditional two-parameter fit and the three-parameter fit.

The three-parameter fit is the more exact case for all EM data. But for distributions with a steep slope (small σ), definition of the threshold value is both unnecessary and expensive. It is unnecessary because projections of allowed use currents for small-σ distributions are generally sufficient for intended designs. It is expensive because prohibitively large sample sizes are required to determine the threshold value with an appropriate degree of certainty. But for large-σ distributions, the conventional two-parameter projection yields a dismal value for the allowed use current. At the same time, determination of the threshold value can be accomplished by sample sizes that can be achieved, and the failure times are short enough that the test duration is short. Allowed use current density values obtained using the threshold value of failure time are reasonable.

7.5 THERMAL CONSIDERATIONS FOR ELECTROMIGRATION

So far, we have dealt with the general behavior and analysis of electromigration failure in simple isolated lines. The temperature of these lines is produced by external heating, such as from an oven. Heating can also be produced by the current flowing in the line itself, if the current is great enough, and it can be produced by nearby lines and devices on the chip. In this section we will examine both direct and indirect Joule heating, for which two additional concepts will be required. The first concept is thermal resistance, and the second is the thermal coefficient of resistance of a metal.

7.5.1 Self-Heating

Every conductor with resistance R carrying a current I dissipates energy in the form of heat. Collisions between the conduction electrons and atoms of the conductor cause the atoms to vibrate with higher frequency and greater amplitude than would be the case with no current flowing. The line experiences this increased atomic vibration as a rise in temperature. The energy per unit time, or power, dissipated by the current is given by

$$P = I^2 R, \tag{7.41}$$

where P is the power in watts, I is the current in amperes, and R is the resistance in ohms.

The temperature of the conductor depends upon the environment around the conductor because heat is conducted away by whatever is in contact with it. If the surrounding material is a good thermal conductor, such as another metal, the temperature of the current-carrying conductor will not increase much. If the conductor is suspended in a vacuum, the temperature can increase to a very high value and possibly melt the conductor. In the most usual situation, the applied current is constant in time, such that the power dissipation in the wire is also constant. In order for the temperature to remain constant, therefore, thermal power conducted away must be constant as well. The relationship between the temperature of the line and the power dissipated in it is given by

$$\Delta T = R_\theta P \qquad (7.42)$$

where ΔT is the temperature increase over the ambient temperature (temperature of the surroundings) in °C or K, R_θ is the thermal resistance in watts/(m-K), and P is the power in watts. The thermal resistance determines how fast the power is conducted away. Calculation of thermal resistance can be very complex, depending on the geometry involved, and often requires computer simulation through finite element analysis (FEM). However, considerable intuition and quantitative information can be obtained for lines in integrated circuit chips from a relatively simple approach.

Thermal and electrical resistances are both conceptually and mathematically similar. Consider a layer of insulating material, such as glass, sandwiched between two layers, one of metal and one of silicon (Si). Figure 7.25 shows a rectangular metal line with thickness, h, width, w, and length, l. A current is passed through the line at a sufficiently high magnitude to generate measurable heating. The metal line rests on a line of oxide with thickness t_i, width, w, and length, l, which in turn rests on a large slab of Si held at room temperature. Assume that the Si volume is large enough to act as a thermal reservoir (heat sink), such that large amounts of heat added or removed from it will have a negligible effect on its temperature. If the metal is held at temperature, T, and the Si has a temperature of T_o = room temperature, and if $T > T_o$, heat will flow from the metal line to the Si according to

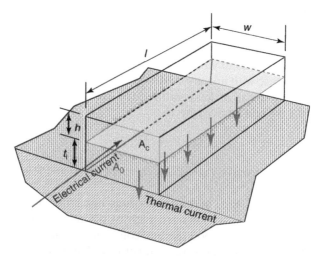

Figure 7.25. Sketch of a metal line over an oxide layer with the same footprint, both on the surface of a Si substrate. The electrical current flowing along the line through the area A_e produces heat in the line that flows downward through the oxide layer with footprint A_θ into the Si substrate.

Equation 7.42. The heat per unit time, or thermal power, is P, and $\Delta T = (T-T_o)$ is the temperature difference between the metal and the Si, The thermal resistance, R_θ, is given by $\Delta T/P$ and has units of K/W. If the temperature of the line can be measured and if the power dissipated by the line is known, the thermal resistance is the slope of the temperature versus power curve.

However, P may be difficult to measure for many configurations. Although $P = I^2 R$, and I^2 is known because it is the applied current, R is temperature dependent and increases with temperature for the primary conductors, Cu, Al, Ag, and Au. Thus, any heating from the current will increase the resistance of the line, and unless the resistance increase is known, the power can not be calculated. This increase in resistance with temperature is a result of increased electronic scattering caused by the increased atomic vibration at higher temperatures. It is characterized by the thermal coefficient of resistance (TCR), β, in Equation 7.43.

$$\beta = \frac{1}{R_o} \frac{dR}{dT} \qquad (7.43)$$

β is easily measured with an oven and an ohm meter by measuring the resistance of a conducting line at several different temperatures and finding the slope of the resulting points (which generally describe a straight line). If we replace the slope, dR/dT, with the ratio of differences, $\Delta R/\Delta T$ (where $\Delta R = R - R_o$ and $\Delta T = T - T_o$ and where both ΔR and ΔT are positive), Equation 7.43 can be rewritten as shown in Equation 7.44.

$$\beta = \frac{1}{R_o} \frac{\Delta R}{\Delta T} = \frac{R-R_o}{R_o \Delta T}, \qquad \text{so} \qquad R_o \beta \Delta T = R - R_o.$$
$$\text{Then} \quad R_o + R_o \beta \Delta T = R \quad \text{and} \quad \therefore \quad R = R_o(1 + \beta \Delta T). \qquad (7.44)$$

The equation on the right of the second line says that the resistance at any temperature, T, is found from the initial resistance, R_o, increased by an amount proportional to the temperature change, where β is the proportionality constant. The resultant resistance value depends on the value of resistance chosen for the starting resistance, R_o. Because resistance varies with temperature, the value must correspond to the temperature T_o in $\Delta T = T - T_o$. The value of β will be different for different values of R_o. Values for β and R_o given in this chapter are taken at 30°C, but not everyone uses this reference. Two common choices are room temperature and 0°C.

Referring back to Figure 7.25, the electrical cross sectional area, A_e, is equal to the cross sectional area of the line, $A_e = wh$, and the thermal cross sectional area is equal to the bottom surface of the line, $A_\theta = wl$, since the heat in the line flows down through the glass into the Si substrate. The thermal resistance is easily calculated (if the thermal conductivity of the glass is known) by the formula $R_\theta = (1/K_i)L_\theta/A_\theta$, where K_i is the thermal conductivity (the inverse of thermal resistivity), L_θ is the length of the thermal conductor (= insulator thickness, $\pm_i$), and A_θ is the cross-sectional area through which the thermal current flows ($A_\theta = wl$).

TABLE 7.3. Thermal and Electrical Analogs

Electrical Equations	Thermal Equations
$V = IR$	$\Delta T = PR_\theta$
$R = \rho l / tw$	$R_\theta = (1/K_i) t_i / lw$

This expression is very similar to that for calculating electrical resistance in the metal line, $R = \rho L / A$, where ρ is the electrical resistivity of a conductor, L is the conductor length, and A is the cross-sectional area through which the current flows. The similarity in form is evident from Table 7.3 where the expressions defining thermal resistance and electrical resistance are summarized.

The temperature of the line with a current I flowing in it can be calculated from Equation 7.42, by inserting $I^2 R$ for P and Equation 7.44 for the electrical resistance:

$$\Delta T = PR_\theta$$

$$\Delta T = (I^2 R)(R_\theta) = I^2 R_\theta * R_o (1 + \beta \Delta T)$$

$$\Delta T (1 - I^2 R_\theta R_o \beta) = I^2 R_\theta R_o \qquad (7.45)$$

$$\Delta T = \frac{I^2 R_\theta R_o}{1 - I^2 R_\theta R_o \beta}.$$

If the numerator and denominator of the right-hand side are divided by R_θ, and if we convert the numerator from current to current density, j, the temperature change in the wire due to Joule heating is given by Equation 7.46:

$$\Delta T = \frac{I^2 R_o}{\frac{1}{R_\theta} - I^2 R_o \beta}; \quad I^2 R_o = \frac{I^2}{A^2} A^2 \rho_o \frac{L}{A} = j^2 \rho_o AL = j^2 \rho_o V$$

$$\therefore \Delta T = \frac{j^2 \rho_o}{\frac{1}{R_\theta V} - j^2 \rho_o \beta}. \qquad (7.46)$$

The difference between the two terms in the denominator of Equation 7.46 represents a balance between the thermal conductance and the power production. The thermal conductance is essentially fixed by the thermal conductivity of the insulator surrounding the wire and the geometric aspects of heat flow. The power production depends both on the current or current density (depending on which form of the equation is addressed) and the magnitude of the thermal coefficient of resistance. For low values of the current density, thermal conductance will be much larger than the power dissipated, and temperature changes will be small. But as j is increased, the magnitude of the second term in the denominator will eventually grow close to the magnitude of the first term, the difference will be very small, and the temperature change will become very large, until the conductor is melted or burns up.

Equation 7.46 is generic in the sense that it has not yet been applied to a specific geometric arrangement or set of materials, other than that of a conductor with a cross-sectional area and a thermal resistance between the conductor and a heat sink some distinct away. It is not at all restricted to the arrangement in Figure 7.25, and holds, for example, for a circular wire embedded in rubber and surrounded by water. However, once the thermal resistance is given a specific form, the expression will be specific to the materials and geometric arrangement for which the thermal resistance is calculated. For example, for the arrangement shown in Figure 7.25, the thermal resistance in Table 7.3 would be inserted in Equation 7.46 to produce

$$\Delta T = \frac{j^2 \rho_o}{\frac{1}{\frac{t_i}{K_i} lwt} - j^2 \rho_o \beta} = \frac{j^2 \rho_o}{\frac{K_i}{tt_i} - j^2 \rho_o \beta}. \tag{7.47}$$

Structures such as that shown in Figure 7.25 can be made, and may be useful for studying thermal conductivity of the insulator material, but as a practical matter, the oxide underlying the metal line would not normally be etched away to conform to the width of the line. Rather, the oxide would either be grown or deposited as a uniform layer, and the line would be formed on top. Such a structure is shown in Figure 7.26.

In this scenario, heat generated in the line would flow to the sides and down to the substrate, as well as directly down beneath the line. This thermal spreading allows additional heat flow compared to the case shown in Figure 7.25. Bilotti [19]

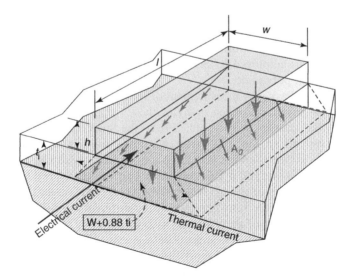

Figure 7.26. Similar to Figure 7.25 except that the conductor now rests on a sheet of oxide, which enables the heat flow to spread laterally and increase the conductance.

has shown that for lines on a film of dielectric, the thermal width, w, in the expression for R_θ in the table above is increased by a factor of $(1 + 0.88\ t_i/w)$ to produce a thermal resistance of

$$R_\theta = \frac{t_i}{lwK_i\left(1 + 0.88\frac{t_i}{w}\right)}. \tag{7.48}$$

Schafft [20] used Bilotti's expression for the thermal conductance of a line on an oxide film over a Si substrate (Figure 7.26) to calculate the increase in temperature of that line as a function of the current density in the line. Schafft calculated the thermal profile of a NIST structure, including the transition region where the line tapers from the narrow dimension to the wider lines that lead to bond pads. Assuming that the line is so long that the end effects are insignificant, the temperature of the line is obtained from Equation 7.47 by substituting Equation 7.48 for R_θ to produce Equation 7.49.

$$T = T_o + \frac{j^2\rho}{\frac{K_i}{tt_i}\left(1 + 0.88\frac{t_i}{w}\right) - j^2\rho\beta}. \tag{7.49}$$

Verification of the model can be accomplished by measuring the temperature of lines with a range of widths on the same metal level. Examination of Equation 7.49 shows that all of the parameters except K_i can be determined, either from independent measurement (such as for the TCR and sheet resistance of the metal) or from layout definitions and test conditions. The model is then fit to the data, allowing the value of K_i to vary. The value of K_i that produces the best fit will be the effective thermal conductivity of the insulator. (Line width can be checked by measuring the resistance of narrow and wide lines and comparing the ratio with that expected from design widths and sheet resistance, to achieve better accuracy.) This approach was employed by Schafft to measure the thermal conductivity of thin film SiO_2 and polyimide.

One might well ask why measurement of the thermal conductivities of various insulators is of importance when many of these values are available in common references. The answer is two-fold. First, the materials for which data is commonly available in the usual reference texts are generally bulk materials, i.e., made in thicknesses that are very large compared to the dimensions of the film thicknesses on IC chips, and necessarily made by processes that differ from those used to fabricate layers of similar material on a chip. One might therefore expect that the properties of the thin films would be different from those of the bulk materials. Second, many of the materials used on the more advanced chips simply are not available in bulk form, and no data is available. A third reason is that many chips are now built with several different types of insulation layers, and an effective or composite conductance can be obtained that will be more accurate.

The data obtained by Schafft's approach can be verified by comparison with other techniques for measuring thin film thermal conductivities, such as the 3-ω method [21]. The advantage of Schafft's method is that it can be implemented using simple structures that can be measured by standard pad connections with wafer probers available in a manufacturing test environment.

Structures such as that shown in Figure 7.26 are commonly made in the industry at each level of metal, but they are always subsequently covered by a layer of insulator for protection. The covering insulator layer could be removed or omitted for Al metallizations to produce the shape in Figure 7.26, and useful thermal measurements could be made. But for Cu metallizations, exposure to air at elevated temperatures causes the lines to oxidize and the measurements can become compromised. So almost always, the lines to be tested for EM are embedded in and surrounded by dielectric. The dielectric to the sides and above the line provides additional paths for conducting away thermal energy, and additional modifications to Equation 7.49 are needed to adequately calculate the temperature of the line for high current densities. Harmon et al. [22] extended the work done by Schafft and Bilotti by employing the Schwartz–Christoffel transform technique used by Bilotti to model the fully embedded line to produce a formula that would encompass two-dimensional heat transport from a line with a rectangular cross section. Their formula is the sum of three separate conductances, G_b for the bottom conductance, G_s for the side conductance, and G_t for the top conductance, and is shown below in Equation 7.50. For uniform dielectric layers (all having the same thermal conductivity), such as SiO_2, Lee [23] has shown that exchanging a value of 2.8 for 0.88 in Equation 7.49 above provides a reasonably accurate approximation to Equation 7.50.

$$G_s = G_b + G_s + G_t \tag{7.50}$$

where

$$G_b = 2\frac{k_i}{\pi} \ln\left[\frac{1}{3} + \frac{2}{3}\cosh\left(1 + \pi 3^{-\frac{3}{2}} + \frac{\pi \omega}{2t_i}\right)\right],$$

$$G_s = 2\frac{k_i}{\pi} \ln\left[-1 + C^2 + 2C\sqrt{(C^2 - 1)}\right],$$

$$G_t = 2\frac{k_i}{\pi} \ln\left[\left|\frac{1 + \sqrt{(1 + (t_i + h)/t_c)}}{1 - \sqrt{(1 + (t_i + h)/t_c)}}\right|\right],$$

$$C = 1 + h/t_i.$$

k_i is the thermal conductivity of the insulator, t_i is the underlying insulator thickness, h is the metal line thickness, and t_c is the overlying insulator thickness.

The situation can be much more complex. For example, in Cu technologies, a special dielectric layer, such as Si_3N_4, is deposited over the Cu lines as a capping layer to contain the Cu in the line. The dielectric between the lines may be a low dielectric constant (low-k) dielectric like fluorinated silicon glass (FSG), or a porous material, and the dielectric underlying the first metal level can be boro-phosphor-silicate glass (BPSG). SiO_2 may be used for upper metal levels instead of low-k dielectric. Since each of these dielectrics have a different thermal conductivity, the thermal resistance will vary with direction from the line, and with the metal level. Application of the Harmon–Gill model will then yield an effective thermal conductivity for each of the three terms in Equation 7.50. Chen [24] was able to use the Harmon–Gill model to extract the effective thermal conductivity of composite dielectrics for a couple of these complex situations.

From a practical perspective, most lines in an actual product chip are part of an array of wires at the same metal level, and are sandwiched between additional levels of metal over and under them. For Cu technologies, even when the density of lines is relatively low around a particular line, metal fill shapes, inserted to help maintain film thickness uniformity during chemical mechanical polishing opera tions, occupy the space. This means that the actual thermal conductance will be greater than that for a line surrounded by oxide alone. Modeling efforts would have to include these metal shapes and lines to produce reasonably accurate values of conductance for selected situations. Since there are a large number of different layout possibilities, detailed modeling is unwieldy and impractical. A useful compromise is to measure the conductance at each metal level for a range of line widths and extract values for the effective thermal conductivity. From a reliability perspective, having the thermal conductance in product chips be greater than that obtained in modeling and measurement of test structures is an advantage because it provides additional margin against failure.

7.5.2 Wafer-Level Electromigration

Wafer-level reliability (WLR) electromigration testing is another case where the thermal conductance of the structure under test must be known. WL EM is performed before chips are cut out of a wafer, with the current applied to the bond pads through the W probe pins. For several methods of WL EM, the only source of heating is the self-heating due to the applied current.

WL EM requires a direct application of the principles of temperature measurement in embedded lines. Whereas most electromigration testing is done with chips mounted into ceramic packages and stressed in ovens, WL EM is done on a wafer prober with the chips still in wafer form and with the wafer at room temperature. The heating is produced by high currents, and the temperature is determined by Equation 7.49, or a similar expression. Since the heating is produced by the current, the lifetime of the line depends upon only one variable, the current.

WL EM has several advantages over the conventional oven-based testing. First, the testing is done while the chips are still in wafer form. This eliminates the time and cost involved in dicing the wafer into chips and mounting and wirebonding the chips into ceramic packages. It also eliminates the need for the specialized equipment used for electromigration testing, which, in itself is very expensive to buy, operate, and maintain. Second, WL EM can be conducted very rapidly, producing failure times on the order of seconds or minutes, in contrast to the hours, or even weeks required for conventional package testing in ovens. This speed of testing translates into larger volumes of data and either greater certainty in the projected reliability of the parts, or greater allowed use current densities.

The most significant disadvantage of WL EM is that both the heat and the current are supplied by the current, and some method is needed to evaluate the current density exponent and activation energy independently. A second serious disadvantage is that for the higher temperatures the temperature of the line is not uniform along its length. The nonuniformity is caused by local differences in thermal conductance. For example, the ends of the line are contacted by larger metal lines above or below the line of interest, and the thermal conductance of those lines is generally greater than that of the line under test. When the line under test is at its test temperature, the ends of the line are likely to be cooler than the center, and a temperature gradient will be present. Diffusivity is greater in a region of high temperature compared to that in a region of lower temperature, such that when a mild flux divergence is present within a region with a temperature gradient, a void is likely to form there and become the failure location.

Since the failure location in conventional testing is usually at or near a via, the difference in failure modes becomes an issue of uncertainty in defining the true failure time of the device. For example, in the case of an aluminum metallization with refractory conductive layers both over and under the aluminum layer, a line will still conduct current if the Al layer is completely voided out because the refractory layers form a shunt around the void. However, although the conductive path around a void in the line will employ both refractory layers, a void located at the very end of a line over a via will prevent current from the via from reaching the upper refractory path. All of the current will have to use the lower refractory layer. The resistance increase per unit length of void will be greater if only one refractory path is available and because failure time is defined in terms of a percentage resistance increase, the failure time will be earlier.

In addition to differences in failure time, the distribution of failure times for WL EM may be different from those obtained using the same structures in conventional oven-based tests. The combined uncertainties introduced by differences in the shape of the failure distributions and in the values of the median time to failure make comparison between the two methods difficult. As of now, there is no consensus for equating failure data from WL EM with that from conventional oven testing.

While there may not be universal acceptance of using WL EM data for reliability projections for products, there is no doubt that it is useful for line monitoring or for comparing wafers produced with different process modifications. There are at least three different methods presently accepted for WL EM. They are 1) the SWEAT (standard wafer electromigration accelerated test), 2) the isothermal, and 3) the constant current methods. The constant current method seeks to keep the current in the test structure constant during the test. The isothermal method seeks to keep the temperature in the structure constant during the test, and the SWEAT method attempts to keep the failure time constant.

Problems exist with all three methods. In the constant current method, the same current is applied to the whole population of structures tested. For the narrowest lines, which have the greatest variation in on-wafer line width, this approach results in a substantial variation in current density from line to line, and this in turn causes substantial variation in the temperature reached by the line. Thus determination of the actual stress conditions is difficult. For the isothermal method, the temperature from one structure to the next can be controlled reasonably well by monitoring the resistance of the line, but not all parts of the line may be at the same temperature and acceleration, as discussed above. For the SWEAT method, control of the temperature and current depend upon a calculation of the projected median time to failure, so that both vary over the duration of the test.

7.5.3 Indirect Joule Heating

Although electromigration rules usually restrict the magnitude of the allowed DC current to be lower than that which would cause significant Joule heating, the same is not true for alternating current (AC), where the current flows first in one direction, then in the other. Pure AC (that is, either a pure sinusoidal current wave or a symmetric pulsed current wave) does insignificant damage to interconnect lines tested over long durations [25]. Naturally, the period of the cycle must be short in comparison to the time in which DC electromigration damage will occur, but this is generally not a concern with modern high speed circuits. The reason AC current seems to do no damage is that any damage done during the first part of the cycle appears to be reversed during the second part of the cycle, when the current flows in the direction opposite to that in which it flowed in the first part of the cycle.

Since no electromigration damage is observed as the result of AC, designers would like to be able to use very high current densities in some applications. However, even though electromigration damage does not accumulate, AC still produces heat so that the quantity that limits the magnitude of current density that can be used is Joule heating. Clearly, the current density must not be allowed to become great enough to melt the AC line. But long before this occurs, a hot AC line can heat up the surrounding wires sufficiently to accelerate EM failure in nearby DC lines. As a general rule, current densities below $\sim 25\,\mathrm{mA/um^2}$ cause less than one degree Celsius of Joule heating. But heating increases as the square of the current density, as shown in Equation 7.46, so that once it becomes noticeable it increases rapidly to dangerous levels. Furthermore, lifetime in a DC line decreases between 20 and 25% for a mere 5°C temperature increase (depending on the activation energy of the EM). Therefore, we can set a limit on the magnitude of the AC current if we assume that the nearest neighboring DC line reaches the same temperature as the AC line. (Neighboring lines are generally at a lower temperature than the hot AC line because of the effects of radial spreading.) Thus the ability to predict the temperature of an AC line becomes important to design specifications.

7.6 CLOSING REMARKS

We have covered an overview of electromigration in integrated circuit chip metallizations. This has included discussion of the phenomenon of EM itself, and how it occurs in a metal, as well as how it is measured and how the metallizations system, which includes the primary conductor, cladding or redundant layers, and the surrounding dielectric, affect the measurements. The methodology for testing specific structures, analyzing the data, and using that analysis to determine the currents that are allowed under use conditions was described, including some of the statistical concepts that support the analysis. Differences in the behavior of Cu and Al metallizations were discussed.

The statistical discussion was centered on a uniform population of parts that failed in the same way. We also looked at cases where failure could occur in different modes, such that the failure population displayed a staggered or bimodal failure distribution, and the implications of minimum failure time was used to introduce the concept of a three-parameter lognormal distribution. Joule heating resulting from high currents was introduced, and we used the concepts of thermal resistance to show some simple dependencies of Joule heating on materials properties and structure geometries. The application of thermal conductance to wafer-level EM testing was briefly mentioned.

Despite this somewhat lengthy overview, this chapter has only lightly touched on the complexities that can be encountered in electromigration testing. Different technology configurations often create structures very different from all of those that have been tested before. And new advances in materials and processes are altering the effects of EM on progressively small wires. It is wise to keep in mind that the electron mean free path is about 400 Å in Cu at room temperature, and that minimum line widths in the most advanced technologies are less that 0.1 μm. This means that the mean free path is comparable to the line width and suggests that effects as yet not encountered are likely to appear in the near future. While this fact is bound to be a challenge to the microelectronics industry, it also represents an opportunity for additional pioneering work to those in the metallizations reliability community.

REFERENCES

1. J. M. Poate, K. N.Tu, J. W. Mayer, editors, *The Properties of Thin Solid Films*. John Wiley and Sons, New York: 1978, Ch 8, Section 8.5.

2. P. S. Ho. Motion of inclusion induced by a direct current and a temperature gradient. J. Appl. Phys., 41(1): 1970, p 64.

3. See reference 1, Ch 8.

4. J. R. Black. Mass transport of aluminum by momentum exchange with conducting electrons. IEEE 6th Ann. Int. Reliability Symp. IEEE: 1968.

5. C.-Y. Li, P. Borgeson, and T. D. Sullivan. Stress-migration related electromigration damage mechanism in passivated, narrow interconnects. Appl. Phys. Lett. 59(12): 1991, p 1464.

6. J. M. Poate, K. N. Tu, J. W. Mayer, editors, *The Properties of Thin Solid Films*, Wiley, 1978, pp 15–24.

7. I. Blech. Electromigration in thin aluminum films on titanium nitride. J. Appl. Phys. 47(4): 1976, pp 1203–1208.

8. P.-C. Wang, Thermal and electromigration stress distributions measured by X-ray microdiffraction. Ph.D. Thesis, Columbia University, 1997.

9. A. Witvrouw et al. Incubation, time-dependent drift and saturation during Al-Si-Cu electromigration: Modelling and implications for design. IEEE IITC: 1998, p 27.

10. H. S. Rathore et al. Electromigration and current-carrying implications for aluminum-based metallurgy with tungsten stud-via interconnections. SPIE Vol. 1805, Submicrometer Metallization: 1992, 251–262.

11. C.-K. Hu. Electromigration failure mechanisms in bamboo-grained Al(Cu) interconnects, Thin Solid Films 260(1): 1995, pp 124–134.

12. C. K. Hu et al. Electromigration and stress-induced voiding in fine Al and Al-alloy thin-film lines. IBM J. of Res. and Develop. 39(4): 1995, 474.

13. D. Edelstein et al. Full copper wiring in a sub-0.25 μm CMOS ULSI technology, Electron Devices Meeting, Technical Digest, International, IEEE: 1997, pp 773–776.

14. X. Federspiel et al. Dynamics of resistance evolution during electromigration stress. 1998 IEEE IRW Final Report. 1998.

15. A. Fischer et al. Experimental data and statistical models for bimodal EM failures. Proc. 38th Ann. Int. Reliability Physics Sym., IEEE: 2000, pp 359–363.

16. J. Gill et al. Investigation of via-dominated multi-modal electromigration failure distributions in dual damascene Cu Interconnects with a discussion of the statistical implications. Proc. 40th Ann. Int. Reliability Physics Sym., IEEE: 2002, pp 298–304.

17. R. Filippi et al. Paradoxical predictions and a minimum failure time in electromigration. APL 66(15): 1995, pp 1897–1899.

18. B. Li et al. Application of three-parameter lognormal distribution in EM data analysis. Microelec. Reliabil. 46(12): 2006, pp 2049–2055.

19. A. A. Bilotti. Static temperature distribution in IC chips with isothermal heat sources. IEEE Trans. Electron. Devices ED-21: 1974, p 217.

20. H. Schafft. Thermal analysis of electromigration test structures. IEEE Trans. Electron Devices ED-34(3): 1987, 664–672.

21. S. M. Lee, D. G. Cahill. Heat transport in thin dielectric films. JAP, 81(6): 1997, pp 2590–2595.

22. D. Harmon, J. Gill, T. Sullivan. Thermal conductance of IC interconnects embedded in dielectrics. 1998 IEEE IRW Final Report; 1998, pp 1–9; Gill et al. Predicting thermal behavior on interconnects. 1999 IEEE IRW Final Report: 1999, pp 54–60.

23. T. C. Lee et al. Comparison of isothermal, constant current and SWEAT wafer level EM testing methods. 2001 IEEE IRPS 39th Ann. Proc.: 2001, pp 172–183.

24. F. Chen et al. Measurements of effective thermal conductivity for advanced interconnect structures with various composite low-k dielectrics, Proc. 42nd Ann. Int. Reliability Physics Sym., IEEE: 2004, pp 68–73.

25. J. Tao et al. Modeling and characterization of electromigration failures under bidirectional current stress, IEEE Trans. on Electron Devices, 43(5): 1996, pp 800–808.

26. T. D. Sullivan. Reliability in copper metallization and low-*k* dielectrics for ULSI interconnects. IEEE IEDM Short Course, Reliability for Logic and Memory Technologies. A. Oates, Organizer, J. L. Hoyt and D. B. M. Klaassen, Chairpersons, 1998.

INDEX

Reliability Wearout Mechanisms in Advanced CMOS Technologies. By Strong, Wu, Vollertsen, Suñé,
LaRosa, Rauch, and Sullivan
Copyright © 2009 the Institute of Electrical and Electronics Engineers, Inc.

Printed and bound by CPI Group (UK) Ltd, Croydon, CR0 4YY

27/10/2024

14580330-0005